Industrial Hydraulic Control

A textbook for Fluid Power technicians

Fourth Edition

Peter Rohner
(Dip. Tech. Teaching)

JOHN WILEY & SONS
BRISBANE • NEW YORK • CHICHESTER • WEINHEIM • TORONTO • SINGAPORE

This edition published 1995 by
JACARANDA WILEY LTD
33 Park Road, Milton, Qld 4064

Offices also in Sydney and Melbourne

First published 1984 by
AE Press (Stanley Thornes and Hulton (Aust.) Pty Ltd)
Second edition 1986
Third edition 1988

National Library of Australia
Cataloguing-in-Publication data

Rohner, Peter 1939–.
Industrial hydraulic control.

4th ed.
Includes index.
ISBN 0 471 33524 X

1. Fluid power technology. 2. Hydraulic Control. I. Title.

629.8042

Cover photograph by courtesy of Robert Bosch (Australia) Pty Ltd

All the experiments described in this book have been written with the safety of both teacher and student in mind. However, all care should be taken and appropriate protective clothing should be worn when carrying out any experiment. Neither the publisher nor the author can accept responsibility for any injury that may be sustained when conducting any of the experiments described in this book. No legal responsibility is accepted for the information contained in this publication.

Illustrated by Peter Rohner

Printed in Singapore

10 9 8 7 6 5 4 3 2

Foreword

Effective training is the key to the growth and development of any industry and nowhere is this as pronounced as in the field of hydraulics.

The diversity of hydraulic product designs and control techniques complicates the learning process. However, these problems can be clarified by a thorough understanding of first principles. This textbook effectively tackles the subject of hydraulics by explaining the underlying principles of fluid statics as they relate to practical systems. It offers an up-to-date guide to the details that are so important to all levels of understanding: design, circuit analysis, and troubleshooting.

The book develops logically with the explanation of the operation of various hydraulic components (pumps, valves, actuators) with an impartial evaluation divorced from the design and development philosophy of individual manufacturers.

Each subject is comprehensively illustrated with practical details of application, relevant calculations, sizing and design parameters.

Chapters 9 to 12 cover details relating to how the system accessories are applied and explains the importance of hydraulic fluids. These are often overlooked points that can make or break even the best selection of major components.

Complete systems are considered in the final sections as an interesting conclusion that ties together the material from earlier chapters.

This extensively illustrated manual should prove to be an invaluable aid for teaching theoretical and practical aspects of hydraulics. It will also be a handy source of information for technicians and engineers working with hydraulics in the field.

Roy W Park B E (Hons)
Managing Director *Moog Australia Ltd*

Contents

Preface

Hydraulic systems are used in industrial and mobile applications such as construction and mining, transportation, general manufacturing, stamping presses, steel mills, agricultural machines, aviation, space technology, deep-sea exploration, marine technology, amusement industries and offshore gas and petroleum exploration.

The impressive, and ongoing growth of this versatile and exciting technology continues to create numerous job opportunities for properly trained and accredited people in all areas of fluid power: engineers, technicians, sales and service personnel are badly needed, whilst there is also a shortage of trained fluid power teachers.

This **fourth edition** includes four completely new chapters. **Cartridge valve technology** (logic valves) in chapter 13, **Power saving pump controls**, as used on mobile and industrial systems (load-sensing and constant power controls), in chapter 14, **Proportional and servo hydraulic control systems** in chapter 16 and **PLC electronic programmable controllers for hydraulic systems** in chapter 19. PLC control is a powerful and nowadays widely used technology for the control of sequential hydraulic systems. Because of its simplicity and versatility, the step-counter method is frequently used to program, and design ladder logic for, PLC controllers. Since the step-counter method has long been applied in pneumatic control design, it seemed to be applicable to introduce it in chapter 18 as a natural bridge to PLC circuit design.

These four totally new chapters present highly complex, but these days relevant technology in an easy to understand way. With their addition, the text becomes a versatile all-round book for technical and further training at TAFE colleges and technical universities.

The purpose of this textbook is to facilitate an understanding of the fundamental principles of industrial hydraulics and to provide a practical working knowledge of the commonly encountered components for designing, installing and maintaining industrial and mobile hydraulic systems. As such, the textbook is aimed at those students who are preparing themselves for industry in design, maintenance or sales.

The book is part of a comprehensive learning and teaching package, consisting of a textbook, a separate workbook, a teacher's book (workbook with solutions) and over 340 overhead projectorial masters.

Modern teaching methodology is applied throughout the book to stimulate and facilitate learning. Basic formulae, where introduced, are coloured with a pink background, components functions are described, and simple circuits show component application and integration. Pumps, valves, actuators, filters, etc. are depicted as sectioned illustrations together with their matching I.S.O. graphic symbol. The pressure control chapter, because of its complexity, is rounded off with a comprehension and review summary.

Numerous worked design examples are given to illustrate important design criteria for the sizing and selection of pumps, motors, valves, accumulators and piping systems. Nomograms, as well as calculative methods are used for calculations on component and system sizing.

The appendix section contains a concise formulae collection, all applicable I.S.O. fluid power symbols, unit of measurement, prefixes for fractions and multiples of base units, an extensive conversion table and a comprehensive index with over 800 listings.

Peter Rohner (Dip. Tech. Teaching), *Senior Lecturer Fluid Power Control — Royal Melbourne Institute of Technology (RMIT University)*

Acknowledgments

I am very grateful to my wife Heidi and my children Rahel and Michael, without whose patience and support I would never have been able to finish this book.

I would like to thank my colleagues Sid Stribling, Graham Williams, Chris Bourne and Dr Marian Tumarkin from the Fluid Power Department at the Royal Melbourne Institute of Technology (RMIT University) for their numerous comments and suggestions. Barry Perriman deserves special credit for spending many hours of his valuable spare time on vetting manuscript and galley proofs and for making substantial practical contributions, drawn from his extensive experiences as a maintenance and sales engineer.

The cover photograph was obtained from Mr. Wilfried Bork, Chefredakteur of Ölhydraulik + Pneumatik (Zeitschrift für Fluidtechnik) with permission from the owners, ROBERT BOSCH GmbH, Stuttgart, Germany.

Chapter 17 (Faultfinding of hydraulic circuits) was reproduced by kind permission of Castrol Australia Pty Ltd.

The following materials have been reproduced by courtesy of the firms and organisations listed:

Appendix 4, Conversion table	— Applied Measurement Australia Pty Ltd
Figures 55, 71, 72, 151–153, 155, 156, 169, 173, 287–292, 298	— Robert Bosch (Australia) Pty Ltd
Figure 134	— Hagglunds Australia
Figures 65, 67, 68, 79, 120, 121, 136, 138, 139, 233, 234	— Rexroth Australia Pty Ltd
Figure 207	— Moog Australia Pty Ltd
Figures 66, 137	— P&G Hydraulic Installations (Abex Denison)
Figures 191–193	— The Shell Company of Australia Ltd

Please note

Colour coding of fluid pressure used throughout this book:

- High or system pressure
- Return line to tank, pump inlet pressure, or lowest pressure
- Reduced pressure
- Gas pressure equal to system pressure

1 Physical principles in hydraulics

Personnel who operate, service, or design fluid power systems should have a thorough understanding of the physics and properties of fluids and their behaviour under different circumstamces.

Liquids and gases flow freely, and for that reason both are called fluids (from the Latin *fluidus*, meaning flow).

A fluid is defined as a substance which changes its shape easily and adapts to the shape of its container. This applies to both liquids and gases. Their characteristics are discussed throughout this book.

Transmission of force by fluids

When one end of a bar of solid material is struck, for example with a hammer, the main force of the blow is transmitted straight through the bar to the opposite end. The direction of the blow determines the direction of the major force transmitted, and the more rigid the bar, the less force is either lost in it, or transmitted at angles different to the direction of the blow.

When a force is applied to the end of a column of a confined liquid (fig. 1) that force is transmitted straight through the column to its opposite end, but also — equally and undiminished — in every other direction, sideways, downwards, and upwards. This physical behaviour is defined by Pascal's Law (fig. 2). Pascal's discovery has opened the way to the use of confined fluids for power transmission and force multiplication.

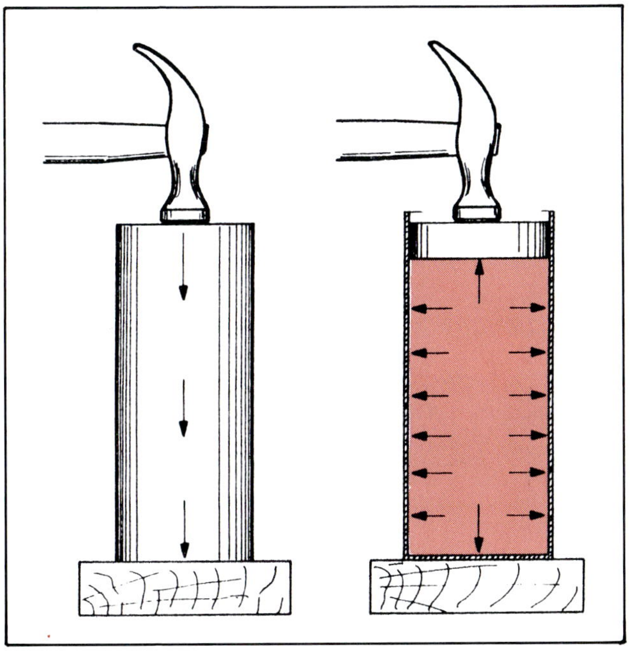

Fig. 1. Transmission of force through a solid material and through a static fluid.

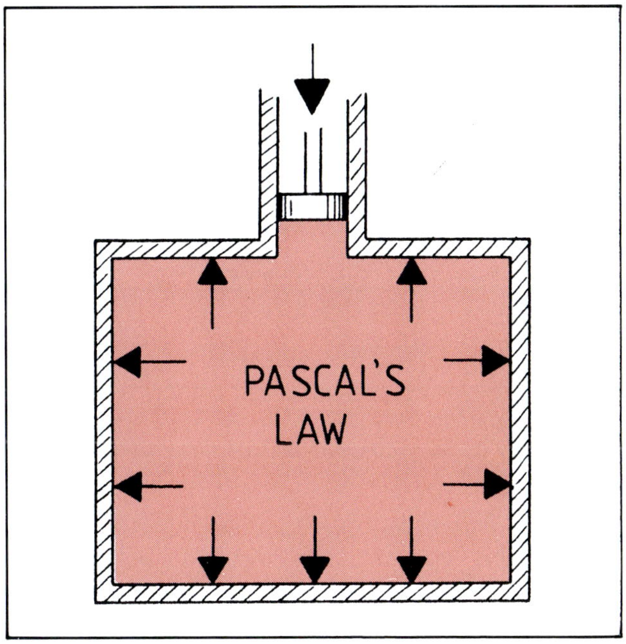

Fig. 2. Pascal's Law states that pressure applied to a static and confined fluid is transmitted undiminished in all directions and acts with equal force on equal areas and at right angles to them.

Atmospheric pressure

The weight of the air (every gas has a mass) causes so called atmospheric pressure. Atmospheric pressure at sea level is approximately 101.3 kPa. Atmosphereic pressure can be measured with a barometer. Torricelli's mercury barometer is shown in fig. 3. The atmospheric pressure acts upon the open surface of the mercury container and thus supports the mercury column in the vacuum tube. The distance "h" is proportional to the atmospheric pressure and varies with the altitude.

Vacuum pressure

Any pressure below normal atmospheric pressure is termed vacuum pressure. It can be measured with the same barometer, depicted in fig. 3. Pressure readings below the calibration mark "A" denote vacuum

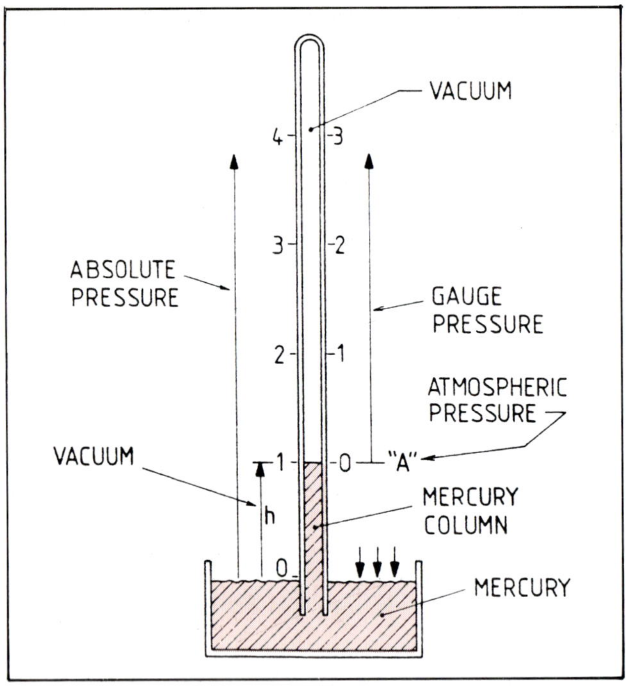

Fig. 3. Mercury barometer. Atmospheric pressure acting onto the mercury in the receptacle supports the mercury in the vacuum tube.

pressure. At absolute zero pressure, the mercury column in the vacuum tube will disappear entirely since no pressure will be present to support a mercury column. Therefore it can be said that any pressure reading on the scale is absolute pressure, or an absolute pressure measurement.

Gauge pressure versus absolute pressure

Most pressure gauges used in hydraulic systems however, have a pressure calibration based on atmospheric pressure. This means that atmosphere pressure is regarded as zero pressure, and any pressure above atmospheric pressure is thus a positive pressure reading. Most pressure gauges of this nature do not show pressure readings below atmospheric pressure. Pressure readings on such gauges are called gauge pressure, and pressure readings on gauges which start from absolute zero pressure are called absolute pressure (fig. 3).

Pressure in liquids

Under static conditions and without any external forces, the pressure at any point within a fluid system is proportional to the height of the fluid column above that point (fig. 4). Torricelli called the pressure at the bottom of the fluid column (tank) the "head pressure". All that is required to work this pressure in Pa is the specific weight of the fluid in the tank and the force that is produced by its weight. For example, a water column of 10 m height with a base area of 1 m^2 weighs 10 000 kg (specific weight for water is 1000 kg per 1 m^3).

The Law of Newton states:

Force = Mass × Gravitational Acceleration

Thus it can be calculated that the force in Newtons exerted onto the resting surface of the water column is:

Force (N) = 10 000 kg × 9.81 = 98 100 Newtons

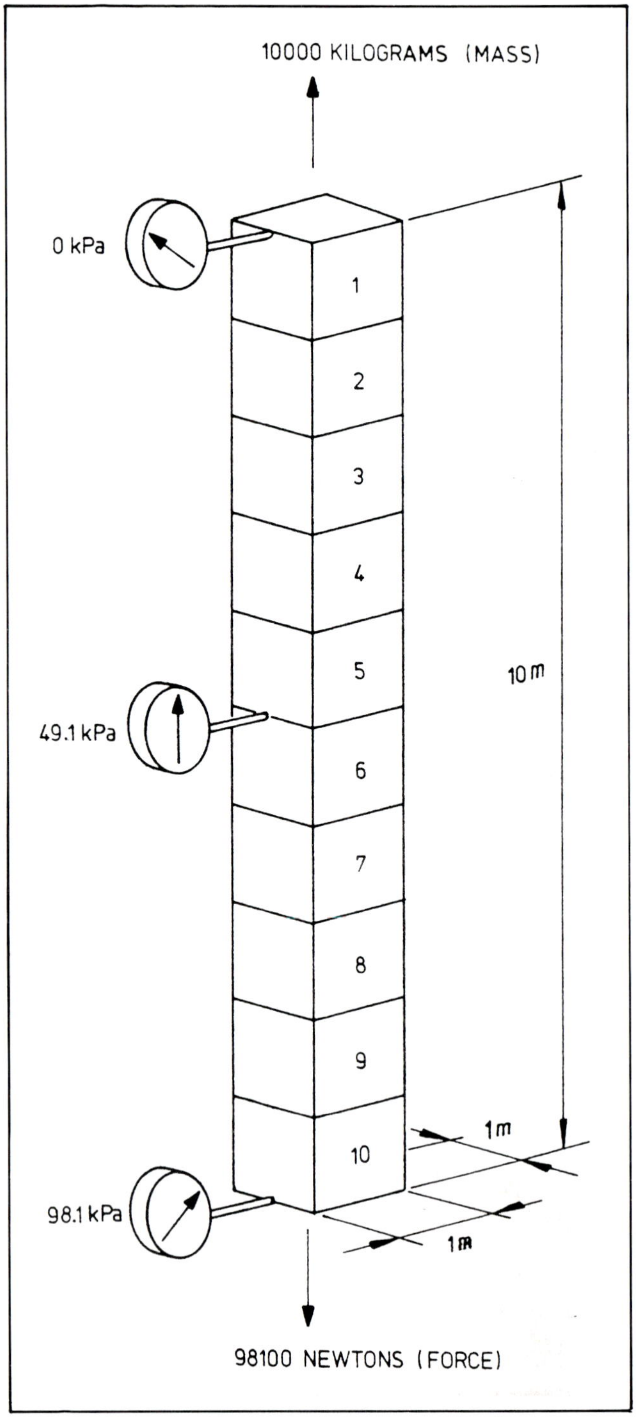

Fig. 4. A water column of 10 m with a base area of 1 m^2 produces a force of 98 100 Newtons but its mass is 9.81 times less.

Pressure in the S.I. system is expressed in Pascals. Since the Pascal is a very small unit, the prefixes kilo and mega are used to express pressure in industrial and mobile hydraulics. European countries predominantly use the unit "bar" to express pressure, whereby one bar equals 100 kPa or 100 000 Pa.

The relationship between Pascals and Newtons is (fig. 4):

$$1\ \text{Pa} = \frac{1\text{N}}{1\text{m}^2},\ \text{or Pressure (Pa)} = \frac{\text{Force (N)}}{\text{Area (m}^2\text{)}}$$

It can therefore be said that for every metre depth, the pressure within the fluid column (water column) increases by 9.81 kPa or 9810 Pa. This pressure changes of course, when a fluid with a different specific weight is used.

Example

The pump of an industrial hydraulic system is fitted with a raised reservoir to prevent cavitation. The reservoir provides a 6 metre mineral oil column and the specific weight of mineral oil is 910 kg per m^3.

Calculate the pressure exerted onto the intake port of the pump.

$$\text{Force} = 6 \times 910\ \text{kg} \times 9.81 = 53\ 560\ \text{Newtons}$$

$$\text{Pressure} = \frac{\text{Force}}{\text{Area}} = \frac{53\ 560\ \text{N}}{1\ \text{m}^2} = 53\ 560\ \text{Pa} = 53.56\ \text{kPa}$$

Flow and pressure drop

Whenever a hydraulic fluid is flowing, a condition of unbalanced force causes the fluid motion. Hence, when hydraulic fluid flows through a pipe with a constant diameter, the pressure will always be lower downstream than at any point upstream (fig. 5). The pressure differential or pressure drop is caused by the friction in the pipe. Whenever fluid is flowing through a system, heat is created by friction. Thus, a part of the hydraulic energy is permanently lost as heat energy, which is measurable as pressure drop, shown in the successive vertical head pipes in fig. 5.

Fluid flow

The fluid in a hydraulic system is in general under pressure and completely fills the pipes of the system. This pressure may result from friction within the piping system, from gravity or the weight of the fluid, or from load resistance to the pump flow.

When a hydraulic fluid flows through straight piping at low velocity, then the particles of the liquid move in straight motion parallel to the direction of flow, so that heat loss due to friction is minimal. Such flow is referred to as laminar flow. Detrimental factors such as high flow velocity, sharp bends and elbows in the piping, rough internal pipe surfaces, etc. cause cross currents within the flow and turbulence starts to develop. Turbulent flow causes significant increases in friction and a pressure drop; thus heat is produced (input energy is wasted).

Although friction can never be eliminated entirely, it can be controlled to some extent if the aforementioned detrimental factors are avoided or reduced (fig. 6).

Flow through an orifice

An orifice is a hole, narrower than the pipes to which it is fitted. The orifice is generally used to control flow (Chapter 6), or to create a pressure differential (Chapter 5).

As long as there is flow through an orifice, there must be a pressure drop across the orifice, which means that the pressure downstream — that is, in the direction of flow (fig. 7) — is less than the pressure upstream. This pressure loss is permanent since the friction (work) has been transformed into heat, and this heat cannot be regained. As soon as the flow is stopped the Law of Pascal must be applied for the now static condition, and

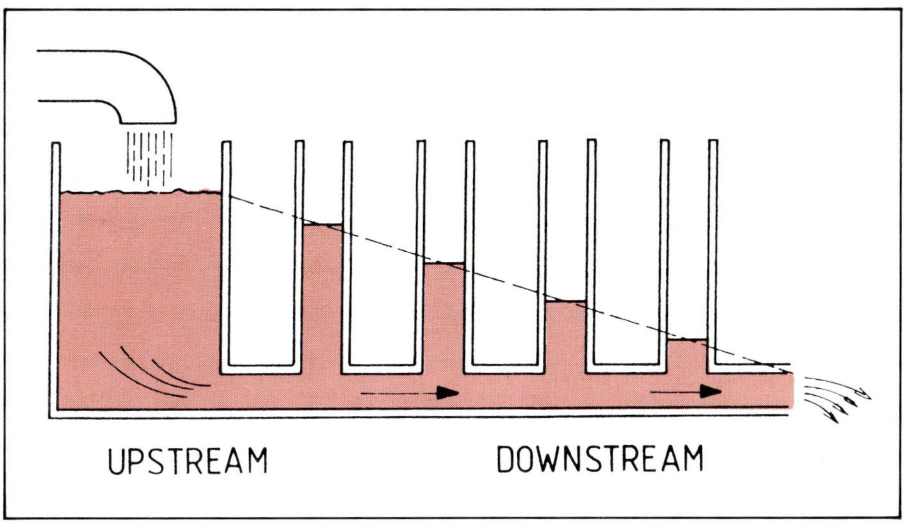

Fig. 5 Friction in pipes causes heat and pressure drop.

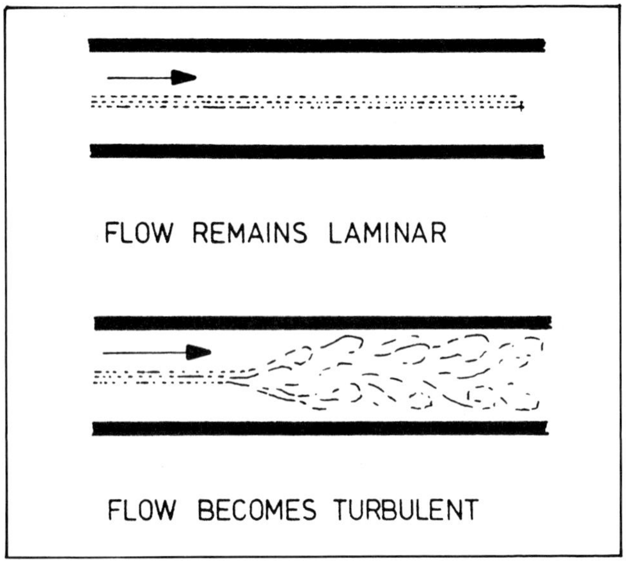

Fig. 6 Flow behaviour in pipes.

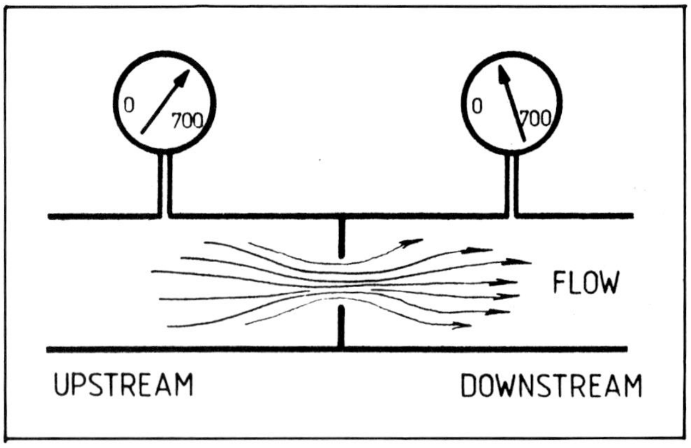

Fig. 7 The orifice causes friction and thus a permanent pressure drop.

pressures on both sides of the orifice equalise (fig. 8).

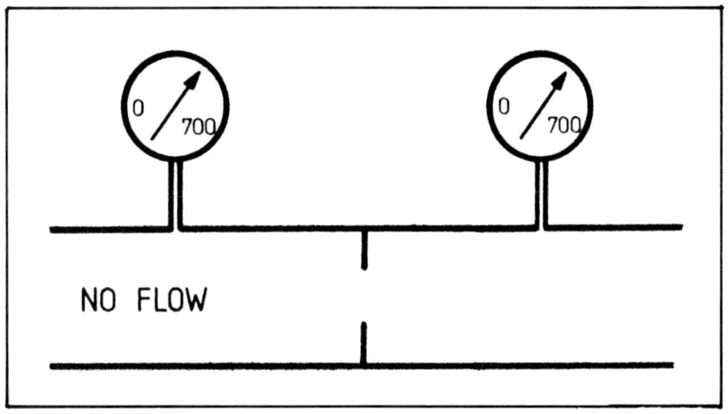

Fig. 8 Without fluid flow there is no pressure drop.

Flow rate and flow velocity

The average speed with which the particles of a fluid move past a measuring point is called fluid velocity or speed. The fluid velocity is measured in metres per second (abbreviated m/s). Flow velocity must be carefully controlled and considered when the hydraulic system is designed, as an excessive flow velocity causes turbulent flow with its resulting high pressure drop.

Flow rate is the measure of flow volume that streams past a measuring point in a given time. Flow rate is measured in litres per second (abbreviated L/s) or cubic metres per second. (The latter is less frequently used.) Flow rate has a direct bearing on the speed with which a hydraulic actuator moves a load and is therefore governed by the design concepts of the machine (for further details see Chapter 6).

Force transmitted by a liquid

The conservation of energy is a very fundamental principle, and states that energy can neither be created nor destroyed. At first sight, the multiplication of force as depicted in fig. 9 may give the impression that something small is turned into something big. But this is wrong, since the large piston on the right is only moved by the fluid displaced by the small piston on the left. Therefore, what has been gained in force must be sacrificed in piston travel distance (fig. 10).

The concept of force multiplication explained here is also termed "hydraulic lever", and is used to demonstrate how energy is transferred within a basic hydraulic system. The design of a hydraulic lifting jack is depicted in fig. 11.

Example

Calculate the pump stroke if the piston area ratio is 100:1, the large piston diameter is 150 mm, and the pumping piston must be moved up and down 400 times to lift the large piston a distance of 130 mm.

Step 1

The volume required to displace the large piston is equal to the volume to be pumped by the small piston:

$$0.15\ \text{m} \times 0.15\ \text{m} \times 0.7854 \times 0.13 = 0.0022973\ \text{m}^3$$

Step 2

$$\text{Area small piston} = \frac{0.15\ \text{m} \times 0.15\ \text{m} \times 0.7854}{100} = 0.0001767\ \text{m}^2$$

Step 3

$$\text{Stroke} = \frac{\text{Volume}}{\text{Area}} \quad \therefore \quad \text{Total stroke} = \frac{0.0022973\ \text{m}^3}{0.0001767\ \text{m}^2} = 13\ \text{m}$$

Step 4

$$\text{Single Stroke} = \frac{\text{Total Strokes}}{\text{No. of Strokes}} = \frac{13}{400} = 0.0325\ \text{m} = 32.5\ \text{mm}$$

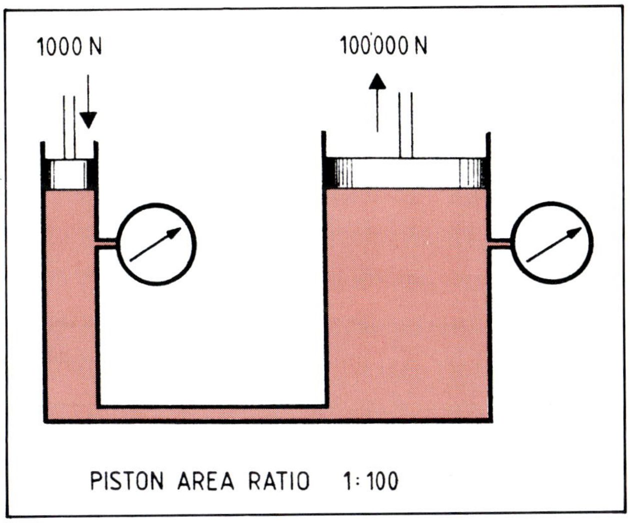

Fig. 9 Force multiplication with hydraulic lever.

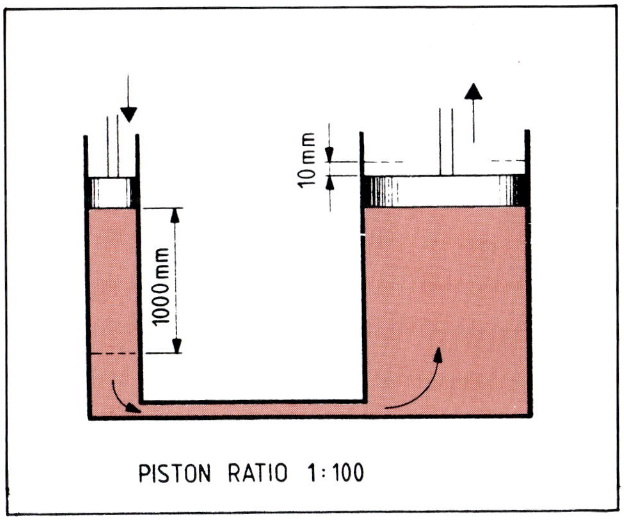

Fig. 10 Force multiplication requires a sacrifice of distance

Gas laws

Since gases too are used in hydraulic systems (in accumulators, for example), it is important to mention at least some of the basic behavioural characteristics of gases. Robert Boyle discovered by experimentation and direct measurement, that when the temperature of an enclosed sample of gas was kept constant and the pressure doubled by means of a piston (fig. 12), the

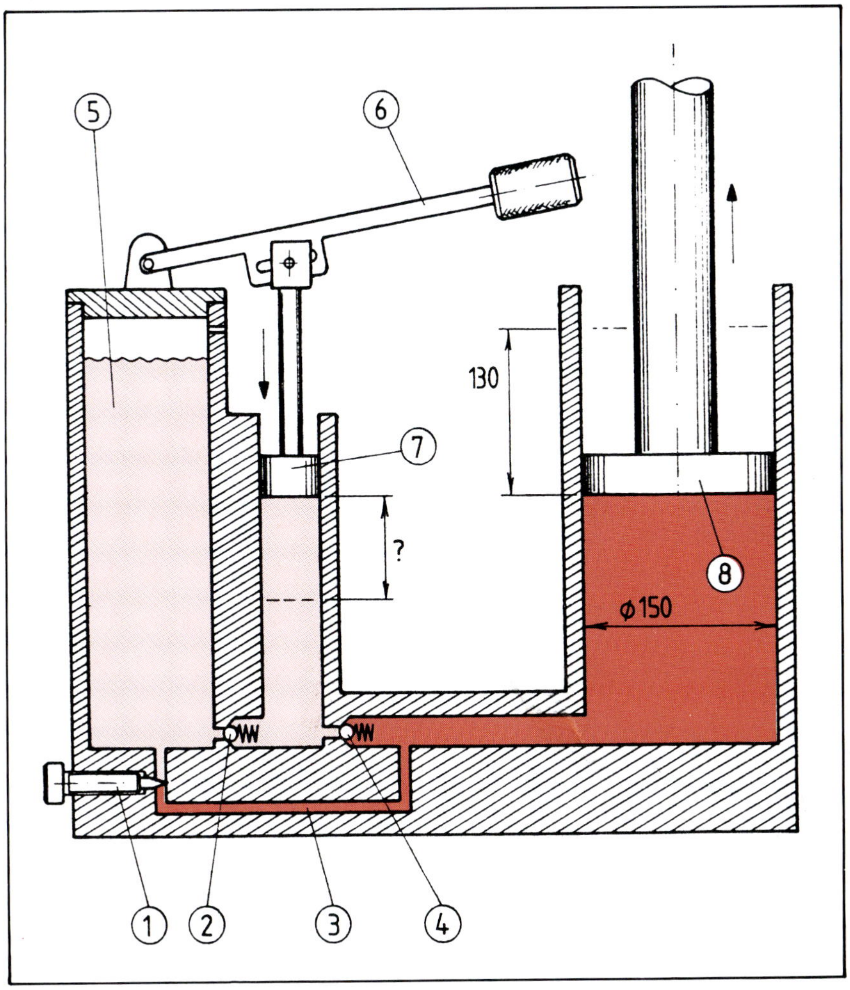

(1) Lowering valve.
(2) Check valve.
(3) Return line.
(4) Check valve.
(5) Fluid reservoir.
(6) Pumping lever
(7) Pumping piston.
(8) Lifting piston.

Fig. 11 Hydraulic lifting jack.

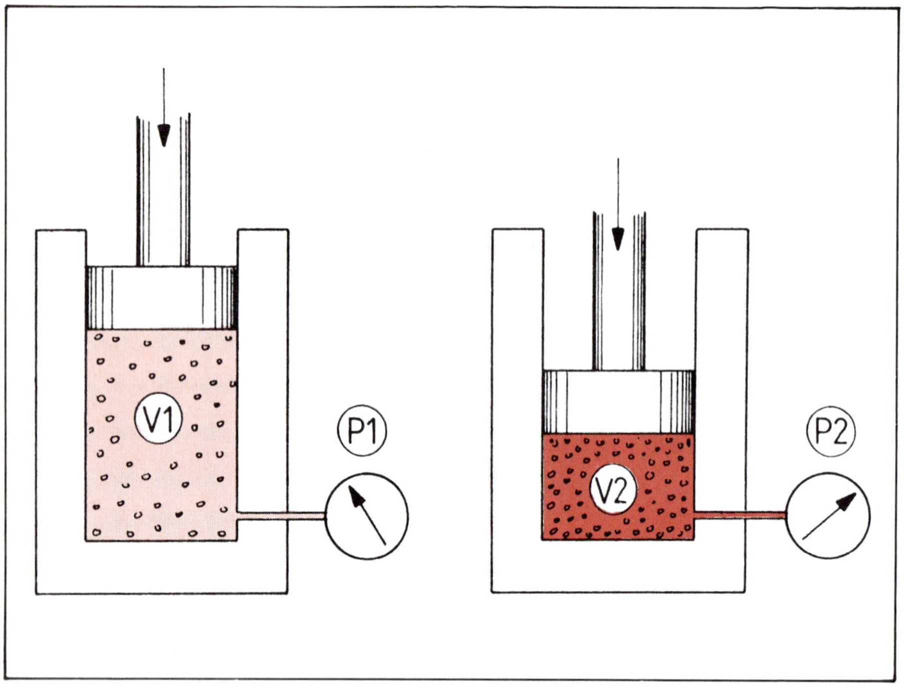

Fig. 12 Temperature maintained — volume decreased.

volume was reduced to half the previous volume. As the pressure decreased (the piston retracted), the volume increased again. He concluded that the product of volume and pressure of an enclosed gas at a constant temperature remains the same (constant). Thus, he derived Boyle's Law:

$$V_1 \times p_1 = V_2 \times p_2 = \text{constant, or } \frac{V_1}{V_2} = \frac{p_2}{p_1}$$

It must be remembered that for calculation purposes one has to convert all given gauge pressures to absolute pressures.

Example

The accumulator in a hydraulic system with a precharge pressure of 900 kPa is filled with hydraulic fluid until the gas pressure shows 2700 kPa. How much hydraulic fluid has been pumped into the accumulator if the volume of the accumulator is 0.4 m^3?

$V_1 = 0.4\ m^3$ $p_1 = 900\ kPa + 101.3\ kPa$

$V_2 = ?$ $p_2 = 2700\ kPa + 101.3\ kPa$

$$\frac{V_1}{V_2} = \frac{p_2}{p_1} \qquad V_2 = \frac{p_1 \times V_1}{p_2}$$

$$V_2 = \frac{1001.3\ kPa \times 0.4\ m^3}{2801.3\ kPa} = 0.143\ m^3$$

Therefore, the compressed gas volume is 0.143 m^3 and, the total volume being 0.4 m^3, the remaining volume of 0.257 m^3 has been filled with fluid.

Theoretically this is what would occur with an ideal gas. But because density changes with temperature (that is, gases expand when heated and contract when cooled, see fig. 13), there will be some minor changes in practice with regard to pressure and volume. (A more detailed explanation is given in Chapter 8.)

Whereas Boyle's Law describes the change of state of a gas at a constant temperature, experimental work by Charles (in which pressure was maintained constant while temperature varied, and the change in volume was measured) enabled him to formulate another gas law.

Charles' Law states: at constant pressure the volume of a gas varies in direct proportion to a change in temperature. Expressed in a formula this is:

$$\frac{V_1}{V_2} = \frac{T_1}{T_2}, \text{ or transposed } \quad V_2 = \frac{V_1 \times T_2}{T_1}$$

To solve problems of this nature absolute values of temperature and pressure must be used.

Absolute zero on the Kelvin temperature scale is equivalent to −273°C. One Kelvin temperature unit is equal to 1°C. Thus 100°C equals 373 Kelvin.

Example

A balloon with a gas volume of 0.1 m^3 at a temperature of −14°C is heated to a temperature of 90°C. What is its increased gas volume if the pressure remains constant?

$$V_2 = \frac{V_1 \times T_2}{T_1} = \frac{0.1\ m^3 \times (90° + 273°)}{(-14° + 273°)} = 0.14\ m^3 \quad \text{(fig. 13)}$$

Gay-Lussac's Law supplements the gas laws of Charles and Boyle. He observed that if a volume of a gas is kept

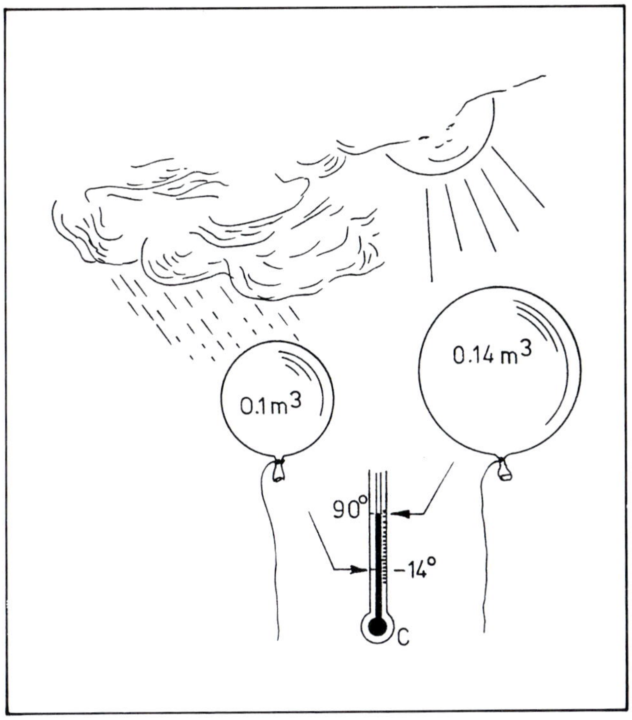

Fig. 13 Pressure maintained — temperature increased.

constant, the pressure exerted by the confined gas is directly proportional to the absolute temperature of the gas. Expressed in a formula this is:

$$\frac{p_1}{p_2}=\frac{T_1}{T_2}, \text{ or transposed } p_2=\frac{p_1 \times T_2}{T_1} \quad \text{(fig. 14)}$$

The combination of the three gas laws of Boyle, Charles, and Gay-Lussac results in the general gas law:

$$\frac{p_1 V_1}{T_1}=\frac{p_2 V_2}{T_2} \quad \text{or} \quad \frac{pV}{T}=\text{constant}$$

Transmission of power

There are four basic methods of transmitting the power of a prime mover to a machine. These methods are electric, mechanical, hydraulic, and pneumatic. Each is used to transmit power and modify motion, and each method has its special capabilities and limitations. Hydraulic systems integrate the various methods to accomplish a most effective and efficient form of power transmission.

Advantages of hydraulic power transmission

- Freedom of location of input and output power converters such as prime movers, pumps, and actuators.
- Great efficiency and economy due to low friction losses and high system reliability (efficiency is approx. 70%–80%)-
- Safety and overload protection by means of relief valves.
- Emergency power stored in an accumulator.
- Infinitely variable control of output force, output torque, output speed, and actuator position.
- Extremely high output forces and force multiplication by means of the "hydraulic lever".
- Low inertia and ease of shock absorption during actuator motion, reversal, start, and stop.
- Hydraulic systems are self-lubricating and power can be diverted to alternative actuators.

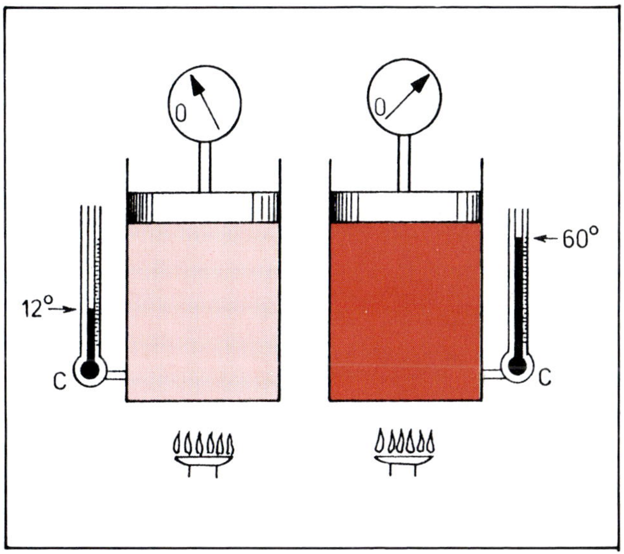

Fig. 14 Volume maintained — temperature increased.

The concept of power transmission

Power is the measure of a defined force moving through a given distance at a given speed. To understand this fundamental concept, the term force must be explained.

Force may be defined as any cause which tends to produce or modify motion. Due to inertia, a body at rest tends to stay at rest and a body in motion tends to maintain that motion until acted upon by an external force. Force is measured in Newtons.

The concept of pressure must also be explained. Pressure is force per unit area and is expressed in Pascals (Pa). Both force and pressure are primarily measures of effort. A force or pressure may be acting upon a motionless object without moving the object, if the force or the pressure is insufficient to overcome the inertia of the object.

Work is a measure of accomplishment. For example, the piston of a hydraulic actuator exerts a force on an object and moves the object over a given distance. Thus, work has been accomplished. The concept of work, however, makes no allowance for the time factor. The work performed by the hydraulic actuator while moving the object from point A to point B in 8 seconds is

$$\text{Pressure (p)} = \frac{\text{Force}}{\text{Area}}$$

$$\text{Force (F)} = \text{Pressure} \times \text{Area}$$

$$\text{Area (A)} = \frac{\text{Force}}{\text{Pressure}}$$

$$\text{Power (P)} = \text{Force} \times \text{Velocity}$$

$$\text{Force (F)} = \frac{\text{Power}}{\text{Velocity}}$$

$$\text{Velocity (v)} = \frac{\text{Power}}{\text{Force}}$$

$$\text{Power (P)} = \text{Pressure} \times \text{Flowrate}$$

$$\text{Pressure (p)} = \frac{\text{Power}}{\text{Flowrate}}$$

$$\text{Flowrate (Q)} = \frac{\text{Power}}{\text{Pressure}}$$

Fig. 15 Relationships of force, pressure, area, work, and power.

precisely the same as when it moves the object from point A to point B in only 2 seconds, but the task performed in 2 seconds is obviously much greater. To explain the difference of performance one must resort to the definition of power.

Power is work performed per unit of time. Thus it can be said that power is the rate at which energy is transferred or converted into work. Power is expressed in Watts. Mathematical relationships of force, pressure, area, work, and power are shown in fig. 15.

The basic concept of a hydraulic system

Hydraulics is the engineering science of liquid pressure and liquid flow. Hydraulic power transmission systems are concerned with the generation, modulation, and control of pressure and flow, and in general such systems include:

- Pumps which convert available power from the prime mover to hydraulic power at the actuator.
- Valves which control the direction of pump-flow, the level of power produced, and the amount of fluid-flow to the actuators. The power level is determined by controlling both the flow and pressure level.
- Actuators which convert hydraulic power to usable mechanical power output at the point required.
- The medium, which is a liquid, provides rigid transmission and control as well as lubrication of components, sealing in valves, and cooling of the system.
- Connectors which link the various system components provide power conductors for the fluid under pressure, and fluid flow return to tank (reservoir).
- Fluid storage and conditioning equipment which ensure sufficient quality and quantity as well as cooling of the fluid (fig. 16).

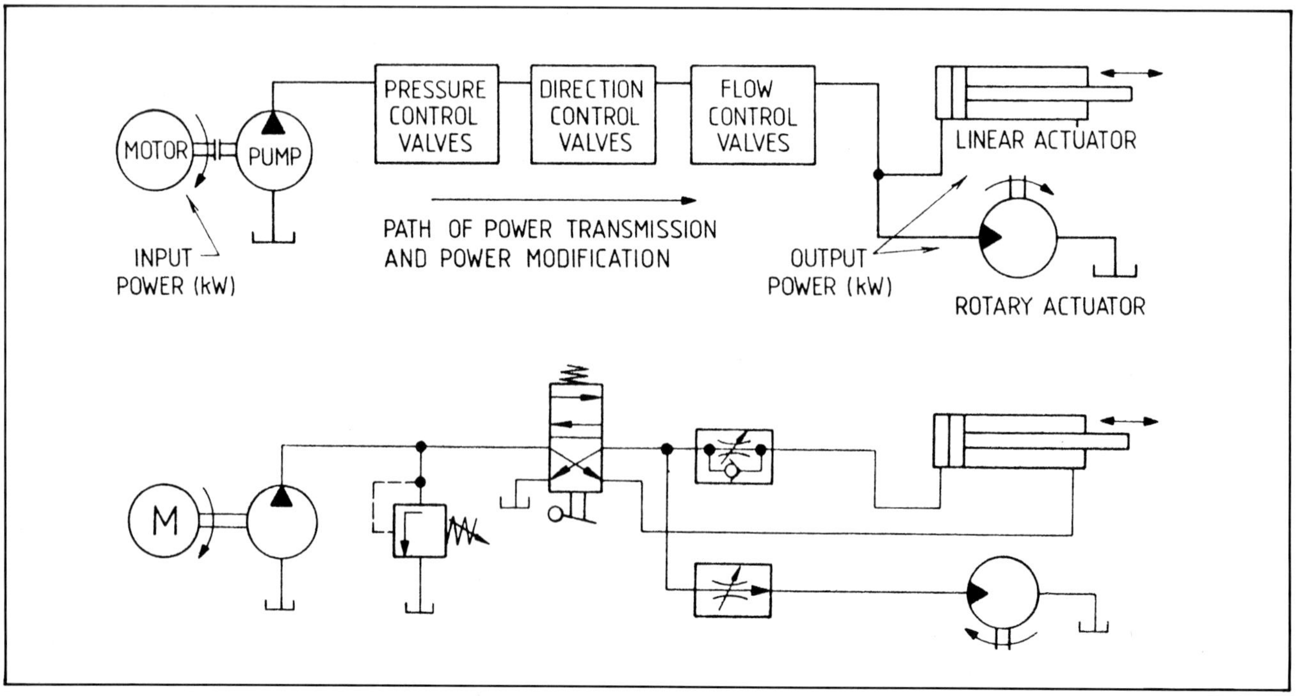

Fig. 16 Basic hydraulic system.

2 Directional control valves

A valve is a device which receives an external command (either mechanical, fluid pilot signal, or electrical), to release, stop, or redirect the fluid that flows through it.

Directional control valves in particular, as their name implies, control fluid-flow direction. They are applied in hydraulic circuits to provide control functions which:

- Control direction of actuator motion;
- Select alternative control circuits;
- Perform logic control functions.

Directional control valves are classified according to their design characteristics:

1. Internal valve mechanism (internal control element) which directs the flow of the fluid. Such a mechanism can either be a poppet, a ball, a sliding spool, a rotary plug, or a rotary disc.
2. Number of switching positions (usually two or three). Some valves may provide more than three, and in some exceptional cases up to six switching positions.
3. Number of connection ports (also called number of ways). These ports connect the hydraulic pressure lines to the internal flow channels of the valve mechanism and often also determine the flow rate through it.
4. Method of valve actuation which causes the valve mechanism to move into an alternative position. Fig. 25 shows an array of valve actuators.

Valve symbols

Symbols are an ideal way of drawing and explaining the function of fluid power components and of directional control valves in particular. Symbols exist for most of the more commonly used valves. They are standardised to provide and safeguard an internationally agreed form of circuit drawing. These standards should be closely adhered to, and only deviated from if they are not available for a particular valve, or if a valve consists of several standardised parts. Most fluid power valves are drawn according to I.S.O. standards (International Standards Organisation) or C.E.T.O.P. standards (European Fluid Power Standards Committee).

Valve switching positions

For each of the switching positions provided by a valve, the graphic symbol shows a square, sometimes called a box. This means the valve on the left in fig. 17 provides two switching positions and the valve on the right three. The non-actuated position of a valve is assumed by its moving parts (valve mechanism) when the valve is not connected, not pressurised, and not actuated. In most hydraulic control circuits all valves are depicted in this non-actuated position, and by convention two-position valve symbols should normally be connected on the right-hand square (fig. 17).

Three-position valves should be connected to the centre square, which again depicts the valve's non-actuated position (sometimes also called the neutral

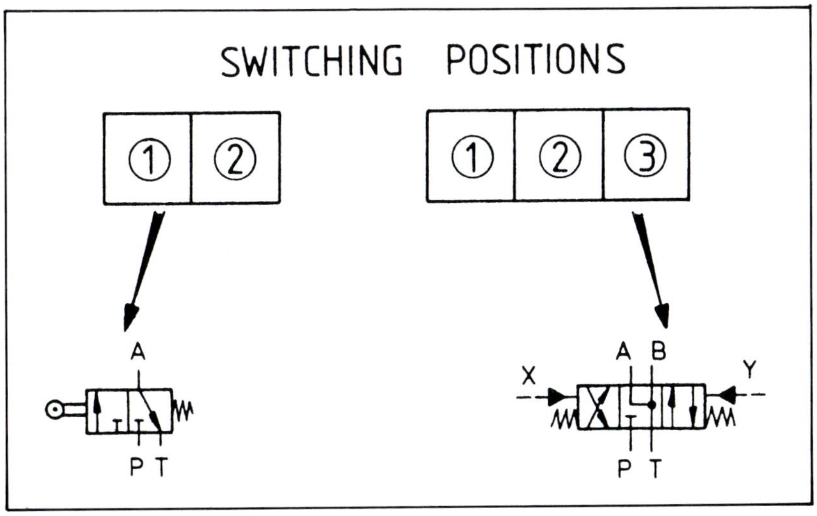

Fig. 17 Squares represent switching positions.

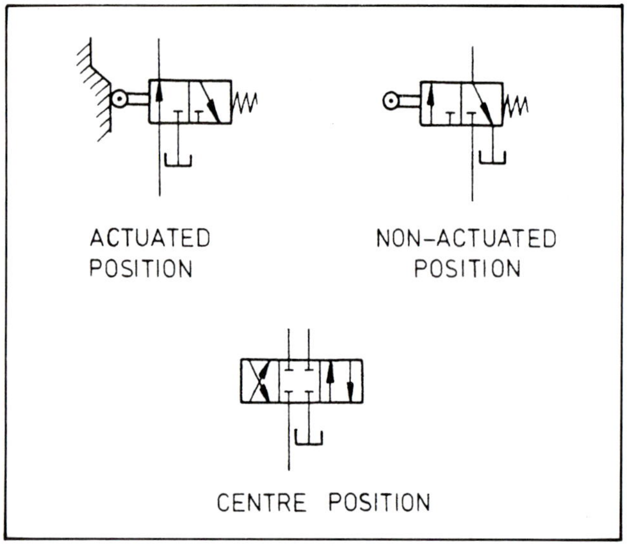

Fig. 18 Valve connection methods.

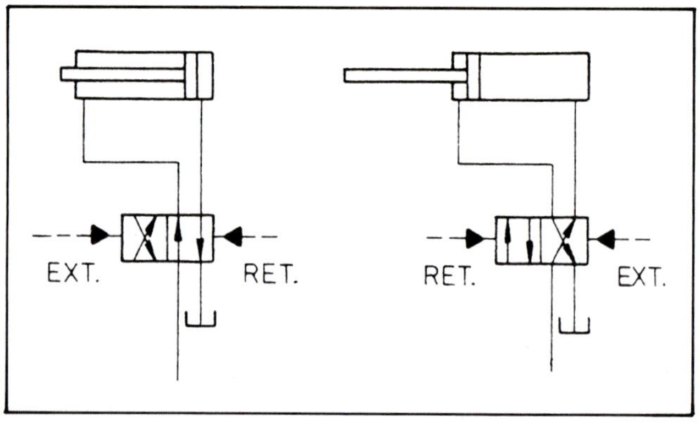

Fig. 19 Flow path configurations depend on initial actuator position.

position). However, it is the practice of some circuit designers to draw valves in circuit diagrams in their actuated or so-called initial position. This position is assumed when the valve is installed in the machine and levers, cams or machine parts are pressing onto the valve's operating element, so that the valve's mechanism is no longer in the non-actuated position. Since this is not a standard practice, it is imperative to state on the circuit diagram that all valves are shown in their initial position. Also, each of the valves which are actuated must be shown with a cam, and all pressure lines are to be connected to the left-hand square (fig. 18).

Two-position valves which are not spring biased and can freely assume both switching positions (bi-stable), may be drawn in whatever flow-path configuration

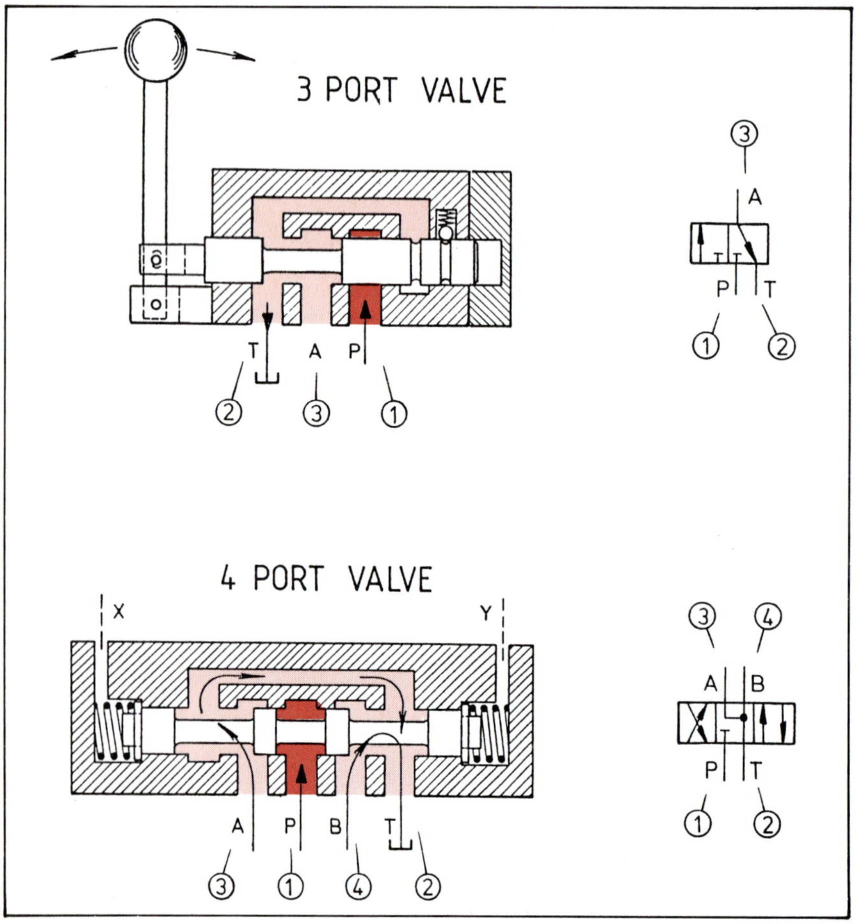

Fig. 20 Valve ports.

demanded by the actuator position. But here again, the pressure lines must be connected to the right-hand square (fig. 19).

Valve ports (ways)

Directional control valves are frequently described by the number of working ports. Two-port valves (two-way valves) are used either to open or close the flow path in a single line (fig. 41). This makes the two-port valve an on/off valve.

Three-port valves have three working ports (fig. 20). If a three-port valve has two inlet ports and one outlet port it is used as a selector valve to guide two flow lines to a common outlet (fig. 47). If the three-port valve has two outlets and one inlet it can likewise be used as a selector or shunt valve to determine which outlet will receive the flow.

Four-port valves are mainly used to control double-acting linear actuators or reversible hydraulic motors (fig. 19). One line (P) connects the valve to the pump, the ports A and B are connected to the actuator and port T connects the common return line (actuator exhaust fluid) to tank (reservoir).

Valve mechanism

Directional control valves consist of a valve body or valve housing and a valve mechanism usually mounted on a sub-plate. The ports of the sub-plate are threaded to hold the tube fittings which connect the valve to the fluid conductor lines. The valve mechanism directs the fluid within the valve body to the selected output ports, or stops the fluid from passing through the valve.

External signal commands (electrical, manual, pilot pressure), and internal signal commands (pilot pressure, spring force) may be applied to shift the valve mechanism (figs. 22 and 36).

Spool, rotary disc, or rotary plug (fig. 21) mechanisms are predominantly used for directional control valves, whereas poppets and balls are the preferred mechanisms for check valves and shuttle ("or" function) valves (figs. 40–45).

Valves which are spring biased are said to have a normal position. Thus, any external command will move the valve mechanism into the actuated position. In the case of a three-position valve as shown in fig. 25, the valve's normal position is the spring-centred position.

Normally closed and normally open valves

These terms are generally applied to three-port valves. The valve shown in fig. 22 (top), a two-position valve with three ports, permits no flow from pressure port P to output port A in its normal (spring biased) position. In the case of a normally open valve (fig. 22, bottom), the supply is permitted to stream to output port A while the valve is in its normal position. Both valves, when actuated, change their flow mode to the flow configuration depicted in the left-hand square of the valve symbol.

Methods of valve actuation

The term actuation in relation to hydraulic valves refers to the various methods of moving the valve mechanism. Valves can be actuated by five basic methods: manually, mechanically, electrically, hydraulically, or pneumatically.

In machine applications any of these methods, or combinations thereof, may be used to gain optimum control. Manual methods use hand or foot actuators such as levers, push buttons, knobs, and footpedals.

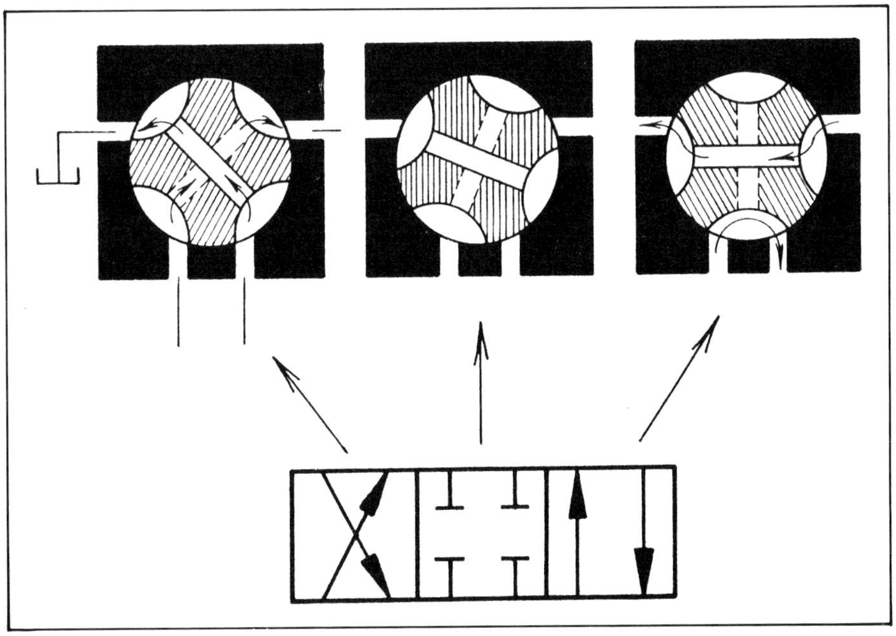

Fig. 21 Rotary plug valve mechanism.

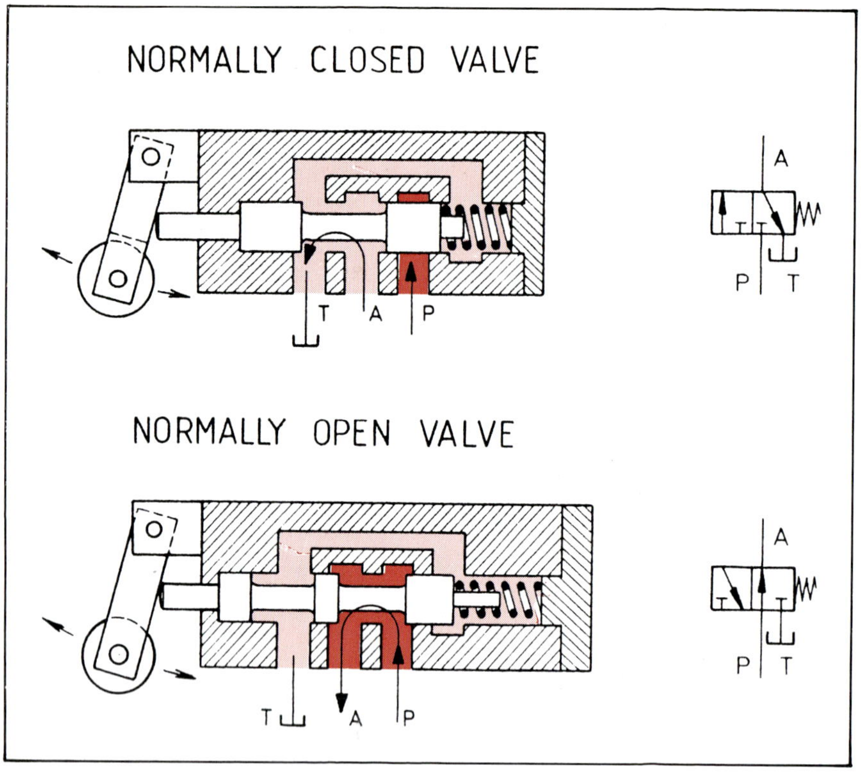

Fig. 22 Valves with spring bias (normal position).

SYMBOL	PORTS	POSITIONS	COMMENTS
	2	2	N/C (IF SPRING RESET)
	2	2	N/O (IF SPRING RESET)
	3	2	N/C (IF SPRING RESET)
	3	2	N/O (IF SPRING RESET)
	4	2	
	5	2	
	3	3	FULLY CLOSED CENTRE
	4	3	FULLY CLOSED CENTRE
	4	3	TANDEM CENTRE
	4	3	FULLY OPEN CENTRE
	4	3	PRESSURE CLOSED, A & B TO TANK

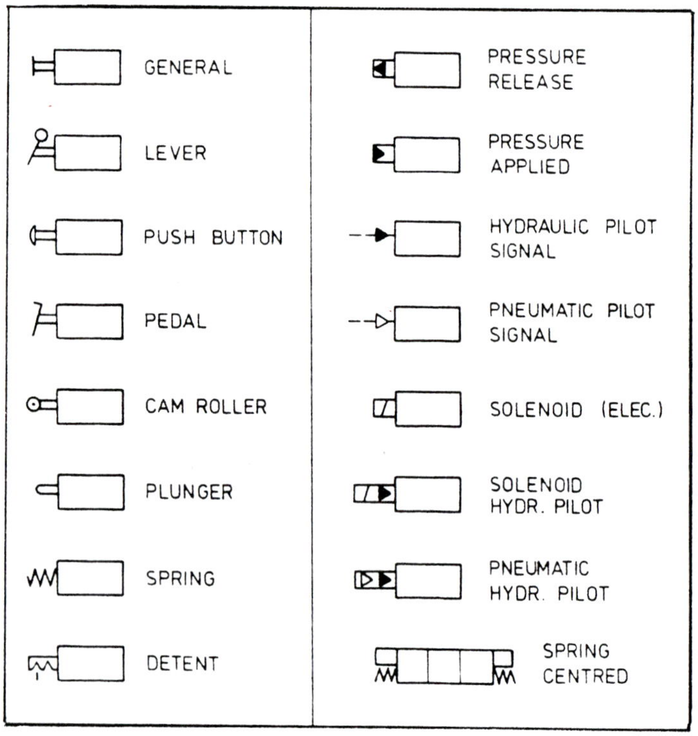

Fig. 24 *above:* Commonly used valve actuation methods.

Fig. 23 *left:* Commonly used valves.

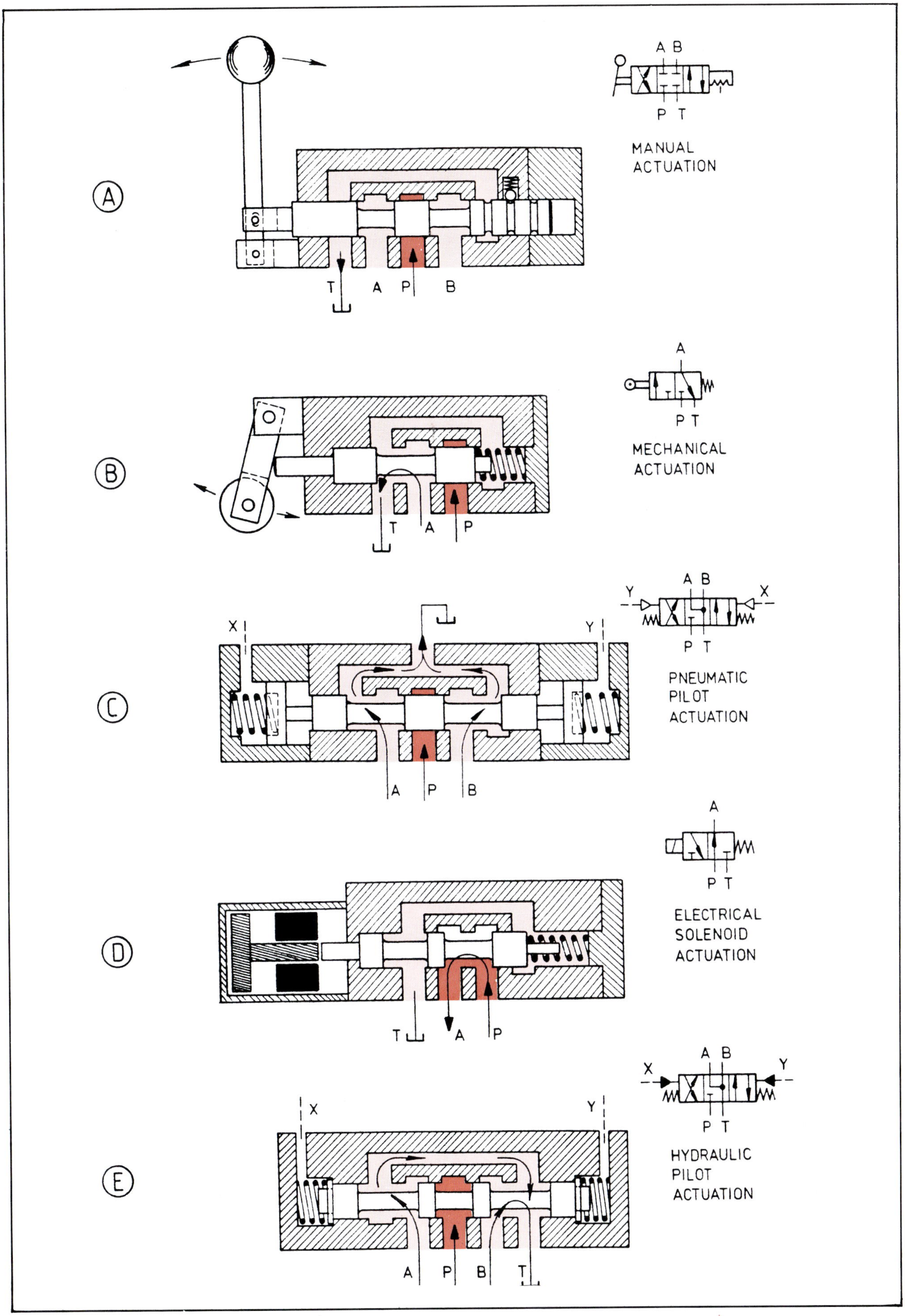

Fig. 25 Basic valve actuation methods applied to spool valves.

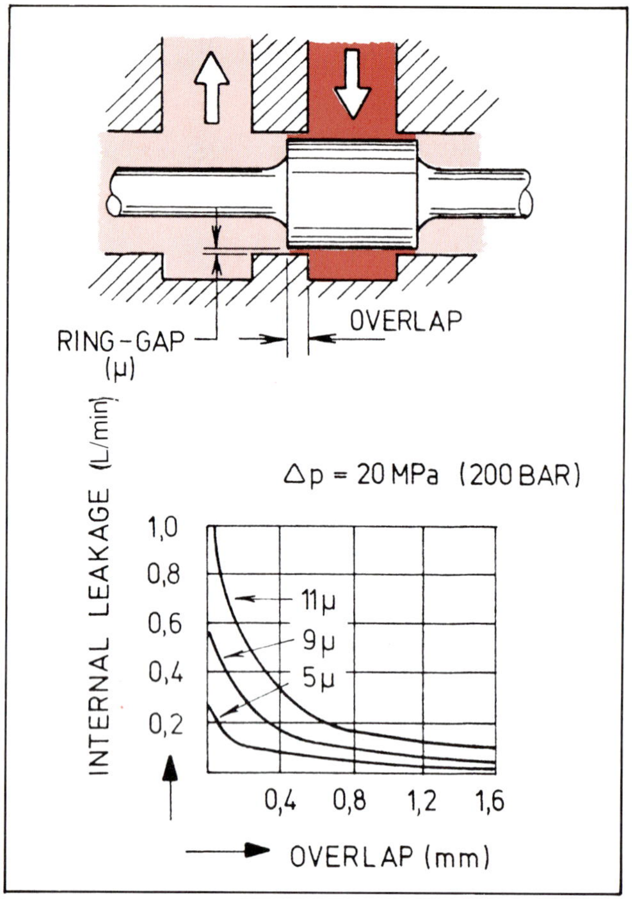

Fig. 26 Internal leakage on spool valves.

Electrical methods use A.C. or D.C. solenoids, either of the air-gap or oil-immersed type. These solenoids may move the valve mechanism directly, or provide a hydraulic pilot signal which in turn operates the valve mechanism (fig. 18).

In hydraulically and pneumatically operated valves, the fluid-pilot signal acts onto a piston, which in turn moves the valve mechanism to its alternative position (fig. 25C and E).

Spring-actuated valves

The terms spring-biased or spring-offset, and in the case of three-position valves spring-centred, refer to the application of springs which return the valve mechanism to its normal position. Since such springs will return the valve to its normal position as soon as the signal command is cancelled, such commands must be maintained as long as the valve is to remain in its non-normal or actuated position.

A valve without a spring bias is actuated entirely by its external controls, and can therefore "float" between its two extreme positions as soon as the external control is cancelled. To avoid this, a "detent" mechanism or friction pads may be built into the valve (fig. 25A), or either of the two control signals (which can be electric, hydraulic, or pneumatic) must be maintained throughout the cycle (fig. 25E).

Sealing and internal leakage

Within the spool valve pressurised cavities are sealed from non-pressurised cavities by means of a minimal clearance ring gap. This ring gap is between the spool land and the valve bore and may vary from 5 to 10 μm. The internal leakage which occurs past this ring gap depends on the:

- clearance tolerance of the ring gap (fig. 26);
- viscosity of the hydraulic fluid;
- overlap length of spool land and bore (fig. 26);
- differential pressure of the pressurised fluid and the non-pressurised cavities.

Valve centre condition

The centre position of a directional control valve is designed to satisfy specific requirements or conditions of the hydraulic circuit and its actuators. In order to satisfy a variety of circuit conditions, these valves come also in a variety of centre conditions. The valve body is generally of uniform design, but the valve spool differs (compare fig. 27 with figs. 28–30).

The fully open centre

This interconnects all ports and the pump flow is redirected to the tank at low pressure. At the same time, the actuator piston is free to move with the load. This means it can "float" to whatever position external forces drift it (fig. 27). The disadvantage of an open centre is that other actuators tied to the same pump (flow source) are not flow supplied while the valve is centred.

The fully closed centre

This blocks all ports and the actuator may be held fixed in an intermediate position for a short period only. Spool valves leak internally and if the valve is subjected to system pressure for more than a few minutes, pressure will build up on both actuator lines and the piston rod will extend. The fully closed centre valve does not provide flow re-direction when the valve is centred, but it does allow independent operation of other actuators connected to the same pump (fig. 28).

The tandem centre

This condition blocks port A and port B and may therefore be used to stop an actuator in any intermediate position, but for a short period only. It also allows the pump flow to be re-directed to the tank when no actuator movements take place.

The tandem centre valve also lends itself to be series connected to other tandem centre valves. Thus actuators may be operated individually or simultaneously (fig. 52). With this arrangement pump flow

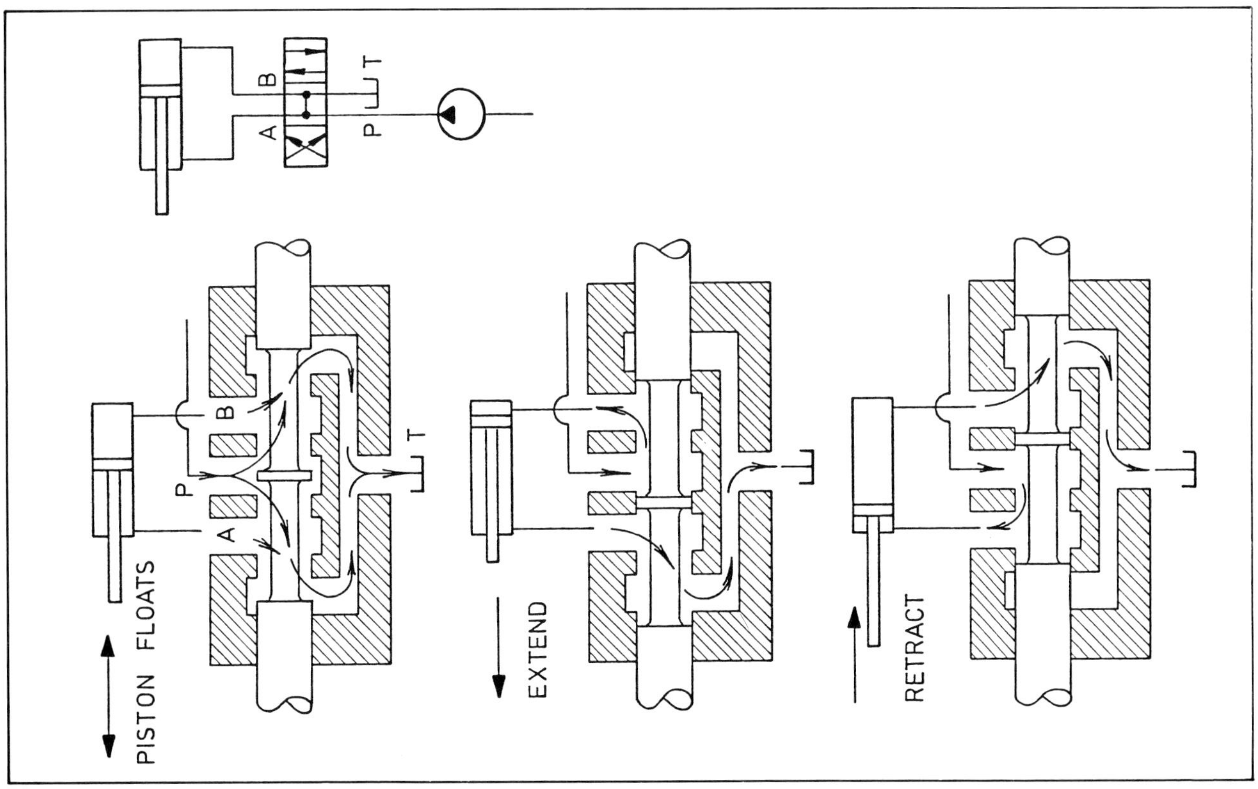

Fig. 27 Fully open centre.

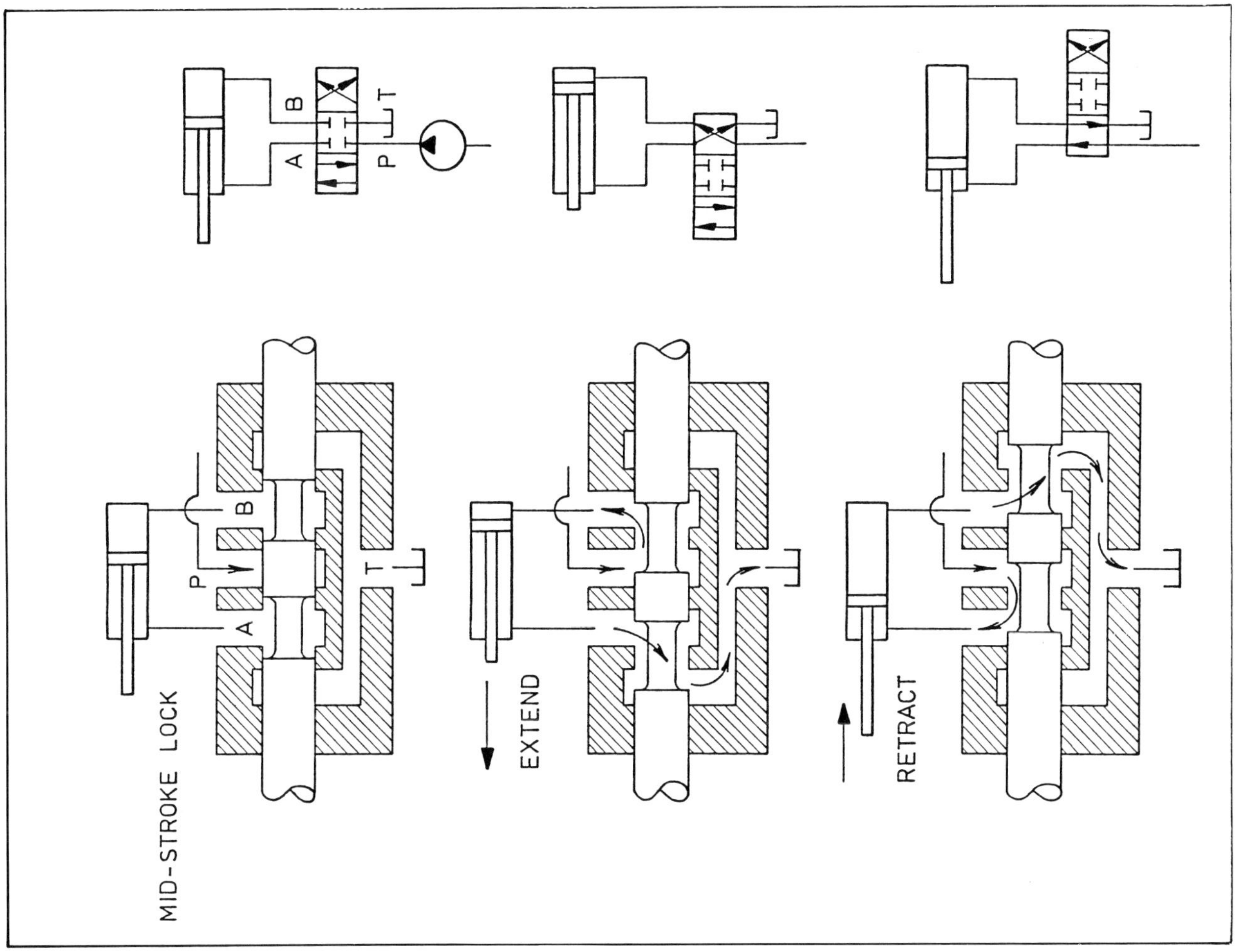

Fig. 28 Fully closed centre

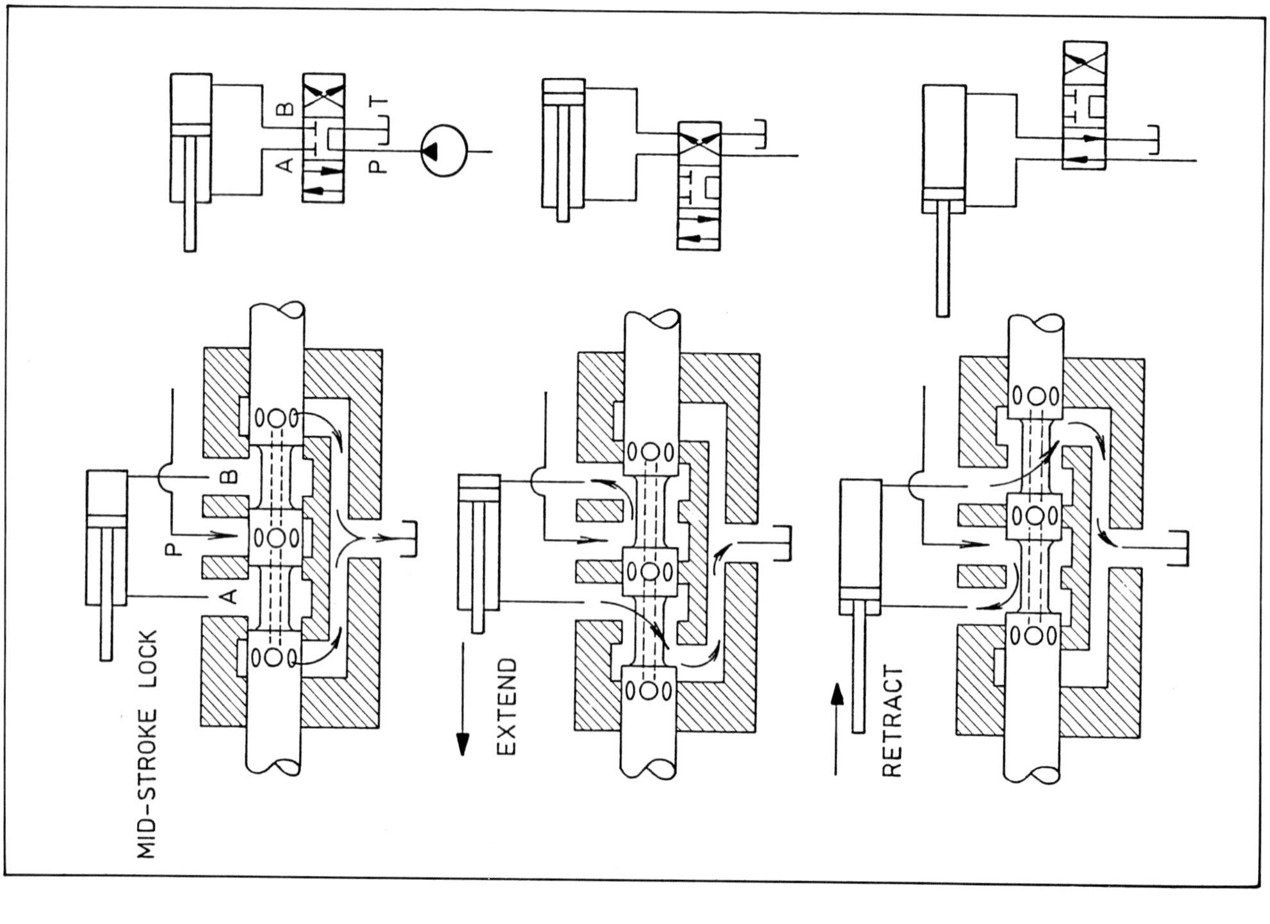

Fig. 29 Tandem centre.

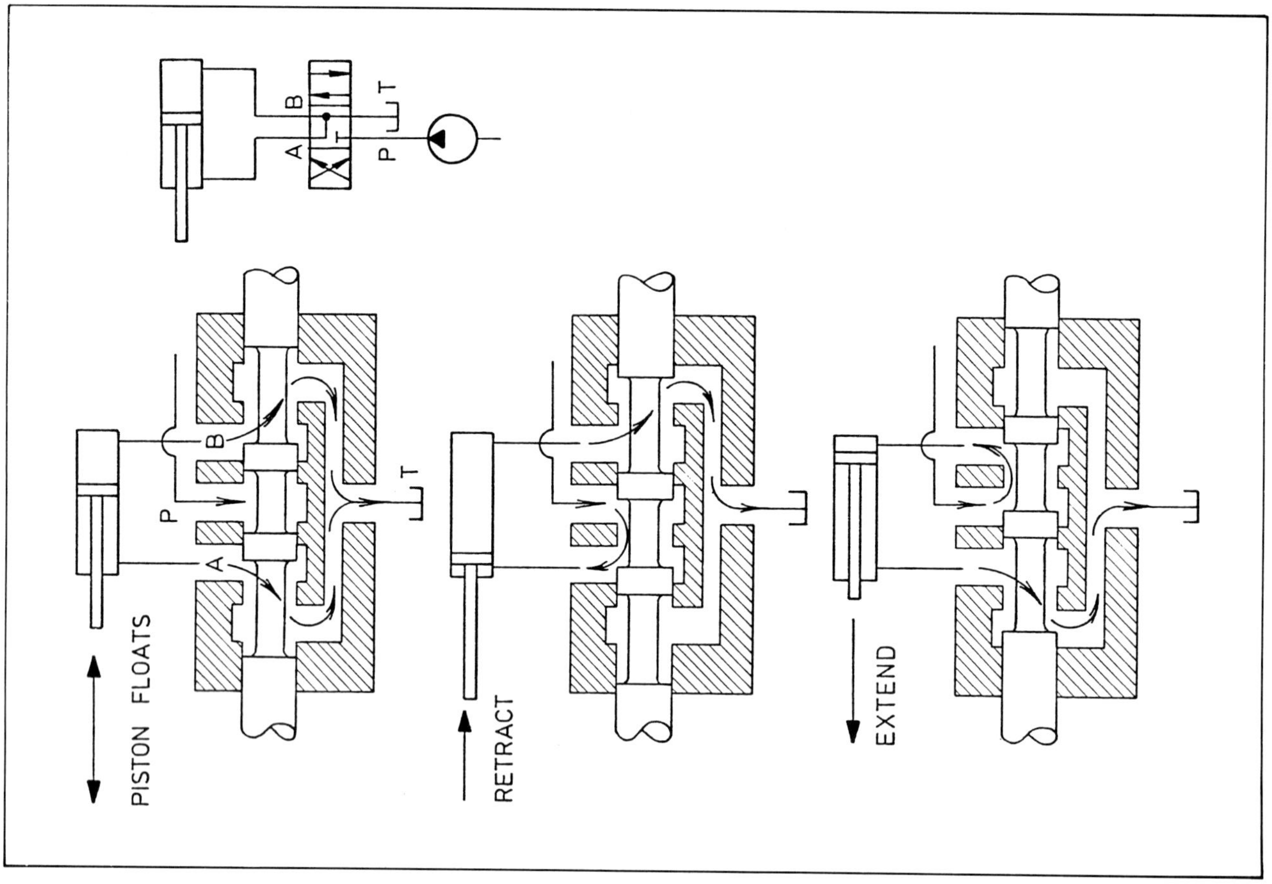

Fig. 30 Float centre.

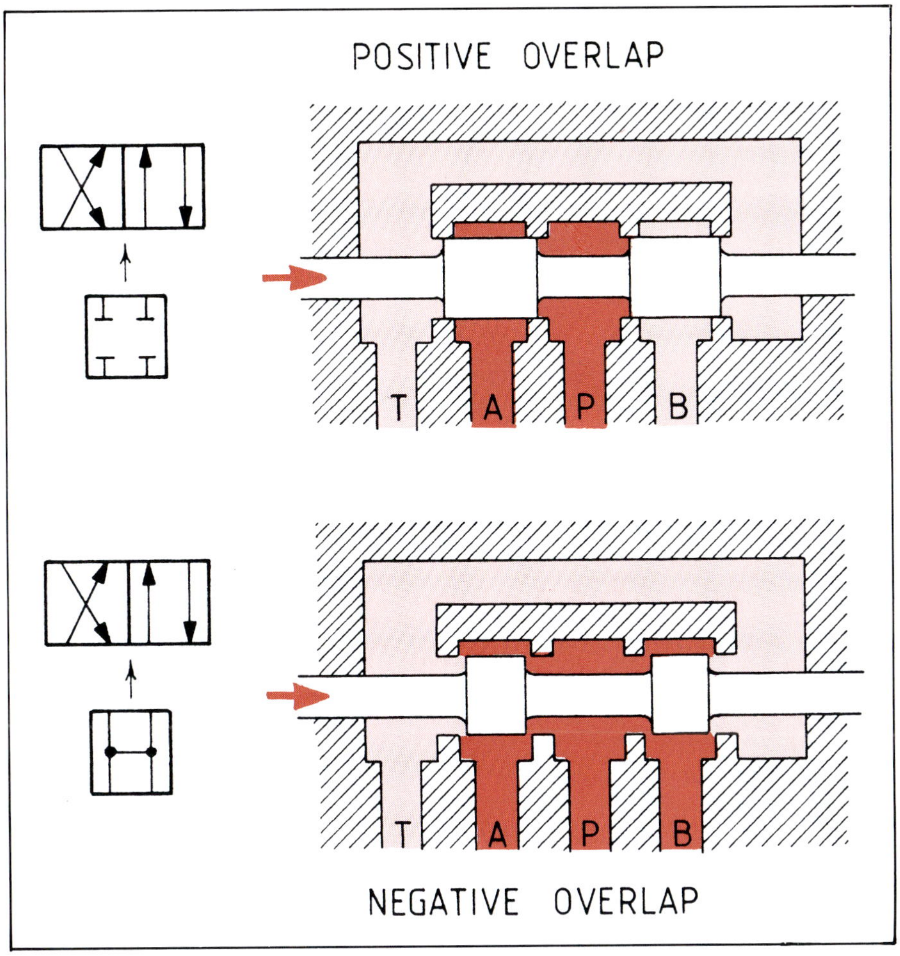

Fig. 31 Different spools for positive or negative overlap.

can still be re-directed to the tank. (See fig. 29 for the tandem centre valve.)

The "float" centre

This condition connects ports A and B to the tank while the pressure supply port P is blocked. This condition also allows the actuator to float, as already described for the fully open condition. Since the pressure supply port is blocked, independent operation of other actuators tied to the same pump is possible (fig. 30).

A further advantage of the float centre condition is that pressure in the actuator lines cannot build up, since both ports A and B are drained to the tank. This permits the float centre valve to be used to stop a load in an intermediate position, provided a pilot-operated check valve is connected to it (see circuit application fig. 43).

Cross-over condition

Directional control valves with only two selectable valve positions are in reality three-position valves with a special centre condition. This centre condition is known as the transition or cross-over condition. For most two-position valves this cross-over is a fully closed or fully open condition. The cross-over condition is shown between interrupted lines (fig. 32) and is only drawn into the valve symbol when a special condition is required for the functioning of the circuit.

The fully closed centre cross-over is achieved with a positive spool overlap (fig. 31). For a short period all valve ports are sealed against each other. Thus system pressure on the actuators is prevented from collapsing during the cross-over. The fully closed centre cross-over may however cause undesirable pressure peaks in the system which vary with the amount of fluid flow and switching time.

The fully open centre cross-over is achieved with negative spool overlap (also called underlap) (fig. 31). For a short period all valve ports are connected to each other. This results in smooth, pressure-peak free switching during the cross-over. However, undesirable actuator movements may occur with certain load conditions.

Figure 32 illustrates a typical machine tool application for a fully closed cross-over condition. A fully open cross-over condition for this application would not be suitable since the actuator with its load would momentarily lunge downward during the position change of the valve.

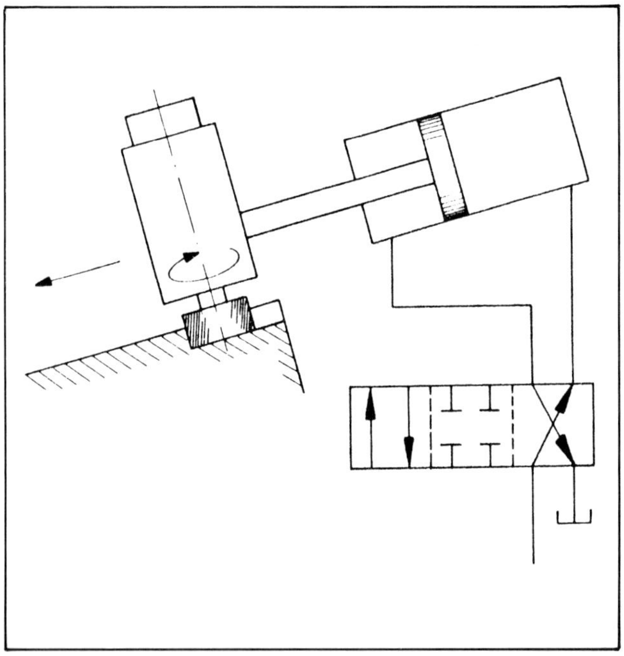

Fig. 32 Fully closed cross-over condition (positive overlap).

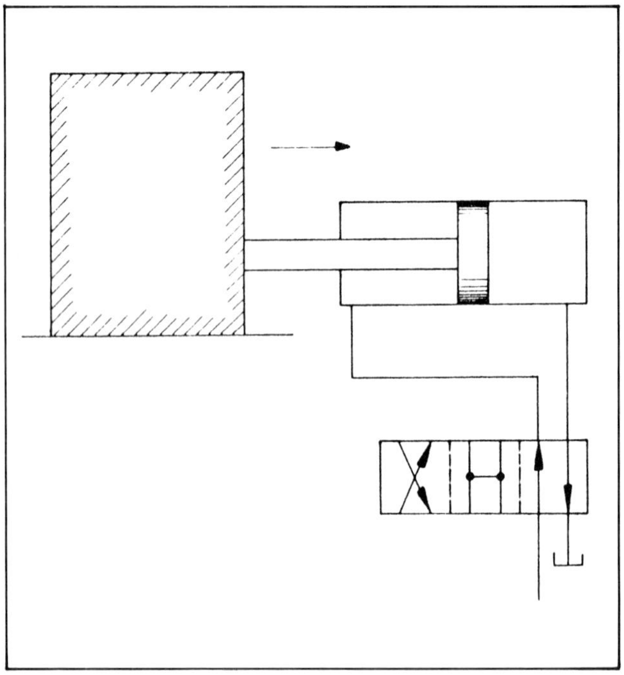

Fig. 33 Fully open cross-over condition (negative overlap).

Large capacity control valves

The force needed to move the spool in a large-capacity directional control valve can be very considerable, and if the valve is to be solenoid actuated, it would require an extremely large solenoid. A small solenoid or pneumatically operated master valve is therefore mounted on top of the large capacity valve (slave valve) and it directs hydraulic pilot signals to either end of the slave valve spool.

See fig. 34 for a cross-sectional drawing of an electrically controlled (solenoid) pilot valve, which in turn operates a directional control valve, and see fig. 35 for its graphic symbols.

Pilot choke control

During slave-spool position change-over in pilot operated directional control valves, pressure shock can occur, as large fluid quantities are forced to change direction quickly. Pilot choke control allows adjustment of the speed at which the slave spool shifts. Thus, pressure shocks are reduced or avoided. A choke control block containing two flow control valves is "sandwiched" between the master valve and the slave valve (fig. 36). Pilot choke control can be "meter-in" or "meter-out".

Pilot pressure sources

The detailed symbol for a pilot operated directional control valve in fig. 35 shows that the pressure supply for the pilot valve is either internally connected to the slave valve, or the flow to the pilot can come from an external source. Certain control situations make it necessary to provide the valve with external pilot pressure. This may be the case when internal pilot pressure is either too high, too low, or is fluctuating and unreliable. When external pressure is required, the internal line must be plugged (fig. 34) and pilot fluid enters the valve through a separate port, usually labelled "X".

Back pressure check valve

Pilot operated directional control valves which have the P port connected to the tank by the slave spool centre condition, and are internally pilot pressure supplied (fig. 39), require a back pressure check valve to retain sufficient pilot fluid pressure. Some valves have a 450 kPa back pressure check valve built into the valve. These check valves may be located in the P port passage with an effect similar to the arrangement shown in figs. 38 and 39. (For check valve description and illustration, see fig. 40.)

Check valves

The check valve is a very special type of directional control valve as it only permits fluid flow in one direction and blocks flow in the reverse direction (fig. 40).

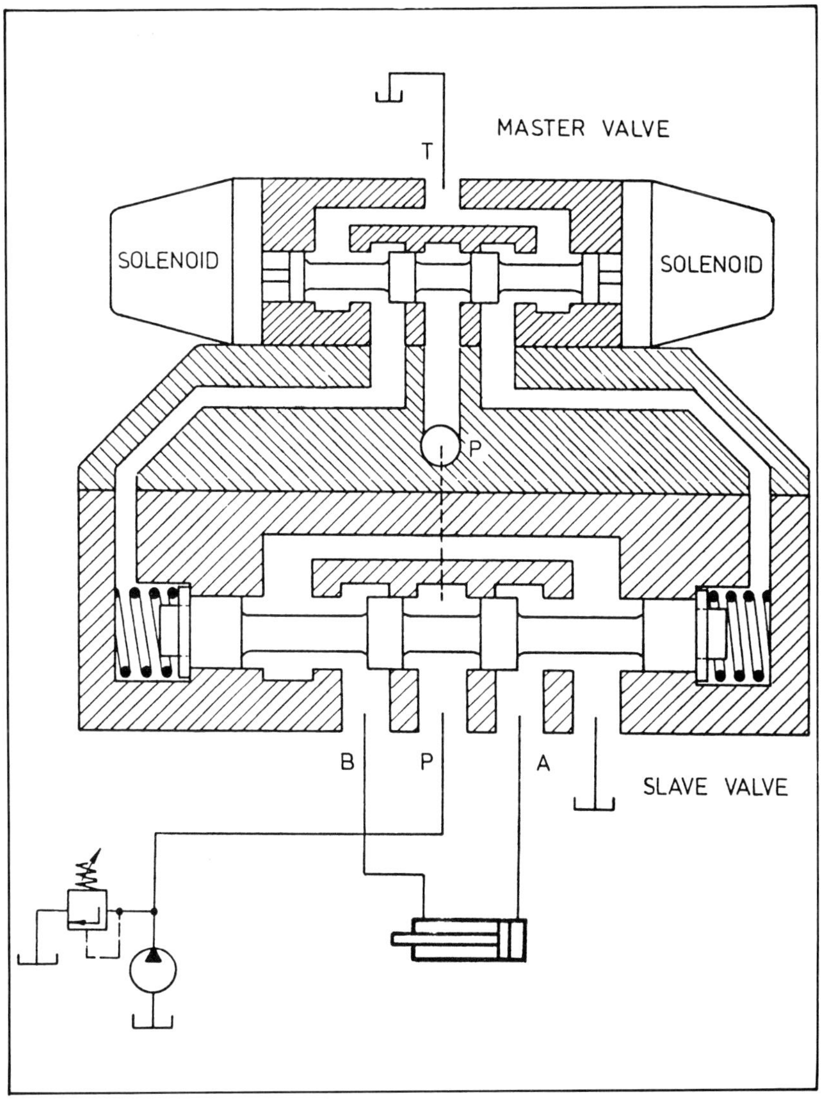

Fig. 34 Pilot operated valve: the master valve sits on top of the slave valve.

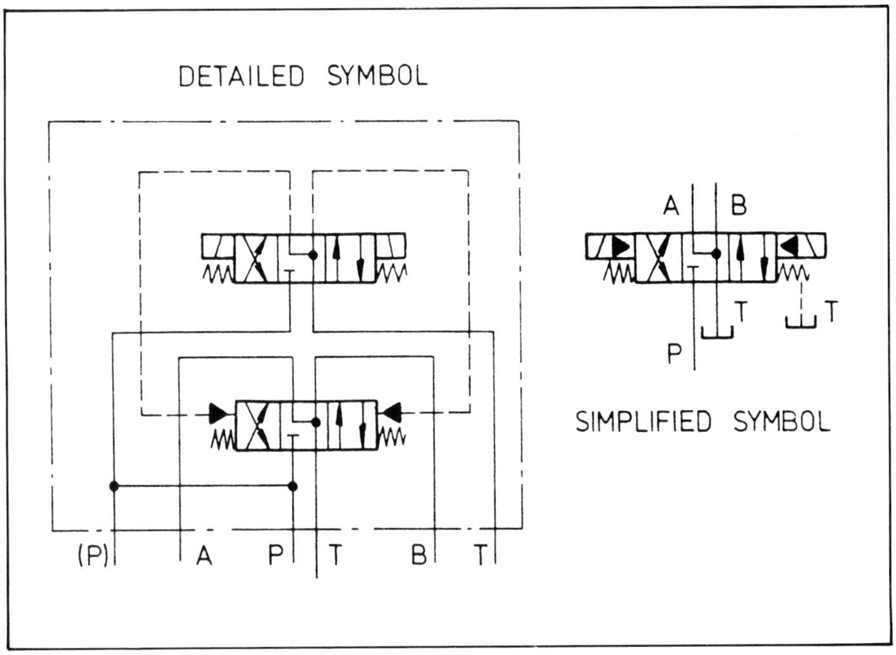

Fig. 35 Symbols for solenoid controlled pilot operated valve. (Also available with pneumatically pilot operated master valve.)

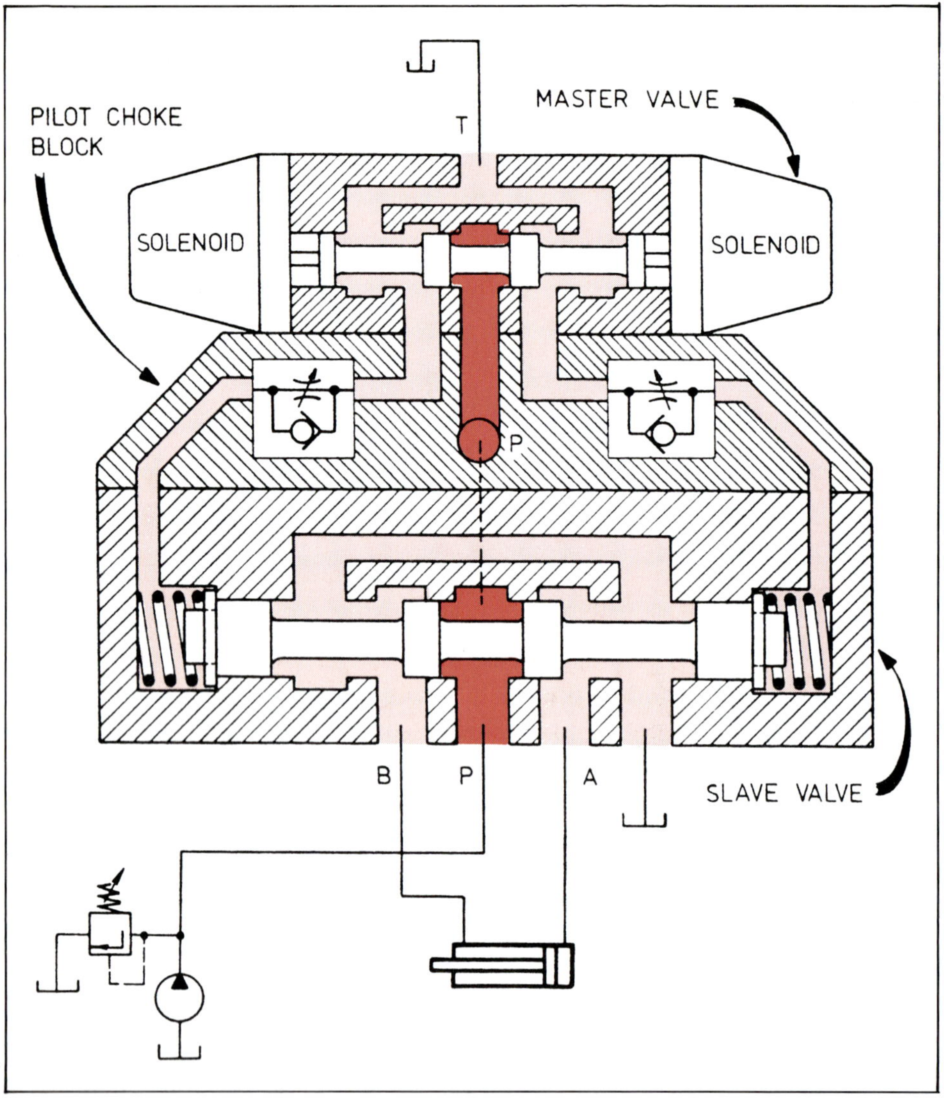

Fig. 36 Pilot choke control to reduce and control the speed of spool shift.

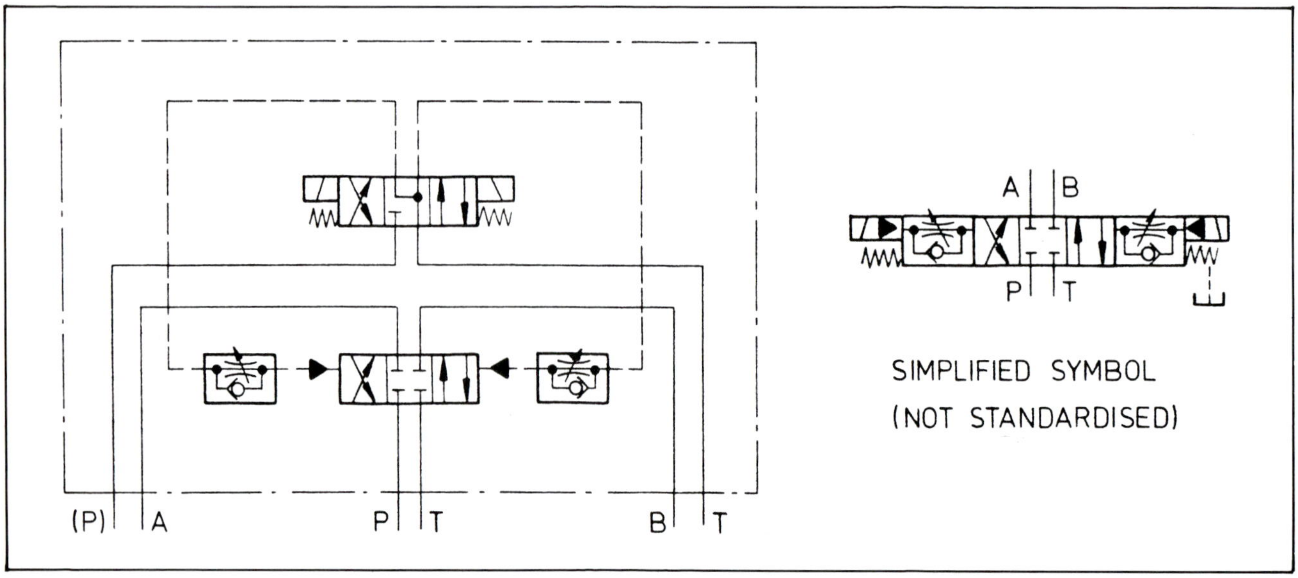

Fig. 37 Detailed symbol for pilot operated directional control valve with inbuilt choke control.

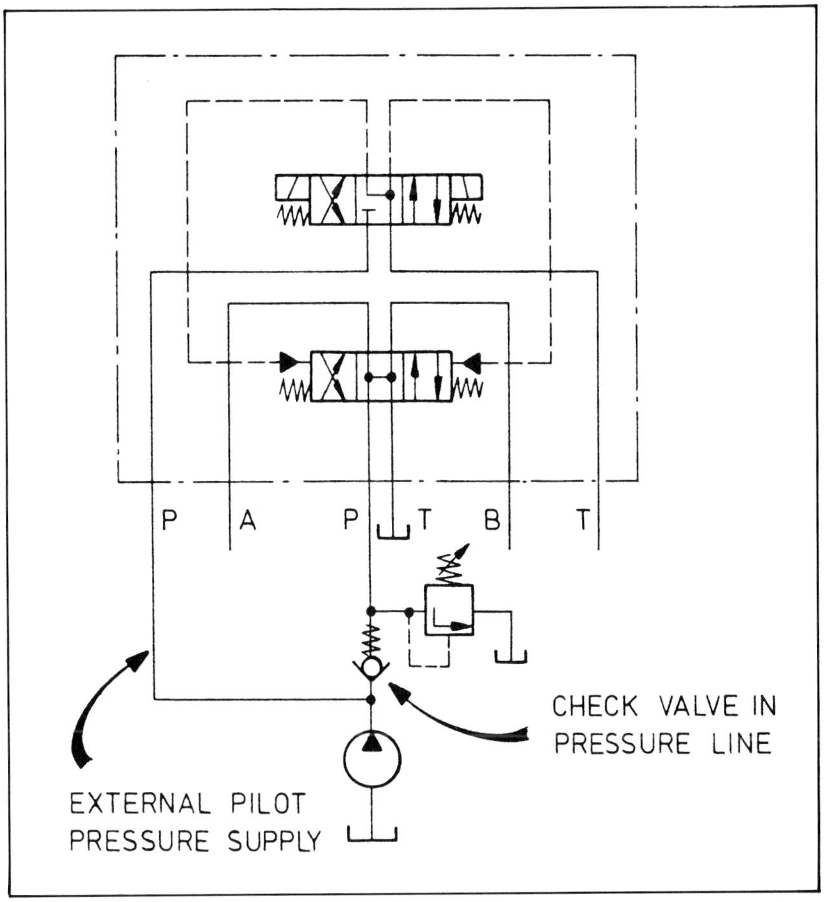

Fig. 38 450 kPa check valve in pressure line creates pilot pressure whenever the pump is running. Some valves may have the pilot check valve already located in the the D.C.V. body.

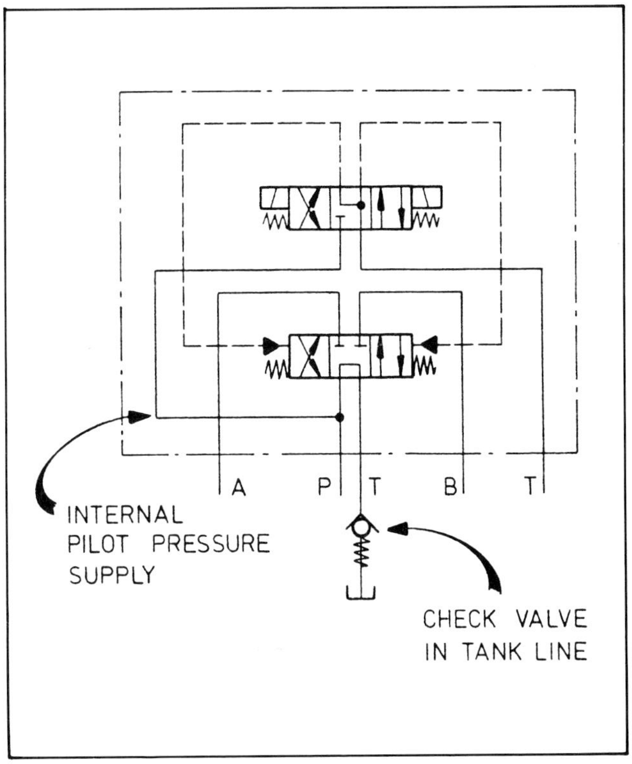

Fig. 39 450 kPa pilot pressure check valve maintains pilot pressure.

Its valve mechanism is either a ball or a poppet which fits onto a valve seat and has extremely good sealing characteristics compared with the spool mechanism discussed previously. The more the pressure increases against the poppet, the harder the poppet is pressed against its seat, thus ensuring a tight seal. A typical application for check valves is shown in fig. 41, where they control the flow of fluid direction in a hydraulic lifting jack such as is used to lift cars.

Pilot operated check valves

In certain applications control of reverse flow through the check valve is desirable. A pilot operated check valve provides this function (fig. 42). A typical application of pilot operated check valves is shown in fig. 43.

The pilot operated check valves prevent the pressurised fluid that is trapped in the cavities on either side of the piston, from leaking past the spool lands of the directional control valve. Thus, the piston is hydraulically locked or fixed in its position. When the actuator is signalled to extend, the pilot line to the left-hand check valve is pressurised, and the poppet is pushed off its seat. Thus, the fluid on the rod side of the piston can exhaust to the tank.

For the retraction function of the actuator, the pilot

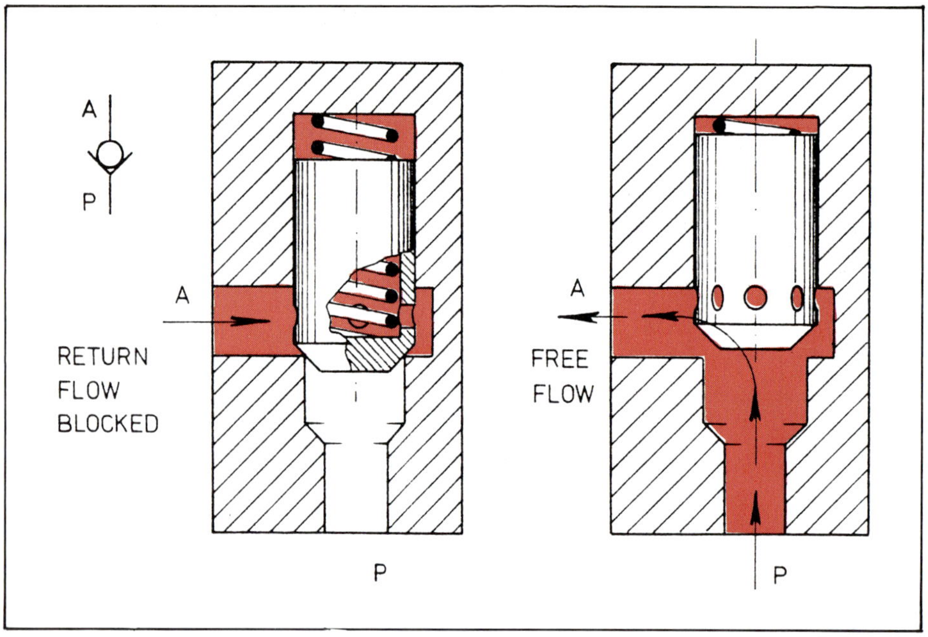

Fig. 40 Poppet type check valve.

pressure builds up in the line leading to the right-hand check valve. Thus, the right-hand check valve is pushed open and the actuator can retract.

The pilot piston of the check valve must have a larger area than the poppet seat (fig. 42) to ensure that the valve can be opened even if the pressure in the trapped fluid increases due to load or heat induced pressure build-up. The ratio of these areas, known as the opening ratio, may range from 1.5:1 up to 7:1.

A special application of a pilot operated check valve is the pre-fill valve. A large pilot operated check valve is used to fill the actuator of a press by gravity flow in order to allow rapid extension of the actuator during the approach stroke. On the return stroke, the pilot opens the check valve and the actuator pushes the fluid back into the overhead tank (fig. 44).

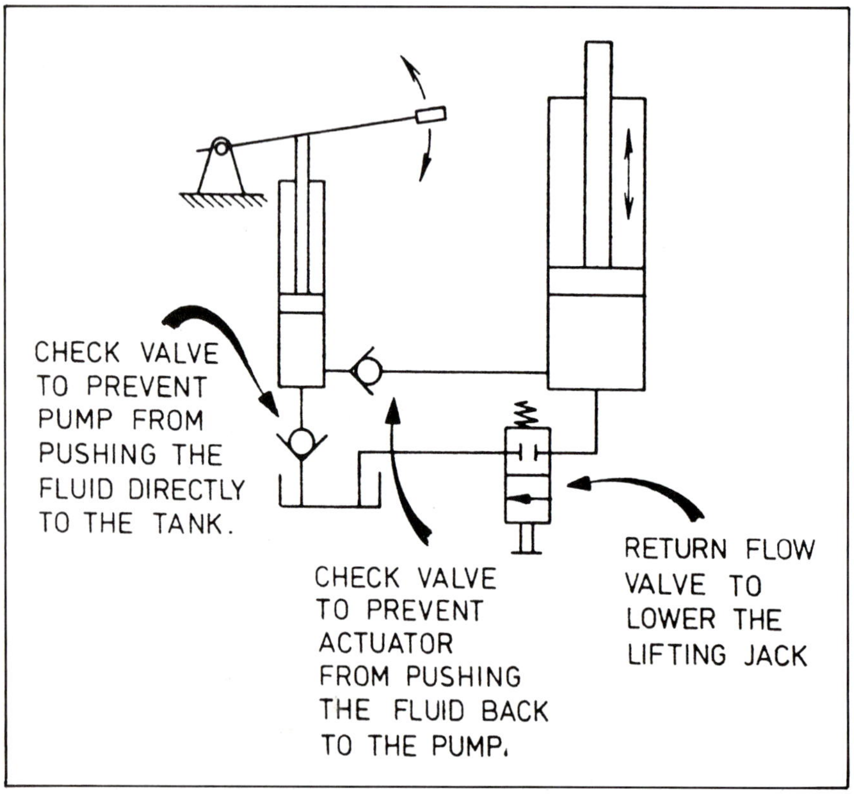

Fig. 41 Typical check valve application in a lifting jack.

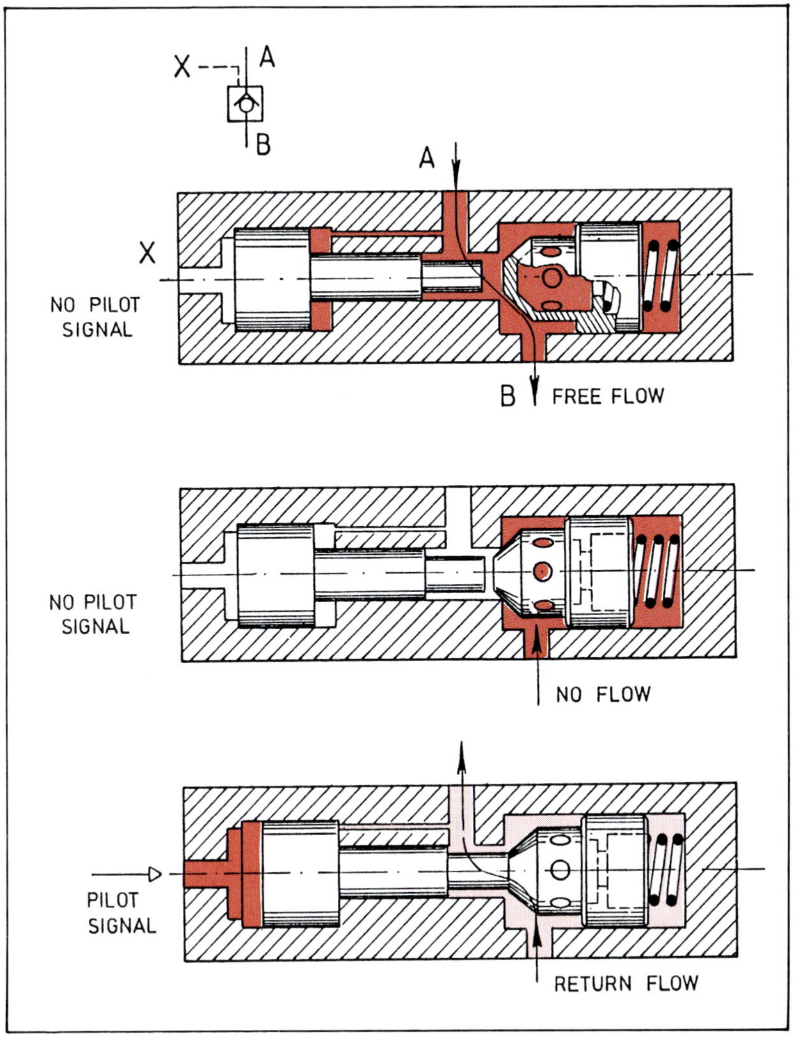

Fig. 42 Pilot operated check valve.

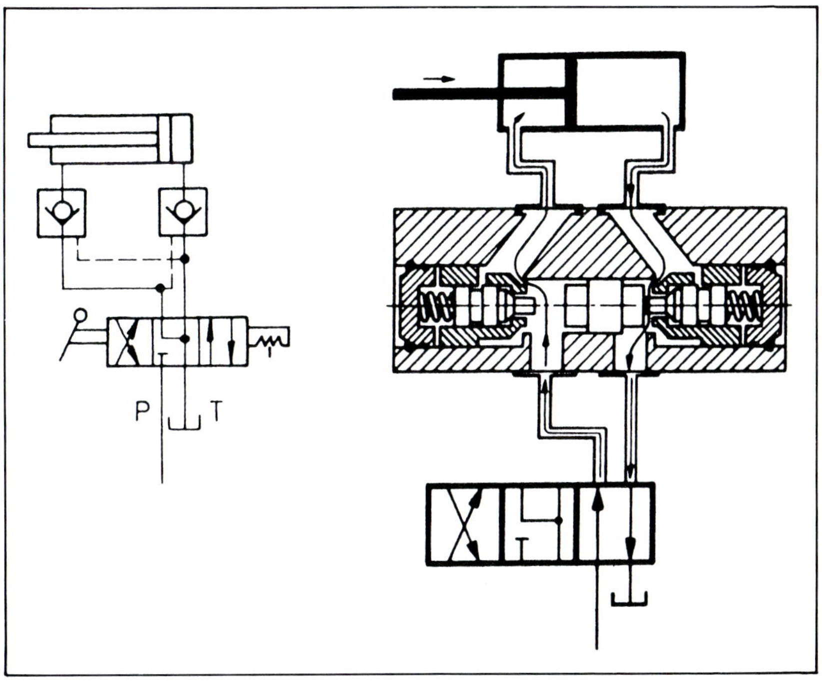

Fig. 43 Typical application for pilot operated check valves.

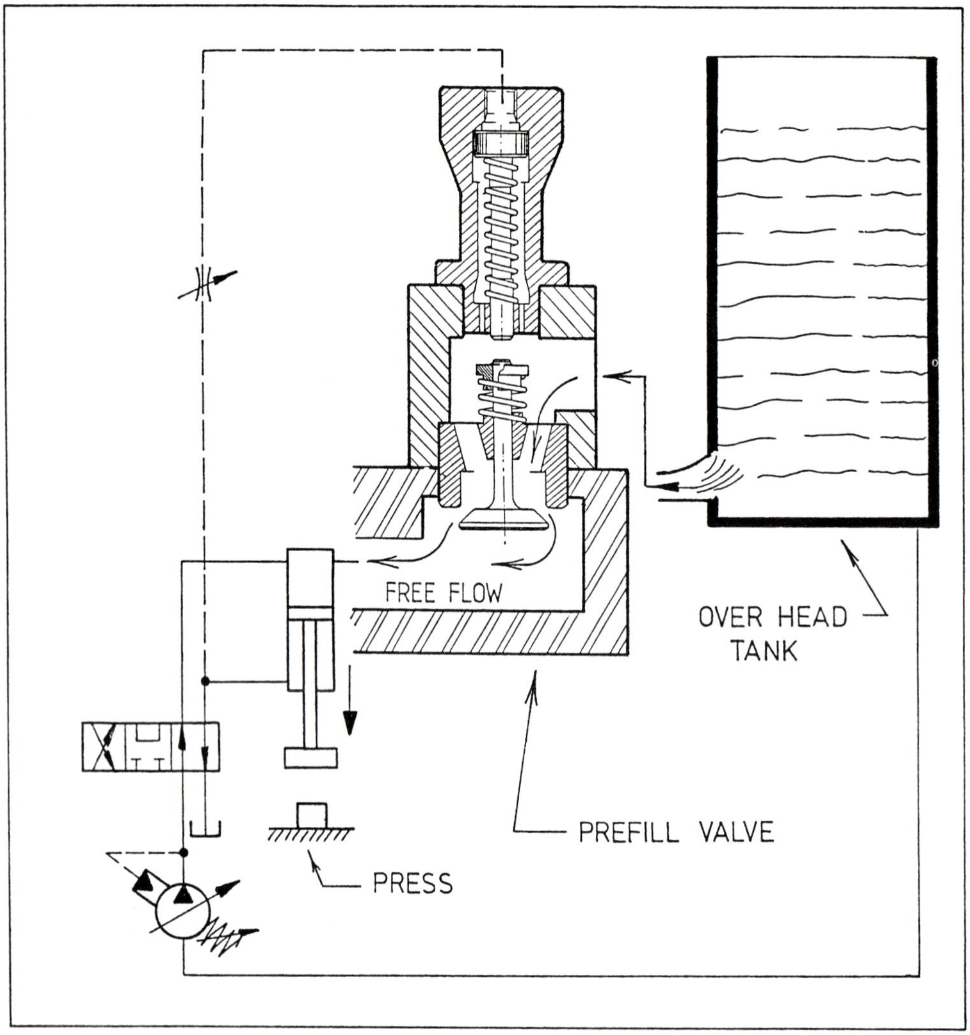

Fig. 44 Pilot operated check valve used as pre-fill valve in a press circuit.

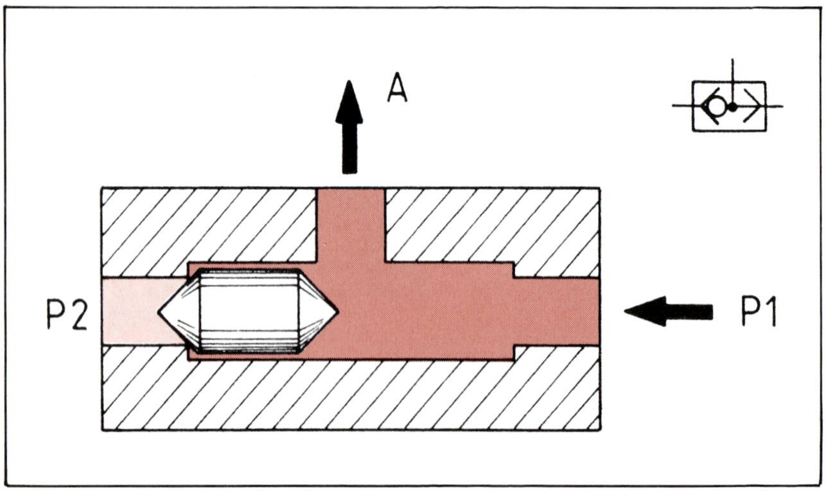

Fig. 45 "or" function valve (shuttle valve).

"Or" function valve (shuttle valve)

The "or" function valve is a special type of check valve. It has two valve seats but only one valving element, which can either be a double ended sealing cone or a sealing ball. The "or" function valve has two inlet ports and one outlet port (fig. 45). Signal P1 may pass through to port A but is prevented from passing to port P2, conversely, signal P2 can pass through to port A but is prevented to pass to port P1.

The "or" valve can for some applications be replaced by two check valves facing each other. But where the outgoing signal A is used as a pilot signal to actuate a directional control valve, this application would be wrong since the pilot could not exhaust back to the source. (See fig. 50 for locked-in pilot signal.) Applications for "or" function valves are shown in figs. 46 and 47.

Directional control valves for logic functions

Directional control valves may be used to perform logic control functions. One such function is the "or" function as depicted in figs. 46 and 47, where the actuator may be signalled to extend either from start 1 push button valve "or" from start 2 push button valve. The shuttle valve prevents the hydraulic signal from exhausting through the tank port of the other start valve. The "or" function may also be achieved by means of a two position, three-port valve, normally closed, and pilot operated, as shown in fig. 47.

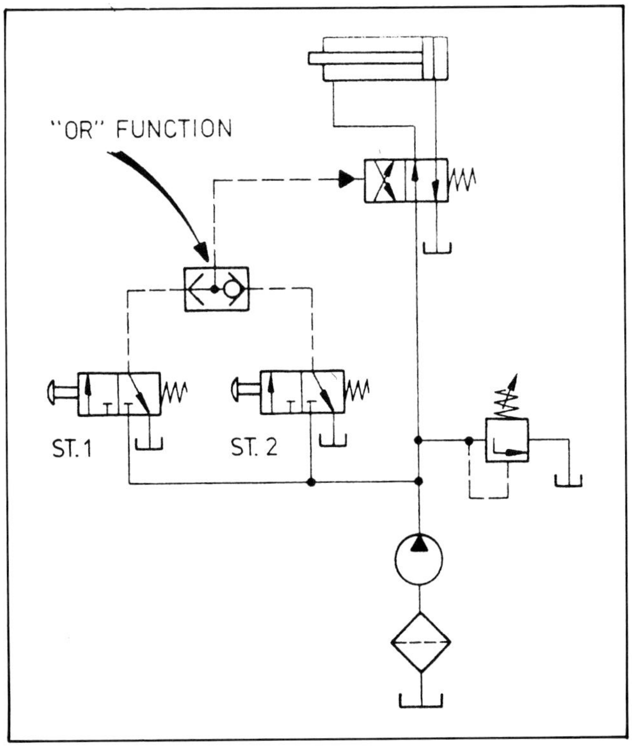

Fig. 46 "or" function with shuttle valve.

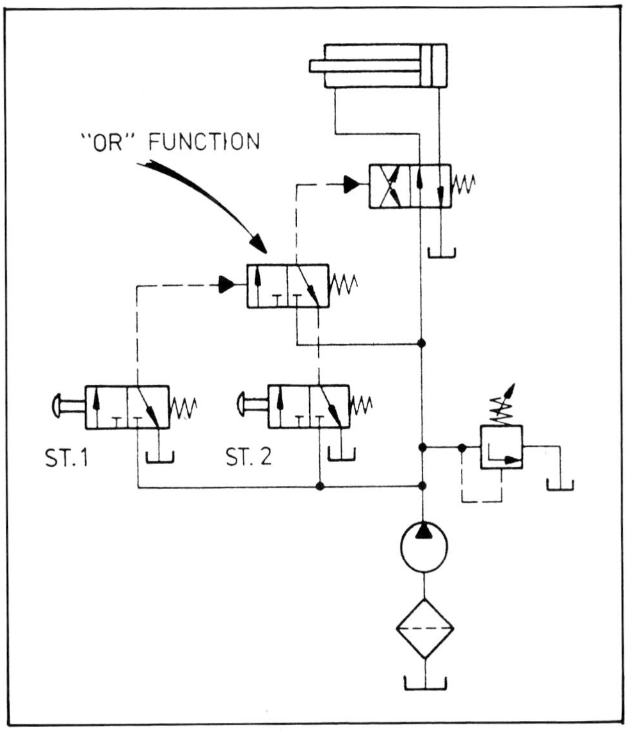

Fig. 47 "or" function with two position, three port directional control valve.

The "and" function is another logic function which is easily achievable with directional control valves (fig. 48). The "and" function may also be achieved by arranging the signal valves in series order as shown in fig. 49. The actuator is signalled to extend if the start valve is actuated "and" the actuator is fully retracted. This is signalled by the limit valve a_0 at the end of the retraction stroke (figs. 48 and 49).

Locked-in pilot signals

Pilot signals must be able to exhaust when the opposite pilot signal is applied and the valve spool moves (figs. 50 and 51). Two-position valves, as depicted in fig. 50 on the right-hand side, do not allow the trapped pilot signal to exhaust and are therefore not suitable for pilot signal control.

Sequence control with directional control valves

The actuator circuit depicted in fig. 51 extends when the piston is fully retracted and the start valve is actuated. (The "and" function is formed by the limit valve a_0 and the start valve.) The actuator retracts automatically when the piston rod is fully extended. The retraction signal is given by the limit valve a_1.

The hydraulic pilot operated directional control valve can only select two positions and maintains these positions until the reverse signal is given. This type of directional control valve would be of the "no spring" or "no spring, detented" type.

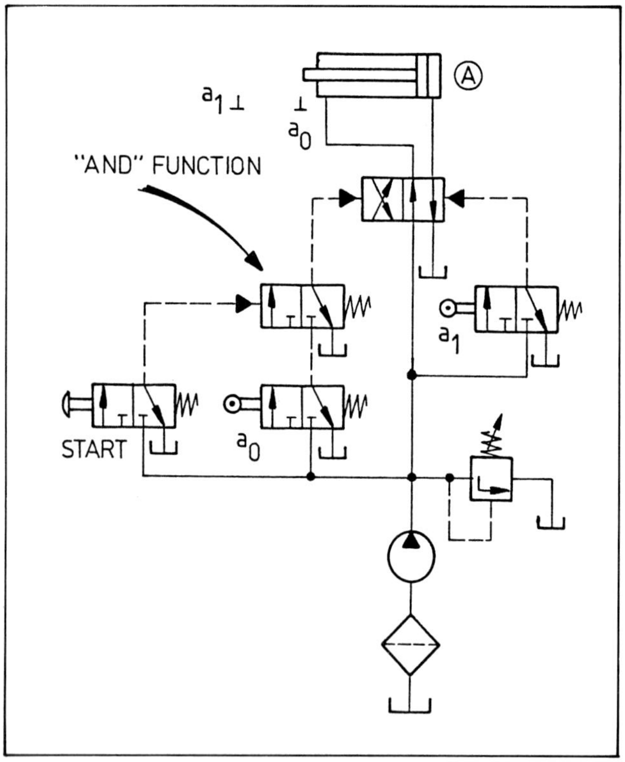

Fig. 48 "and" function with directional control valve.

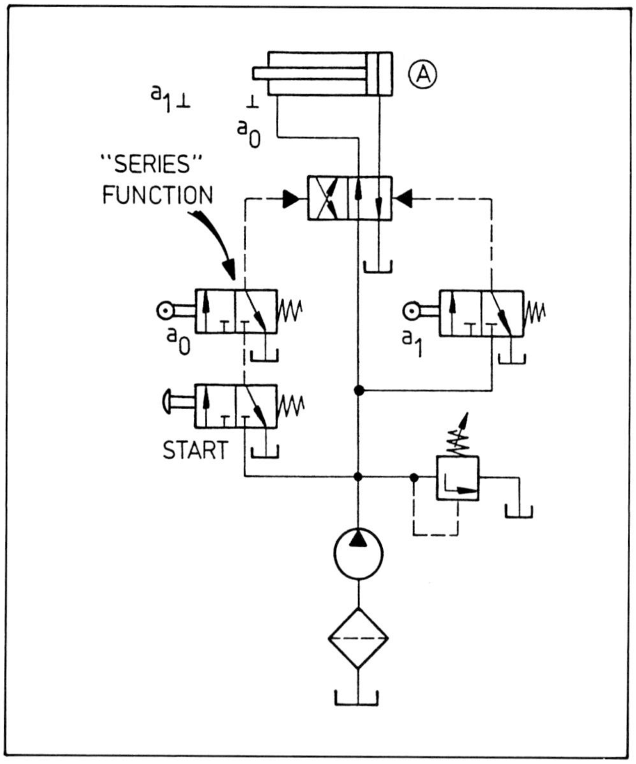

Fig. 49 "and" function by series arrangement.

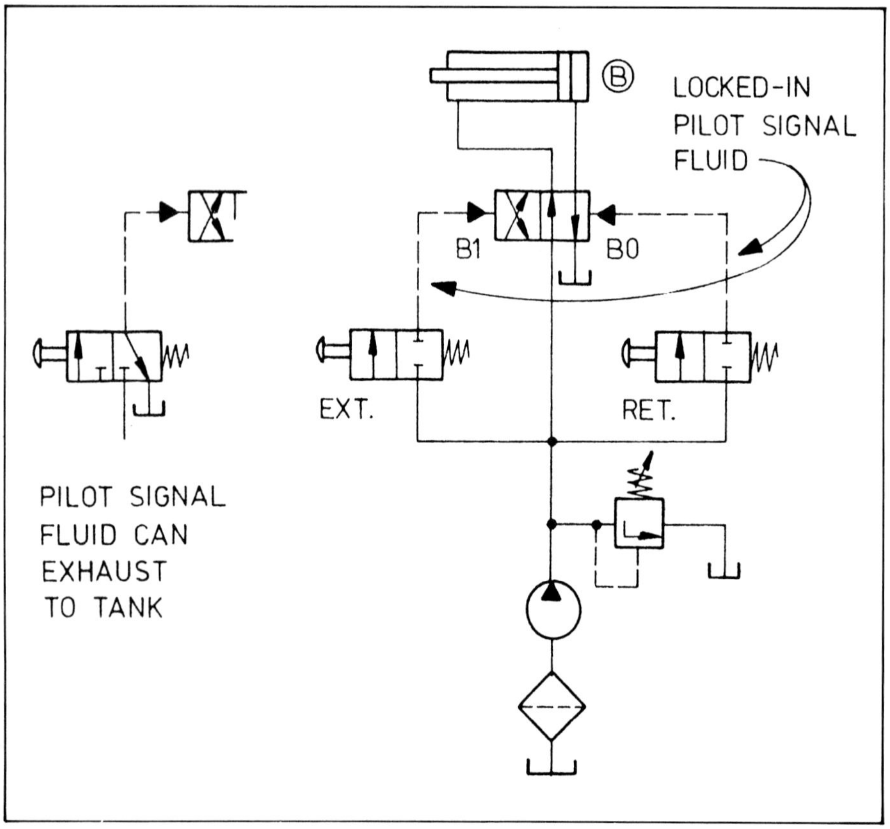

Fig. 50 Prevention of locked-in pilot signals.

Directional control valves in series

With tandem centre directional control valves in the power circuit, actuators can be operated simultaneously or individually as required (fig. 52). However, it must be noted that the pressure for the actuators B and C is diminished, since each tandem centre valve creates an extremely high pressure drop of approximately 300 kPa. This means that the input pressure for actuator C is reduced to only 1400 kPa, when the input pressure into the directional control valve for actuator A is measured to be 2000 kPa! Furthermore, it must be noted that actuators B and C depend for their position and speed on the flow displacement of the previous actuator.

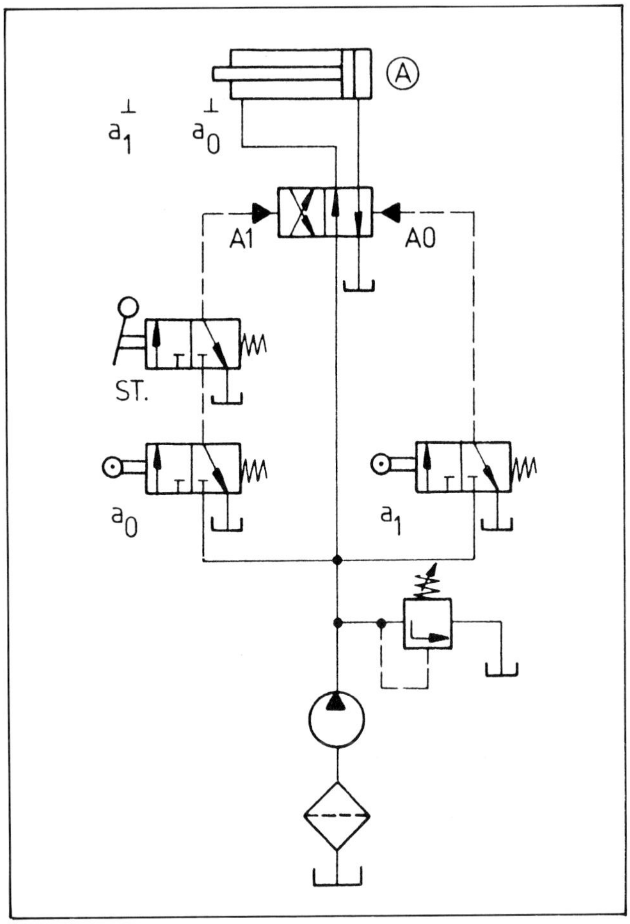

Fig. 51 Sequence control entirely with hydraulic directional control valves.

Fig. 52 Directional control valves in series connection.

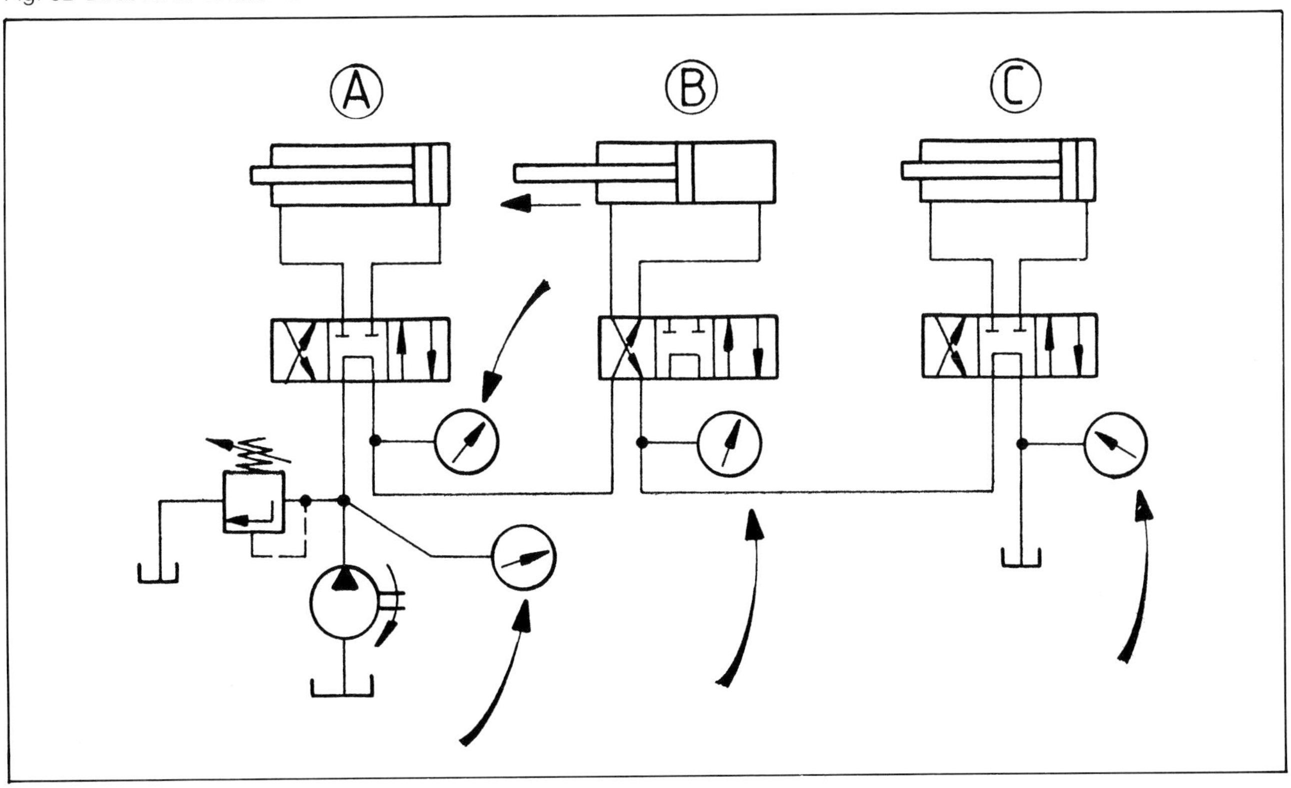

3 Hydraulic pumps

Pumping theory

The positive displacement hydraulic pump is a device for the conversion of mechanical energy into hydraulic energy.

When driven by its motor (prime mover) it basically performs two functions. Firstly, it creates a partial vacuum at the pump inlet port. This vacuum enables the atmospheric pressure to force fluid from the reservoir (tank) into the pump. Secondly, the mechanical action of the pump traps this fluid within the pumping cavities, transports it through the pump, and forces it into the hydraulic system.

Cavitation

If the pump could "pull" a perfect vacuum on the pump inlet port, there would be some 101.3 kPa (~ 1 bar) pressure (absolute) available to push the fluid into the pump. (See function of mercury barometer, chapter 1.) However, to avoid cavitation damage to the pump this "priming" pressure must be much less.

There is approximately 10% air (by volume) in solution in the hydraulic fluid. When the vacuum pressure at the pump intake exceeds the "vapour" pressure of the fluid, air comes out of solution and forms "vapour" bubbles. These bubbles are carried through the pump and implode (collapse rapidly) when exposed to high pressure on the outlet side. Such implosion causes cavitation. Cavitation (easily recognisable by its sharp crackling noise) causes metal erosion on the pump outlet and therefore shortens the working life of the pump considerably. Some simple remedies to prevent cavitation are:

- A raised reservoir (tank). This means a column of hydraulic fluid (head) charges the pump with positive pressure and vacuum pressure in the inlet line is thus avoided.
- Minimal "lift" on the pump intake to keep vacuum pressure (suction pressure) low or at least above the vapour pressure level (fig. 53).
- Pump intake line large enough to keep the flow velocity below 1 m/s, and short enough to cause minimal pressure drop.

Entrained air

If the fittings and lines on the pump inlet are not perfectly tight, air at atmospheric pressure can be drawn into the oil stream and will be carried through the pump (similar to the air bubbles mentioned under cavitation). This air/oil mixture causes excessive pump noise and metal erosion but is somewhat different from cavitation.

Entrained air when compressed at the pump outlet side, forms an air cushion within the hydraulic fluid which does not dissipate in the fluid but passes into the system. The foaming of the hydraulic fluid as a result of entrained air causes loss of actuator control and overheating of the system. Some simple remedies to prevent entrained air are:

- Tightening of loose joints and replacement of porous line fittings in the pump intake line.
- Keep the oil level in the tank well above the intake to prevent "whirlpool" action.
- Avoid foaming from return line into the tank (see construction of hydraulic tank).
- Maintain good sealing on the pump shaft seal to avoid air being drawn into the pump.

Pumps create flow, not pressure

It is often assumed that pumps create pressure, but the sole purpose of pumps is to create flow. Pressure is force per unit area and is created by resistance to flow. A pump is a mechanism designed to produce the flow necessary for the development of pressure. But it cannot of itself produce pressure, since it cannot provide resistance to its own flow.

To illustrate this principle, the leaking actuator shown in fig. 54 does not create any pressure loss since the remaining fluid within the actuator still raises the piston. Pressure would only be lost, if the entire flow from the pump escaped past the piston. The piston leak does however, affect the piston speed, since ⅔ of the pump flow is returning to tank without performing any work. Thus, the intended piston speed is also reduced by ⅔. (See also Chapter 6, fig. 124, bleed-off flow control.)

Pump ratings

Maximum tolerable operating pressure (kPa or bar), flow output (L/min) at maximum drive speed of the pump drive shaft (r.p.m.), and geometrical displacement per pump shaft revolution (m^3) are the three main factors by which pumps are rated. These ratings are determined by the manufacturer. If exceeded for some length of time this may result in reduced pump service life, or

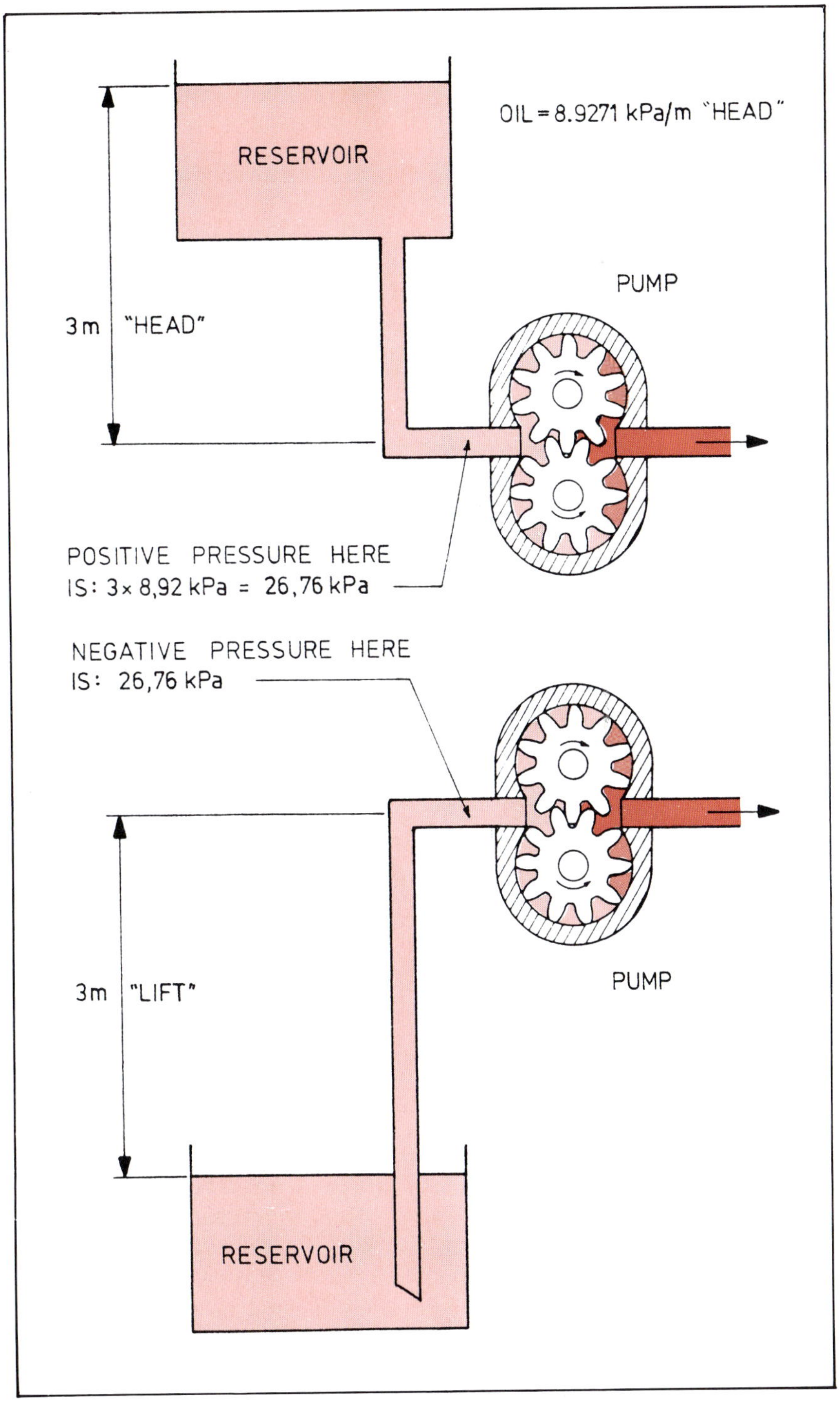

Fig. 53 Alternative pump priming conditions.

serious damage to the pump and the hydraulic system. Pump ratings must therefore be given for differing operating conditions, such as:

- maximum discharge pressure for continuous operation;
- maximum intermittent operating pressure;
- maximum peak pressure for short peaks only (fig. 55).

The flow output (displacement) from a pump may be expressed by its flow output per pump shaft revolution (L), or by its nominal flow rating in L/min. Both ratings are widely used.

Displacement (sometimes also called geometrical displacement) is that volume of hydraulic fluid which gets transported through the pump by a single rotation of the pump drive shaft. Pumps may have a fixed or a

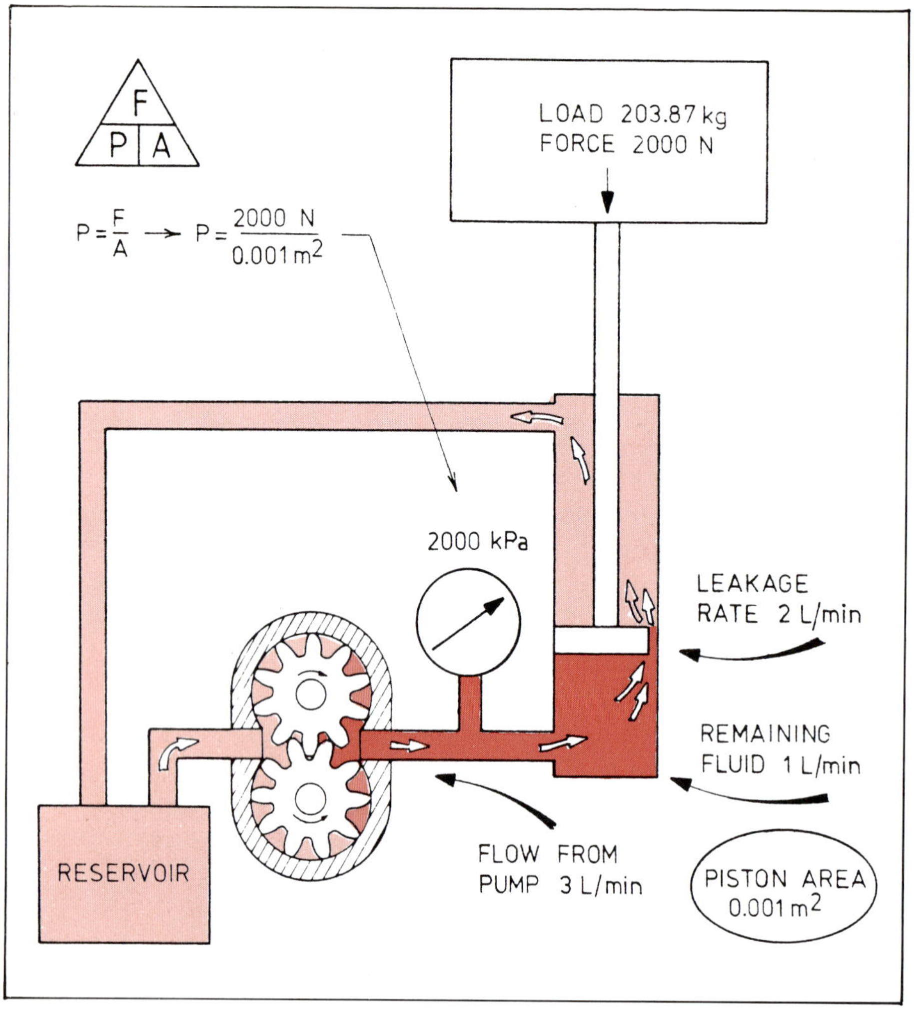

Fig. 54 As long as the leakage rate is less than the pump flowrate, the load will rise and pressure on the pump outlet is developed by the load.

variable displacement, this depends on the pump's design and its application in the hydraulic system. The flow from variable displacement pumps can sometimes also be reversed without changing the direction of pump shaft rotation. (See swash plate and bent axis pumps, figs. 66 and 69.)

Pump selection

Pumps should never be selected on an empirical basis. A number of important factors should be determined and considered before the right pump can be chosen. Some of these factors are:

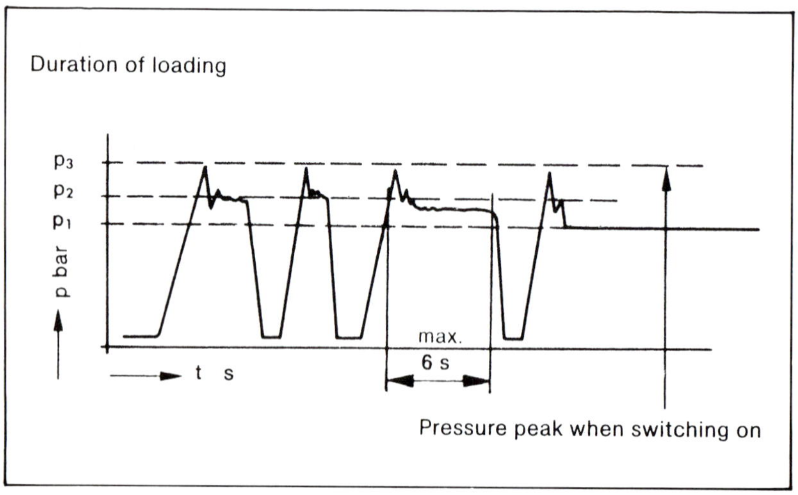

Fig. 55 Various maximum pressure ratings for pumps.

- maximum system pressure required to produce sufficient force output with the actuators;
- maximum (peak) flow required or average flow required, when system is using an accumulator;
- pump performance, operating reliability, ease of maintenance, initial purchasing cost, and pump noise;
- pump-flow control during non-action stages of the system, fixed displacement, variable displacement.

Variable displacement pumps with complex controls gain more and more acceptance for industrial applications. The purchasing cost of these pumps may be up to ten times the price of fixed displacement pumps. However, their many performance advantages off-set the high cost if input power is reduced and system complexity can be simplified.

Pump Principle	Pressure (bar) p max		Speed (RPM)		Q_{max} (l/min)	Pressure Fluctuations	Noise Level dBa	Efficiency tot	Filtration min
	from	to	n min	n max					
Gear	40	100	500	3000	300	pulsating	90	50–80	100
Gear, hydrost. bal.	100	200	500	6000	200	pulsating	90	80–90	50
Internal Gear (gerotor)	50	70	500	2000	100	low pulsation	85	60–80	100
Internal Gear, (crescent)	150	300	500	2000	50	low pulsation	65	70–90	50
Screw	50	140	500	3000	100	free of pulsation	75	60–80	50
Vane	50	100	500	3000	100	low pulsation	80	65–80	50
Vane, hydrost. bal.	140	175	500	3000	300	low pulsation	85	70–90	50
Variable Vane	40	100	1000	2000	200	low pulsation	80	70–80	50
Fixed Vane	100	140	500	2000	100	low pulsation	80	70–85	50
Cam	30	50	—	—	200	low pulsation	—	—	—
Axial Piston, swashplate	200	250	200	2000	3000	pulsating	90	80–90	25
Axial Piston, bent axis	250	350	200	2000	500	pulsating	90	80–90	25
Radial Piston	350	650	200	2000	100	pulsating	90	80–90	50
In-Line-Piston	350	500	50	1000	300	pulsating	—	—	50

Fixed displacement versus variable displacement pumps

Both fixed and variable displacement pumps are of the positive displacement type.

In fixed displacement pumps the amount of flow which is displaced by each pump shaft rotation cannot be varied. Thus, the pump's displacement can only be varied by changing the speed of the pump. Since industrial hydraulic systems generally use constant speed electric motors as prime movers, there are not many applications in which fixed displacement pumps can be used.

Flow control valves can be applied to control the speed of-hydraulic actuators (see Chapter 6). However, flow control valves can also contribute to considerable heat development. If, in fixed displacement pump circuits, the actuators require varying flow rates during operation, then the fixed displacement pump must be sized to deliver the highest flow required. Unfortunately, when less flow is required, the excess flow from the pump must be "dumped" over the system relief valve at maximum system pressure. This converts the unwanted energy directly into heat. For this reason fixed displacement pumps should only be used in constant speed circuit applications, or in circuits where speed control is of very short durations, such as end-cushioning or short-load deceleration.

Nevertheless, there are still applications where fixed displacement pumps do the job, and do it well. But it is imperative that the fixed displacement pump is sized precisely for the speed (flowrate) required.

As a rule of thumb, a fixed displacement pump is adequate only if none of the following statements hold true:

- System pressure must be maintained on a stalled actuator.
- The hydraulic circuit operates over a broad speed range.
- The pump cannot be unloaded by the circuit design during idle periods.
- During a large portion of the cycle the actuator must be operated at a very low speed.

The efficient use of available energy resources has become an important issue. Thus, engineers carefully evaluate the power requirements of new machines and appliances, and they are constantly looking for new ways to reduce energy demands to the lowest possible level.

The use of energy-saving, variable displacement pumps has contributed greatly to overcoming the stigma that hydraulic systems are inherently inefficient. Variable displacement pumps only deliver flow when and as required by the system.

The most important advantage of variable displacement pumps is that heat is not generated by moving oil around the circuit when no actuator work is being done. Even when a fixed displacement pump is being unloaded (with a tandem centre valve, for example), energy is converted to heat simply because the oil is in motion. On the other hand, the variable displacement pump can be controlled to produce this energy only when it is needed. Thus, no heat is developed, and no energy is wasted. In addition, variable displacement pumps often eliminate or reduce the need for flow control and pressure reducing valves, and thus off-set the higher initial cost of such pumps.

Pump efficiency

In theory, a positive displacement pump transfers a quantity of hydraulic fluid equal to its geometrical displacement per pump shaft revolution, and its flow output should therefore be proportional to pump shaft speed. However, the real output is less than the theoretical displacement because of internal leakage or "slip". As the pressure in the hydraulic system increases, the internal leakage past the clearance gaps and seals will also increase, and thus the volumetric efficiency is reduced (fig. 56).

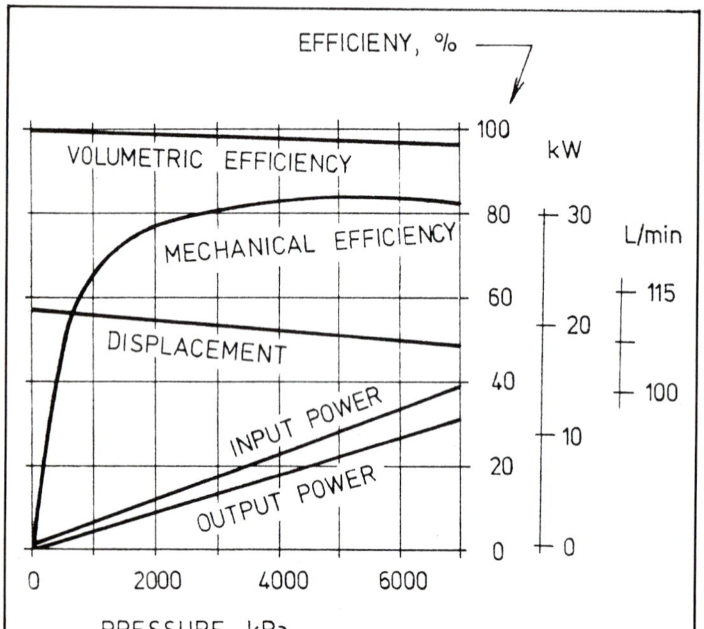

Fig. 56 Efficiency and input power curves for a balanced vane pump.

Gear pumps have a volumetric efficiency of approximately 85 to 96%, for vane pumps this is 85 to 93%, and for piston pumps, which have the highest efficiency, 95 to 98%.

Volumetric efficiency (η_V) determines the internal leak rate at stated revolutions per minute (r.p.m.) and stated pressure (p). All pumps need some internal flow to lubricate moving parts within the pump. Volumetric efficiency is calculated as follows:

$$\text{Volumetric Efficiency } (\eta_V) = \frac{\text{Actual Displacement} \times 100}{\text{Theoretical Displacement}} \ (\%)$$

Overall efficiency (η_0), a term often used in pumping calculations, is made up of volumetric efficiency (η_V) and mechanical efficiency (η_{hm}). (The latter reflects friction losses, see also fig. 56.) Commonly expressed as a percentage, the formula to calculate overall efficiency is:

$$\text{Overall Efficiency } (\eta_0) = \frac{\text{Output Power} \times 100}{\text{Input Power}} \ (\%)$$

Pump calculations

The basic formula to calculate theoretical pump input power (P) can be expressed in a simple triangular arrangement. The formula including overall efficiency (η_0) reads:

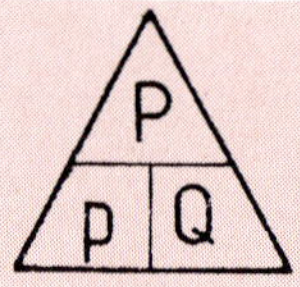

$$P = \frac{p \times Q \times 100}{\eta_0}$$

Example

Calculate the required input power (P) in kilowatts if the pump displaces 105 litres per minute (Q) at a pressure of 4 MPa (p). Overall efficiency (η_0) is 90%.

$$\text{Input Power (W)} = \frac{\text{Pressure} \times \text{Flowrate} \times 100}{\text{Efficiency}}$$

$$\text{Input Power in kW} = \frac{4 \times 10^6 \times 105 \times 100}{10^3 \times 10^3 \times 60 \times 90} = 7.7 \text{ kW}$$

The basic formula to calculate pump flowrate (Q), pump drive speed (n) and geometrical volume (V) (displacement) can be expressed in a simple triangular arrangement. The formulae including volumetric efficiency (η_V) read:

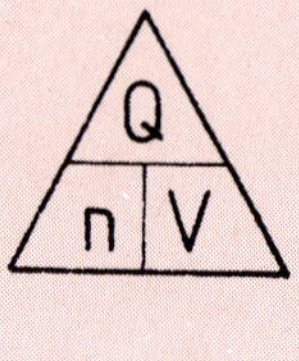

$$V = \frac{Q \times 100}{n \times \eta_V}$$

$$n = \frac{Q \times 100}{V \times \eta_V}$$

$$Q = \frac{n \times V \times \eta_V}{100}$$

Example

Calculate the required geometrical displacement in litres if the pump is driven at 1440 r.p.m. and must deliver a flowrate of 2 litres per second. Volumetric efficiency (η_v) is 98%.

$$\text{Displacement (V)} = \frac{\text{Flowrate} \times 100}{\text{Revolutions} \times \text{Efficiency}}$$

$$\text{Displacement in litres} = \frac{2 \times 60 \times 100 \times 10^3}{10^3 \times 1440 \times 98} = 0.08503\ \text{L}$$

Example

Calculate the flowrate of a pump (Q) if it is driven at 1450 r.p.m. and its displacement (V) is 0.6 L. Assume a volumetric efficiency of 96%.

$$\text{Flowrate (Q)} = \frac{\text{Displacement} \times \text{Revolutions} \times \text{Efficiency}}{100}$$

$$\text{Flowrate in L/s} = \frac{0.6 \times 1450 \times 96 \times 10^3}{10^3 \times 60 \times 100} = 13.92\ \text{L/s}$$

Pump classification

Industrial hydraulic pumps are built in a variety of shapes, sizes, and pumping mechanisms, and are almost without exception of the "positive displacement" type. This implies that the outlet is completely

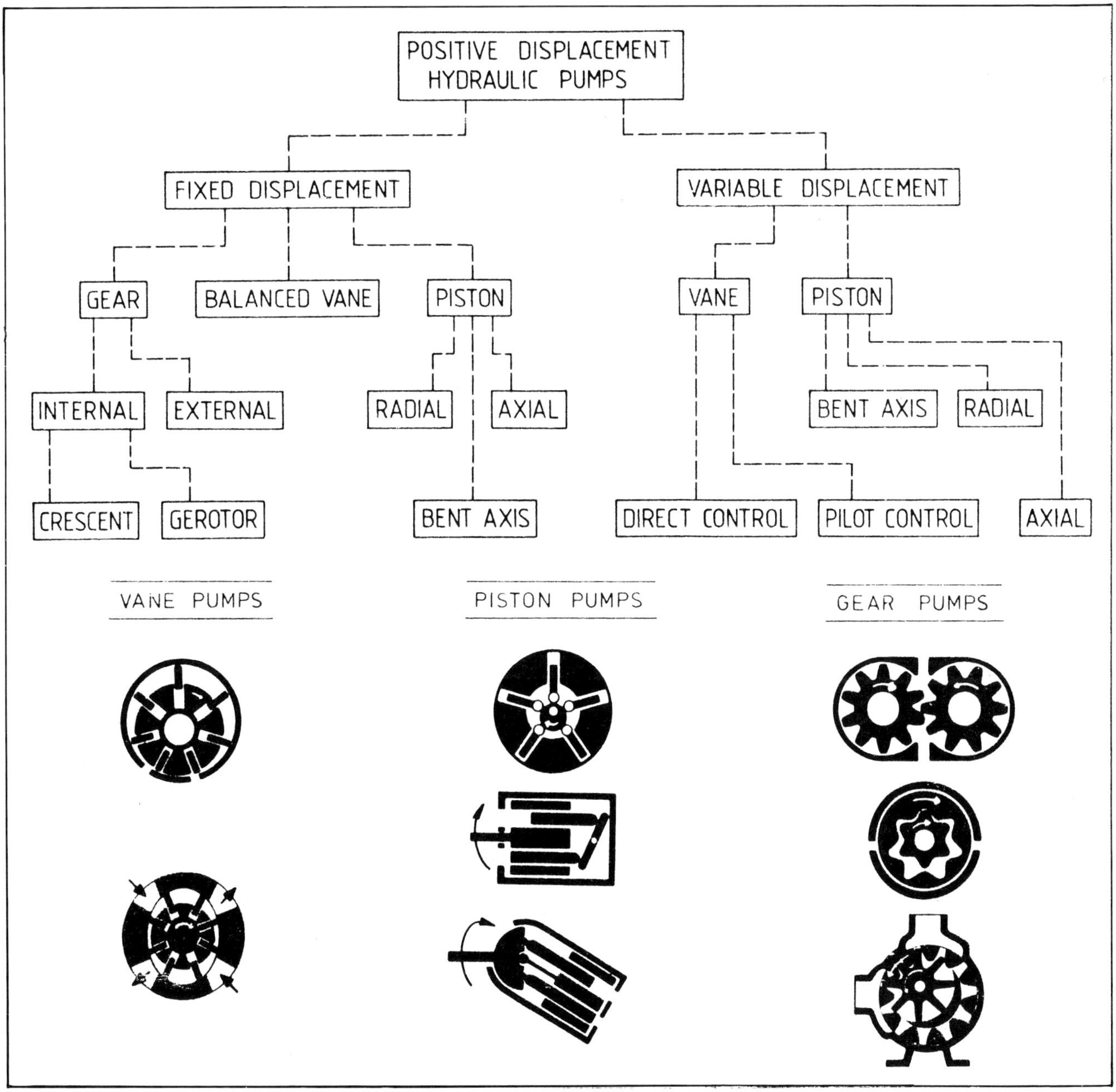

Fig. 57 Classification of positive displacement industrial and mobile hydraulic pumps.

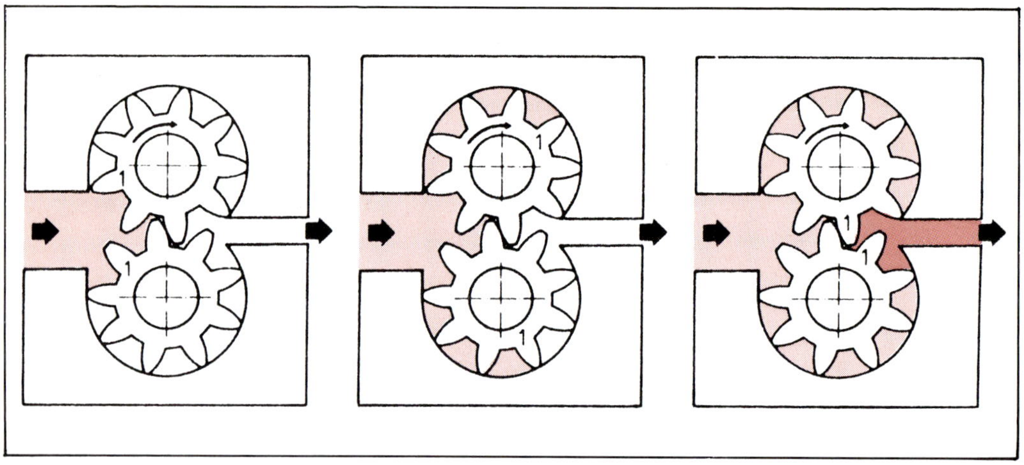

Fig. 58 Fluid transport in an external gear pump.

sealed from the inlet. Thus, theoretically, all the hydraulic fluid which is drawn into the pump must be discharged through the outlet (fig. 58). However, some minor quantities of fluid are lost due to clearance gaps and internal lubrication channels. Hence, a worn pump is less efficient than a well maintained or new pump.

Positive displacement pumps (fig. 57) are classified according to their pumping mechanism into the following groups:

- gear pumps (fixed displacement);
- vane pumps (fixed or variable displacement);
- piston pumps (fixed or variable displacement).

External gear pump

A gear pump consists of a pump housing in which a pair of precisely meshing gears run with minimal radial and axial clearance. One gear (the driver) is driven by the drive shaft attached to the prime mover (the motor). The other gear (the follower) is driven by the meshing gear teeth. As the teeth of the two gears separate, fluid from the pump inlet becomes trapped between the rotating gear cavities and the pump housing (fig. 58). These rotating cavities transport the fluid around the periphery of the pump housing to the pump outlet, where the meshing gears force it through the outlet port into the

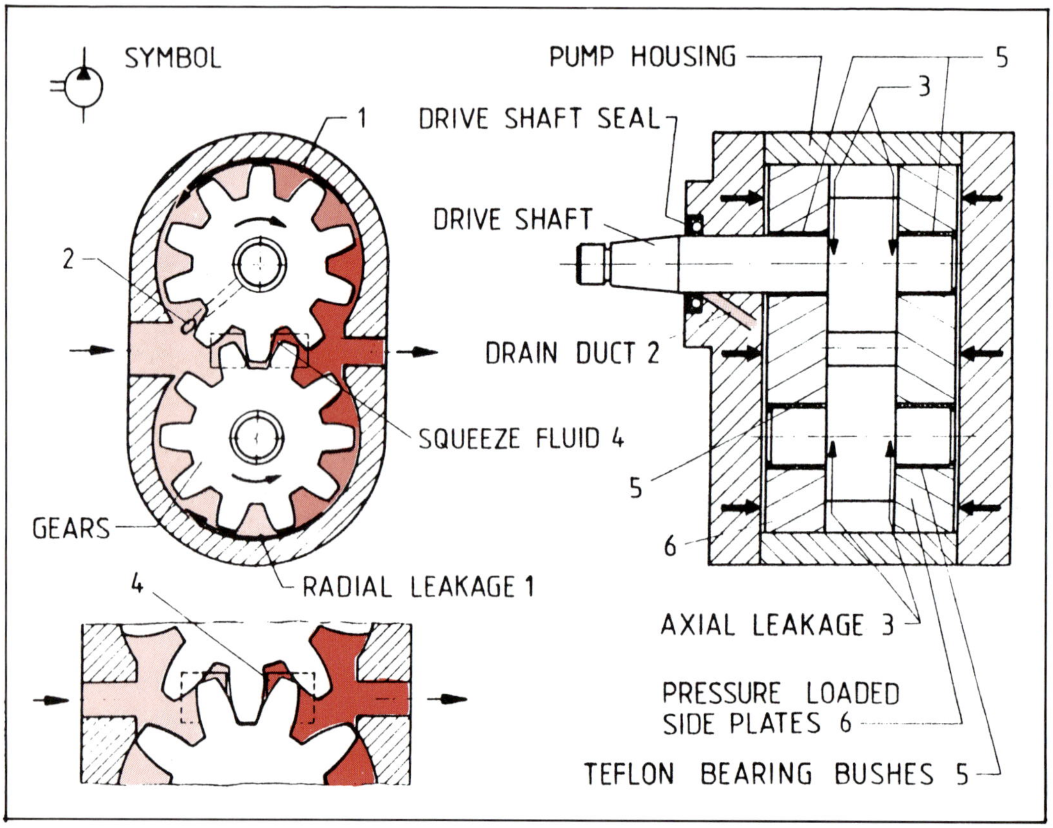

Fig. 59 External gear pump.

hydraulic system. The closely toleranced mesh of the gear teeth provide an almost perfect seal between the pump inlet and the pump outlet, which keeps slippage of the transported fluid to a minimum.

As output flow is resisted, pressure in the pump outlet chamber builds up rapidly and forces the gears diagonally outward against the pump inlet side. Hence, as the system pressure increases, this pressure/force imbalance becomes greater. This imbalance increases the mechanical friction and the bearing load of the two gears. It is therefore imperative that the maximum pressure rating stated by the manufacturer is closely adhered to.

Side plates (pressure plates) or bearing bushes are pressure loaded to prevent excessive leakage of the fluid past the sides of the rotating gears. The more the system pressure increases, the harder these plates are pressed against the sides of the gears. To prevent a pressure build-up behind the drive shaft seal, the inner side of the seal has a drain duct to the low pressure pump inlet side, which avoids leakage past the drive shaft seal. As the gears roll into mesh at the outlet chamber, provision is made to relieve the trapped fluid between the meshing gears. Drain grooves are machined into the side plates to channel the trapped fluid (squeeze fluid) to the pump outlet (fig. 59).

Crescent type internal gear pump

This pump is rapidly acquiring a widespread reputation as a high-pressure, low-noise pump suitable for a variety of hydraulic fluids. It is smaller than the external gear pump for the same flow displacement conditions.

The pump shaft drives the pinion which in turn drives the internal ring gear (rotor). As with external gear pumps, the fluid fills the cavities (voids) formed by the rotating teeth and the stationary crescent. Both the outer gear and the pinion transport fluid through the pump. The crescent seals the low-pressure pump inlet from the high-pressure pump outlet.

Like the external gear pump, the crescent gear pump is also pressure unbalanced. An inherent advantage of this pump is its porting arrangement, where the fluid intake and output zones are much longer than for comparable external gear pumps. Therefore fluid velocities for the filling of the transport cavities is considerably reduced. This not only reduces the noise level but also improves the suction capabilities of the pump (fig. 60).

Gerotor type internal gear pump

The gerotor pumps works in a very similar manner to the crescent pump. The pump shaft is also keyed to the inner rotor and both rotor and rotor ring rotate in the same direction. The inner rotor has one tooth less than the rotor ring. This arrangement forms pumping cavities. Whilst the rotors turn clockwise (fig. 61), pumping cavities are gradually opening up on the pump inlet side until they reach point X and then decrease their volume gradually as they rotate toward the outlet side. The tips of the inner rotor contact the outer rotor ring to seal the pumping pockets from each other.

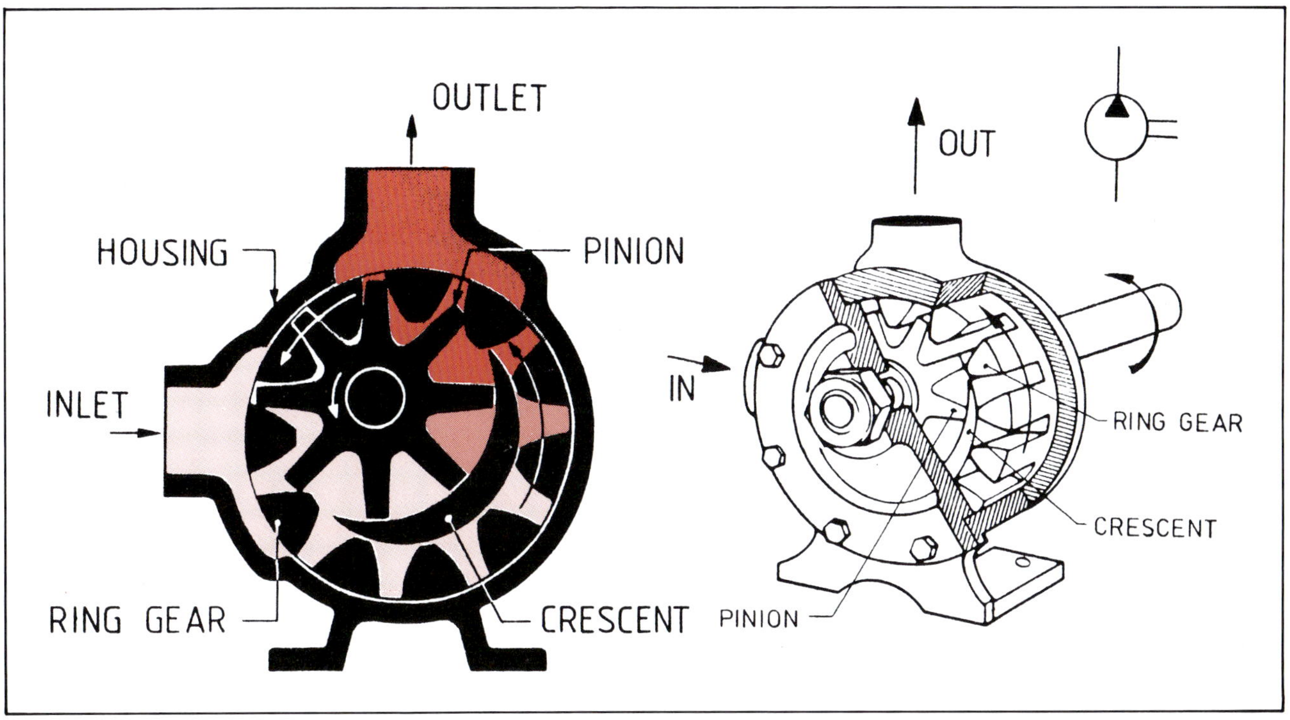

Fig. 60 Crescent type internal gear pump.

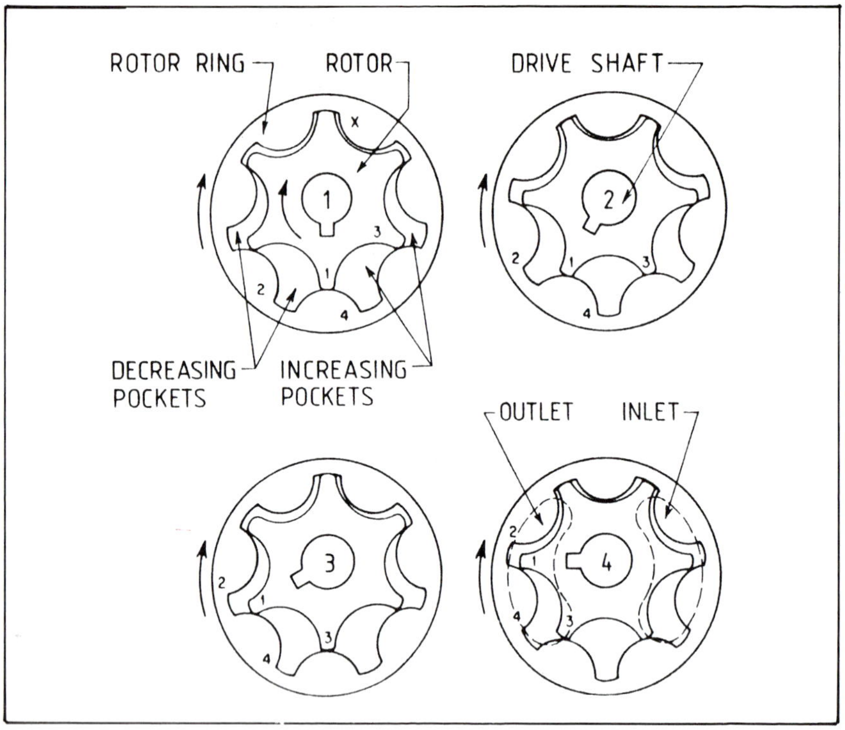

Fig. 61 Gerotor type internal gear pump.

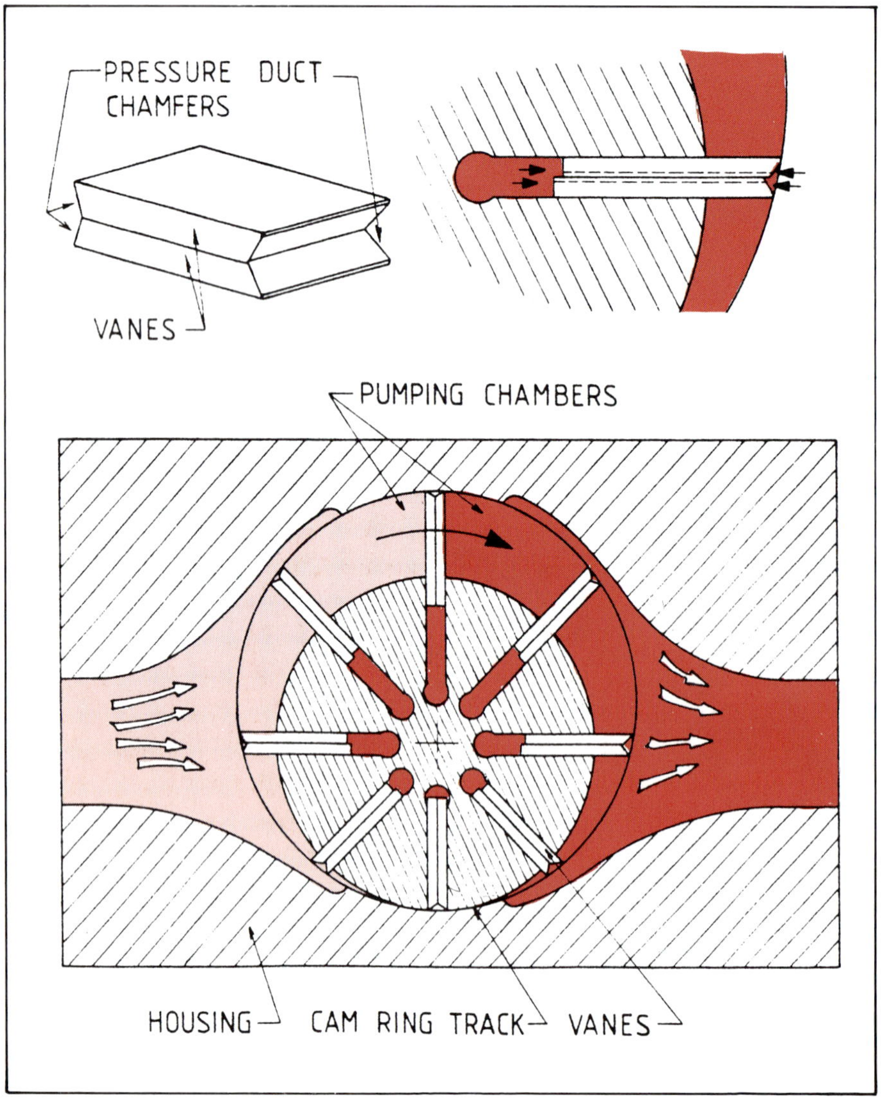

Fig. 62 Function principle of vane pumps.

Fixed displacement vane pump (unbalanced)

In vane pumps an eccentrically arranged slotted rotor with movable vanes inserted into the slots rotates within a circular track. During rotation, the centrifugal force drives the vanes outward against the circular track where they make firm contact and follow the contour of the track. These vanes in conjunction with the rotor and the circular track housing form pumping cavities. As the rotor turns, these cavities transport fluid from the pump inlet to the pump outlet where the diminishing cavities force the fluid into the system.

The pump in fig. 62 is "unbalanced" because all the pressure exposed pumping cavities are on one side of the rotor and the resulting side load may cause the drive shaft bearing to wear prematurely if the pump is used above its pressure rating. System pressure is fed to the underside of the vanes to ensure minimum slippage at the vane tips (fig. 62).

Fixed displacement vane pump (balanced)

Balanced vane pumps use an elliptical track ring (fig. 63) instead of the circular track housing shown in fig. 62. This permits the use of two sets of intake and outlet ports. The diametrically opposed pump outlet openings cancel out the radial pressure loading on the rotor shaft bearing. The resulting force balance permits the pump to be used for higher pressure systems than the unbalanced type. To improve sealing between the vane tip and the elliptical track, pressure from the high pressure outlet is applied to the underside of the vanes (fig. 62).

An extremely high tip contact pressure of the vanes can cause the lubrication film between the vane tips and the elliptical track to break down, resulting in metal-to-metal contact and consequent increased wear.

Some pumps use therefore the so-called "dual vane" construction (fig. 62). Chamfered edges along the tips of the two vanes form a groove. This pressurised groove creates a slightly smaller thrust force than the pressurised underside of the two vanes. Centrifugal force plus pressure-induced force imbalance cause just enough contact thrust to provide good sealing between vane tip and track, without destroying the lubrication seal.

Variable displacement vane pump

The variable displacement vane pump (also known as a variable delivery or variable volume vane pump), pumps the fluid on the same principle as fixed displacement

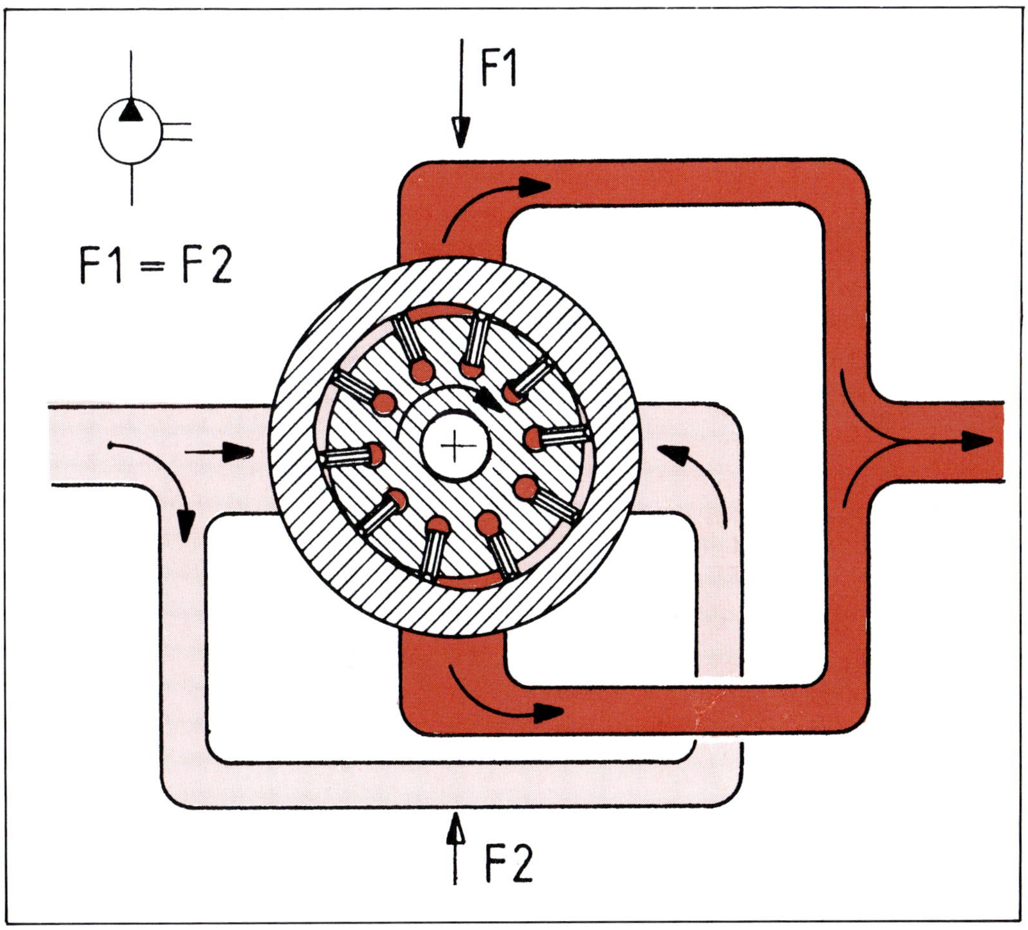

Fig. 63 Function principle of a balanced vane pump.

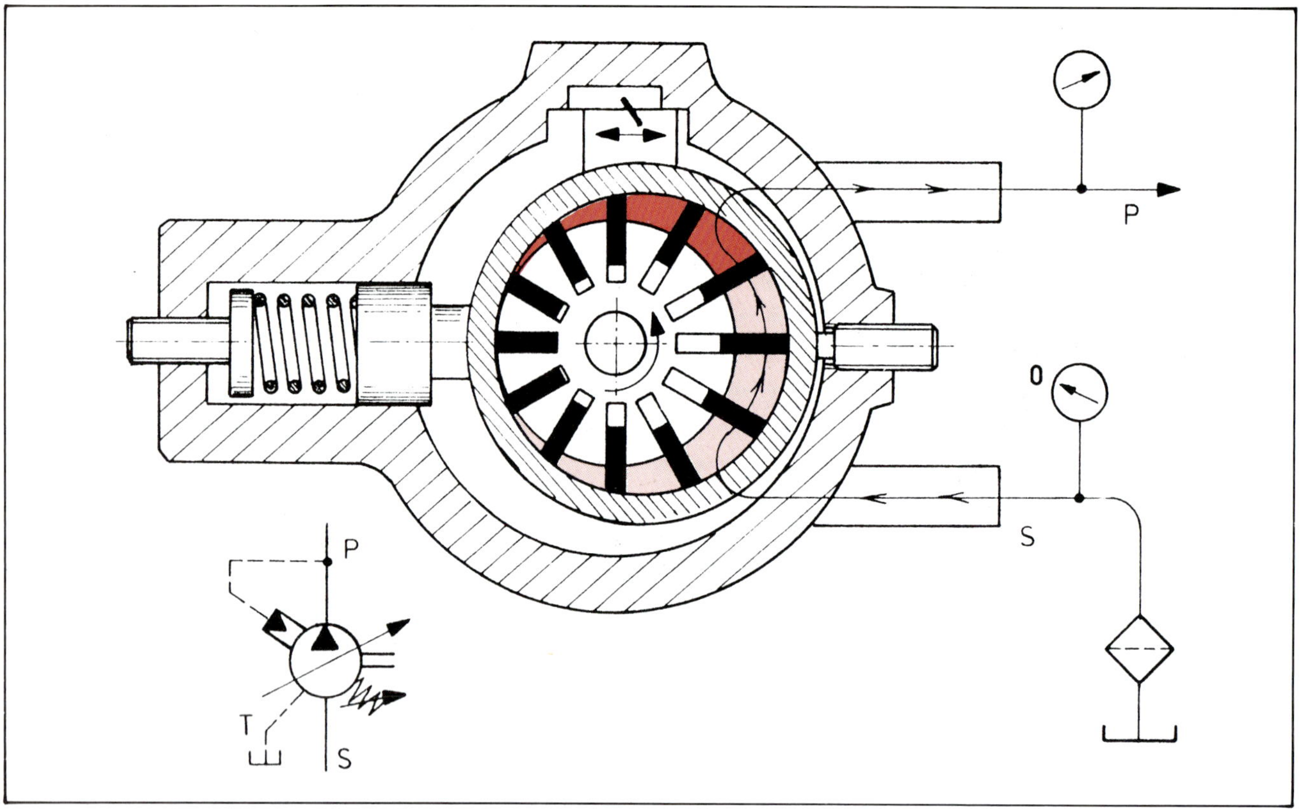

Fig. 64 Function principle of a variable displacement vane pump.

vane pumps. A heavy pressure control spring forces the movable circular track ring into the extreme right hand position (maximum eccentric position). This maximum "throw" position can be adjusted and thus the maximum pumping capacity of this pump is adjustable. Pressure arising from workload resistance opposes the spring force. As system pressure increases the force component acting against the spring also increases. If this force component is greater than that of the spring, the circular track ring moves from its eccentric position into a concentric or "zero" flow position. Thus, when peak pressure has been reached, the pressure control with its spring decreases the flow from the pump so that only leakage fluid is replaced (pumped), and constant maximum system pressure is maintained. Variable displacement vane pumps are pressure unbalanced and are therefore limited to low pressure hydraulic systems (fig. 64).

Piston pumps

Piston pumps are generally regarded as the true high-performance pumps in industrial and mobile hydraulics. Radial piston pumps can handle pressures up to 65 MPa, where vane pumps and gear pumps are struggling to reach pressures of 15 to 20 MPa. Generally, piston pumps are "loafing" at such low pressures, and with modern design, piston pumps often reach an overall efficiency of 95% or more.

Swash plate axial piston pump (fixed or variable displacement)

This kind of pump consists of a pump housing, the swash plate on a fixed or variable tilt angle, a drive shaft, a rotating pumping group, a shaft seal and a control plate with inlet and outlet ports. The rotating pump group, splined to the drive shaft, contains the cylinder block with the pistons.

As the cylinder block rotates, the piston shoes (slippers) follow the stationary (non-rotating) swash plate, causing the pistons to reciprocate. The retracting pistons move past the inlet port slot thus drawing fluid into the expanding pumping chambers. As the cylinder block rotates further, the pistons are forced back into the cylinder block and while moving past the outlet port slot, they force the fluid into the system (fig. 65).

The size and number of the pistons as well as their stroke length determine the pump's displacement. Stroke length again depends on the swash plate angle which is generally about 18° maximum. In fixed displacement pumps, the swash plate is firmly mounted and forms a part of the pump housing. In variable displacement pumps, the swash plate is mounted on a pivoted yoke (swivel yoke).

As the swash plate angle increases so does the cylinder stroke, resulting in increased pump displacement. The swash plate angle can be adjusted manually,

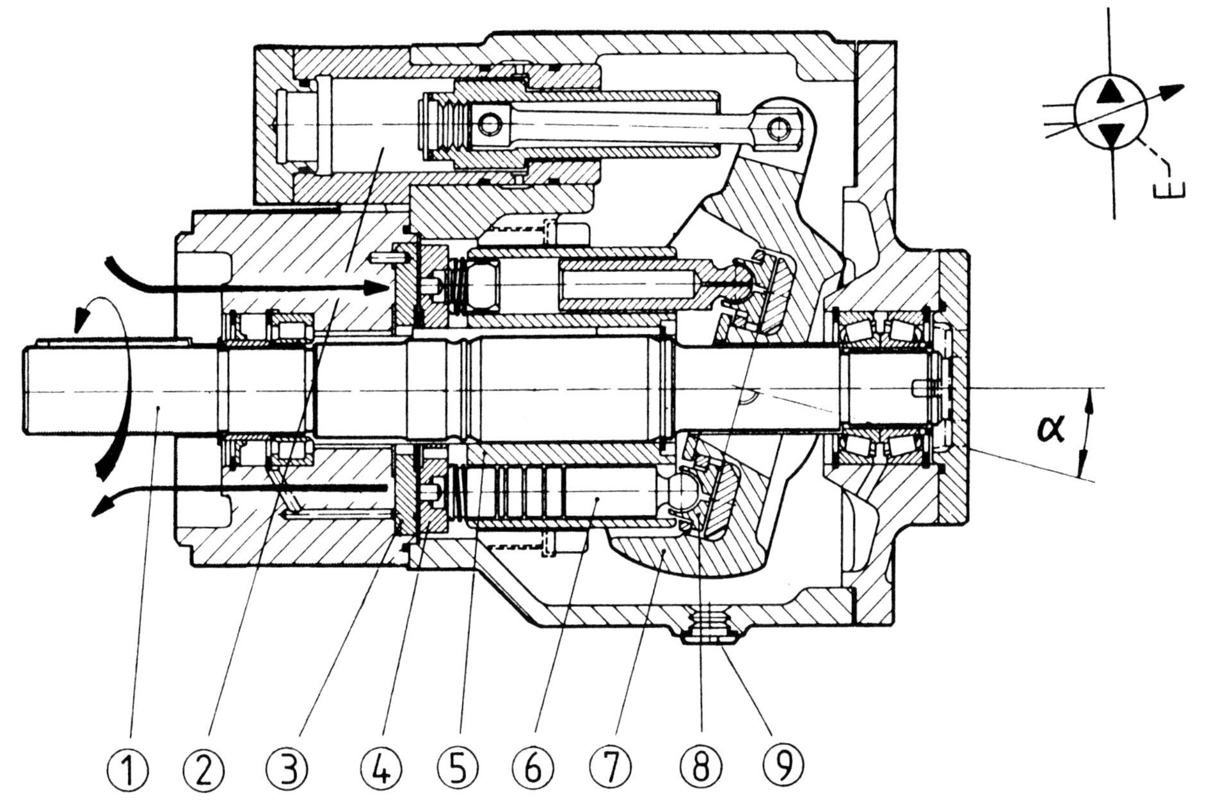

(1) Drive shaft
(2) Pressure controller
(3) Valve plate (fixed)
(4) Valve plate (rotating)
(5) Cylinder block
(6) Pistons
(7) Swash plate (yoke)
(8) Slipper pads
(9) Case drain port

Fig. 65 Swash plate, variable displacement piston pump.

or by means of a pressure controller, or by a sophisticated servo control. The pressure controller maintains a constant output pressure (figs. 65 and 66). When the swash plate is vertical to the drive shaft (zero swash plate angle), the piston stroke also becomes zero, and pump displacement is theoretically zero. Axial piston pumps always have some internal leakage which must be drained to tank through the case drain port.

Bent axis piston pump (variable or fixed displacement)

In the bent axis piston pump the cylinder block rotates at an offset angle to the drive shaft. In the example, a universal drive joint links the cylinder block to the drive shaft so that the drive shaft and the cylinder block rotate with the same velocity, and constant alignment. The piston rods are attached to the drive shaft flange by means of ball joints. Thus, the pistons are reciprocated in and out of their bores as the distance between the rotating drive shaft flange and the rotating cylinder block changes (fig. 67). The universal link is not required to transmit any torque except for the acceleration and deceleration of the cylinder block, and to overcome friction resistance of the revolving cylinder block in the fluid-filled pump housing. The pistons draw fluid into the cylinder bores as they rotate past the inlet port slot, and while extending they force the fluid through the outlet port slot into the hydraulic system.

Pump displacement varies with the offset angle (bent-axis angle). With fixed displacement pumps (fig. 67) the angle is set to approximately 25°. For variable displacement pumps (figs. 68–70) the cylinder block may be swivelled around the centre point to adjust the offset angle. With some pumps the angle may even be changed over centre to reverse the direction of pump flow. The cylinder block angle can be changed either manually (handwheel), with a servo control system, or with a pressure control mechanism. Figure 80 shows such a pressure control mechanism.

Radial piston pump with thrust ring actuation (variable displacement)

The working principle of this pump is shown in fig. 71. The pump has a drum-shaped cylinder block that

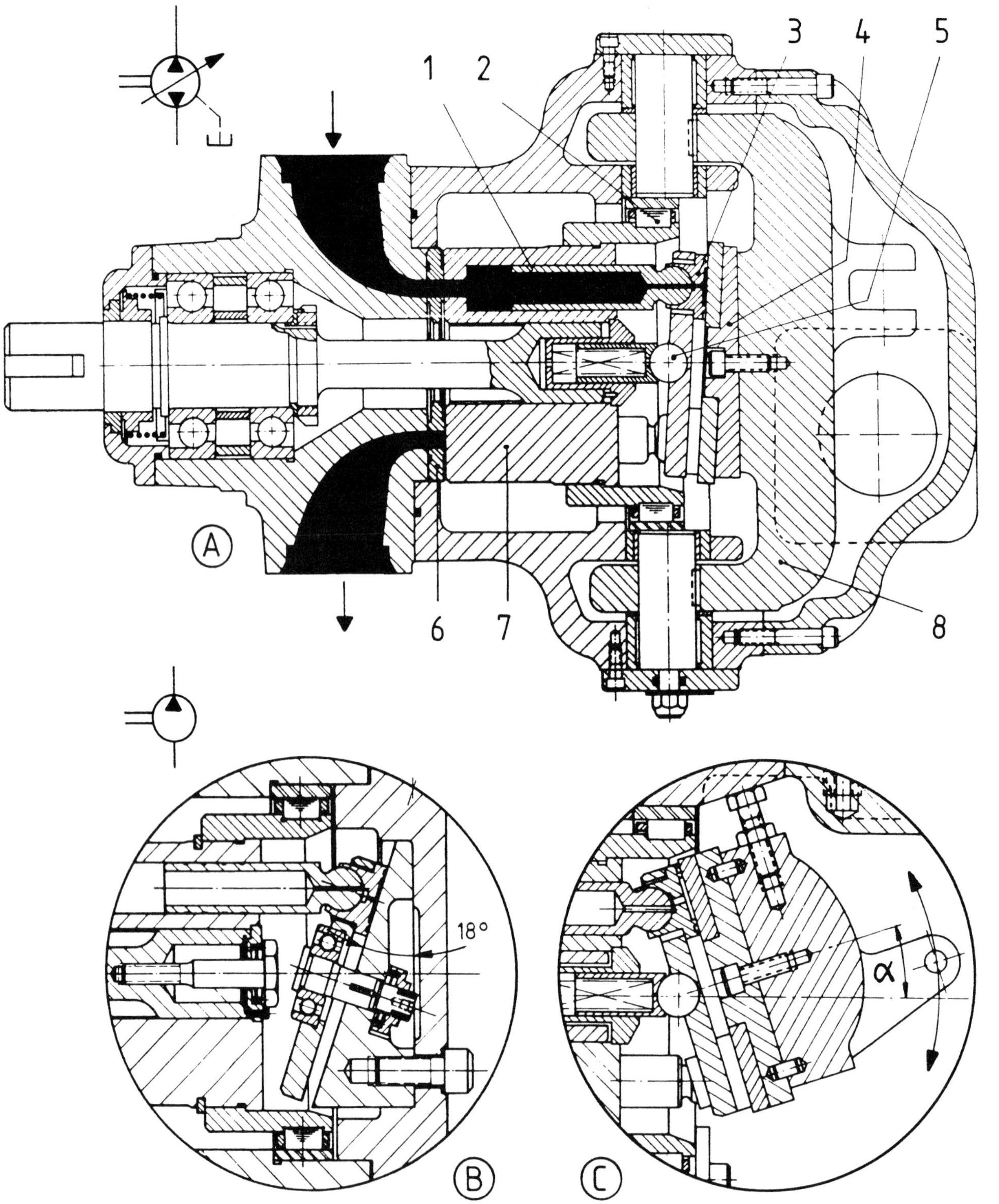

① Piston
② Cylinder block bearing
③ Slipper pads
④ Swash wedge
⑤ Slipper retainer
⑥ Valve plate (fixed)
⑦ Cylinder block (rotating)
⑧ Swash plate (yoke)
Ⓐ Variable displacement pump
Ⓑ Fixed displacement pump
Ⓒ Variable displacement with side view through yoke

Fig. 66 Swash plate, fixed or variable displacement piston pump.

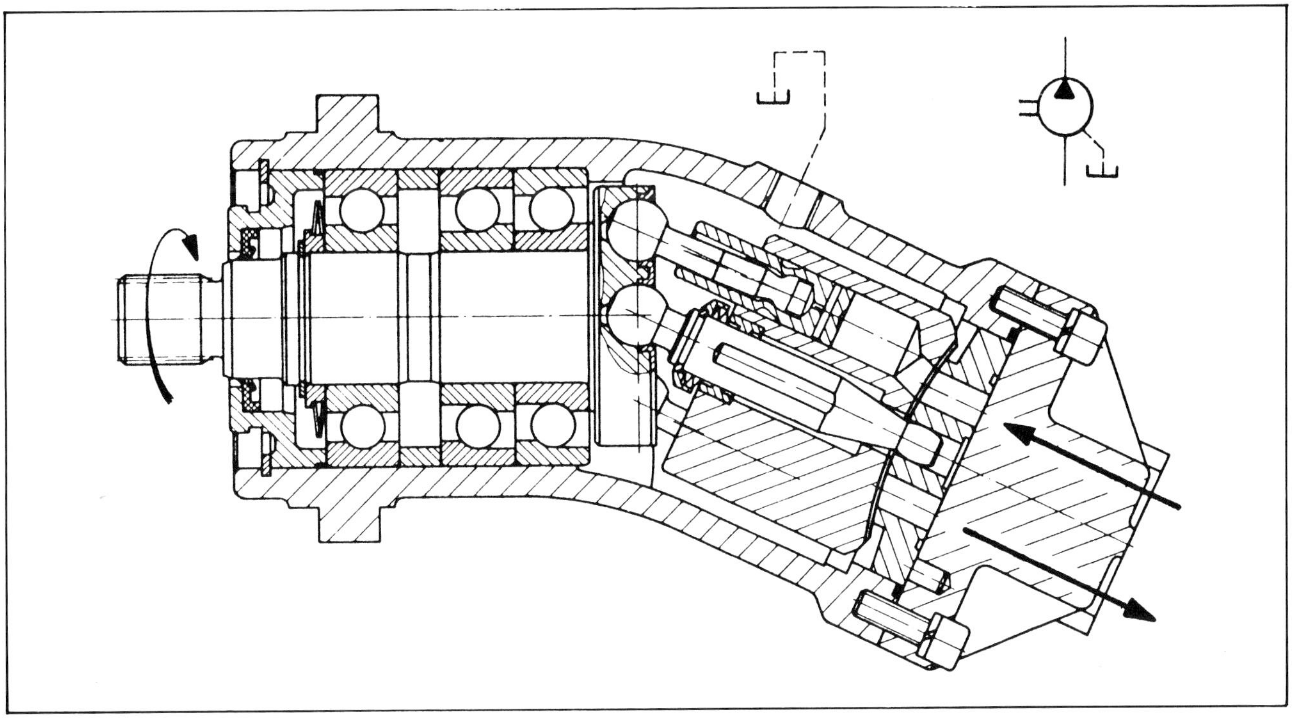

Fig. 67 Bent axis, fixed displacement piston pump.

rotates about a stationary central pintle (or valve shaft). The cylinder block houses the pumping pistons which reciprocate as it rotates. Inside the pintle are the fluid inlet and outlet ducts that connect the pumping chambers to the pump inlet and outlet ports. The cam ring (thrust ring or reaction ring) together with its support block are eccentrically movable to adjust the piston stroke and thus the pump flow.

As the cylinder block rotates, centrifugal force, charging pressure, or sometimes springs, cause the pistons to follow the inner surface of the cam ring. Outward travelling pistons draw in fluid as they move past the inlet crescent of the pintle. Conversely, the inward travelling pistons force the fluid into the system as they move past the outlet crescent of the pintle.

Figure 72 shows a radial piston pump with tandem mounted gear pump driven by a through shaft (or extension drive shaft). This arrangement is useful where two separate supplies are required.

Radial piston pump (fixed displacement)

For high pressure application, for example up to 63 MPa, the pistons are usually moved by an eccentric crankshaft, thus acting radially outwards (fig. 73).

Filling of the cylinders is achieved by the pistons uncovering ports in the cylinders, to allow fluid from the crankshaft to enter. During the pumping stroke, the fluid is expelled from each cylinder via a flat check valve. Such pumps may have three or more cylinders, which can have their output either combined or delivered separately (isoflow).

Piston pump precautions

As all pump lubrication is provided by the hydraulic fluid being pumped, special care must be taken with piston pumps. The pistons, cylinder block, bores, valve plate, slipper pad (shoes), and all mating parts are very finely machined (fig. 65). The finish and trueness of these surfaces control the condition of the pump and determine its volumetric efficiency. Exceptionally fine filtration is required, because even fine abrasives will wear and score these surfaces, thus increasing internal leakage and eventually destroying the pump.

Internal leakage serves an important purpose, since pump lubrication depends entirely on the liquid being pumped. Therefore, initial start-up of piston pumps must never be attempted until the pump housing (case) has been filled with hydraulic fluid. This is often called "priming". However, it does not actually prime the pump for pumping, but assures initial lubrication of bearings and wear surfaces inside the pump. The case drain should always be placed at the highest possible point, and syphoning back of the fluid to tank should be prevented.

Cylinder block loading

To assure high volumetric efficiency, the rotating cylinder block in axial piston pumps must be pressure loaded against the stationary surface of the valve plate. However, this pressure loading must be carefully balanced to ensure an adequate lubrication film of hydraulic oil between the cylinder block and the valve

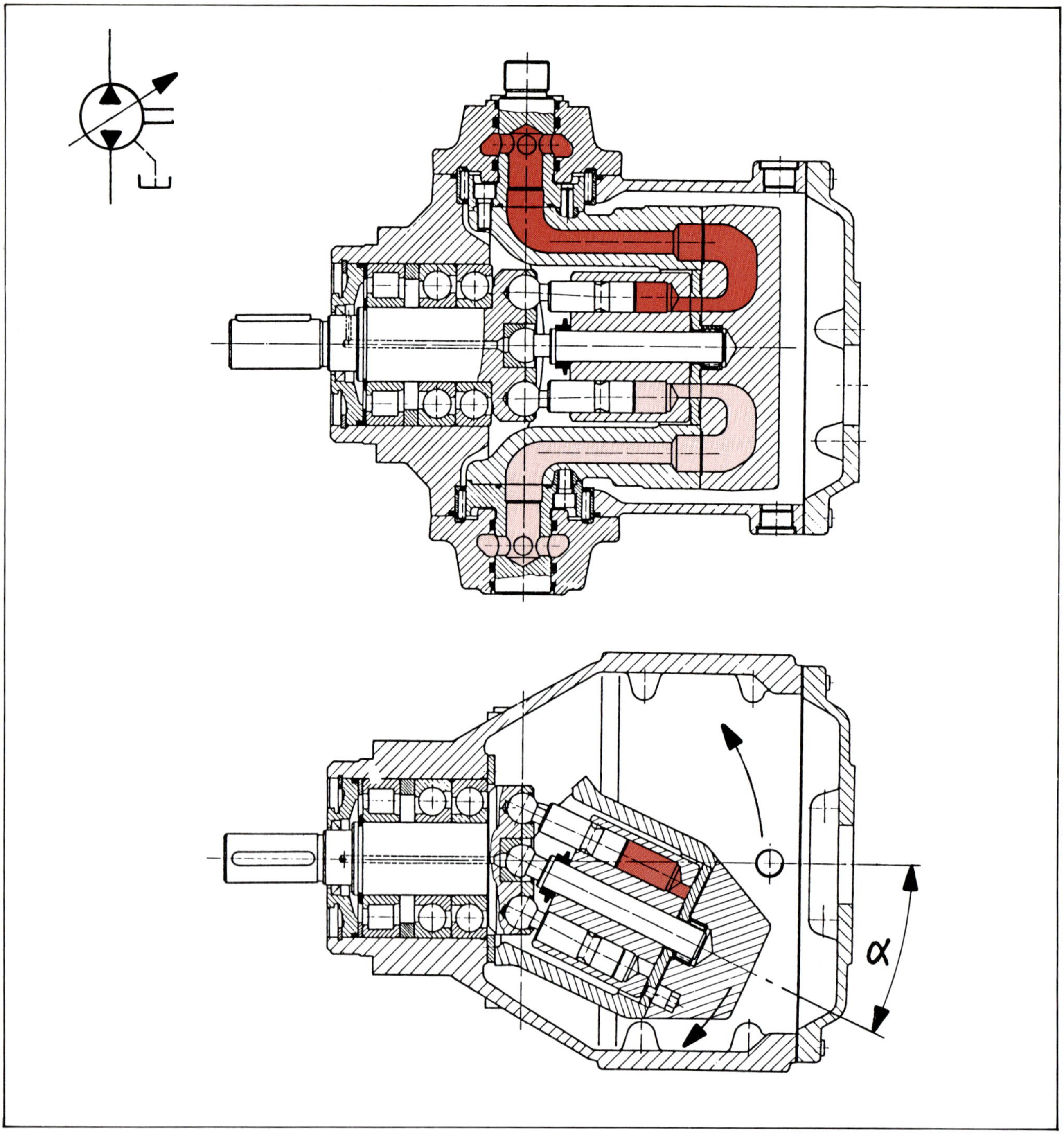

Fig. 68 Bent axis, variable displacement piston pump.

plate. The pressure loading is accomplished by making the total net area of half the number of piston bores slightly larger than the effective area of the pressure kidney on the valve plate. The thrust force so developed, presses the cylinder block firmly against the valve plate (fig. 74). However, this does not guarantee that the two components will always maintain the required firm contact. When the axial piston pump is operated at a pressure higher than the rated pressure, or with extremely high pressure peaks, it is not uncommon for the two components to separate. This can cause wire drawing (groove-like metal erosion), and/or mechanical damage to these critical surfaces of the pump.

Slipper pad bearing failure

Basically, there are four detrimental operating conditions which can cause slipper pad bearing failure.

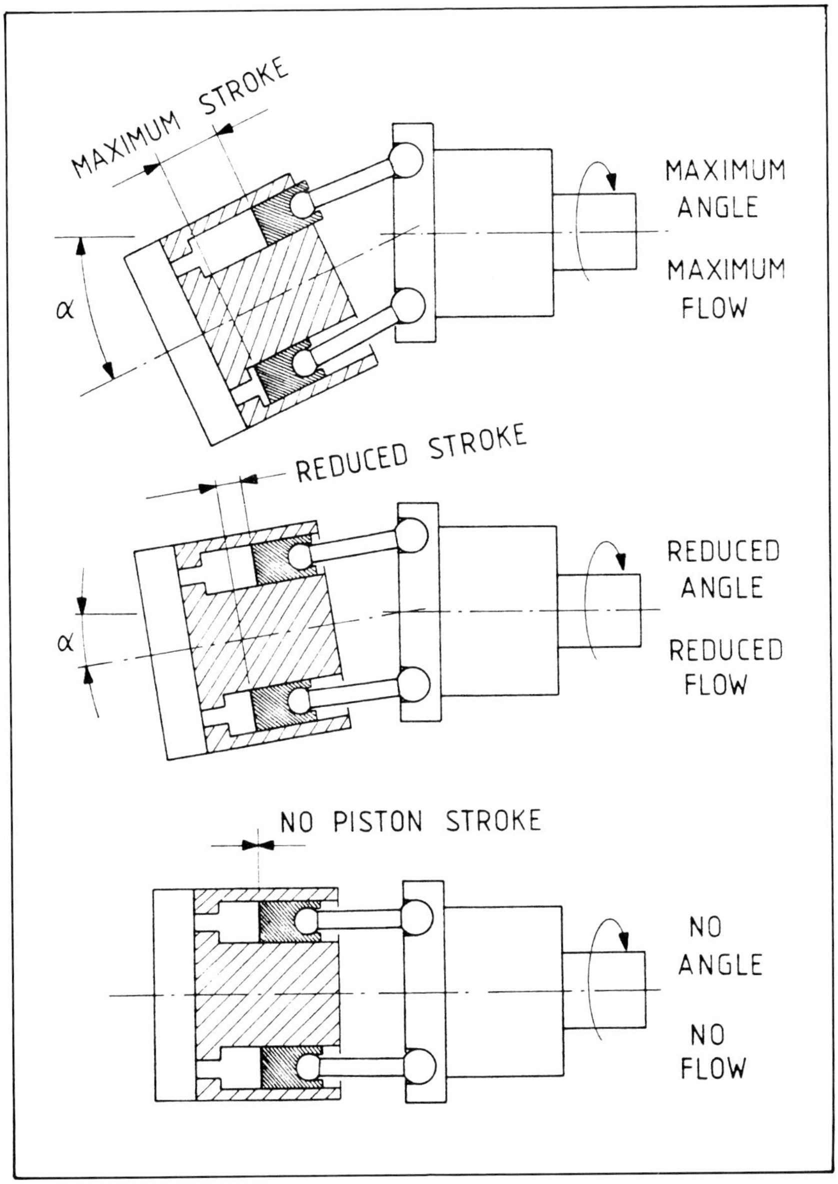

Fig. 69 Function principle of bent axis piston pump.

They are: operating with poorly filtered hydraulic fluid, too much vacuum at the pump suction inlet, excessive case pressure, and excessive pump shaft speed.

Poorly filtered hydraulic fluid accelerates the normal wear of mating surfaces and clogs the critical lubrication passages in the piston ball joints (fig. 66). These passages port pressurised hydraulic fluid between the slipper pad and the swash plate. The slipper pad is designed to have an effective pressurised area which balances the thrust forces of the pistons against it. Thus, a clogging of these passages would reduce or eliminate the vital force balance, and a rapid failure of the slipper pad bearing would be inevitable.

High vacuum suction conditions will ruin the ball joint bearings. The in-line or axial piston pump demands far better inlet conditions than any other hydraulic pump. The ball joint which holds the piston to the slipper pad cannot withstand high tension forces. During suction, the retainer ring extracts the piston by pulling on the slipper pad. If the vacuum on the pump inlet is excessive, the bronze slipper pad is simply pulled off the ball tip of the piston, and the pump continues to operate until the bearing area is totally ruined. Pump inlet supercharging, or a raised reservoir, will avoid such inadvertent conditions. It is also essential to provide adequate case drain conditions.

Excessive pump shaft speed goes hand in hand with excessive speed of the reciprocating motion of the pumping pistons. Needless to say, the piston slipper pads and the ball joints are subject to acceleration and deceleration forces, far beyond the pump's design limits. These forces will eventually destroy the ball joints and the slipper bearing surfaces.

Case drain

The pumping mechanisms of all pumps leak. In pressure controlled pumps and piston pumps this leakage, which can be quite substantial, drains into the pump housing, and must therefore be drained back to tank. Leakage in pressure controlled pumps can range from four litres per minute in small pumps to 20 litres or more in large pumps. Case drain plumbing should be installed with the utmost care, to provide ample drainage capacity, and to ensure that the case pressure remains well below 100 kPa. The case drain line should therefore be full-sized, and always terminate below the minimum fluid level of the reservoir. It is also important that the case drain is connected individually, and not teed into a common return line.

Control of variable displacement pumps

Flowrate and maximum system pressure are the two quantities that can be controlled with variable displacement pumps. With industrial hydraulic systems the pump shaft rotations per minute (r.p.m.) are assumed to be constant, thus flow rate must be constant if the geometrical displacement (swept volume) is kept constant or is not adjustable (as in fixed displacement pumps).

For some variable displacement pumps the geometrical displacement (mL/rev.) may be altered if the swash-plate angle or the bent-axis angle is altered (figs. 65 and 68).

For vane type pumps and radial piston pumps, displacement variation is achieved by moving the cam ring into a more or less eccentric position to the rotor (figs. 64 and 71).

Since variable displacement pumps are used to maintain and limit maximum system pressure, their minimum flow rate will be automatically adjusted by the pump controller to match the internal leak rate of the total system.

Industrial hydraulic systems predominantly use pressure controllers in conjunction with variable displacement pumps but other controllers may be used to regulate flow rate levels, pressure and flow rate combined, or input power (as for mobile systems).

The methods used to gain control are: hand wheels, manual servo, remote hydraulic controls, remote electrical controls, and direct internal hydraulic controls. The control mechanisms vary greatly from one manufacturer to another (figs. 65, 78, 79 and 80).

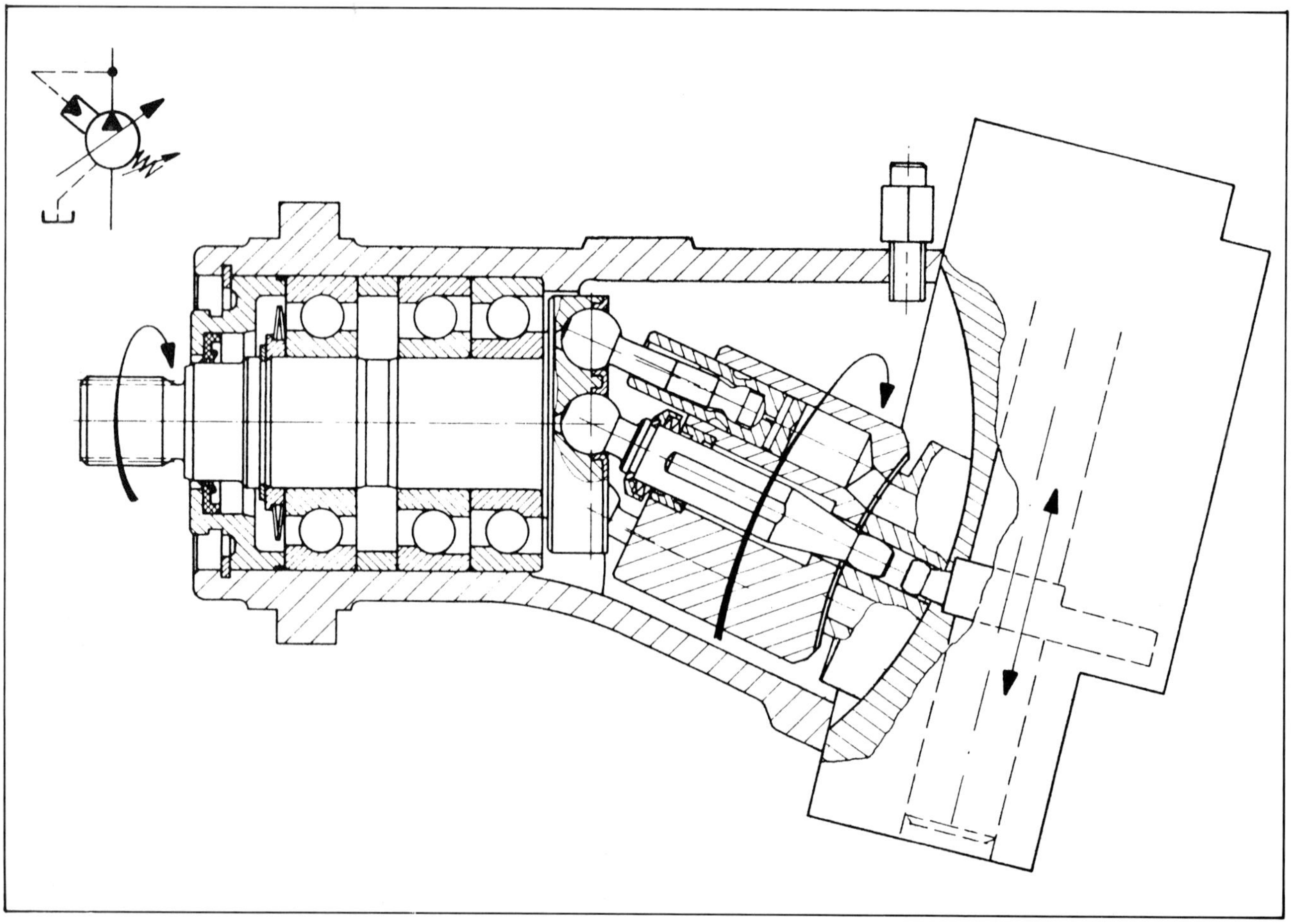

Fig. 70 Bent axis, variable displacement piston pump with pressure control mechanism.

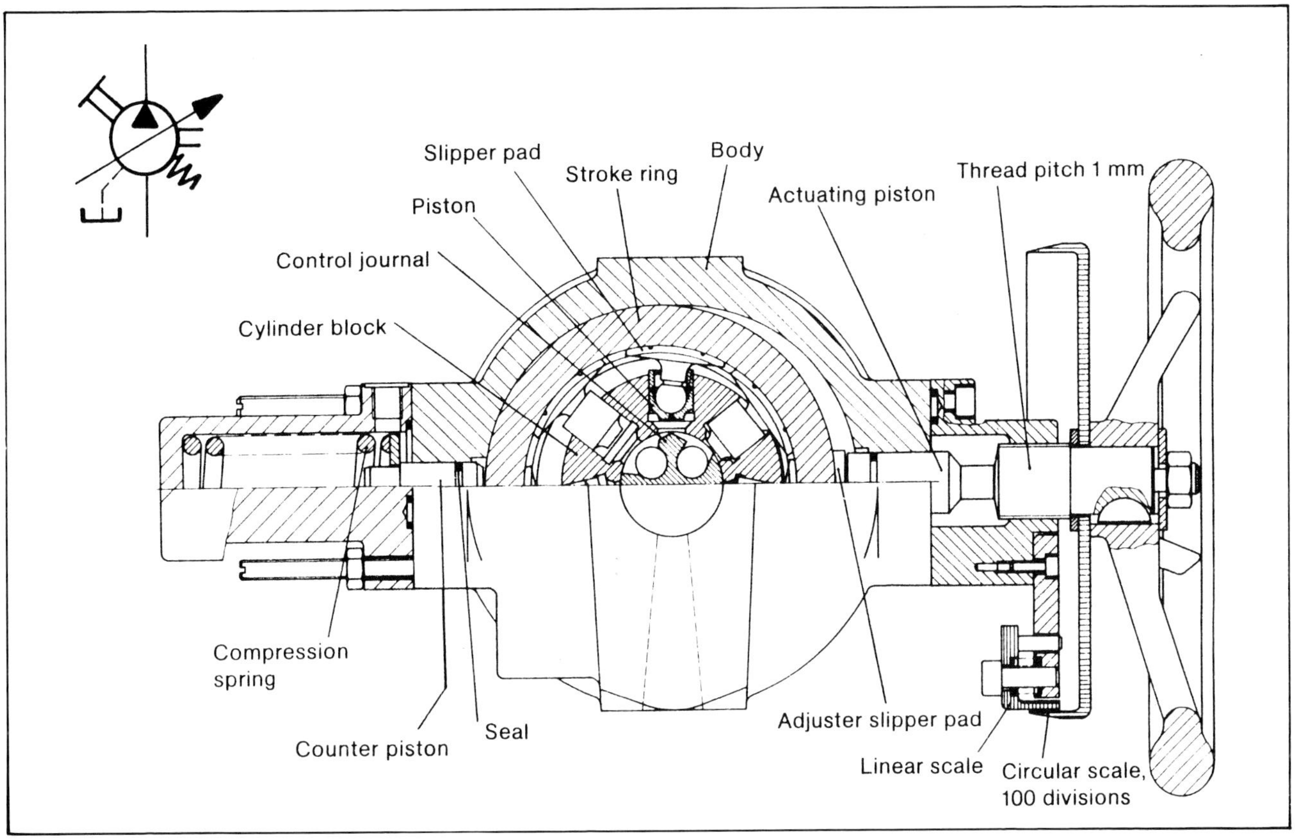

Fig. 71 Radial piston pump, variable displacement with handwheel adjustment (thrust ring type).

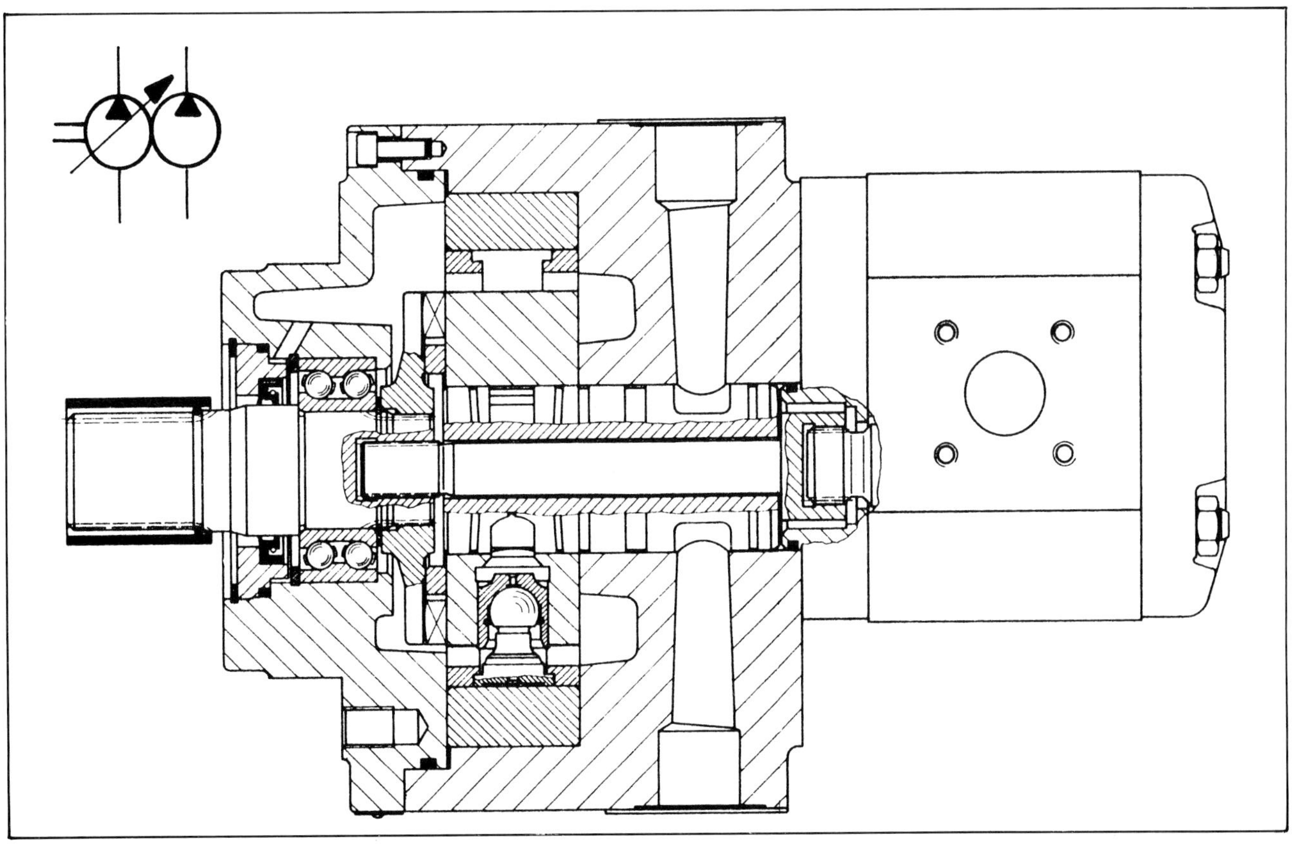

Fig. 72 Radial piston pump with tandem mounted boost pump (the boost pump is of the external gear type).

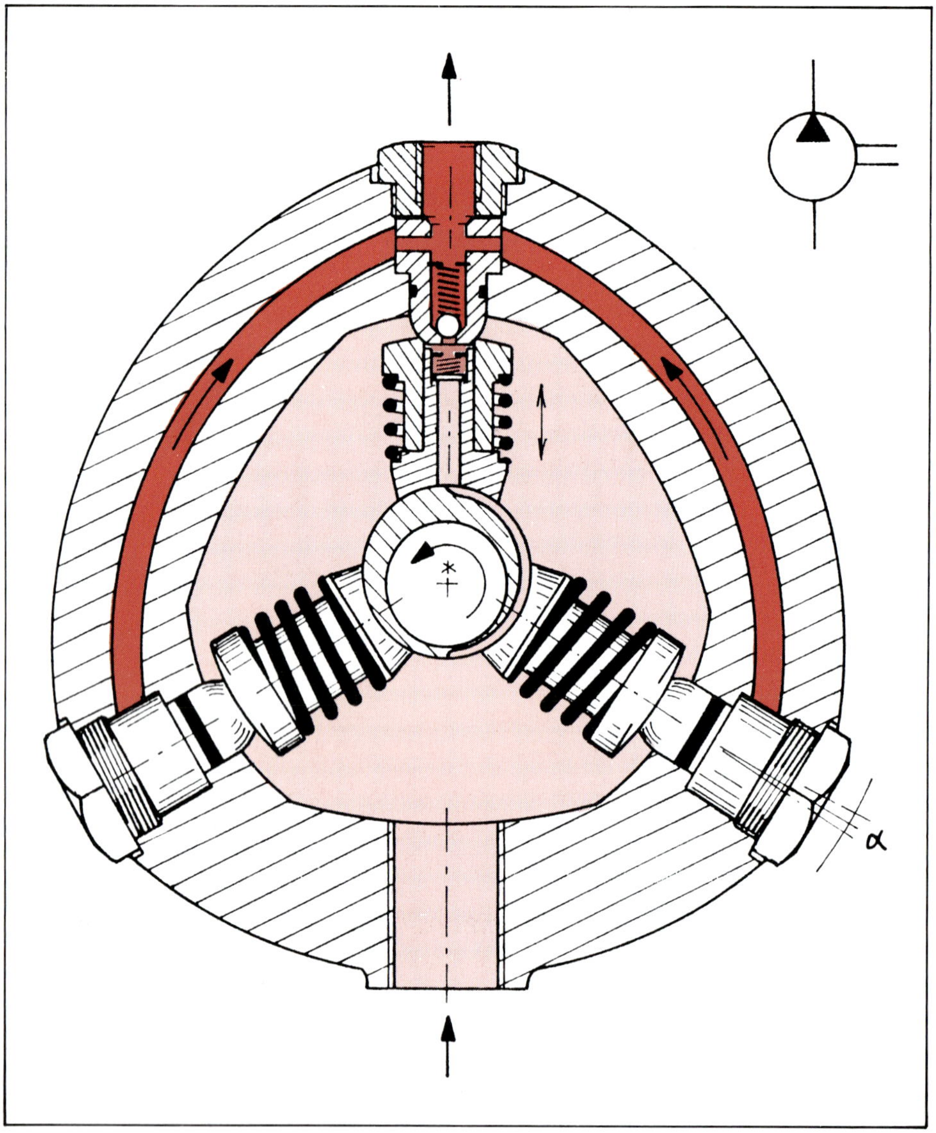

Fig. 73 Radial piston pump (cam actuated type with self-priming).

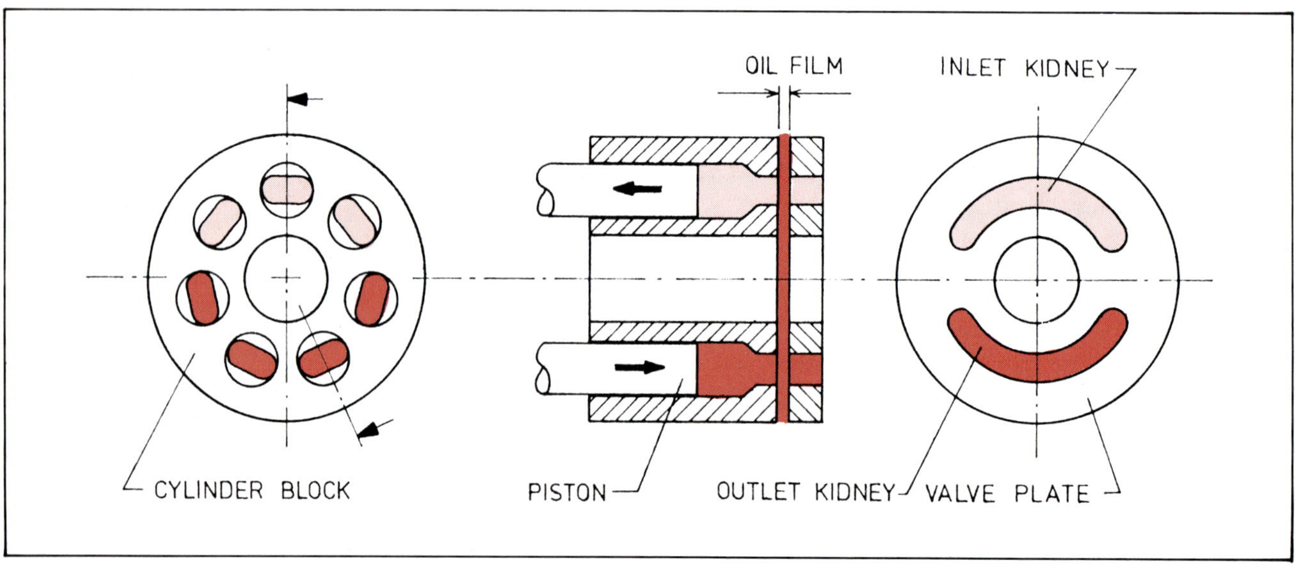

Fig. 74 Cylinder block loading in piston pumps.

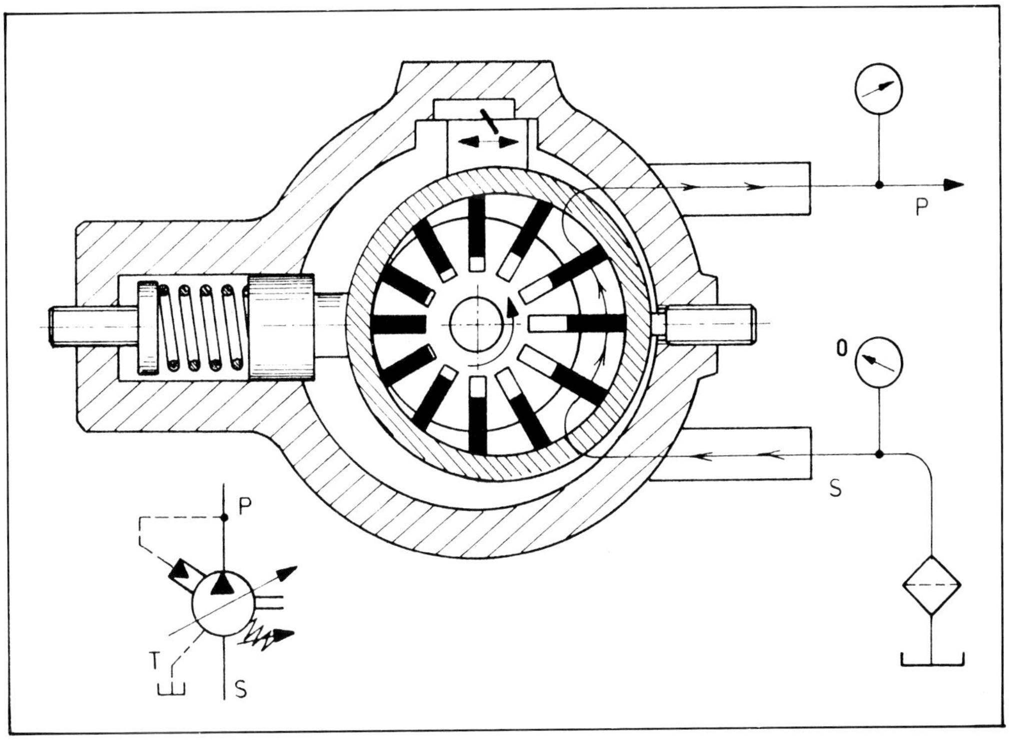

Fig. 75 Variable displacement vane pump.

Ramp type pump control

The pump flow adjusts itself to the flow required by the actuators. If no flow is required by the actuators, but maximum system pressure is reached, then the ramp controller will regulate the flow to almost zero. However, full system pressure will be maintained and any internal leakage (slip) oil in the pump or in the system will automatically be replaced.

The ramp type controller is shown in fig. 75 and its corresponding pressure-flow-power diagram is shown in fig. 76. The ramp shape flow reduction depends on the control spring characteristics. To change sensitivity or response, different springs can be fitted. Ramp controlled pumps do not necessitate a system relief valve to protect system and pump.

Step type pump controls

When the control has been set to the desired pressure threshold, the pump will automatically deliver maximum flow until the pre-set pressure level is reached. It will then maintain such pre-set pressure. Simultaneously, the power input and the flow delivery are reduced to whatever power and flow are required to maintain the pre-set pressure. In contrast to the ramp control, the step controller reduces the flow in a very sharp step (compare figs. 76 and 77).

Systems with step type pump control should be equipped with a pressure relief valve, as step type controllers have a time constant resulting in pressure peaks considerably higher than the nominal setting, when the flow demand is suddenly reduced, for example as shown in fig. 77. They are also very susceptible to oil contamination and could malfunction if oil cleanliness is not monitored and maintained.

Control mechanisms for step type control

Three typical step type controllers are shown in figs. 78, 80 and 81. The step type controller in fig. 75 is used on a vane pump. The cam ring of the vane pump is firmly held between two pistons with an approximate area ratio of 2:1, and system pressure from the pump outlet is acting onto both pistons. Maximum system pressure is pre-set on the adjustable spring of the directional control valve.

When the pre-set pressure is reached, the spool of the control valve is moved against the spring, and the large piston chamber is vented to tank. The small piston will then move the track ring into a more concentric position (to the left) and pump delivery decreases to the flow rate required by the system. The control valve will return to its spring off-set position, as soon as the system pressure drops below the pre-set pressure. Thus, the system pressure is ported back onto the large piston and the track ring is again forced into the full flow position.

Downward regulation (full flow to minimum flow) may cause extreme pressure peaks which could be detrimental to the pump and to the system. A small pressure relief valve will remove such peaks. For maximum tolerable pump pressure, see the manufacturer's specifications. A typical dynamic response test circuit for the pump is shown in fig. 79, with its corresponding

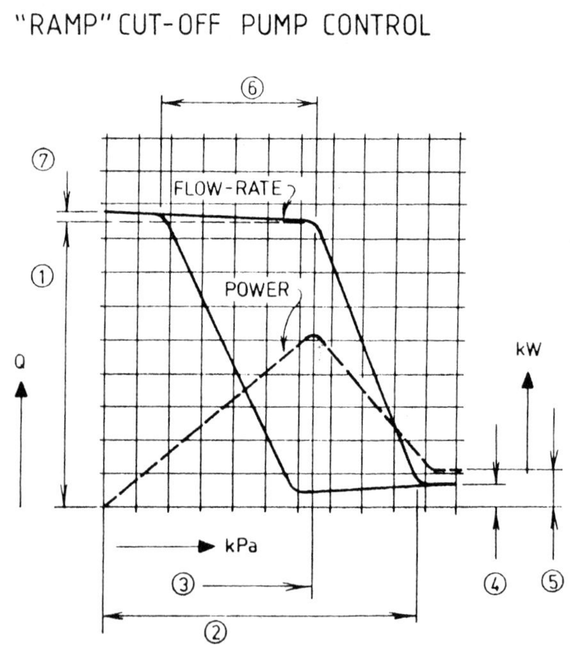

1 Maximum flowrate prior to cut-off
2 Maximum adjusted system pressure
3 Cut-off point
4 Minimum required flowrate to hold system pressure despite internal system leakage
5 Power requirement at minimum flow
6 Adjustment range
7 Decline in volumetric efficiency with increasing pressure

Fig. 76 Pressure-flow diagram for ramp type pump control.

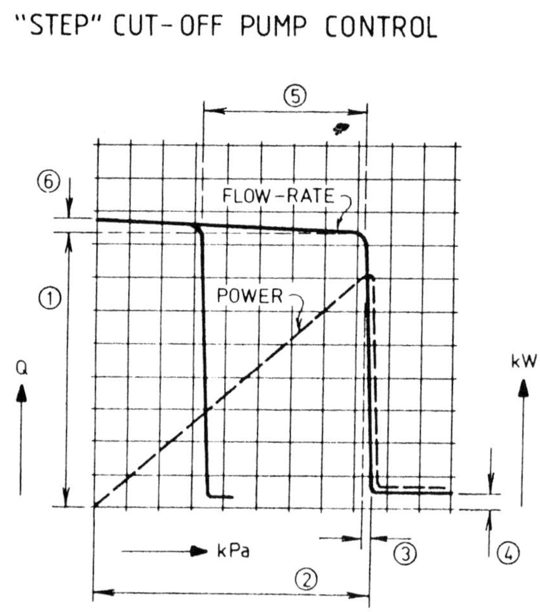

1 Maximum flowrate prior to cut-off
2 Maximum adjusted system pressure
3 Spring constant
4 Minimum required flowrate to hold system pressure
5 Adjustment range
6 Decline in volumetric efficiency with increasing pressure

Fig. 77 Pressure flow diagram for step type pump control.

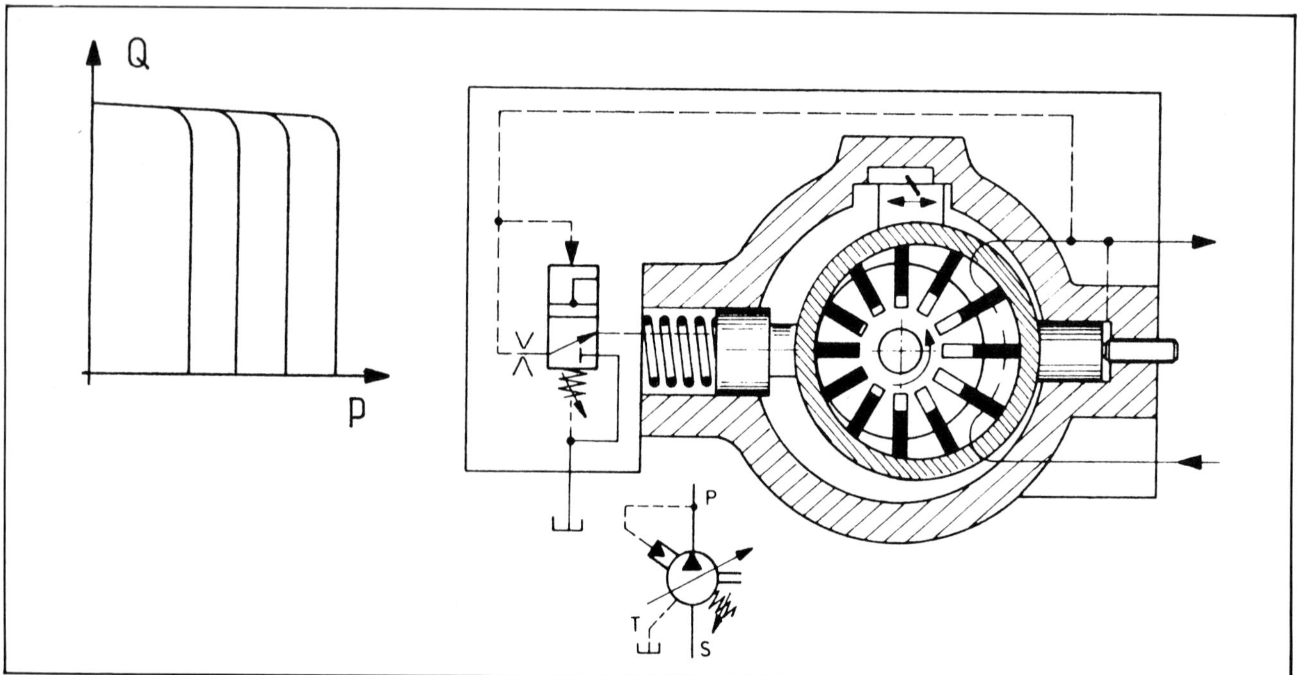

Fig. 78 Vane pump with step type pressure controller.

pressure/time graphs for downward and upward regulation.

The step type pressure controller in fig. 80 controls the swashplate angle of a variable displacement pump. System pressure (P) is ported to the pressure sensing valve and to the annulus area of the adjustment piston. The full piston area is ported to tank. Maximum system pressure is pre-set on the spring of the pressure sensing valve. When the pre-set pressure (P) in the system is reached, the spool of the pressure sensing valve is moved against the spring, pressure P is then ported onto the full piston area of the adjustment piston, and pump delivery decreases to the minimum flow rate demanded by the system. The minimum flow rate is equal to the internal leak rate of the system or whatever flow rate is equired by actuators in motion.

The pressure controller in fig. 81 provides remote maximum pressure control adjustment. System pressure is directed to the large spring loaded piston. This holds the track ring into the full flow position. The pressure relief valve will direct a small amount of fluid to tank when system pressure exceeds its setting. The resulting pressure drop across orifice 1 unbalances the pressure sensing valve and the large piston chamber is vented to tank. The small piston will then move the track ring into a more concentric position (to the left), and pump delivery decreases to system demand.

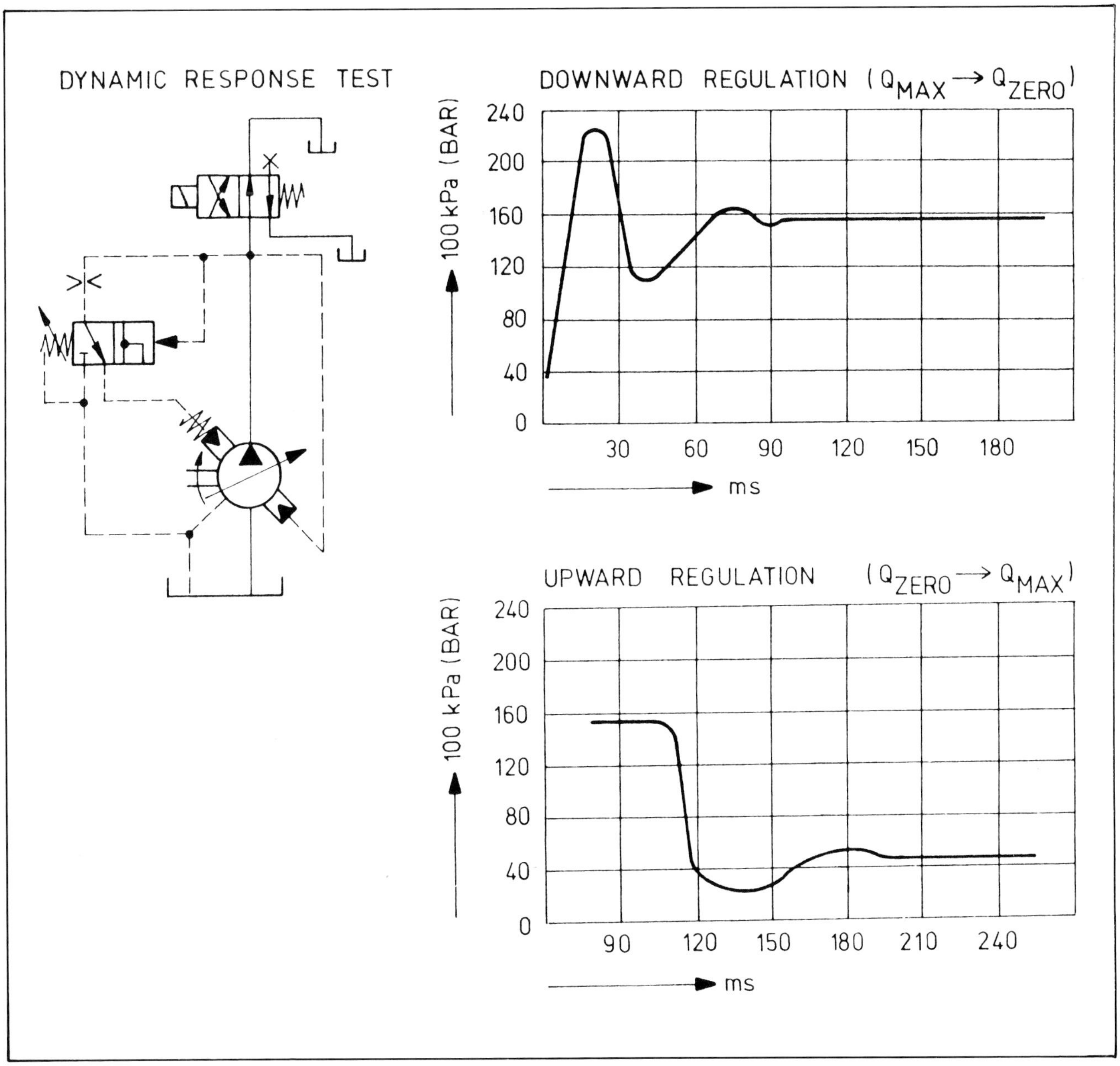

Fig. 79 Test circuit with corresponding pressure-time regulating diagrams for pressure controlled vane pump.

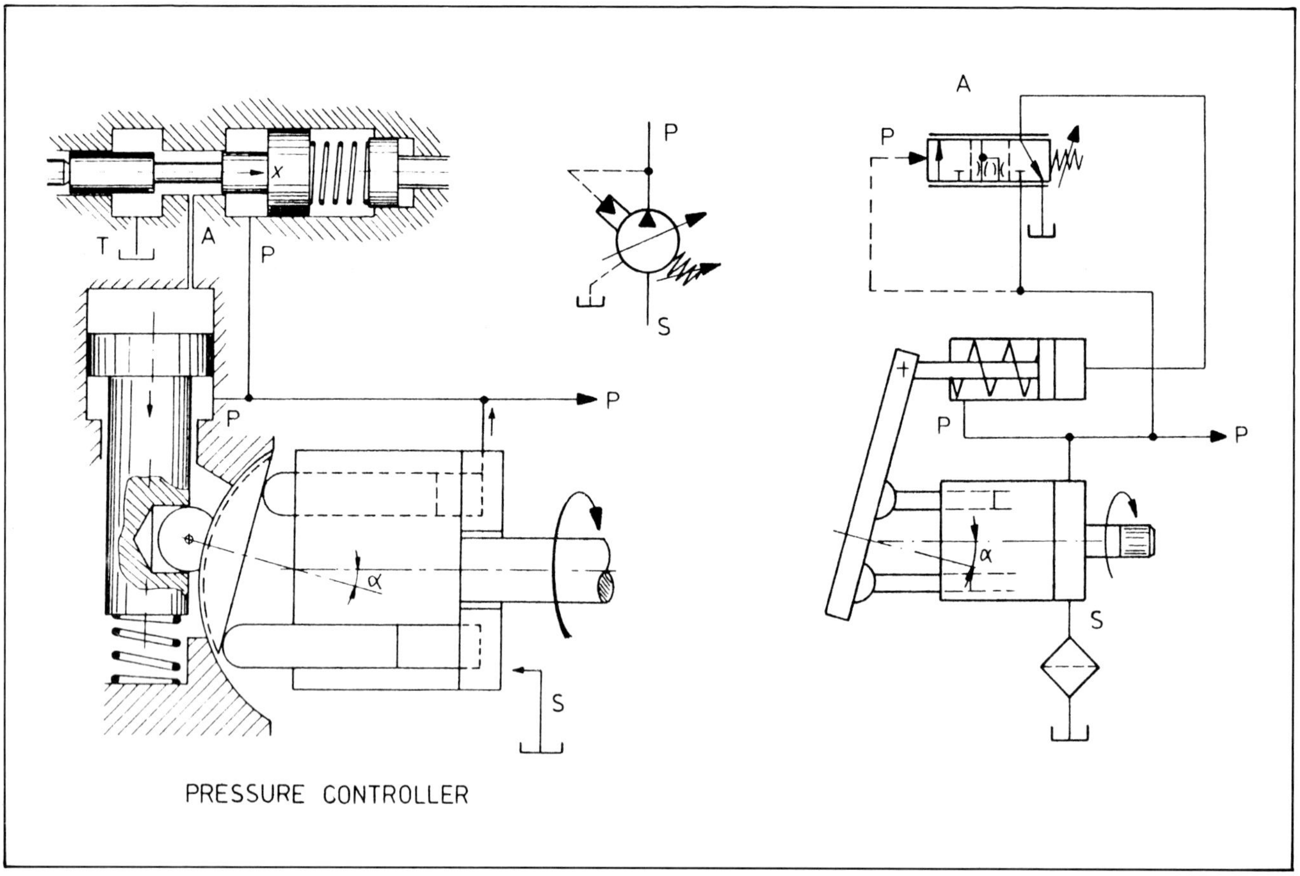

Fig. 80 Step type pressure controller for swash plate piston pump.

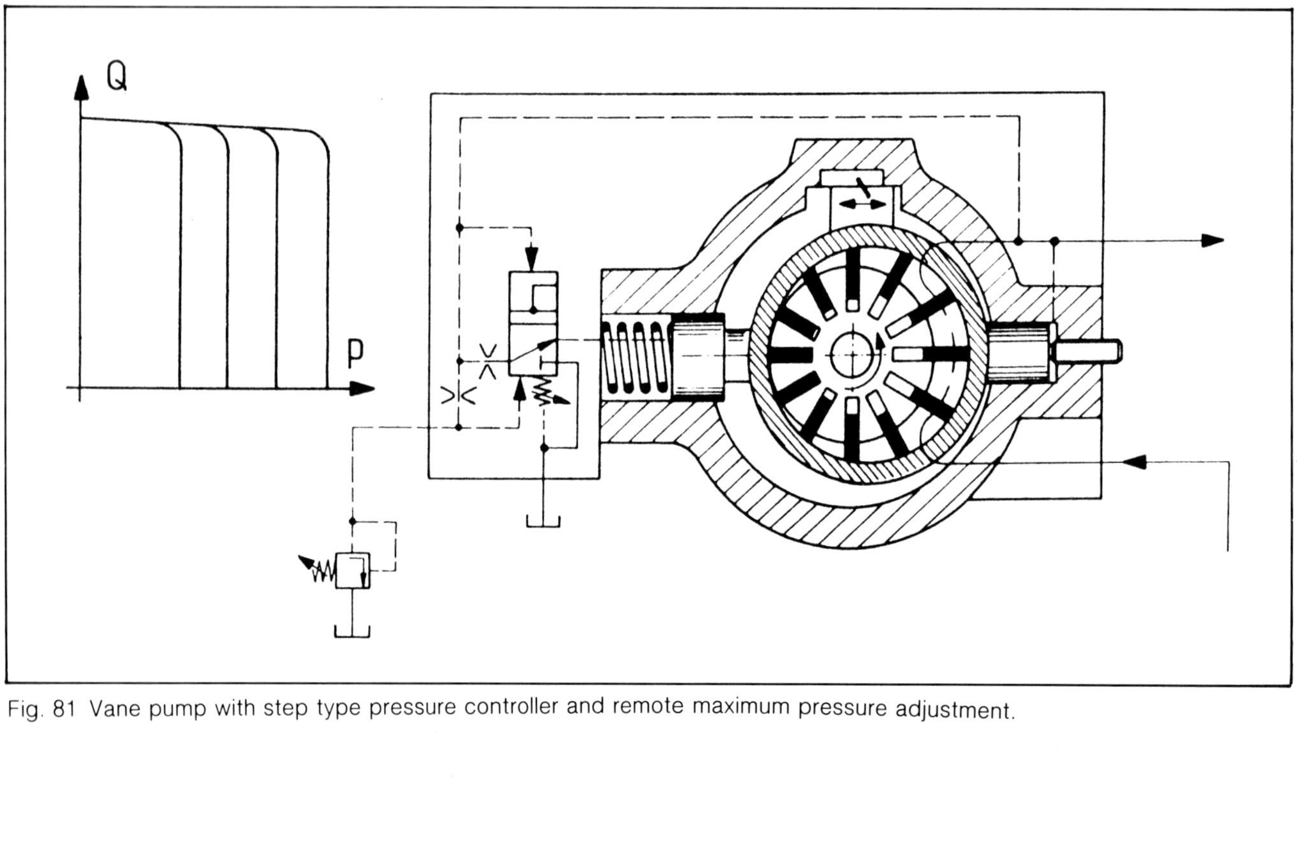

Fig. 81 Vane pump with step type pressure controller and remote maximum pressure adjustment.

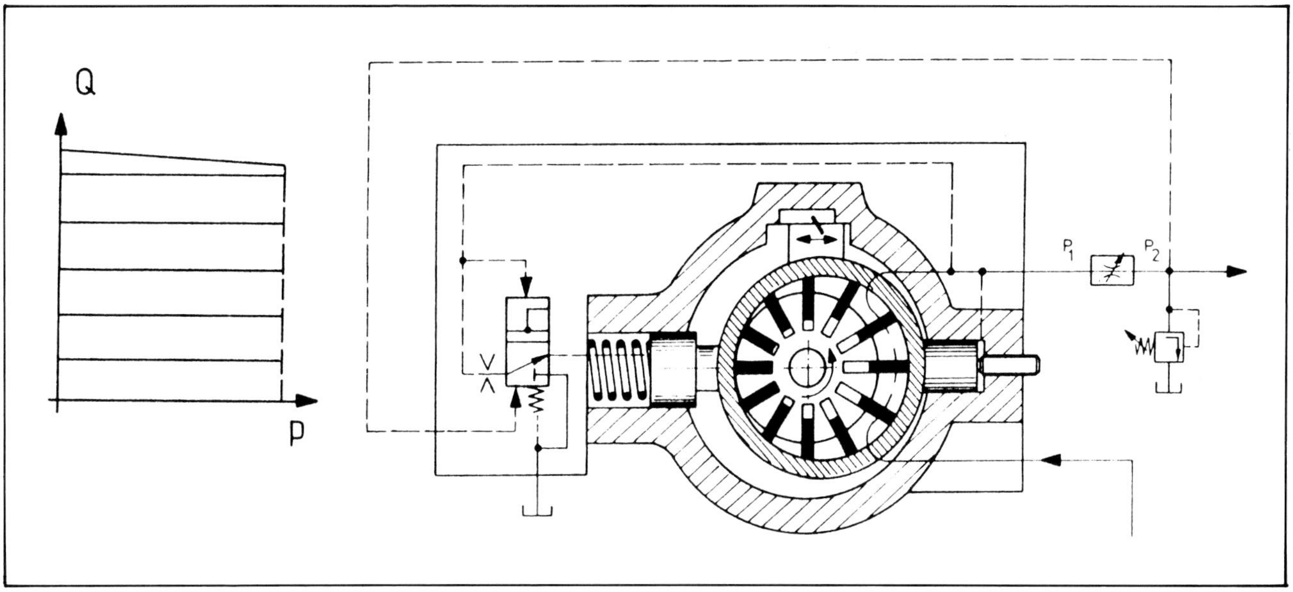

Fig. 82 Vane pump with constant power control.

4 Linear actuators (cylinders)

Hydraulic linear actuators are used to convert hydraulic power into linear mechanical force or motion. Although the actuator itself produces linear motion, a variety of mechanical linkages and devices may be attached to it to produce a final output which is rotary, semi-rotary, or a combination of linear and rotary. Levers and linkages may also be attached to achieve force multiplication or force reduction, as well as an increase or reduction of motion speed (fig. 83).

The main parts of an hydraulic linear actuator are shown in fig. 84. A variety of refinements, additions, and options can be added to this basic actuator (figs. 85, 86 and 89).

The generation of linear thrust force is very simple: fluid under pressure, when delivered to one end of the actuator, acts against the piston area. The piston with the attached piston rod starts to move in linear direction as long as the reaction force is smaller. The developed force is used to move a load, which may be attached either to the piston rod or to the actuator housing (fig. 83). The distance through which the piston travels is known as the stroke.

Actuator types

Single-acting actuators permit the application of hydraulic force in one direction only. These actuators are normally mounted in vertical direction, thus permitting the load to return the piston to its initial position. Where the actuator must be mounted horizontally, an inbuilt spring is used to cause retraction (fig. 85).

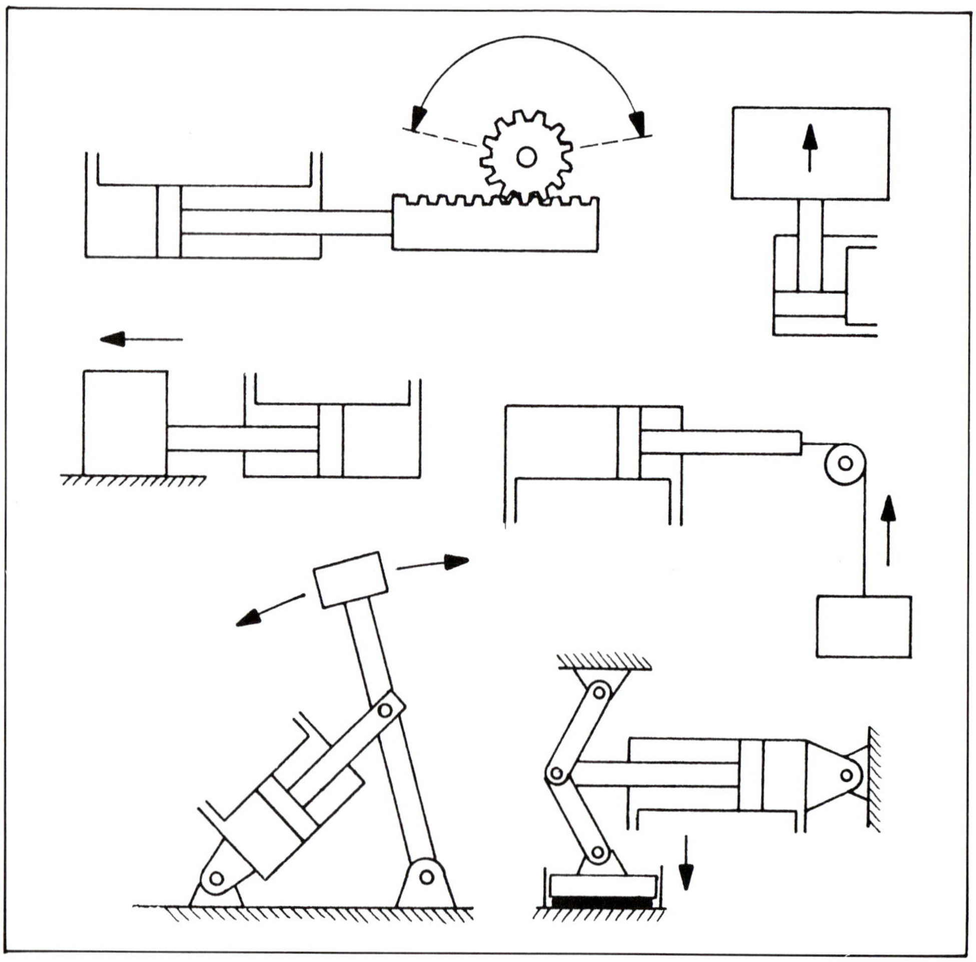

Fig. 83 Linkages and loads attached to linear actuators.

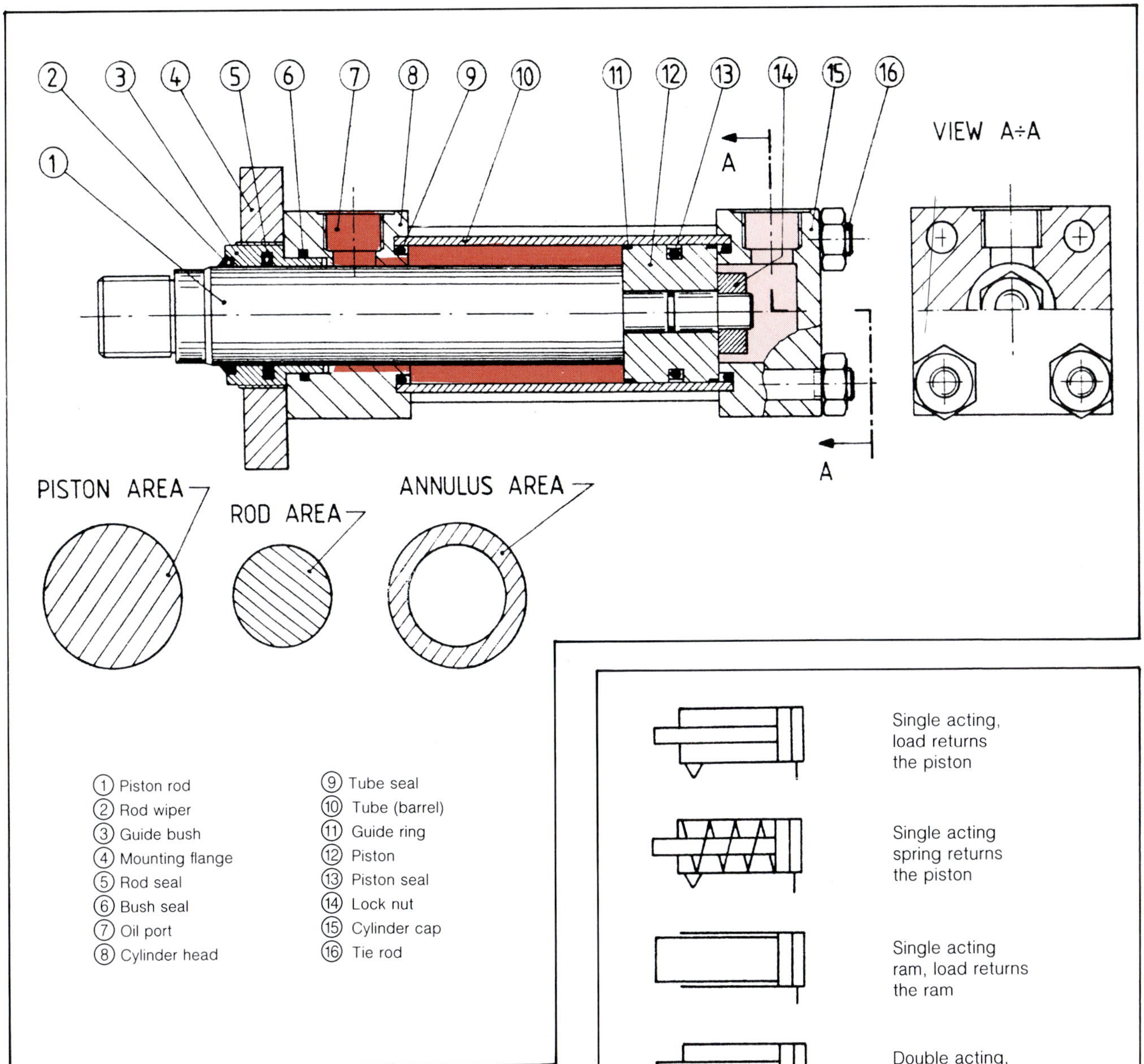

Fig. 84 Typical tie-rod type actuator (cylinder).

Double-acting actuators permit the application of hydraulic force in both directions. However, the retraction stroke develops a smaller force than the extension stroke, since the pressurised fluid acts on a smaller area, known as the annulus area (fig. 84).

Double-ended actuators with rods on both actuator ends are used where the developed force must be equal for both directions (extension and retraction). Since the voids to be filled with hydraulic fluid are equal for extension and retraction, the resulting piston speeds are also equal for both strokes. (Piston speed equals actuator volume divided by the flow rate of the pump.)

Telescopic actuators may be double or single acting. Their collapsed length is only little more than their

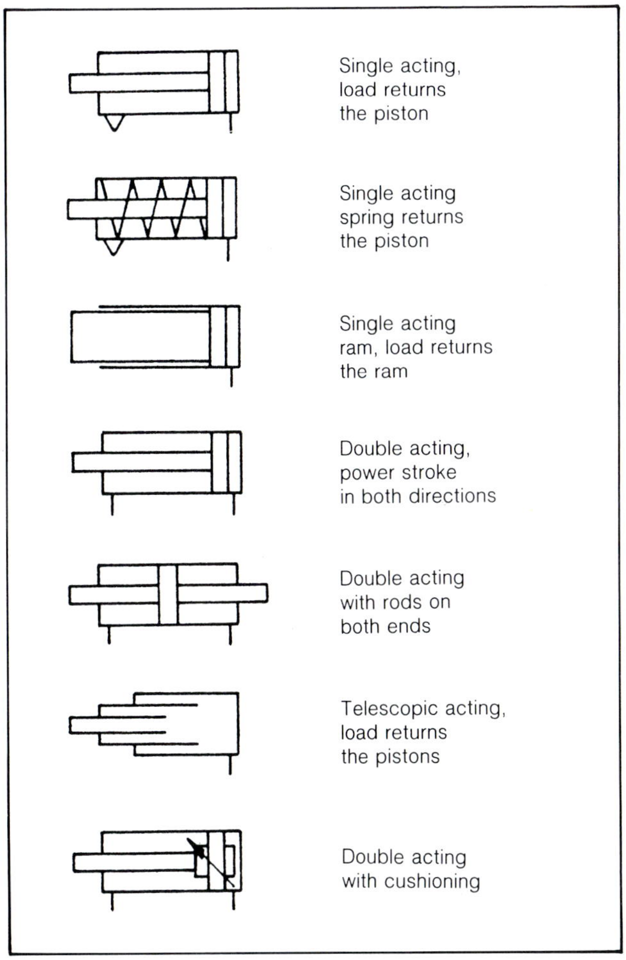

Fig. 85 Graphic symbols for the more commonly used linear actuators.

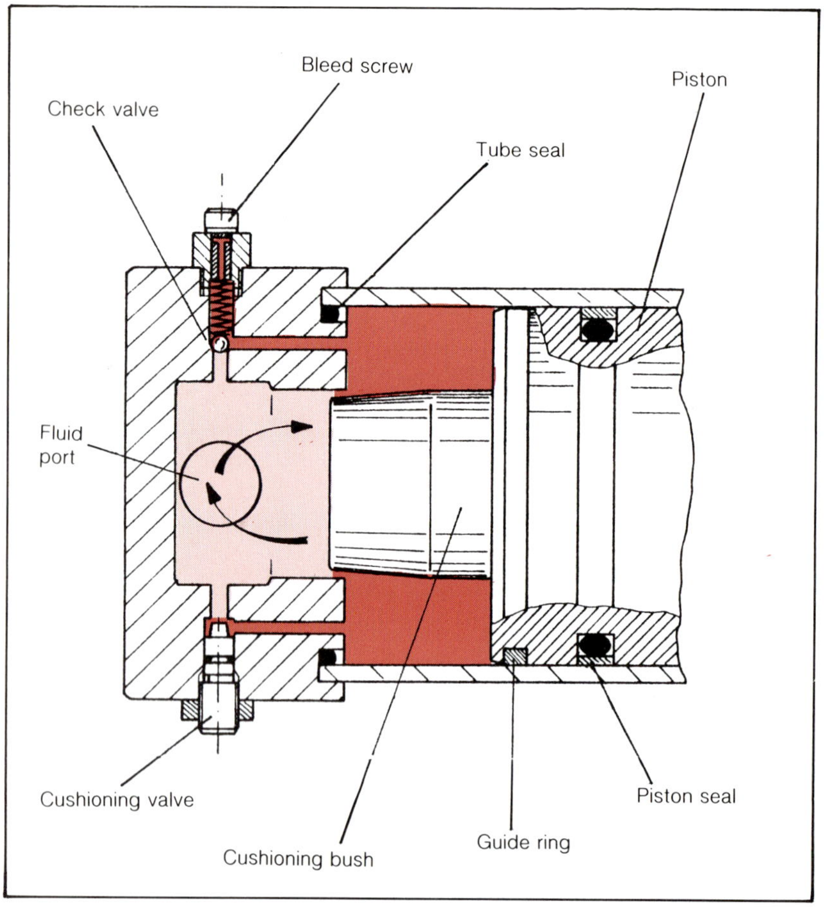

Fig. 86 Typical end position cushioning arrangement.

longest segment. The developed pressure varies with the load and the effective piston area. Thus, the piston segment with the largest piston area extends first. The required pressure increases with each extended segment, since the piston area decreases while the load remains constant. When retracting, the sequence is reversed, i.e. the smallest piston retracts first.

End-position cushioning

Cushioning, or end-position cushioning, refers to braking and deceleration of the final stroke portion until standstill occurs. Cushioning becomes essential above a certain stroke speed. The kinetic energy released on impact at the stroke end must be absorbed by the stroke limit-stops, which are built into the end caps. Their capacity to absorb this energy depends on the elasticity of their material.

An hydraulic braking function (end-position cushioning) must therefore be applied where piston speeds (v) exceed 0.1 m/sec. Figure 82 shows a cross-section of the end-position cushioning mechanism of the end cap. The cushioning of the rod end cap is similar.

The piston is fitted with a tapered cushioning bush. When this bush enters into the bore of the end cap during the final part of the stroke, the main fluid exit begins to shut until it finally closes off completely. This first stage of exit-flow throttling causes an initial speed deceleration. The remaining fluid must now exhaust through the cushioning valve. The degree of cushioning and second stage speed control can be regulated at the cushioning valve (flow control valve).

A check valve is fitted to achieve fast and full force brake-away from the end position. A bleed screw may also be built into the check valve (fig. 86). Some manufacturers have separate bleed screws available as an option. The bleed screw must always be mounted uppermost.

Seals in linear actuators

Seals are grouped into static and dynamic applications. Static seals are fitted between two rigidly connected components. The seals between the piston rod and the piston in fig. 87A and C are static seals, and so are the seals that prevent external leakage past the joint where the tube is mounted onto the cap in fig. 87B, C, and D.

Dynamic seals prevent leakage between components which move relative to each other. Therefore, dynamic seals are subject to wear, whereas static seals are essentially wear free. The seals between the moving piston and the stationary actuator tube (fig. 87A and C) are typically dynamic seals, and so are the seals that

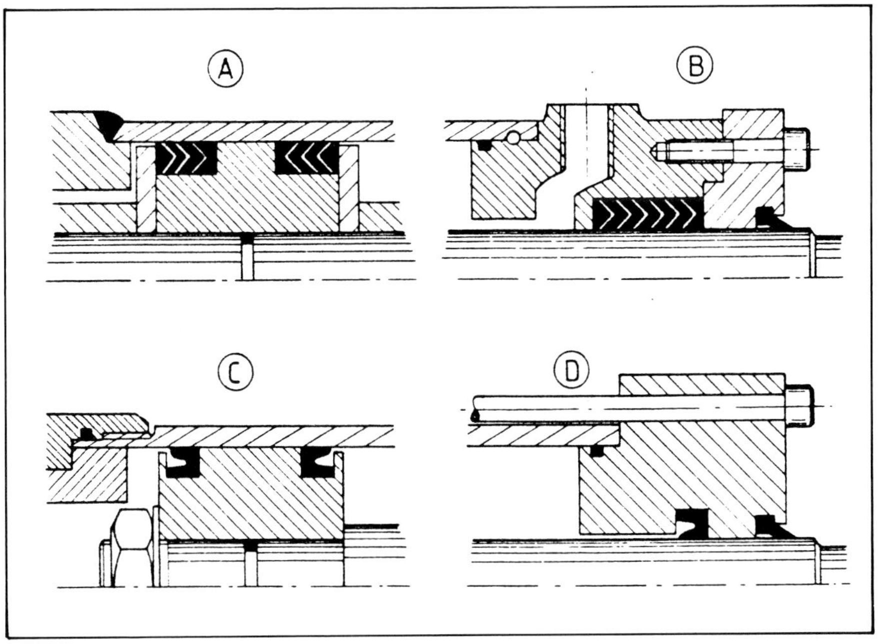

Fig. 87 Various types of seals and tube-to-head connecting methods.

stop hydraulic fluid from escaping past the clearance gap between the moving piston rod and the end cap (head) (fig. 87B and C). The rod-wiper ring may also be regarded as a dynamic seal. It prevents contaminants from being drawn into the actuator when the rod retracts (fig. 87B and D).

O-ring type seals are generally used as static seals whereas dynamic seals may range from simple Chevron type compression packings to complex moulded cup or lip seals. Chevron seals (fig. 87A and B) are less suitable for smooth piston movement at low pressures and low speed, but are an excellent seal for high system pressure and high stroking speeds. Lip-ring seals (fig. 87C and D) provide good sealing even under stationary piston conditions. Lip-ring and glide-ring seals are used for applications with exceptionally low friction requirements. For glide-ring seal applications, the piston is fitted with guide rings (fig. 86).

Various hydraulic fluids require specific seal materials. (Manufacturers of actuators should be contacted for compatibility of seal material and fluid used in the system.)

Actuator construction

The actuator tube is usually of cold-drawn seamless steel with a microfinish honed bore of a surface quality of $R_\alpha \leqslant 1.3\mu m$. Piston rods may be made from high-grade heat-treated steel, surface hardened and hard chrome-plated for some applications. Surface quality of the rod is about $0.2\mu m$. Induction hardening of the surface provides greater protection against mechanical damage and results in a longer life of the dynamic seals. Where the actuator is to be used in an aggressive environment, high tensile stainless steel may be specified for the piston rod, and sometimes also hard chrome-plating to provide a wear-resistant surface. Tube and heads are either screwed or welded to each other or tie-rods or wire retainers may be used (fig. 87).

Actuator failure

Standard type actuators are not designed to absorb piston rod side loading. Thus, actuators must be mounted with care and accuracy, to ensure that the load moves parallel and in alignment with the actuator centreline (fig. 88).

For many applications, the piston rod must be fitted with a clevis or spherical bearing-rod eye, or the actuator must be allowed to swivel around a trunnion mount, permitting it to swing as the direction of the load changes (fig. 89D–G).

Failure of the rod bearing (guide bush) usually occurs when the piston rod is fully extended. If mounting space permits, actuators should be used with a longer stroke than actually required. This gives more bearing distance between the piston and the supporting bearing bush (fig. 88). For various actuator mountings, see fig. 89.

Stop tubes

For actuators with long strokes or applications where the piston is allowed to "bottom-out" on the head end, it is recommended to use stop tubes (fig. 88).

The stop tube is a spacer, placed on the piston rod next to the piston. The stop tube increases the minimum distance between the guide bush and the piston and thus provides more bearing leverage to support side load on the piston rod.

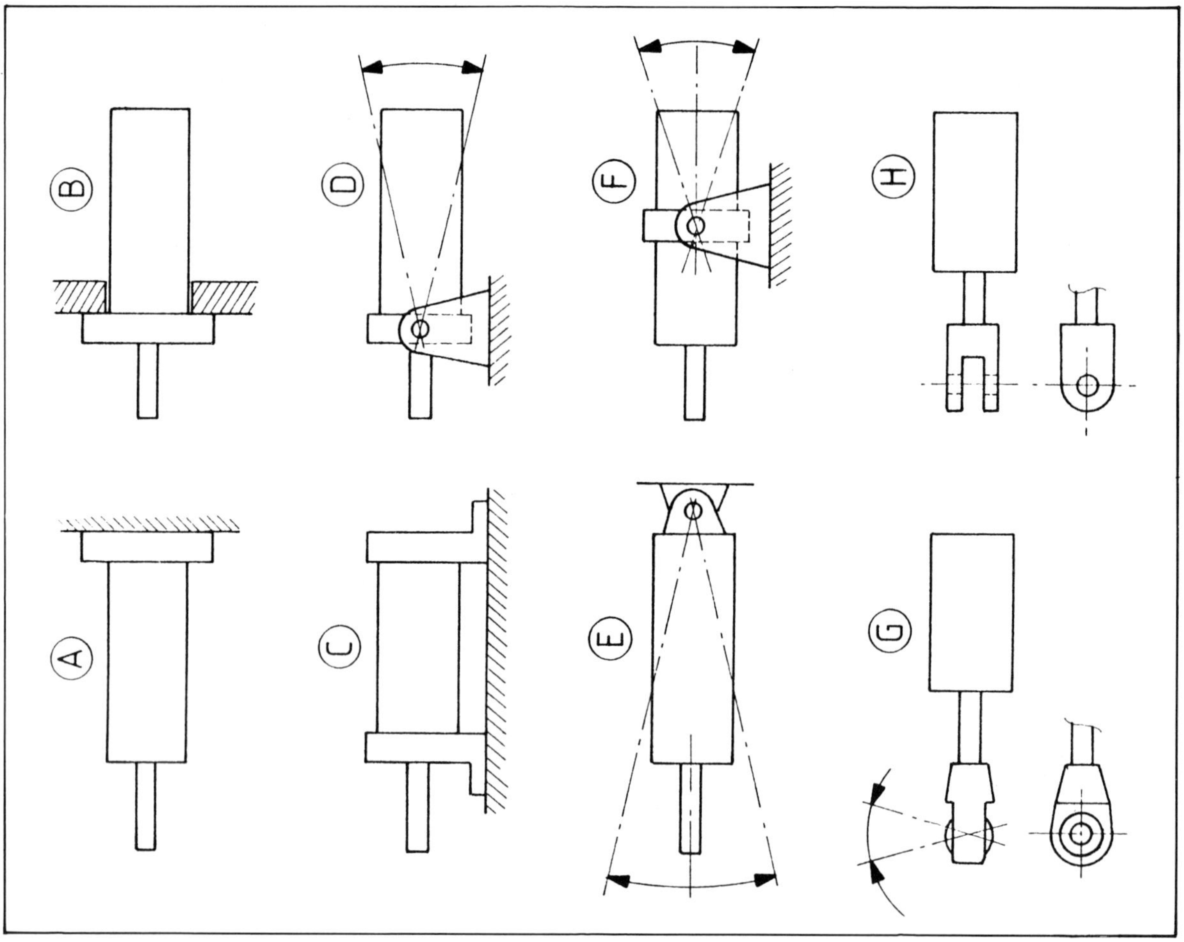
A
B
C
D
E
F
G
H

Fig. 89

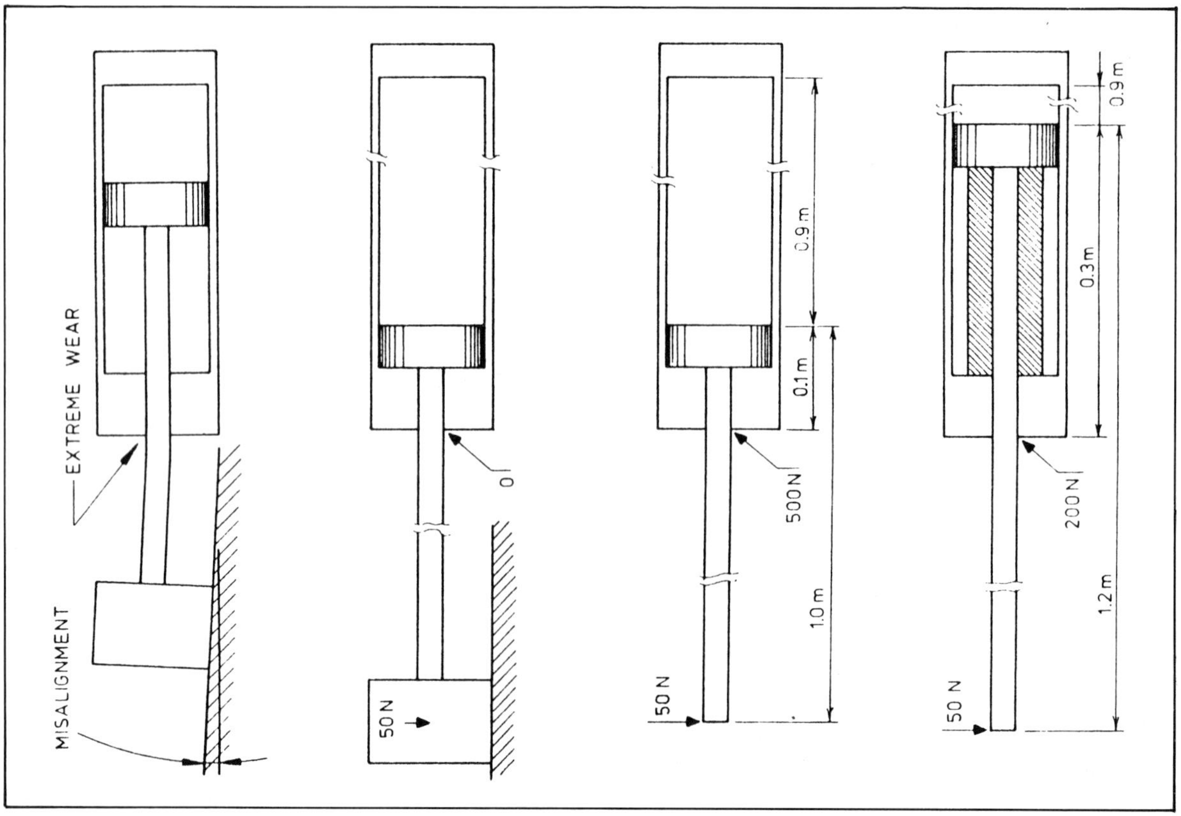
MISALIGNMENT
EXTREME WEAR
50 N
0
0.9 m
0.1 m
500 N
1.0 m
50 N
0.9 m
0.3m
200 N
1.2 m
50 N

Fig. 88

Position of ports and port diameter

With respect to welded actuators the position of ports must be specified before manufacture, but for tie-rod type actuators the positions can be varied by turning the head or the end cap. To accomplish this, the tie rods must be undone on one side and pushed towards the other end. The cap can then be rotated and re-assembled in the desired position.

Some actuator manufacturers offer a choice of port threads and of standard or oversize ports. Many threads may be used, but the most common port threads are the British Standard pipe thread (B.S.P.), the metric I.S.O. thread, and the American national pipe threat (N.P.T.).

Piston speed

For most linear hydraulic actuators, the maximum permissible piston speed is 0.5 m/s. The piston speed is related to the size of the actuator ports, and to the flow-rate to and from the actuator. Thus, piston speed can be controlled if the flow rate is variable (see Chapter 6, flow control valves). Piston speeds greater than 0.5 m/s may for some applications be necessary, but the life of the dynamic seals and the effectiveness of end position cushioning would require special design consideration.

Overhauling load control

Special attention must be given to actuator control, because during some part of the stroke external forces tend to drive the actuator (fig. 90). To prevent such free falling or "run-away" situations, counterbalance or brake valves (Chapter 5) may be used. For positive (non-creeping) load-holding control, see pilot operated check valves (Chapter 2) and their application.

Actuator sizing

The main criteria on which the size of the actuator is based are:

- force output for extension and retraction;
- piston speed for extension and retraction;
- mechanical stability of the actuator.

The force output is expressed in Newtons (N). It depends on the developed system pressure and the effective piston area. It is also affected to some (minor) degree by the friction loss between the moving and stationary actuator components (piston, seals, rod, guide bush, and tube).

The basic formulae to calculate theoretical actuator force output (N), required system pressure (Pa), and effective piston area, can be presented in a simple tri-

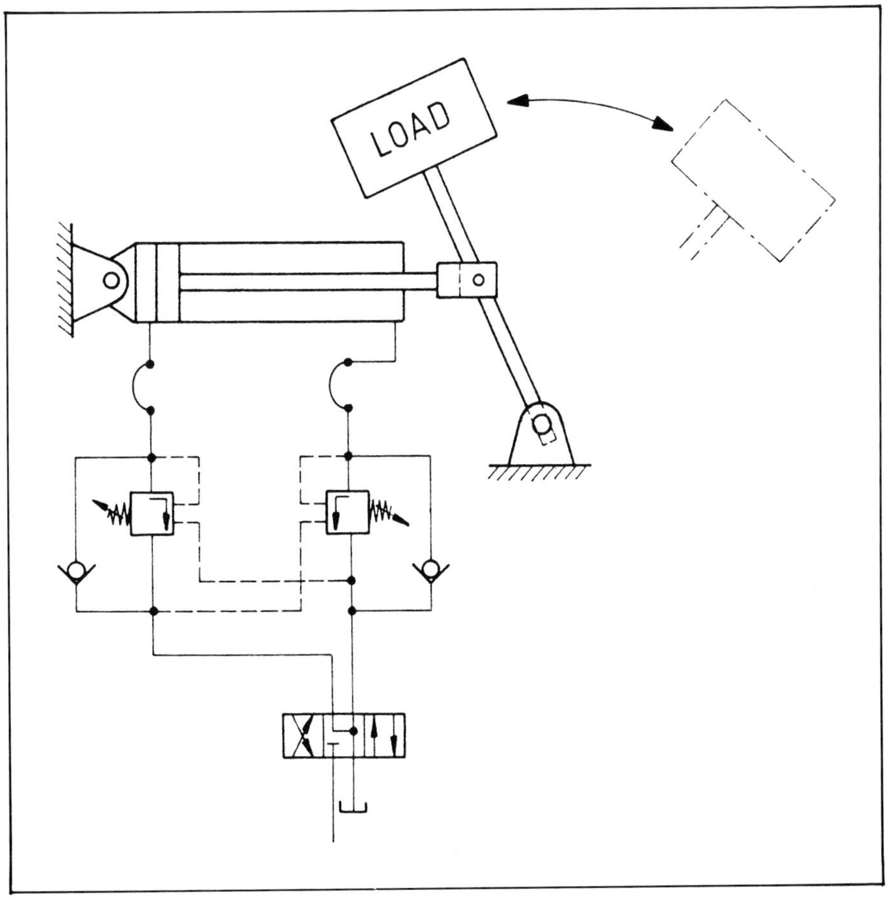

Fig. 90 Overhauling load control for tipping device.

angular illustration (fig. 91). The letter F stands for force, the letter P for pressure, and the letter A for the piston area.
The transposed formulae for each of the three factors are:

- Force = Pressure × Area → $N = Pa \times m^2$
- Pressure $= \frac{\text{Force}}{\text{Area}} \rightarrow Pa = \frac{N}{m^2}$
- Area $= \frac{\text{Force}}{\text{Pressure}} \rightarrow m^2 = \frac{N}{Pa}$

Fig. 91

Three different formulae may be used to calculate piston area. These are:

$$\text{Area} = d^2 \times 0.7854 = \frac{\pi \times d^2}{4} = \pi \times r^2$$

If friction is to be included, the basic formula for force output is:

$$\text{Force} = \text{Pressure} \times \text{Effective Area} \times \frac{\text{Efficiency}}{100}$$

$$N = \frac{Pa \times m^2 \times \eta_{hm}}{100}$$

Therefore:

$$\text{Force}_{(EXTENSION)} = \frac{Pa \times d^2 \times 0.7854 \times \eta_{hm}}{100}$$

$$\text{Force}_{(RETRACTION)} = \frac{Pa \times (d_p^2 - d_r^2) \times 0.7854 \times \eta_{hm}}{100}$$

where

d_p = diameter of the piston (m),
d_r = diameter of the rod (m), and
η_{hm} = hydro-mechanical efficiency (%)

The basic formulae to calculate theoretical actuator speed (v) and required flow rate ($\frac{m^3}{s}$) or (Q) for a piston area (m^2) or (A) can also be presented in a simple triangular illustration (fig. 92).

- Flow rate (Q) = Speed × Area → $\frac{m^3}{s} = \frac{m \times m^2}{s}$
- Speed (v) $= \frac{\text{Flow Rate}}{\text{Area}} \rightarrow \frac{m}{s} = \frac{m^3/s}{m^2}$
- Area (A) $= \frac{\text{Flow Rate}}{\text{Speed}} \rightarrow m^2 = \frac{m^3/s}{m/s}$

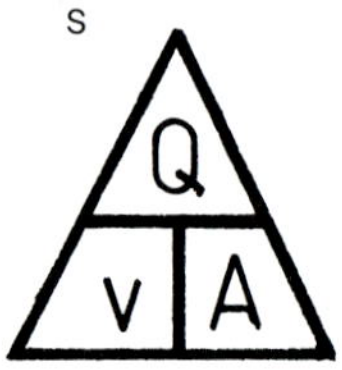

Fig. 92

Note: volume can be also given in litres or cm^3, but the basic unit (m^3) must be used for calculations.

Example

Calculate the force output for a linear actuator and the required pump flow rate for the following specifications:

Piston diameter = 50 mm
Rod diameter = 25 mm
Stroke = 600 mm
Piston speed = 12 mm/s
Efficiency (η_{hm}) = 95%
Pressure = 4000 kPa

$$\text{Force}_{(EXT)} = \frac{4000 \times 10^3 \times 0.05^2 \times 0.7854 \times 95}{10^3 \qquad 100} = 7.46 \text{ kN}$$

$$\text{Force}_{(RET)} = \frac{4000 \times 10^3 \times (0.05^2 - 0.025^2) \times 0.7854 \times 95}{10^3 \qquad 100}$$

$$= 5.60 \text{ kN}$$

$$\text{Flowrate}_{(EXT)} = \frac{0.012 \times 0.05^2 \times 0.7854}{s}$$

$$= 0.000024 \text{ m}^3/\text{s} = 0.024 \text{ L/s}$$

$$\text{Flowrate}_{(RET)} = \frac{0.012 \times (0.05^2 - 0.025^2) \times 0.7854}{s}$$

$$= 0.0000177 \text{ m}^3/\text{s} = 0.0177 \text{ L/s}$$

Piston rod buckling

Column failure, or buckling of the piston rod, will occur if the actuator stroke in relation to the piston rod diameter is, at a required force output, out of (safe) proportion. Piston rod buckling is calculated according to the "Euler formula", where the piston rod is regarded as the buckling member (fig. 93).

$$\text{Euler formula: } K = \frac{\pi^2 \times E \times I}{S_K^2}$$

Note: Under this condition the piston rod buckles!
The maximum safe operating load (in Newtons) is:

$$F = \frac{K}{S}$$

K = critical load (N)
S_K = free buckling length (m) (fig. 93)
S = safety factor (usually 2.5–3.5)
E = elasticity modulus (Pa) → for steel: $2.1 \times 10^{10} \times 9.80665$
F = force
d = rod diameter (m)
I = moment of inertia (m^4) → $\frac{\pi \times d^4}{64}$

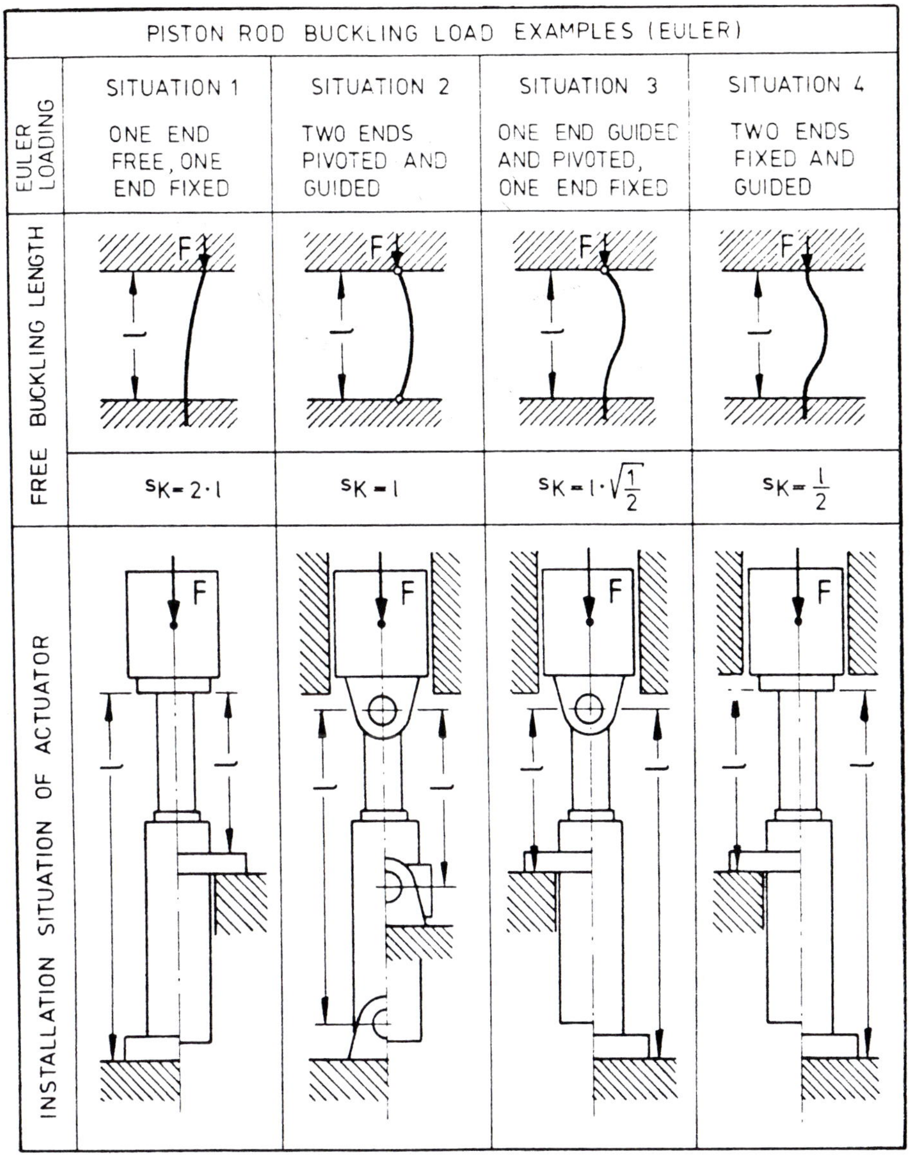

Fig. 93 Piston rod buckling load examples (Euler).

Example

Piston diameter = 100 mm
Rod diameter = 50 mm
Stroke (L) = 900 mm
^{s}K (situation 2) = L

$$K = \frac{\pi^2 \times 2.1 \times 10^{10} \times 9.80665 \times \pi \times 0.05^4}{0.9^2 \quad 64}$$

$K = 769847.34 \rightarrow$ rod buckles!

Safe loading: $F = \dfrac{K}{3.5}$

$$F = \frac{769847.34}{3.5 \times 10^3} = 219.96 \text{ kN (kiloNewtons)}$$

For ease of calculation, the formula for safe operating load can be simplified by grouping all constants in the formula together:

$$F = \frac{K}{S} \rightarrow F = \frac{(\pi^2 \times 2.1 \times 10^{10} \times 9.80665 \times \pi) \times d^4}{S \times S_K{}^2 \times 64}, \text{ or}$$

$$F \approx \frac{10^{11} \times d^4}{S \times S_K{}^2}$$

In hydraulic applications, forces expressed in Newtons usually give extremely high values. It is therefore advisable to use either kilo Newtons (kN) or mega Newtons (MN).

Regenerative actuator control

Regeneration is achieved by using suitable valving which connects the exhaust flow from the extending piston rod with the in-rushing fluid on the piston end. In this way, the exhaust fluid, which would normally return to tank, joins the fluid flow from the pump, thus causing the piston rod to extend with increased speed.

If the full piston area (A1) has a ratio of 2:1 to the annulus area (A2), then the piston rod will extend and retract with equal speed (fig. 94). During regeneration, equal pressure is acting onto both sides of the piston. Therefore, the net thrust for extension can be calculated as:

$Force_{(EXT)} = p \times (A1 - A2)$, where $A1 - A2$ = rod area.

Thus, the extension force is the product of the piston rod area and the system pressure. A similar calculation can be made for the retraction stroke. Since the annulus area and the rod area are equal (with a 2:1 ratio), the retraction force can be calculated as:

$Force_{(RET)} = p \times$ Annulus Area (rod area = annulus area).

Thus, the forces for extension and retraction are equal. With a 2:1 ratio, the exhaust volume for the pushed-out fluid during extension is one half the volume formed by the full piston area (A1). Hence, the pump must fill the other half. Therefore, the regenerated volume and the pumped volume are equal ($V_R = V_P$), and the required volume for extension is:

$V_{EXT} = V_R + V_P$, or $V_{EXT} = V \times 2$.

Furthermore, the required volume for retraction is V_R or $V \times 1$ (remember, V_R and V_P are equal).

Piston rod speed is calculated with the formula:

$$v = \frac{Q}{A}$$

where Q stands for flow rate, A for effective piston area, and v for velocity or speed.

As previously shown, the extension volume is twice the retraction volume ($V_{EXT} = V \times 2$; $V_{RET} = V \times 1$; volume

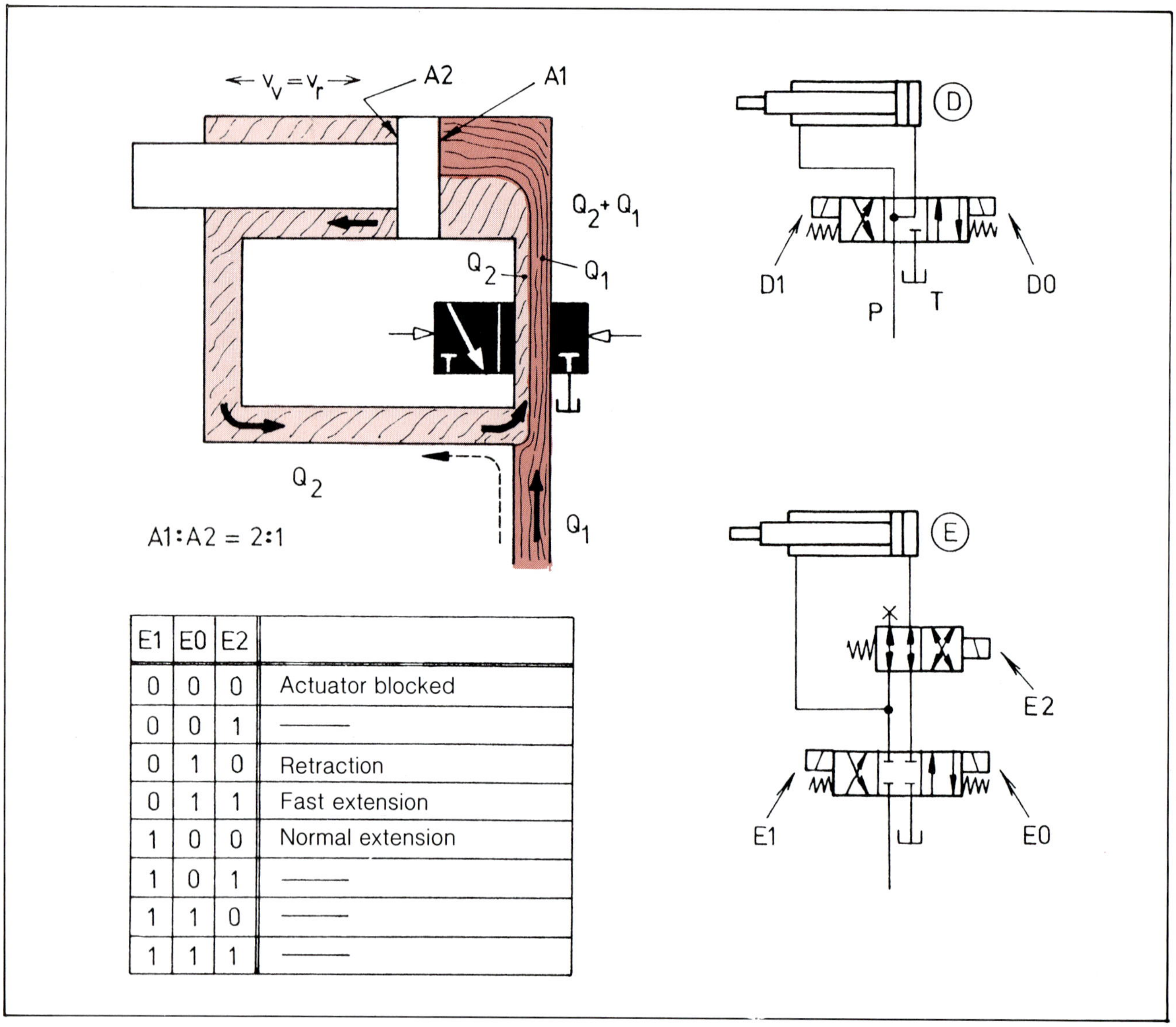

E1	E0	E2	
0	0	0	Actuator blocked
0	0	1	———
0	1	0	Retraction
0	1	1	Fast extension
1	0	0	Normal extension
1	0	1	———
1	1	0	———
1	1	1	———

Fig. 94 Function principle of regenerative actuator with alternative control methods.

ratio 2:1). The flow rate for retraction is equal to the flow rate of the pump. Using the piston rod speed formula one can calculate:

$$V_{RET}=\frac{Q\times 1}{A\times 1} \quad V_{EXT}=\frac{Q\times 2}{A\times 2} \text{ or } V_{EXT}=V_{RET},$$

where
$Q\times 1$ = pump flow rate
$Q\times 2$ = pump flow rate + exhaust flow rate
$A\times 1$ = annulus area
$A\times 2$ = piston area

Therefore, regenerative control with a piston-to-rod area ratio of 2:1 provides equal speed and equal force for extension and retraction.

The control circuit Ⓓ in fig. 94 permits the actuator to extend with normal speed if signal D1 is selected, or to extend rapidly (regenerative flow condition) if the valve's centre configuration is selected. With signal D2 the actuator retracts with normal speed.

The table in fig. 94 explains the effects of the eight possible switching combinations of circuit Ⓔ. The second, sixth, seventh, and eighth combination brings no useful actuator function. Further switching circuits for regenerative actuator control are depicted in fig. 95.

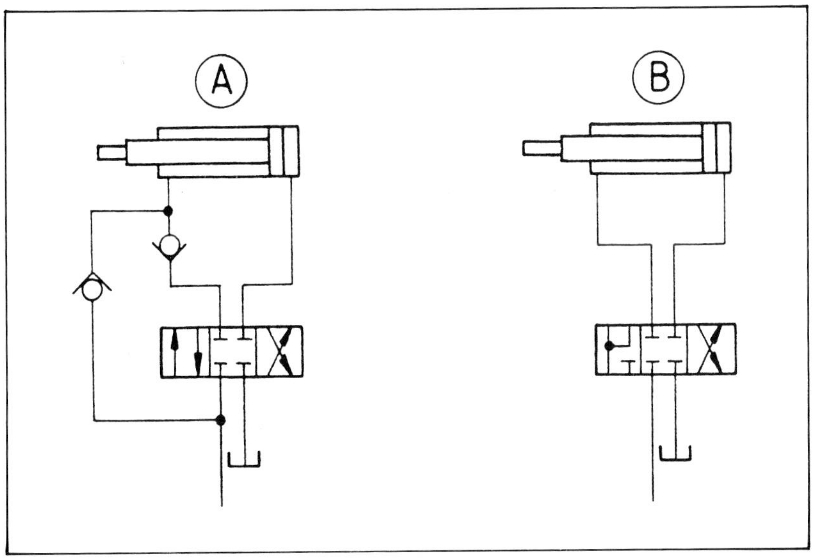

Fig. 95 Circuits for regenerative control.

5 Pressure controls

Hydraulic energy is produced as long as the prime mover (usually an electric motor) drives the pump, and hydraulic pressure develops by resistance to pump flow. Hence, the hydraulic system suffers damage if the pump flow is not stopped or off-loaded (re-circulated) back to tank during non-action periods of the circuit. Non-action periods arise from stalling an actuator, or by reaching the end of the stroke or the circuit sequence, or during the time-delay periods of the circuit sequence.

In order to avoid hydraulic system damage, power wastage, and overheating of the hydraulic fluid, circuit designers use a variety of cleverly designed systems to control maximum system pressure and pump flow during non-action periods (fig. 96).

Pressure control valves are used in hydraulic systems to control actuator force (force = pressure × area), and to determine and (pre)select pressure levels at which certain machine operations must occur. Pressure controls are in the main used to perform the following system functions:

1. To limit maximum system pressure in a hydraulic circuit or sub-circuit, and thus provide overload protection.
2. To provide re-direction of pump flow to tank, while system pressure must be maintained (system unloading).
3. To provide re-direction of pump flow to tank while system pressure is not maintained (system off-loading).
4. To offer resistance to fluid flow at selectable pressure levels (counterbalance force).
5. To provide an alternative flow path for the fluid at selected pressure levels (pressure sequencing).
6. To reduce (or step down) pressure levels from the main circuit to a lower pressure in a sub-circuit.

Pressure control valves are often difficult to identify, mainly because of the many descriptive names applied to them. The function of the valve in the circuit usually becomes the basis for its name. The valves used to accomplish the abovementioned system functions are therefore given the following names, respectively:

1. Relief valves (direct-acting or compound, fig. 99).
2. Unloading relief valve (accumulator charging valve, fig. 104).
3. Offloading valve (fig. 107).
4. Counterbalance valve and brake valve (figs. 108 and 109).
5. Pressure-sequence valves (direct-acting or compound, fig. 110).
6. Pressure-reducing valves (direct-acting or compound, fig. 114).

Direct-acting relief valve

The load which has to be moved by the hydraulic actuator creates a certain resistance to the fluid flow from the hydraulic pump; as this resistance increases, the system pressure increases proportionally. If the actuator reaches a stalling point, then the system pressure will rise almost instantly to such a hazardous level, that damage to the hydraulic system is inevitable. For this very reason, most hydraulic systems are protected by a pressure-relief valve, which limits the maximum permissible system pressure, and diverts some or all of the pump's flow to tank when the pressure setting of the relief valve is reached.

Valve operation

A direct-acting relief valve as shown in fig. 97 may consist of a ball or poppet held firmly onto the valve seat by the spring. The system pressure acts against the pressure-exposed area of the poppet. When the force of the fluid becomes greater (pressure × valve-seat area) than the opposing force of the spring, the poppet is forced off its seat; the relief valve opens, and fluid is released to tank at low pressure. The pressure at which the valve starts to divert flow to tank is called "cracking pressure". As flow to tank increases, the poppet is forced off its seat even more, causing a further compression of the spring.

Thus, when the valve diverts all the pump-flow, the measured system pressure — which is called "full flow pressure" — may be considerably higher than the "cracking pressure". The difference between "full flow pressure" and the "cracking pressure" is called "pressure override" (fig. 98).

Compound-relief valve (pilot operated relief valve)

The "pressure override" caused by the direct-acting relief valve may in some applications be acceptable.

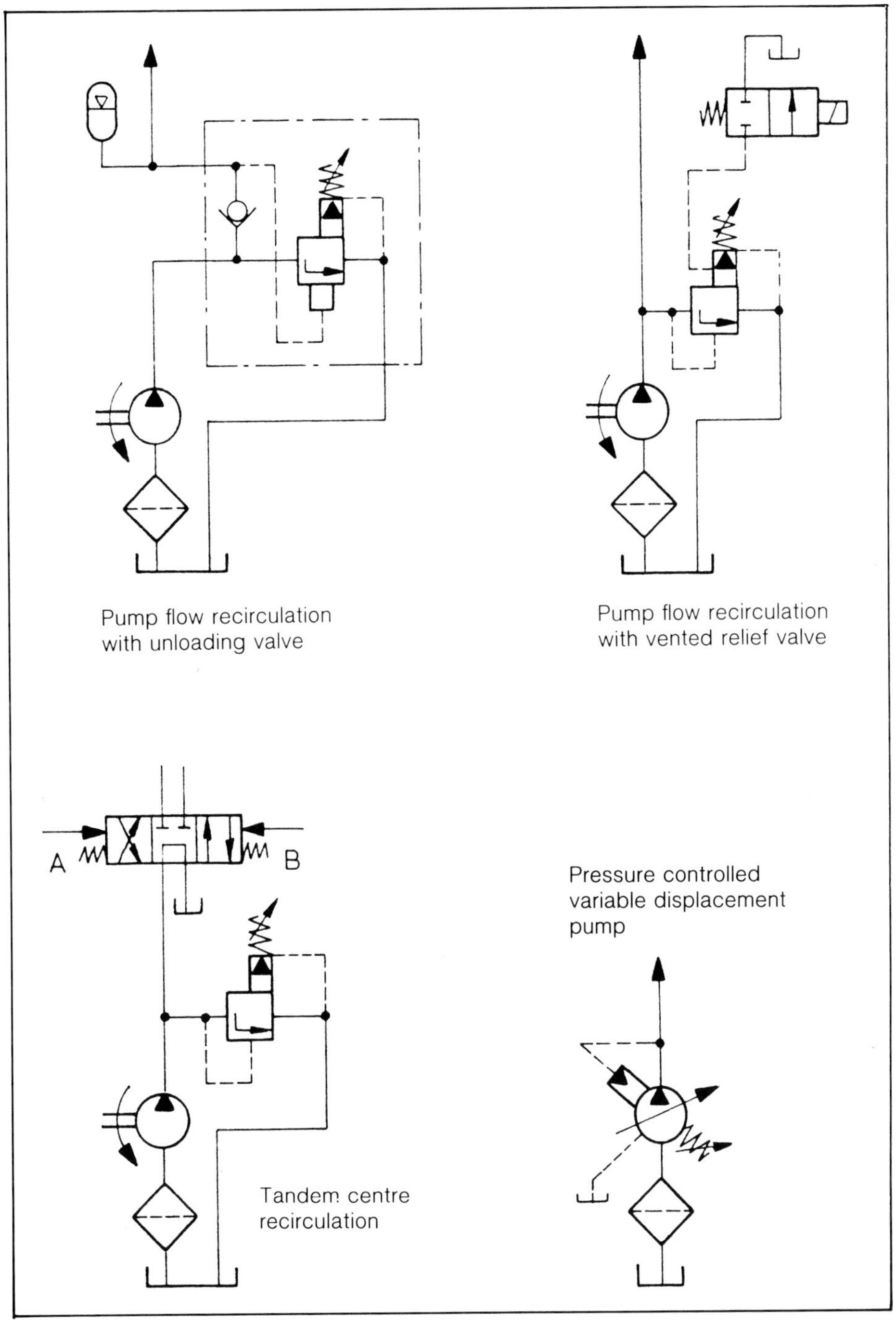

Fig. 96 Four commonly used systems to control maximum system pressure and regulate pump flow during non-action periods.

However, in others it can result in considerable input-power losses due to fluid lost through the valve during the wide "pressure override" band (fig. 98).

Due to the inertia of the valve poppet and the spring, sudden pressure increases on the valve inlet side give rise to pressure peaks beyond the adjusted "full flow pressure" and may well exceed the maximum permissible pressure rating of other components within the circuit. The compound-relief valve minimises the "pressure override" to approximately 100–150 kPa and eliminates pressure peaks almost completely. Thus, the compound-relief valve provides a safe and economical solution, and is the most commonly used type in industrial systems (fig. 99).

Valve operation

The relief valve setting pressure is adjusted with adjustment screw 9. As long as the pressure in the hydraulic system remains below the setting of the relief valve, the static pressure in chambers 2, 4, and 6 remains equal (Pascal's Law).

When the system pressure increases sufficiently to

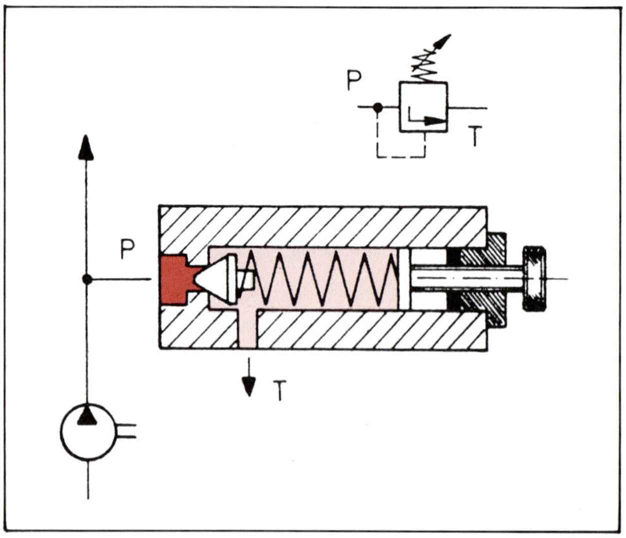

Fig. 97 Direct acting relief valve (simple relief valve).

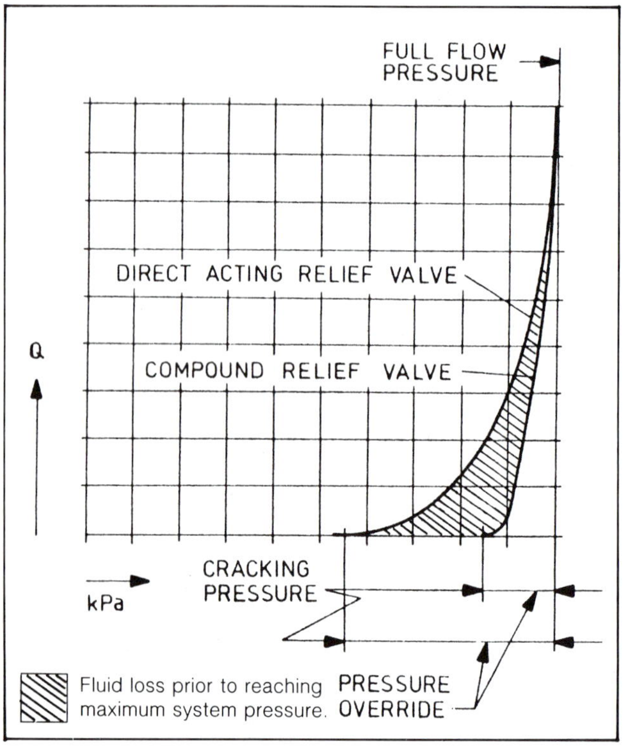

Fig. 98 Q-P graph for pressure-flow relief behaviour, comparing compound and simple relief valves.

force pilot poppet 7 off its seat ("cracking pressure"), fluid starts to flow to tank via internal drain 12 at a very low pressure. The resulting pressure imbalance, due to the pressure drop (Δp) across orifice 3, forces valve piston 11 upwards. This action compresses piston spring 10 and opens tank port T, thus preventing a further rise in pressure. Increased flow through the valve causes the piston to lift further off its seat, but as this only compresses a very light spring, very little "pressure override" is encountered.

As soon as the system pressure sinks below the opening pressure of the relief valve poppet 7, the flow past the pilot relief valve and the orifice 3 stops, and the pressure differential (Δp) disappears. Thus, spring 10 re-seats piston 11 and relief flow from P to T stops (fig. 99).

Venting the compound-relief valve

The compound-relief valve requires pressure build-up during the relief function to keep the pilot relief-valve spring compressed; thus, work is performed by the pump. If the relief valve is required to release flow over a prolonged period, the work performed during that time may lead to considerable pump input power wastage and — worst of all — heat build-up in the hydraulic fluid, which eventually leads to fluid deterioration. The vented compound-relief valve renders the ideal solution to these problems.

Valve operation

Venting-port plug 5 is removed and port V is connected to the venting valve (figs. 99 and 100). The fluid under pressure can, after the venting valve is opened, flow to tank. This flow creates the required pressure drop across orifice 3, which unbalances piston 11 and opens the flow path for the pump flow to be diverted to tank. Since the pressure developed during "venting" is only the pressure drop across orifice 3 (approximately 150–500 kPa), the power used during this time is minimal and the heat build-up negligible.

The venting function is used to offload the pump during non-action stages of the circuit sequence. The venting valve can be controlled either by an electric or a pneumatic signal, and is with some manufacturers' brands mounted directly onto the compound relief valve to reduce unnecessary plumbing and minimise installation cost.

Remote pressure control

As shown in fig. 101, the compound-relief valve may also be controlled from an external location to provide remote pressure selection. A direct-acting relief valve (or a compound-relief valve) is connected to venting port V of the compound-relief valve. Thus, the direct-acting relief valve provides stepless pressure selection, from zero pressure up to adjusted pressure setting of the compound-relief valve.

By means of a directional control valve with three positions (closed centre), three different pressures can be selected (figs. 102 and 103). Signal P2 selects 40 bar (4 MPa). This pressure can be remote-controlled, stepless controlled, or previously fixed. Signal PO selects "venting" of the compound-relief valve (almost identical to the application in fig. 101). If neither of the two signals is present, the centring springs of the valve select P1, which represents the maximum system pressure of the

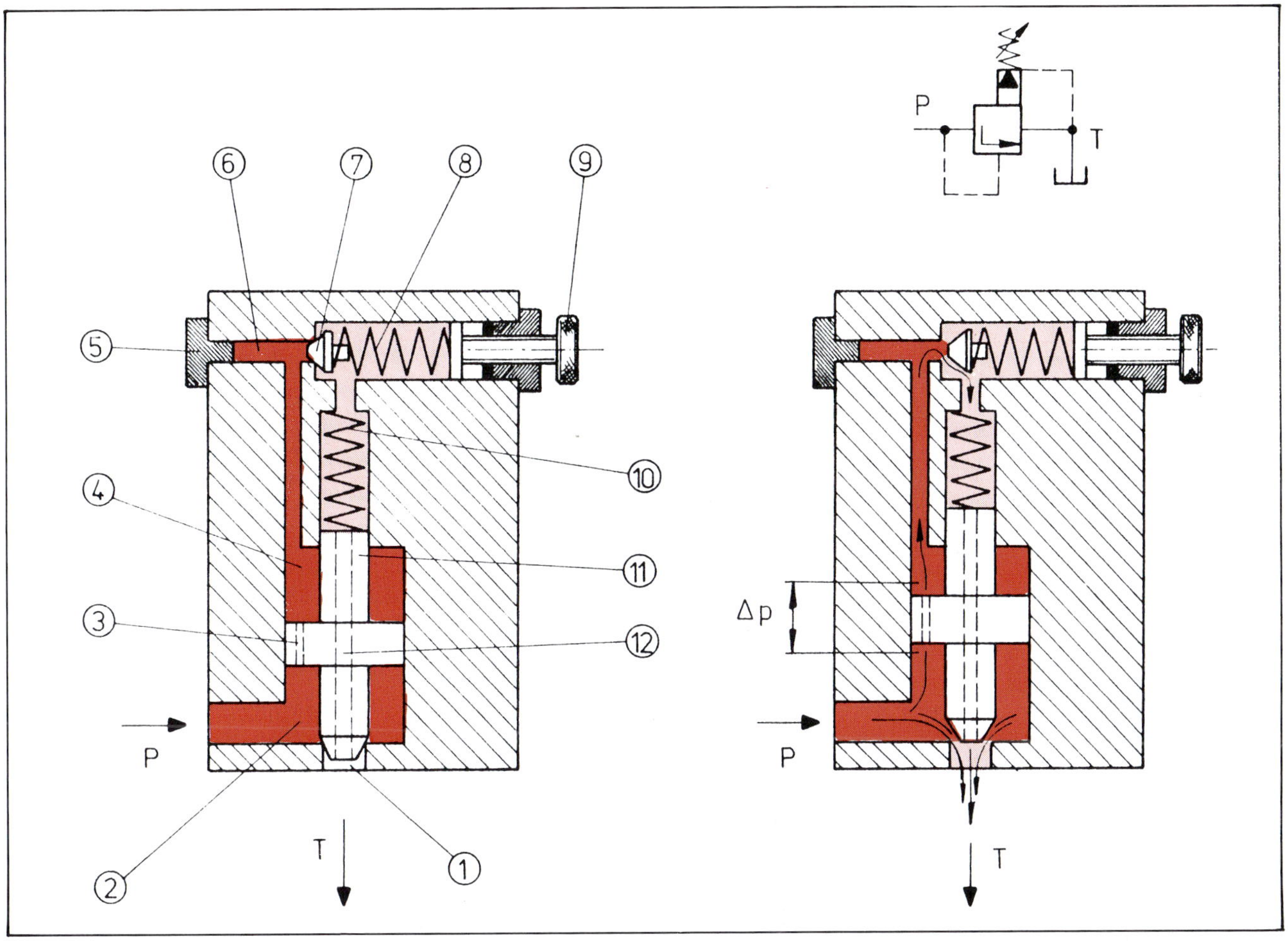

Fig. 99 Compound relief valve (pilot operated relief valve).

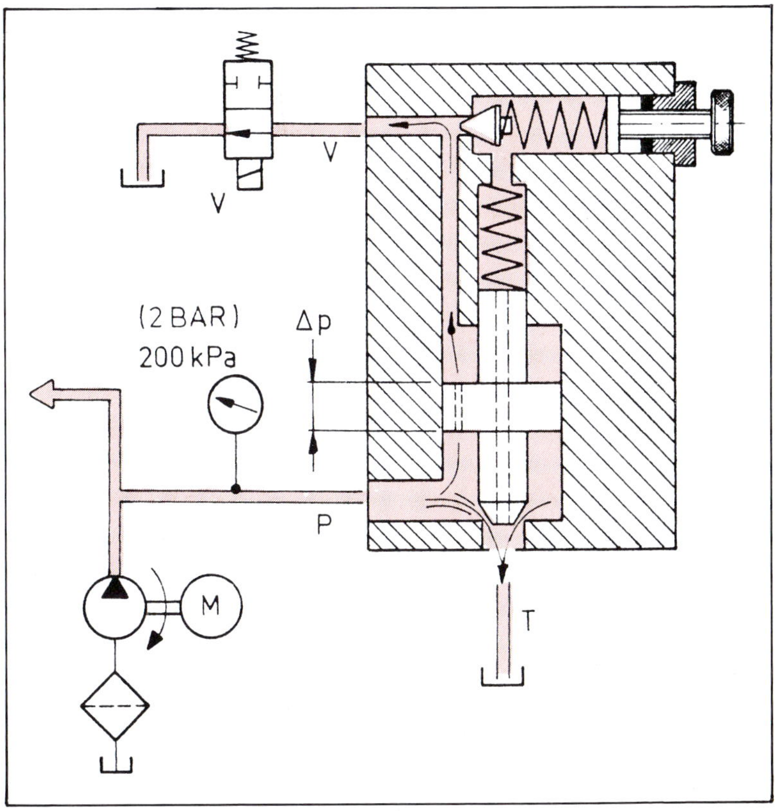

Fig. 100 Vented relief valve: system pressure is almost zero (200kPa) and pump input power during offloading is minimal.

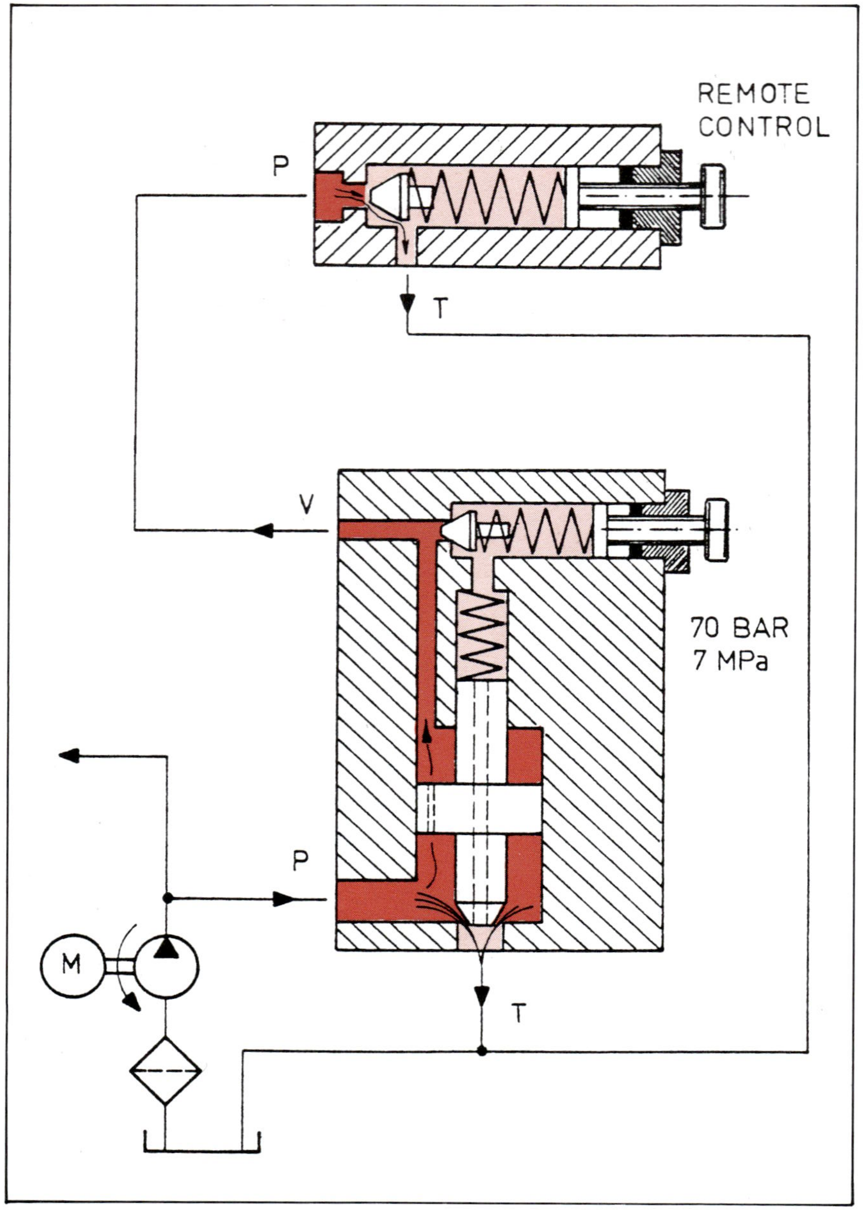

Fig. 101 Remote pressure control: the direct acting relief valve could also be replaced by a compound type relief valve.

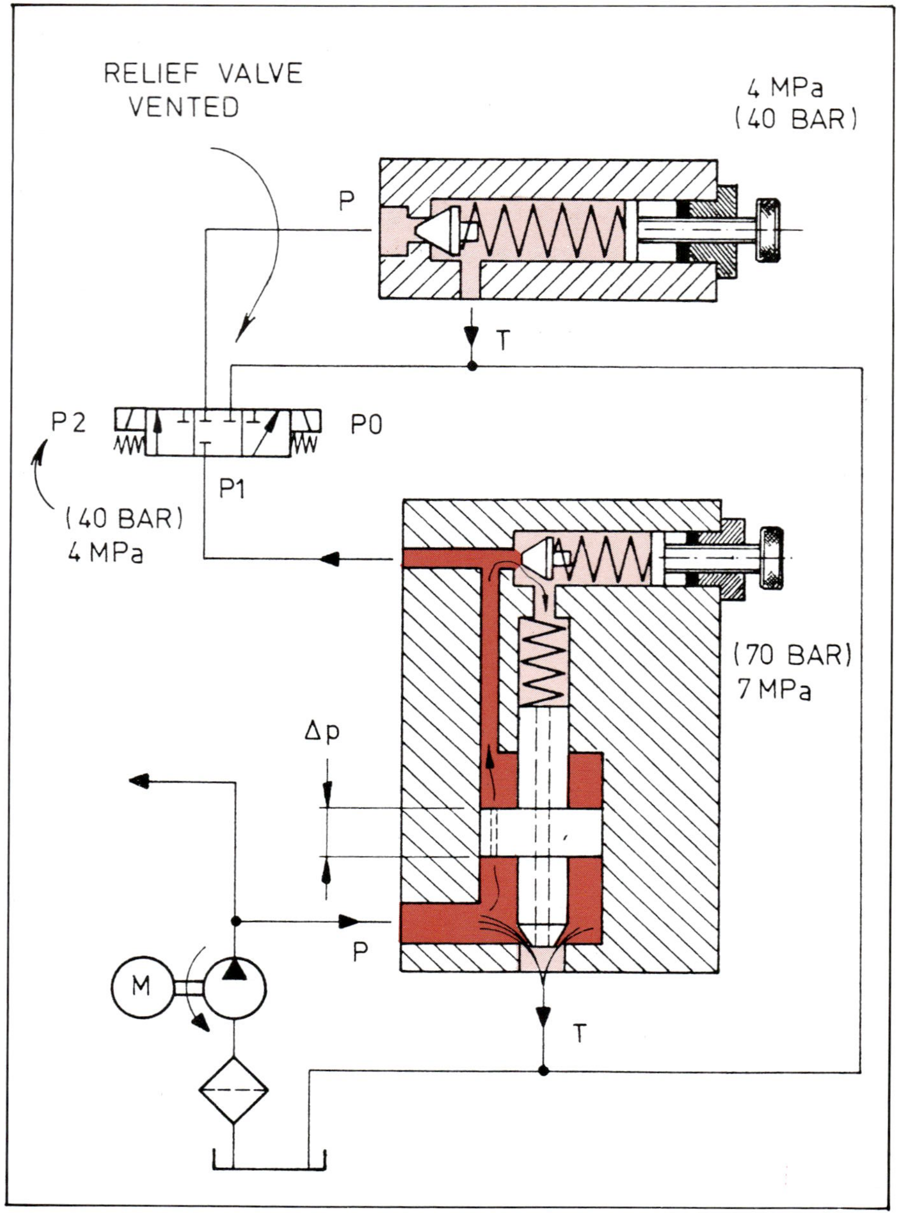

Fig. 102 Multiple pressure control.

compound-relief valve. Various combinations can be devised for such so-called multiple pressure controls, but practical considerations generally limit the number of set pressures to four.

Unloading valve (accumulator charging valve)

The unloading valve (also called accumulator charging valve or differential unloading valve) is in its design closely related to the compound-relief valve (compare figs. 99 and 104). This valve is used to accomplish the following switching and pressure control functions:

- limit maximum system pressure;
- charge the accumulator to maximum system pressure and maintain a working volume and pressure in the accumulator;
- unload the pump when the desired accumulator pressure is reached (e.g. maintain system pressure to actuators).

Valve operation

Figure 104 depicts the flow condition when the accumulator is being charged and the pump flow is directed into the system. As soon as the adjusted maximum system pressure in the accumulator and in the actuators is reached, the pilot relief valve poppet and the main valve piston open, and pump flow is diverted to tank at very low pressure. (For a detailed explanation of valve operation, see the compound-relief valve.) Due to the much

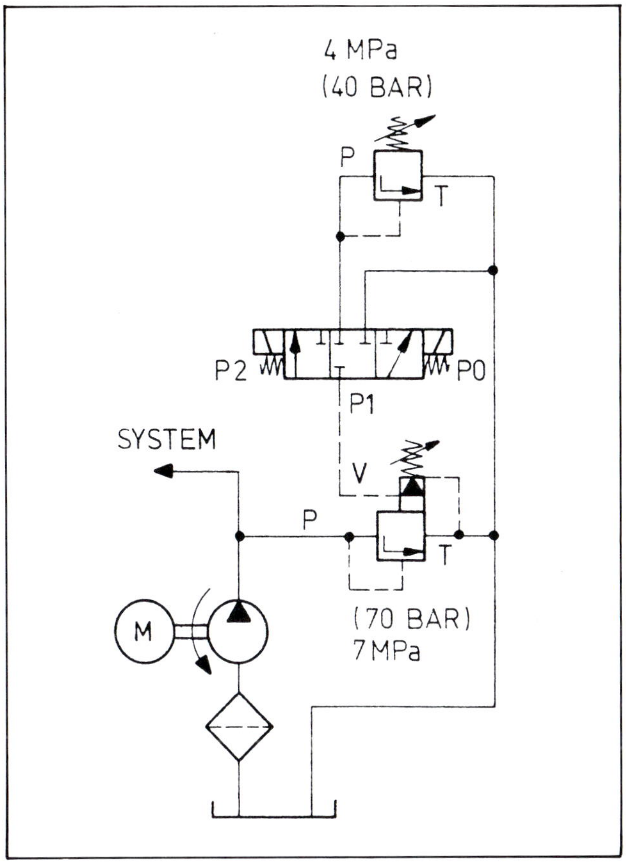

Fig. 103 Graphic symbol circuit for multiple pressure control.

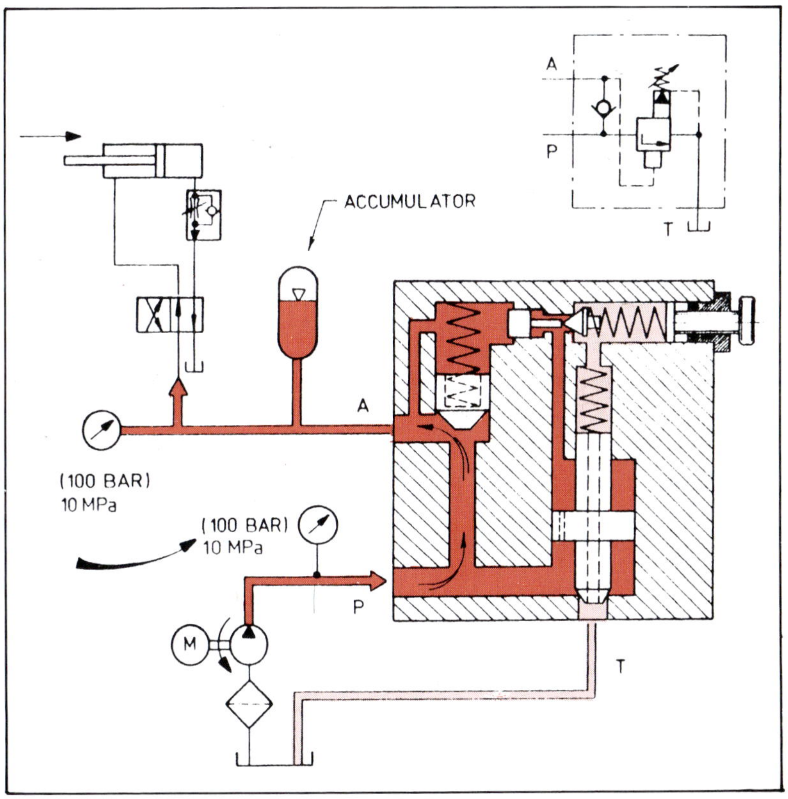

Fig. 104 Unloading valve in accumulator charging position. Note that the tank port(T) is closed and pump flow is directed to the system and to the accumulator. The pilot relief valve is closed and the check valve is forced open.

higher pressure in the accumulator circuit, the check valve closes, and the charged accumulator maintains pressure to the actuators.

While the valve is unloading the pump flow, the pressure in front of the pilot poppet is practically zero. The pilot valve and the main valve piston would therefore immediately re-seat, but for the system pressure from the accumulator, which is now acting via the check valve by-pass onto the differential piston (fig. 105). This pressurised piston forces the pilot poppet completely off its seat and holds it open until the accumulator pressure diminishes to approximately 85% of the maximum system pressure (accumulator charging pressure) as adjusted on the unloading valve. The accumulator pressure diminishes due to volume losses caused by system leakage, actuator movements, or internal leakage of valve spools and seals. The left-hand side of the differential plunger has approximately 15% more circular area than the pressure-exposed area of the pilot poppet (see the diameter indication in the magnified section of fig. 105).

Thus, if the system pressure drops below 85% of the initial maximum system pressure, the pilot poppet will re-seat, and the main valve piston interrupts pump flow from port P to port T. The pump flow now being redirected to port A opens the check valve and streams back into the system, and into the accumulator. Thus, the charging cycle is repeated.

Although system pressure to the actuators is maintained while pump flow is recirculated to tank (with minimum power wastage), the fluctuating system pressure (between 100% and 85% approximately) may prove disadvantageous in certain industrial applications. Thus, other forms of system unloading may have to be selected (see variable displacement pumps, Chapter 3). Unloading valves may also be used in so-called high-low circuits to cut-in and cut-out the low pressure pump (Chapter 11).

Pump unloading with accumulator and electric control

The switching and pressure control functions for the accumulator charging circuit discussed here are identical to the functions achieved by the unloading valve circuit. A normally-open directional control valve actuated by an electric pressure switch is used to vent or de-vent the compound-relief valve (fig. 106).

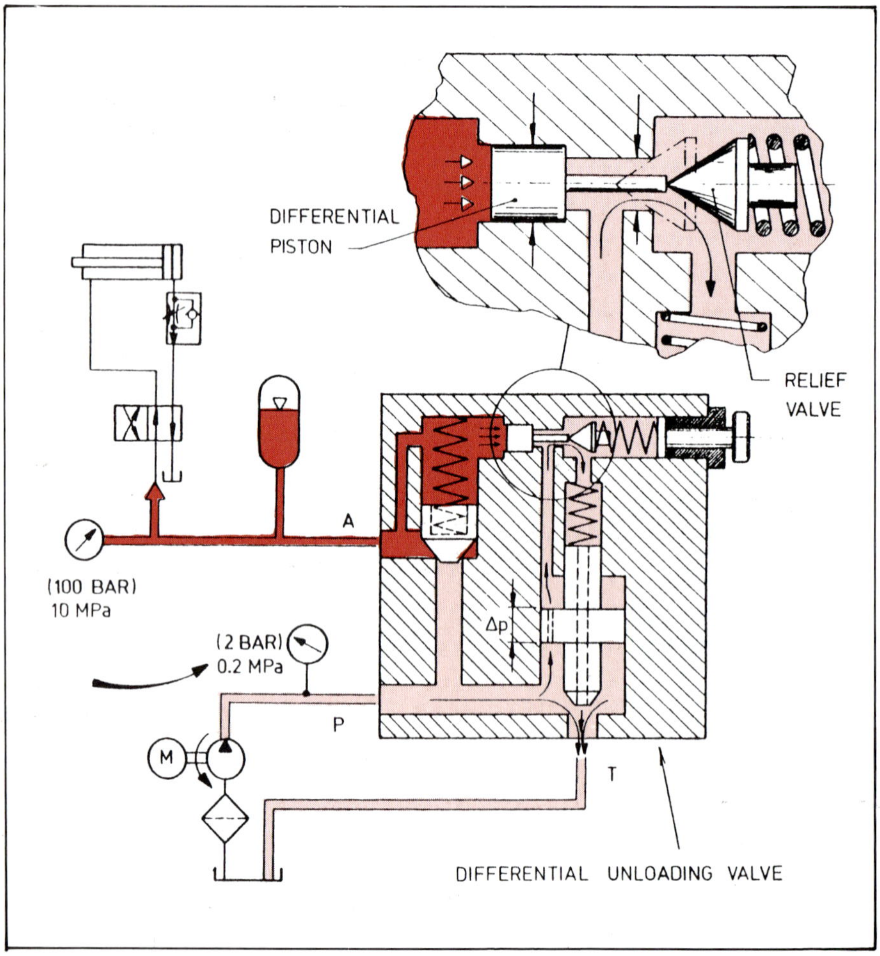

Fig. 105 Unloading valve in pump unloading position. Note that the thank port(T) is open and the pilot relief valve poppet is forced open by the system pressure acting onto the differential piston. The check valve is closed (see also enlarged section and compare the pressure gauge readings with the previous illustration).

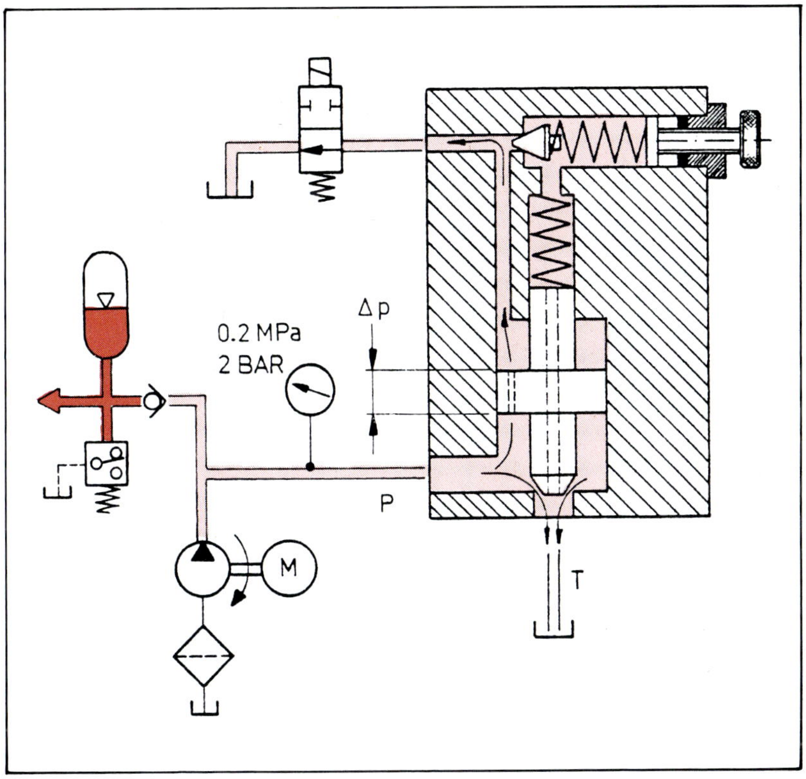

Fig. 106 Pump unloading circuit with accumulator and electric pressure switch, depicting the charging cycle.

Charging operation

The two micro switches of the pressure switch are interconnected to an electric relay in such a way that at low system pressure the solenoid of the venting valve is energised, and the compound-relief valve is no longer vented to tank (de-vent is switched). Pump flow now streams past the relief valve via the check valve to the accumulator.

Unloading operation

When the pressure within the system and in the accumulator reaches the adjusted pressure setting of the pressure switch, the solenoid of the venting valve is de-energised and the compound-relief valve is vented to tank. The check valve now closes, to prevent a return flow of fluid from the accumulator to the pump. Thus, system pressure is maintained and pump flow is unloaded to tank (see fig. 106, and compare with figs. 100 and 104).

Offloading valve

The offloading valve is generally used in conjunction with double pump circuits (fig. 107). The flow from the two pumps is combined to gain more speed while the actuator is traversing. When the high speed is no longer required, or when system pressure rises to the point where the combined pump flow would exceed the input power of the prime mover, the pump with the larger flow is offloaded to tank.

Low pressure operation

The offloading valve is closed under this condition. Flow from the large volume pump passes over the check valve and joins the flow from the low volume pump (the left-hand pump). This condition continues as long as the pressure within the system is below the pressure setting of the offloading valve.

High pressure operation

Figure 107 shows the offloading valve in the open condition, with the large-volume pump recirculating its flow to tank. The check valve is closed, thus the flow from the low volume pump is prevented from also being offloaded to tank. In this configuration, much less power is used than if both pumps had to be driven at high pressure (power = pressure × flow rate). When the motion of the actuator stops, the high-pressure low-volume pump discharges its flow over the system relief valve.

Valve operation

The offloading valve is a direct-acting, remotely controlled spool valve where the spool is held in the closed position by the adjustable spring. When pilot pressure at port X exceeds the pressure setting, the spool is raised and flow from the primary to the secondary port occurs (fig. 107). The hollow spool permits internal leakage (past the spool lands) to drain into the spring chamber, and subsequently to the secondary outlet, and thus to tank.

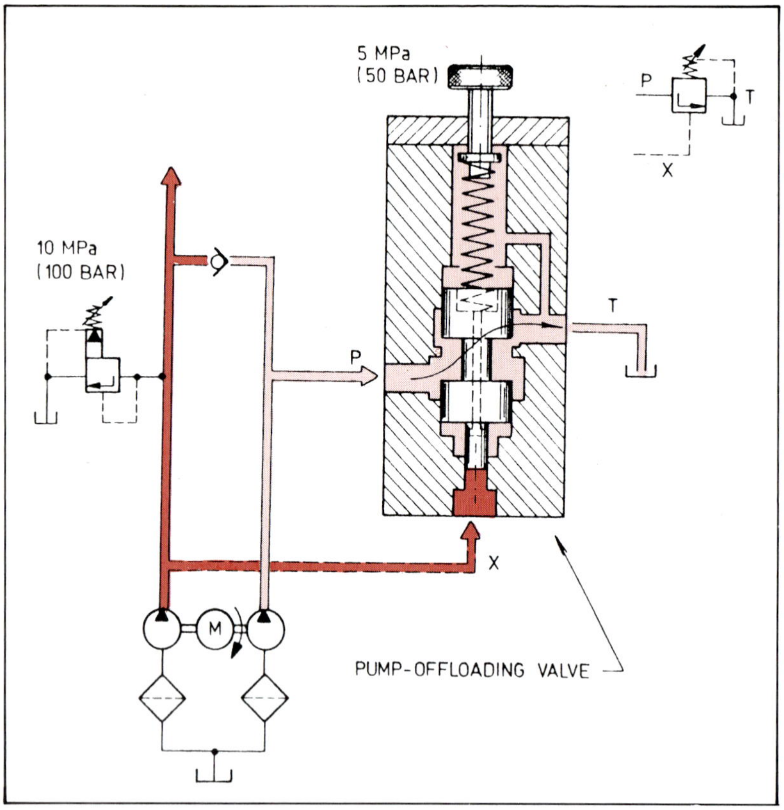

Fig. 107 Offloading valve controlling the high volume, low pressure pump in a "high-low" circuit.

Counterbalance valve (back pressure valve)

The counterbalance valve is applied to create a back pressure or cushioning pressure on the underside of a vertically-moving piston, to prevent the suspended load from "free falling" because of gravity whilst it is being lowered (fig. 108). This counteracting or counterbalancing function has given the valve its name.

Valve operation (lowering)

The pressure setting on the counterbalance valve is slightly higher than the pressure required to support the load from free falling. Due to this back pressure in line A, the actuator piston must be forced down when the load is being lowered. This causes the pressure in line A to increase, which raises the spring-opposed spool, thus providing a flow path to discharge the exhaust flow from line A to the directional-control valve and then to tank. The spring-controlled discharge orifice maintains back pressure in line A during the entire downward piston stroke (fig. 108 shows the load being lowered).

Valve operation (lifting)

Since the valve is normally closed, flow in the reverse direction (from port B to port A) could not occur without a reverse free-flow check valve. When the load is raised again, the internal check valve opens to permit flow for the retraction of the actuator.

Valve operation (suspension)

While the load is held in suspension the valve remains closed. Therefore, its pressure setting must be slightly higher than the pressure caused by the load. Spool valves tend to leak internally under pressure. This makes it advisable to use a pilot-operated check valve in addition to the counterbalance valve, if a load must be held in suspension for a prolonged time (fig. 108).

Brake valve

The brake valve is closely related to the counterbalance valve, and serves — when mounted in the exhaust line of a hydraulic motor — the following functions:

- It prevents the hydraulic motor from overspeeding when an overspinning load is applied to the motor shaft.
- It prevents an excessive pressure build-up during deceleration, and controls the rate of deceleration.

Valve operation

During acceleration, the motor torque is at its peak and therefore the system pressure at the motor inlet is at its maximum. With the required operating pressure acting via port X onto the large spool area, the brake valve spool is forced wide open; thus, exhaust flow from the motor is unrestricted. The area ratio between the large spool surface affected by pressure line X; and the internally connected braking piston affected by the motor exhaust pressure is 8:1.

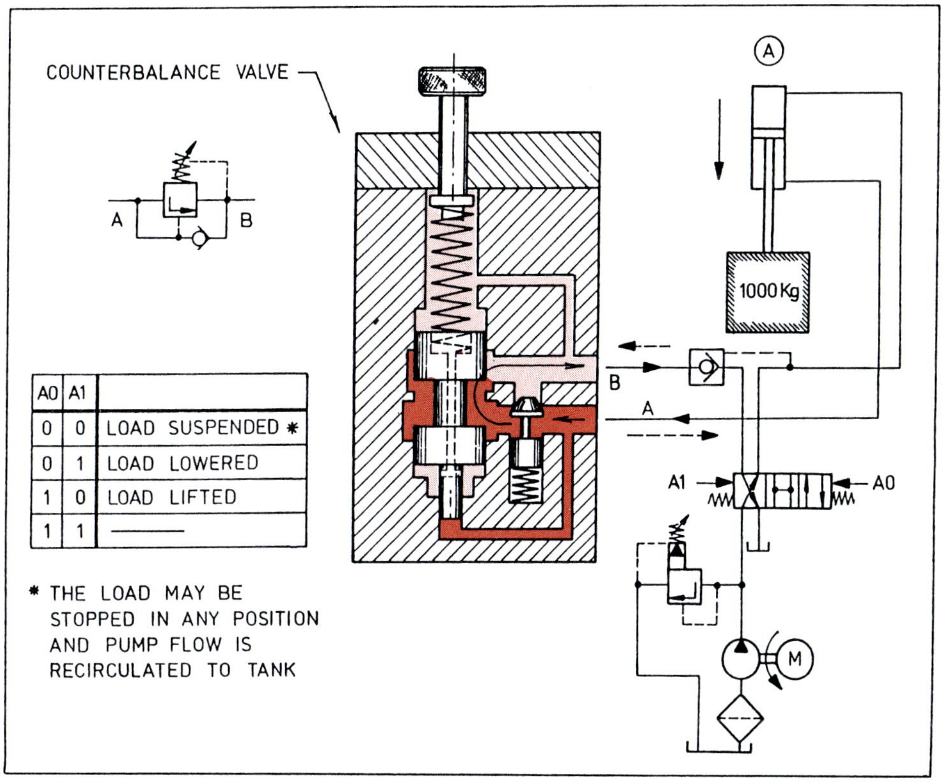

A0	A1	
0	0	LOAD SUSPENDED *
0	1	LOAD LOWERED
1	0	LOAD LIFTED
1	1	———

Fig. 108 Counterbalance valve circuit. The function table shows the required D.C.V. pilot signals for lifting, lowering, and load suspension.

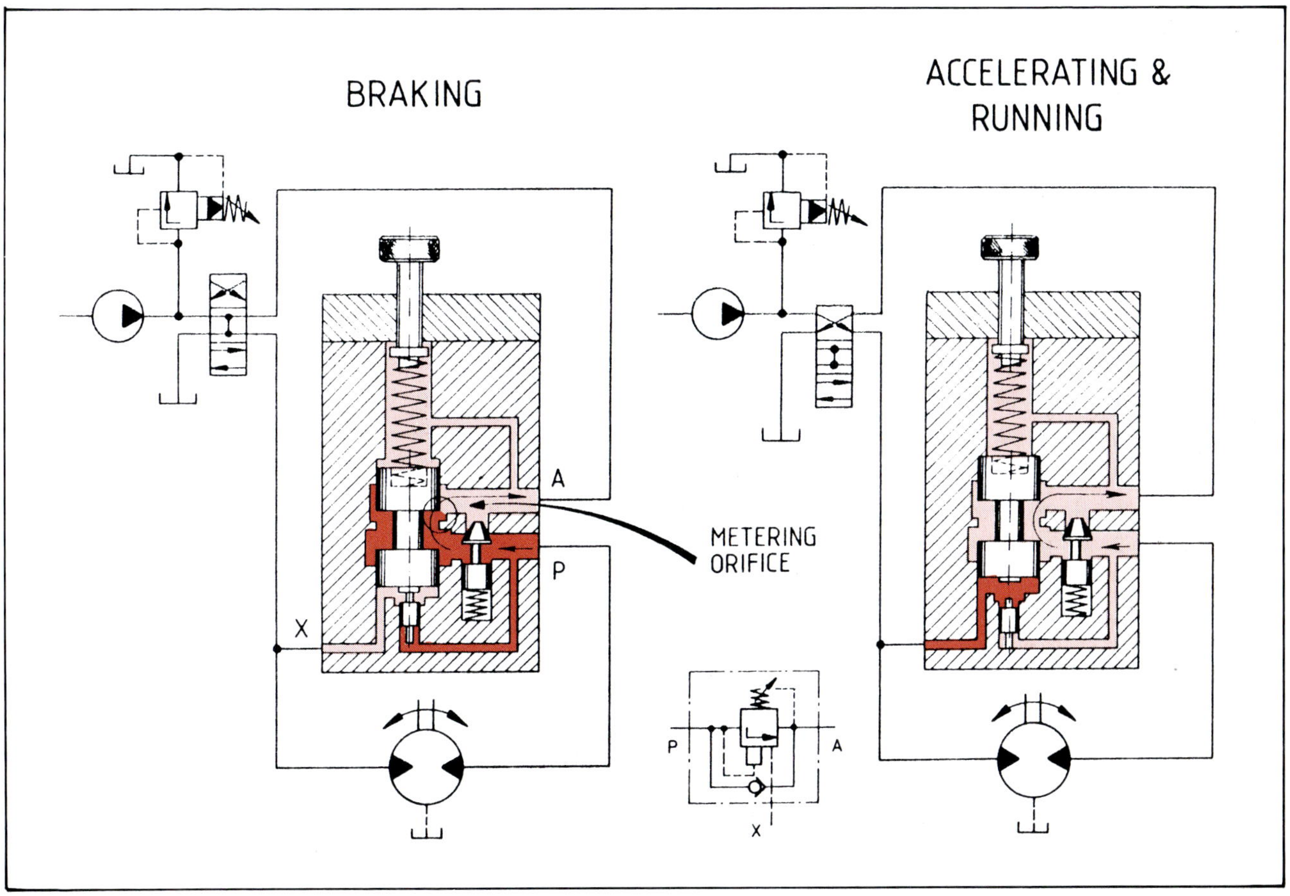

Fig. 109

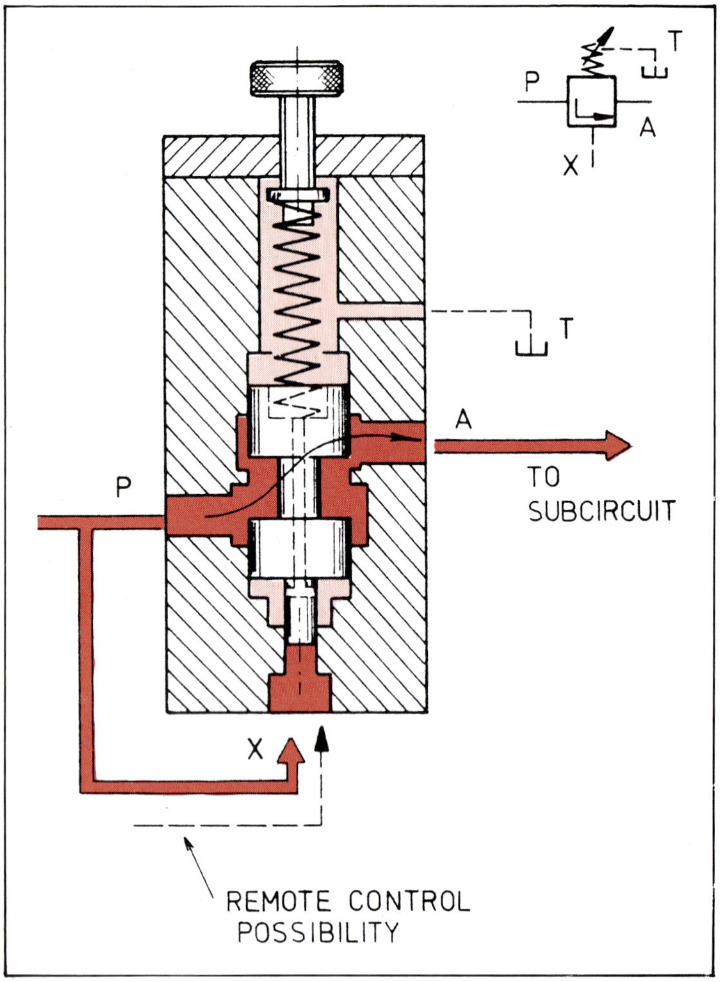

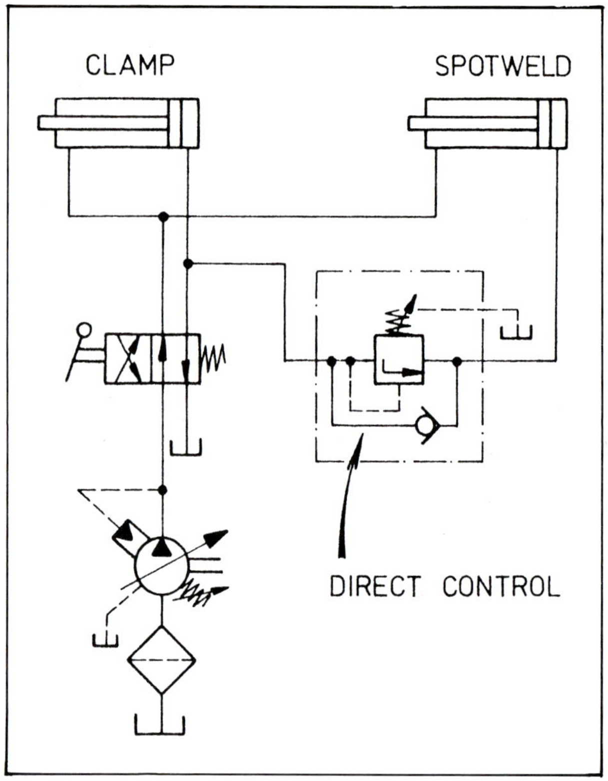

Fig. 111 *above:* Clamp and spotweld circuit. The sequence valve is directly controlled. The circuit provides pressure sequencing for the second sequential step (spotweld) as soon as the pre-set clamping pressure is reached.

Fig. 110 *left:* Direct acting sequence valve.

At operating speed, the created load pressure in line X will hold the brake valve almost completely open, unless external forces try to overspin the motor. If this occurs, the pressure drops off in the motor inlet line and under the large spool area; the spring force tends to reduce the motor discharge orifice within the valve, which increases back pressure, and the motor slows down. This in turn increases the pressure between the reduced discharge orifice and the motor, as well as in the internal pressure line to the braking piston, thus balancing the valve at the required metering position to maintain constant motor speed (fig. 109).

While braking (with the directional control valve in centre position), the load inertia causes the motor to continue its rotation. Thus, the motor becomes a pump which draws fluid from the reservoir and circulates it back through the brake valve. When this happens, the brake valve acts in the same way as during over-spinning, due to external forces. The spool, being balanced between the spring and the force of the braking piston, will gradually reduce the discharge orifice, thus bringing the motor to a coasting stop. The reverse-flow check valve is normally incorporated in order to permit motor reversal.

Sequence valve (single stage)

The sequence valve is closely related to the relief valve in its design and function. It permits hydraulic fluid to flow into a subcircuit, when the pressure in the main circuit has reached the setting of the sequence valve.

A typical application would be a clamp and spotweld circuit (fig. 111). The clamp actuator must be extended first, and as soon as the workpiece is clamped, the spot-weld head actuator must extend. Both actuators are permitted to retract simultaneously. An optional reverse-flow check valve is normally incorporated when the valve is required to operate between a directional control valve and an actuator.

Valve operation

The single-stage, spool-operated sequence valve is a normally closed valve. When pilot pressure at point X exceeds the pressure setting, the spool is raised against the spring, and fluid flows from primary port P to secondary port A. The valve requires an external drain, to permit internal fluid leakage past the spool land to drain back to tank (fig. 110).

It must be noted that sequence valves require a

reverse-flow check valve if return flow is required. Single-stage relief valves may also be remote controlled (figs. 110 and 112), and all sequence valves are externally drained, because the secondary line leads to a pressurised subcircuit.

Sequence valve (two-stage or compound)

Valve operation

Fluid flows without restriction to the main circuit to operate the first phase of the machine function. When the pressure in the main circuit reaches the pressure setting of the sequence valve, pilot poppet 2 is forced off its seat and fluid starts to flow to tank. The resulting pressure imbalance due to the pressure drop (Δp) across orifice 9, forces piston 6 upwards, and fluid can then flow into the subcircuit (fig. 113).

Pressure reducing valve (pilot-operated)

In some fluid power systems it is desirable (and often necessary), to operate a subcircuit at a lower pressure than the main system. Pressure reducing valves are used for this purpose. In contrast to the "normally closed" pressure control valves discussed so far, the pressure reducing valve is "normally open".

The main function of this valve is to limit and maintain a constant downstream pressure (subcircuit pressure), regardless of pressure fluctuations in the main circuit upstream (fig. 114).

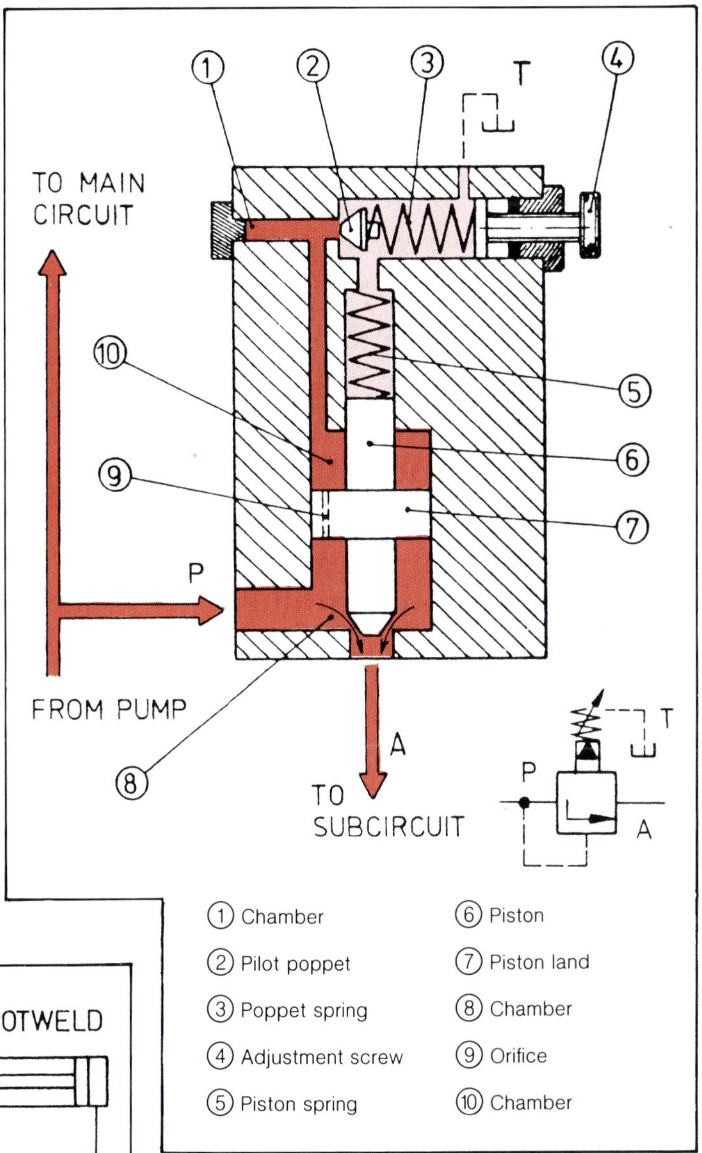

Fig. 113 Pilot operated sequence valve (compound type).

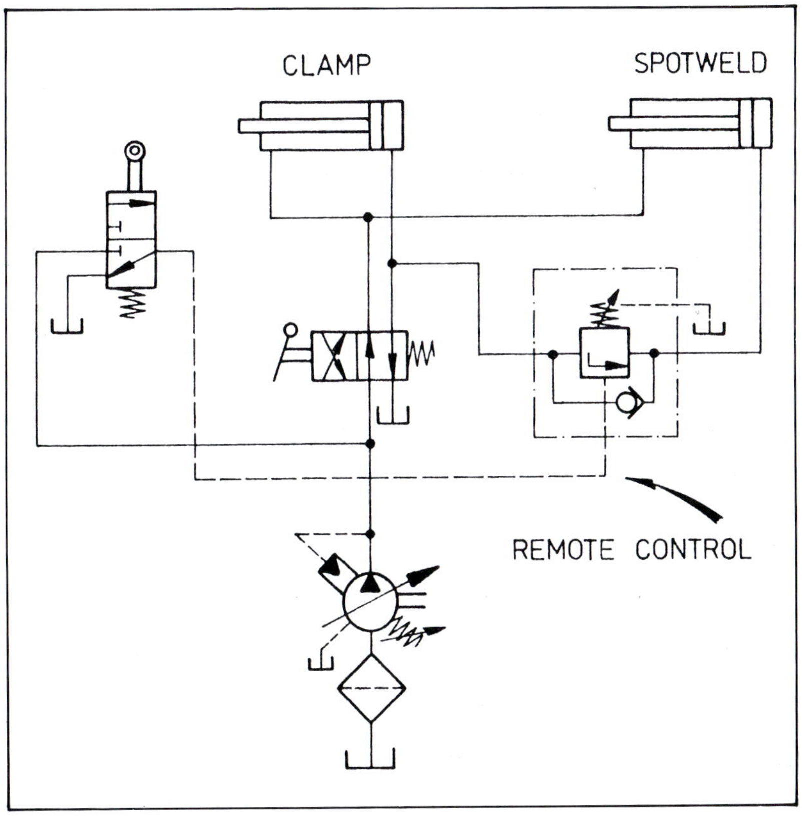

Fig. 112 Clamp and spotweld circuit. The sequence valve is remotely controlled. The circuit thus provides pressure as well as position sequencing for the second sequential step ("and" function).

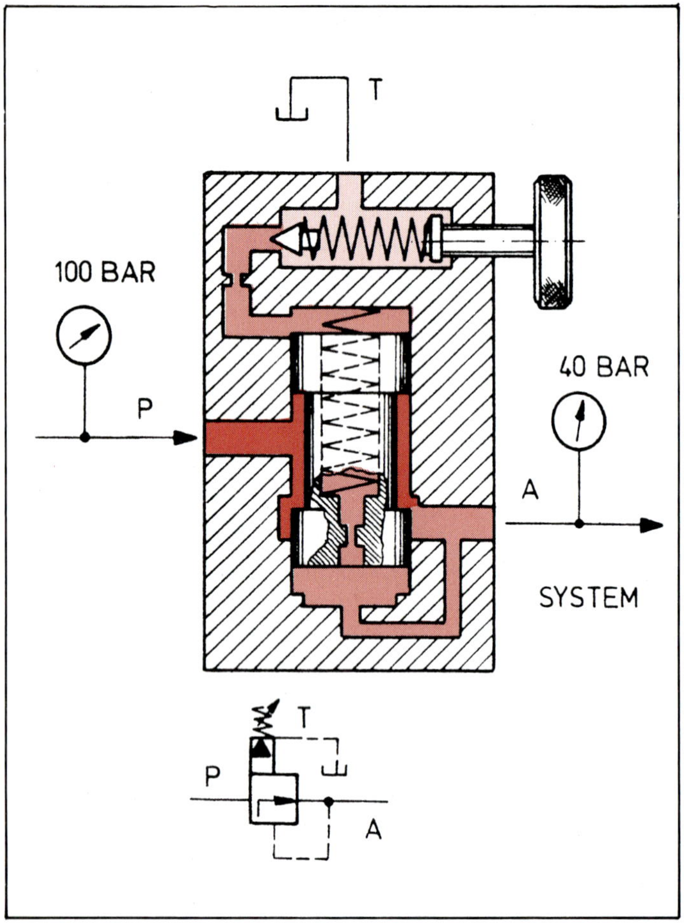

Fig. 114 Pilot operated pressure reducing valve.

Valve operation

The required downstream pressure (subcircuit pressure) is adjusted and set on the pilot relief valve. Below the valve operating pressure, the valve spool is held wide open by the light spring and fluid passes from port P to port A with virtually no restriction. Static pressure equal to the pressure in the subcircuit exists below the valve spool, inside the valve spool, and in front of the pilot relief valve.

When the subcircuit pressure increases and reaches the level of the valve setting, the pilot valve will open and relieve flow to tank. This causes a pressure drop across the internal orifice (Δp) inside the valve spool, so that the higher pressure below the orifice forces the spool upwards against the spring. The spool will therefore remain balanced between the higher pressure below the orifice acting upwards, and the lower pressure above the orifice, plus the light spring acting downwards. In this way the spool reduces or opens the flow passage from port P to port A and limits and maintains a constant pressure in the subcircuit. When no flow is required in the subcircuit, the valve will stay partially open and a fluid flow of some 0.6–1.2 l/min will continuously flow past the pilot valve to tank.

Reverse free flow through the valve is only possible if the pressure in the subcircuit is below the valve pressure setting. If the pressure exceeds the valve setting, the valve will close, thus making reverse flow impossible. Therefore pressure reducing valves are often equipped with a check valve for reverse free flow.

External forces acting onto a linear actuator will increase the pressure between the pressure reducing valve and the actuator. In some systems it is therefore

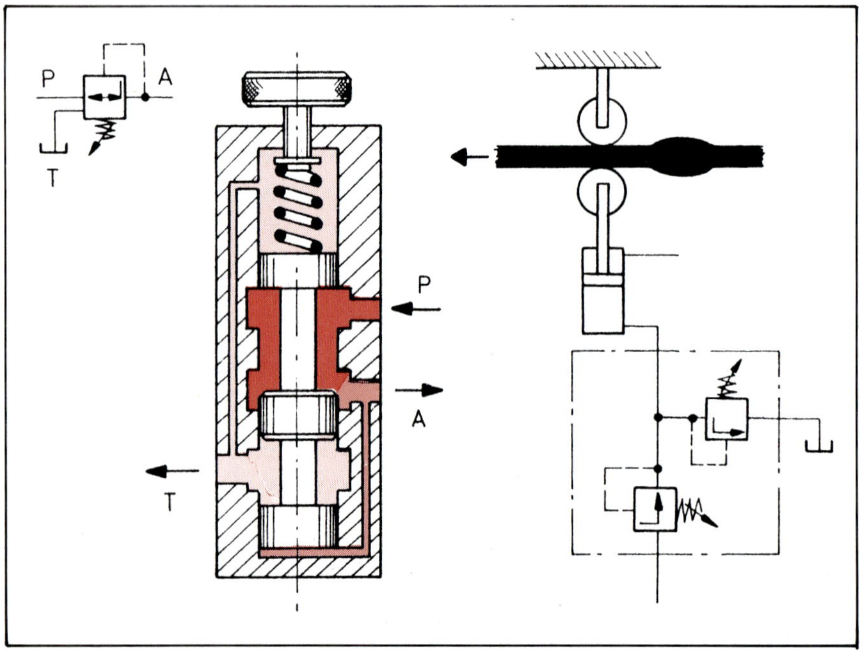

Fig. 115 Direct acting pressure reducing valve with secondary system relief function.

desirable to relieve excess fluid from the secondary system to tank in order to maintain a constant downstream pressure, regardless of such external forces. The pressure reducing valve depicted in fig. 115 fulfils this function but only for a limited flowrate (0.6–1.2 L/min). The control circuit in fig. 115 shows the detailed symbol for this valve and the compound symbol is shown next to the valve.

Pump offload control with directional control valve

An extremely simple and inexpensive way of controlling the pump flow during non-action periods of the actuators, is the use of an open-centre or tandem-centre type directional-control valve. The tandem-centre valve, for example, has both actuator ports blocked in neutral, and the pump is offloaded to tank at a relatively low pressure.

This circuit would still require a pressure relief valve, to protect the system from overloading and limit the maximum system pressure. But compared with a closed-centre system (compare circuits in fig. 116), where the pump flow during non-action periods is pushed over the relief valve, the tandem-centre circuit is definitely power saving (compare pressure gauges on the two systems), and heat build-up in the fluid is minimal. However, most tandem-centre valves have a flow rating (P–T) which is only 50% of the normal maximum for a particular size of valve.

Maximum system pressure control with pump

Variable-displacement, pressure-controlled pumps are often used to control maximum system pressure and conserve pump input power. Such pumps automatically adapt their flow (L/min) to the system requirements. Thus, during non-action stages of the circuit, flow is virtually stopped, and pressure to the actuators is maintained (figs. 117 and 118). "Step type" controlled pumps require an additional system-relief valve, to protect the pump and the system; but "ramp type" controlled pumps (fig. 117) work safely without a relief valve.

System pressure control with the variable-displacement pressure controlled pump has many significant advantages, such as maintained system pressure, minimum power use, constant pilot pressure to directional-control valves, minimal heat development, less control components, etc. However, the initial purchasing cost is higher than the cost of a fixed displacement pump.

Check Valves

Wherever pressure control valves must be placed between the D.C.V. and an actuator one must also install a check valve to provide free reverse flow. (see Figs. 111, 114, 213, 215, 216)

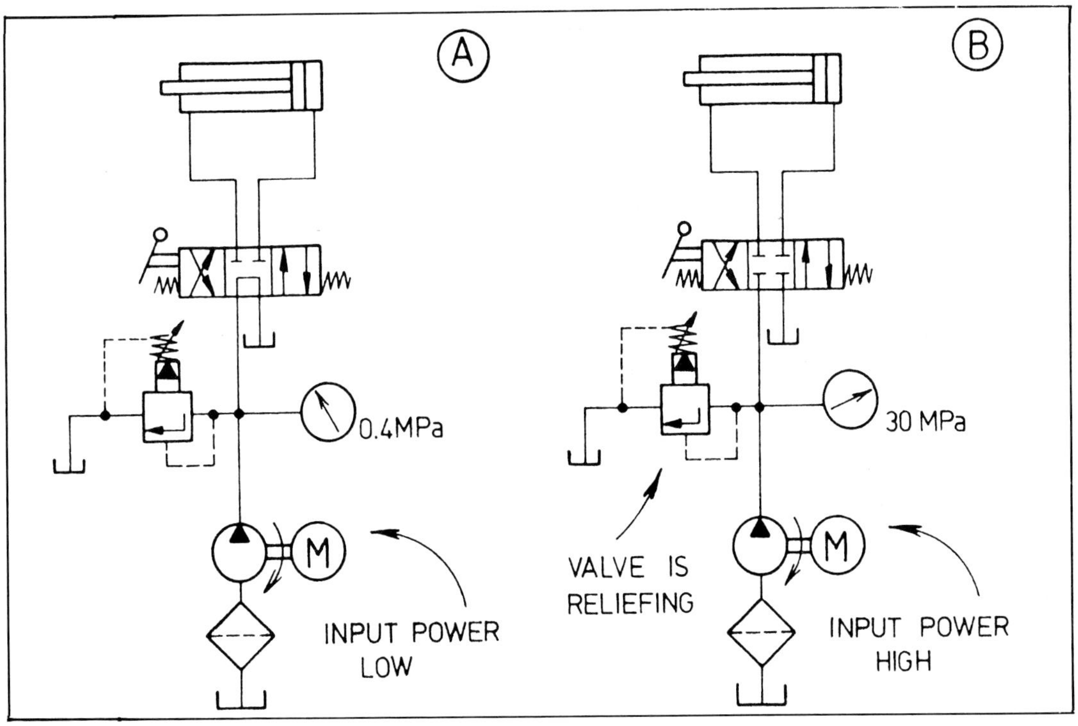

Fig. 116 Pump offload control with tandem centre D.C.V. (circuit A). The fully closed centre D.C.V. in circuit B does not provide pump offload, which means the pump is working hard against the relief valve at 30 mPa.

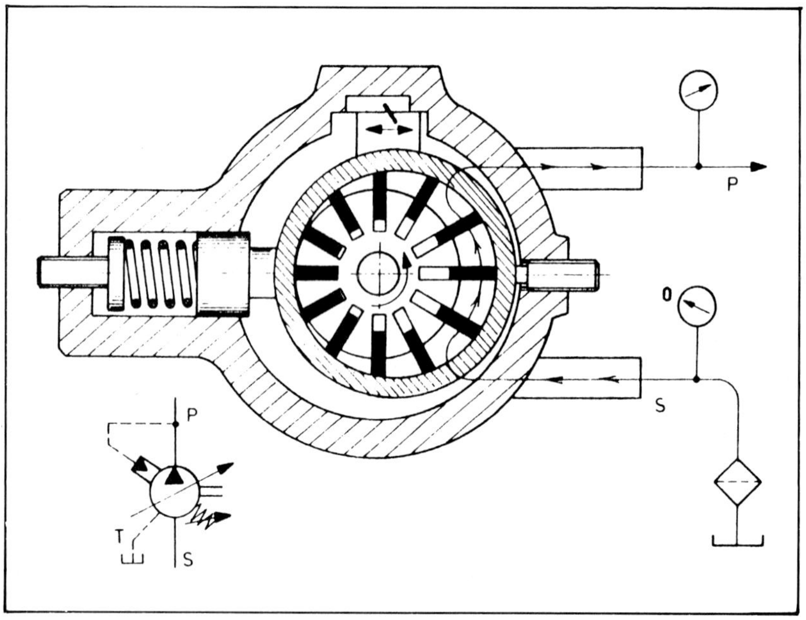

Fig. 117 Variable displacement pressure controlled pump with ramp type pressure control.

Appropriate fields are marked "Yes" if the control mode (top) relates to the control function (left).

*Only for the low pressure pump.

Control function obtained	Control mode (valve)	System relief valve (compound type or simple type)	Pump offload with vented compound relief valve	Pump offload with tandem or open centre D.C.V.	Pump offload with an offloading valve	Differential area unloading valve with accumulator	Unloading with pressure switch, accumulator & vented relief valve	Variable displacement, pressure controlled pump	Pressure reducing valve	Pressure sequence valve	Counter balance valve	Brake valve
	Provide overload protection	Yes	Yes		Yes*	Yes	Yes	Yes				
	Conserve pump input power prevent system overheating		Yes	Yes	Yes	Yes	Yes	Yes				
	Holding system pressure while pump is unloaded					Yes	Yes	Yes				
	Limit actuator output force and motor torque	Yes	Yes			Yes	Yes	Yes	Yes			
	Control actuator operating sequence									Yes		
	Control actuator dynamic braking										Yes	Yes
	Support a vertical load										Yes	
		Pump control							Actuator control			

Fig. 118 Summary of pressure controls.

6 Flow controls

Flow-control valves are used in hydraulic systems to control the rate of flow from one part of the system to another. Flow-control devices accomplish one or more of the following control functions:

- limit the maximum speed of linear actuators and hydraulic motors $\left(\frac{\text{flowrate}}{\text{piston area}} = \text{piston speed}\right)$;
- limit the maximum power available to subcircuits by controlling the flow to them (power = flowrate × pressure);
- proportionally divide or regulate the pump flow to various branches of the circuit.

A partly closed orifice or flow control valve in a hydraulic pressure line causes resistance to pump flow. This resistance raises the pressure upstream of the orifice to the level of the relief valve setting and any excess pump flow must pass via the relief valve to tank (fig. 119).

In order to understand the function and operation of flow-control devices one must comprehend the various factors that determine the flowrate (Q) across an orifice or restrictor. These factors are:

- cross-sectional area of the orifice (mm^2);
- shape of the orifice (round, square, triangular);
- length of the restriction (fig. 122);
- pressure differential across the orifice (Δp);
- viscosity of the fluid (cSt, depending on the temperature).

Thus, the law that governs the flowrate across a given orifice can by approximation be defined as $Q^2 \propto \Delta p$. This implies that any variation in pressure up, or downstream of the orifice changes the pressure differential, Δp, and thus the flowrate through the orifice. The pressure upstream of the flow-control valve is normally kept constant by either the system relief valve or by the pressure controller on a variable displacement pump.

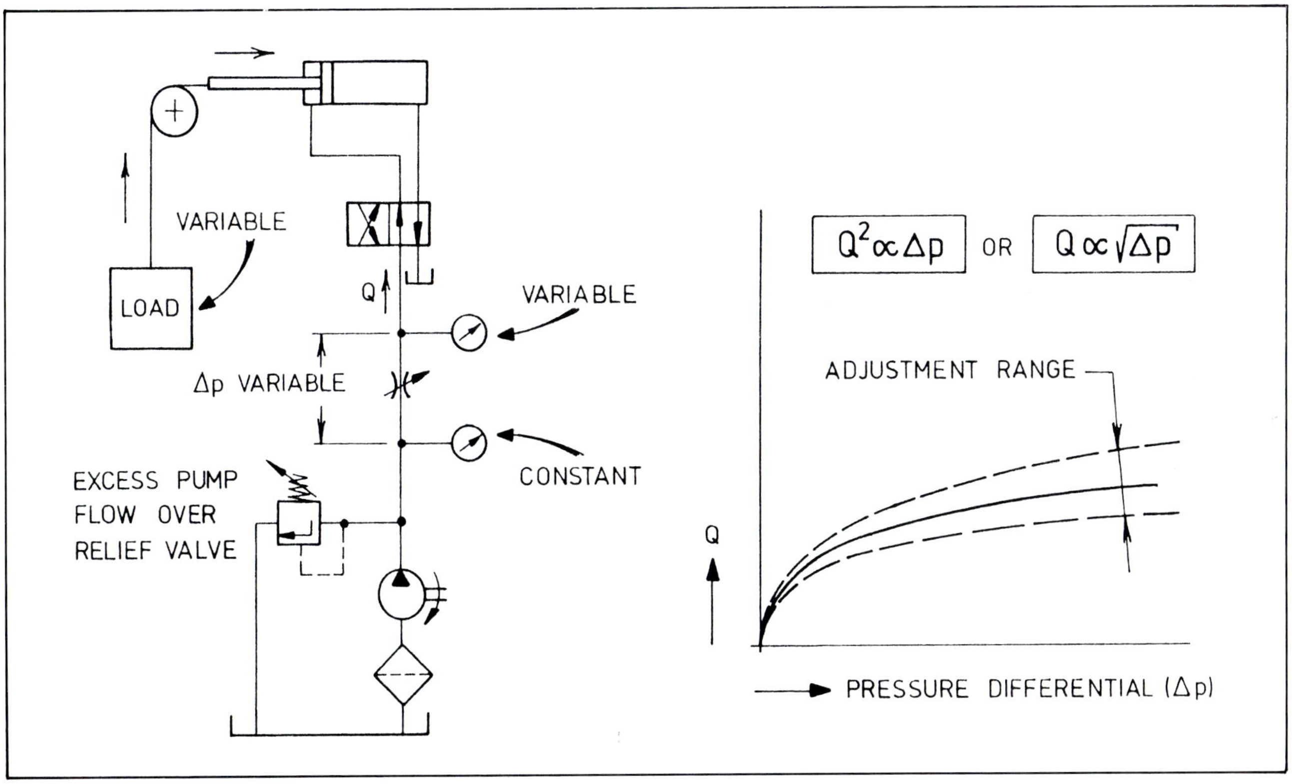

Fig. 119 Simple restrictor type flow control valves should only be used if the pressure differential (Δp) across the orifice remains reasonably constant.

Variations in the pressure differential (Δp) are thus only caused by pressure fluctuations downstream, as a result of varying load forces on the actuator (fig. 119).

Simple restrictor valve (variable throttle valve)

Variable restrictor type flow controls (fig. 120) are used in control circuits where the controlled speed of the actuator is not critical, and thus may vary if the actuator load fluctuates or the viscosity of the fluid undergoes changes.

These valves usually consist of a valve body and a throttling screw for fine adjustment. Where speed control for only one actuator direction is required, a check valve for free flow in the reverse direction is essential. Some restrictor type flow controls have in-built free-flow check valves (fig. 121).

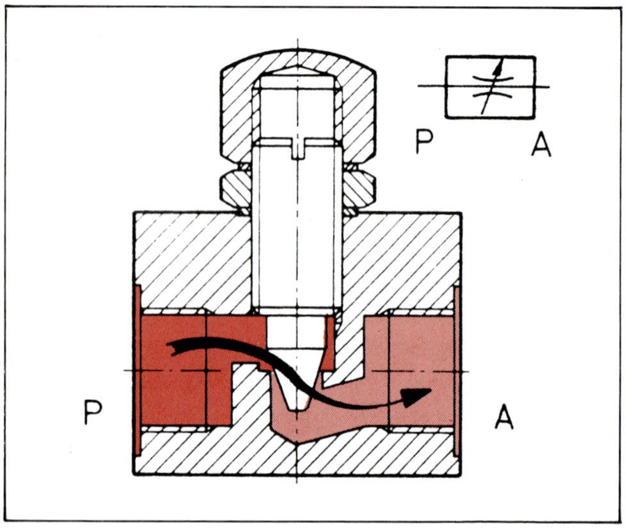

Fig. 120 Simple restrictor flow control valve.

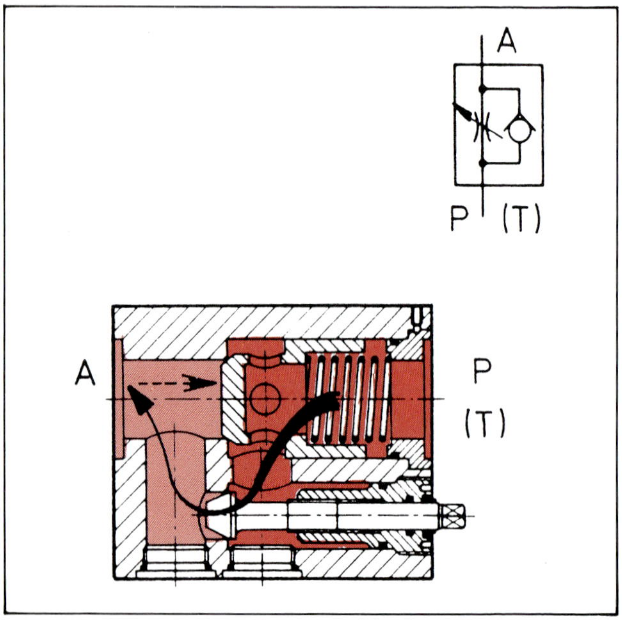

Fig. 121 Simple restrictor with reverse free-flow check valve.

Flow control valve with temperature compensation

Insensitivity to fluid temperature fluctuations is mandatory for satisfactory performance in many hydraulic systems. Fluid temperature affects viscosity, and thus the pressure drop across a given orifice. This is demonstrated in fig. 122.

The long narrow area represents the throttling orifice, which increases the friction of the fluid molecules between each other, and between the molecules and the wall of the narrowed pipe. Such friction, both within and between fluid and pipe, causes a loss of energy from the moving fluid. This causes a loss of pressure and an increase of heat. The friction is directly related to the length of the throttling area. If the throttling length is shortened, the friction is also reduced.

Therefore, the shorter the throttling length, the less sensitive the orifice will be to temperature (viscosity) changes. This implies, that an orifice of "zero" length

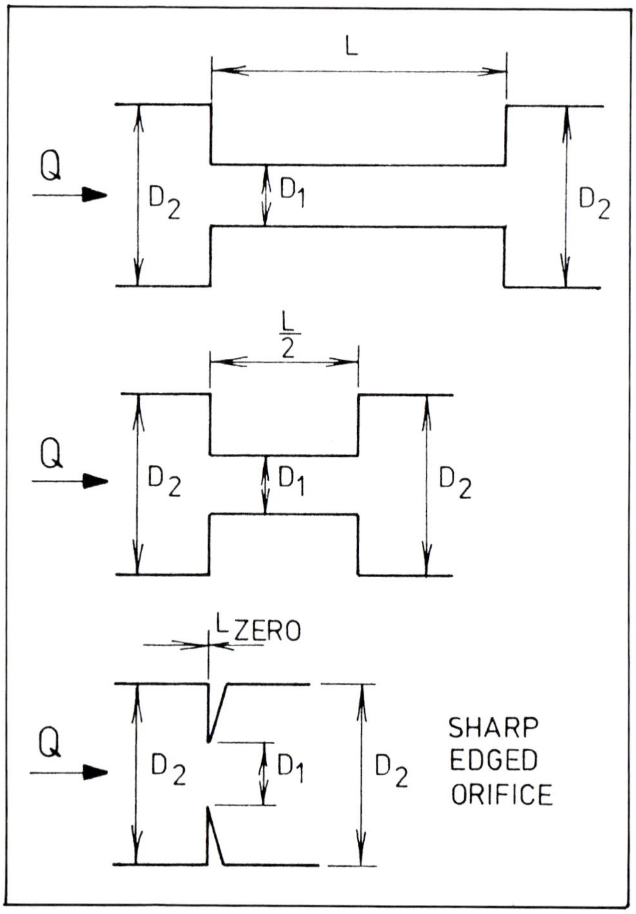

Fig. 122 The sharp edged orifice is less sensitive to fluid temperature variation.

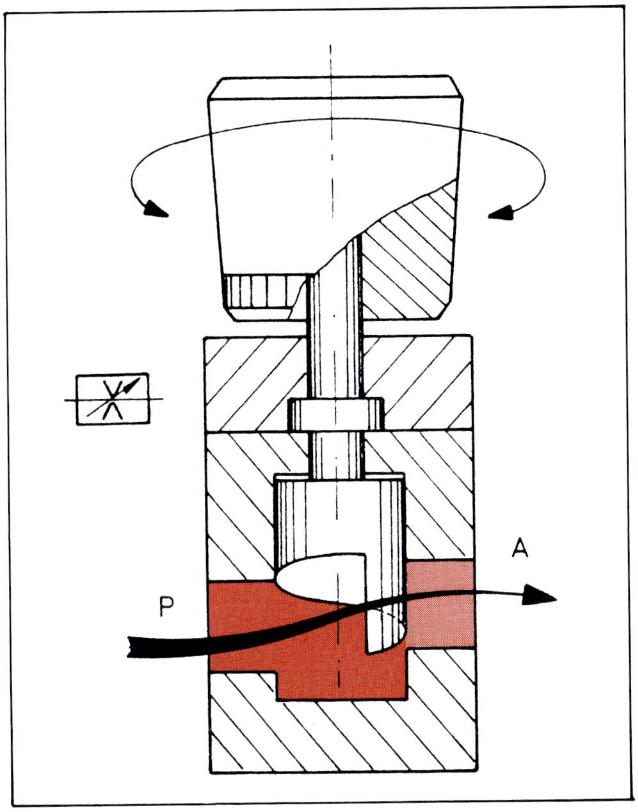

Fig. 123 Temperature compensated flow control valve.

would have no influence on the flow through it. Technologically this cannot be realised. But the so-called "sharp edge" orifice comes close to full temperature insensitivity, and flow deviation is as small as 1 to 1½% within a temperature range of 20–70°C.

A temperature-compensated flow control valve (restrictor valve) is shown in fig. 123.

Flow control valve with pressure compensation

Where accurate and consistent flow-rate control is required, regardless of changes in the pressure differential (Δp), a pressure-compensated flow control should be used. Pressure compensation ensures that the flowrate (Q) through the valve is precisely maintained, even if the upstream or downstream pressure should vary.

Valve operation

Flow passes from port P through the pressure-compensator orifice and then through the control orifice, and leaves the valve at port A. Pressure compensation is based on the use of a pressure-positioned variable orifice (compensator orifice) upstream, in a series arrangement with the control orifice (fig. 124).

The ends of the compensator spool (hydrostat) which have precisely equal areas, are hydraulically connected to the inlet and outlet of the control orifice. Hence, in a static condition, the hydraulic forces will hold the compensator spool in balance, but the bias spring will force it to the far right, thus holding the compensator orifice fully open.

In the flow condition, any pressure drop (Δp) less than the bias-spring force will not affect the fully open compensator orifice; but any pressure drop greater than the bias-spring force will reduce the compensator orifice. Any change in pressure on either side of the control orifice, without a corresponding pressure change on the opposite side of the control orifice, moves the compensator spool. Thus, a fixed pressure differential (Δp) across the control orifice is maintained at all times. By this means, the valve holds the pre-set flowrate across the control orifice constant.

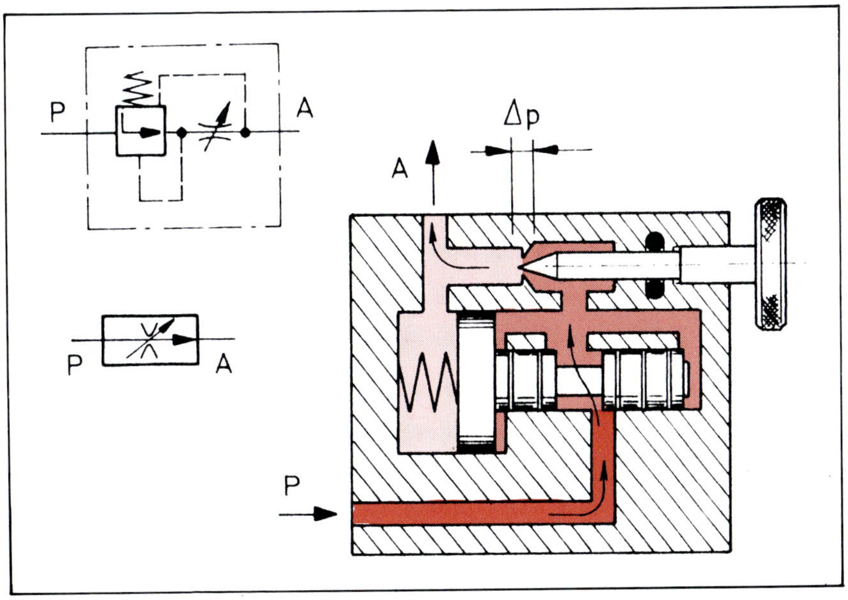

Fig. 124 Pressure compensated flow control valve.

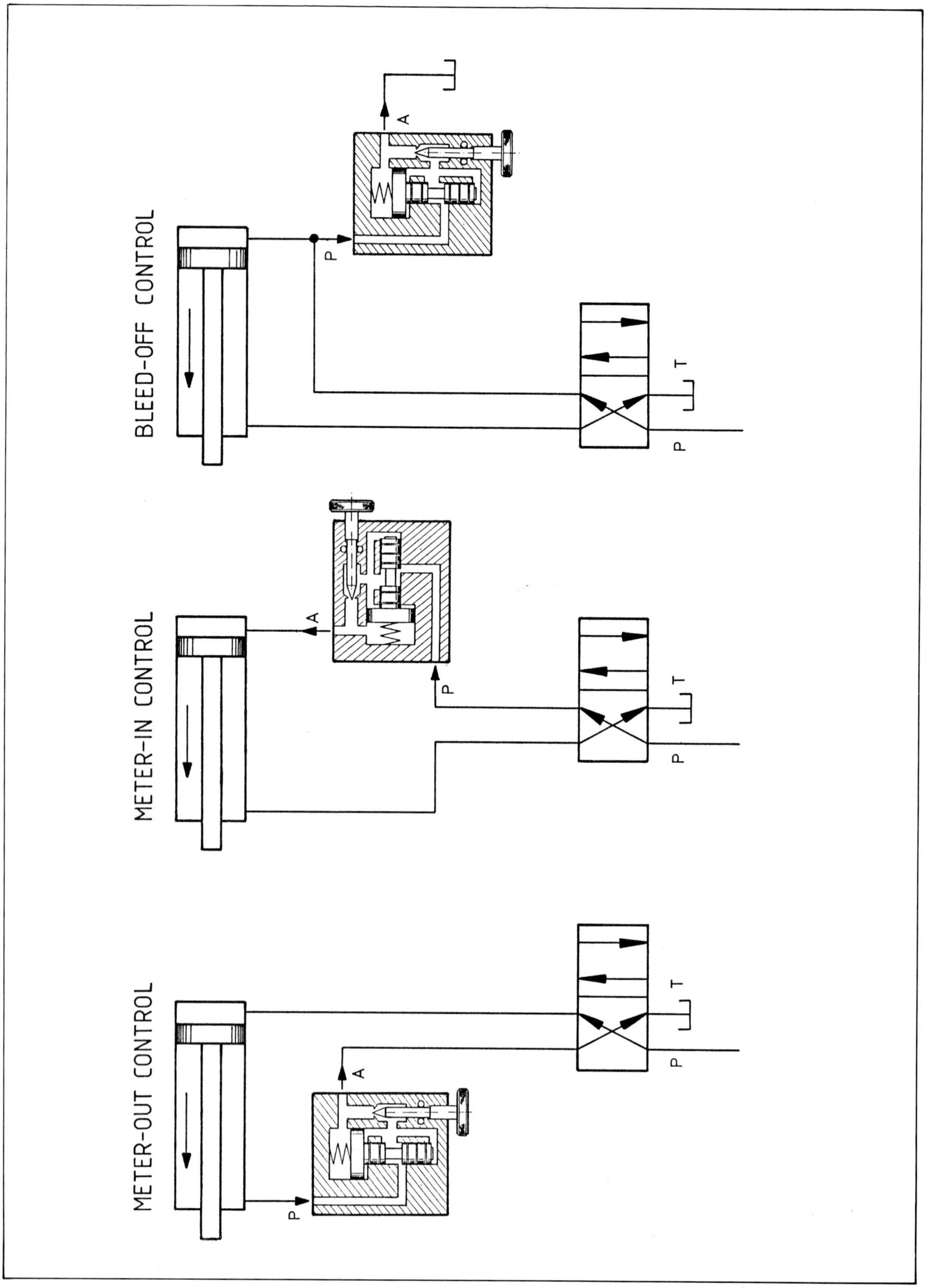

Fig. 125 Flow control methods for restrictor type valves.

Flow control methods for restrictor type valves

Three basic methods are commonly used when applying flow control valves for actuator speed regulation. These methods are: "meter-out, meter-in, and bleed-off". The restrictor type flow control valves shown in figs. 120–124 may be used for all three flow control methods (fig. 125).

A common method of controlling actuator speed in meter-out and meter-in flow control is by means of "sandwich"-type throttle valves, which are mounted between a D.C.V. and its sub-plate. These throttle valves are un-compensated and will therefore give only approximate control if the fluid conditions vary greatly. They are usually made as a pair, in one body, incorporating reverse free-flow check valves.

Meter-out flow control

This speed control method is highly accurate, and used wherever a free-falling load or overhauling load tends to get out of control ("runaway" condition).

The flow control valve is located between the actuator and the directional-control valve, and controls the exhaust fluid from the actuator. If both actuator strokes are to be controlled, the valve can be installed in the tank line of the directional-control valve. But one must beware of the pressure intensification exceeding the T port rating of the directional-control valve if actuators with oversize piston rods are used. If only one stroke is to be speed controlled, a reverse free-flow check valve would be required for rapid retraction.

As a disadvantage it must be mentioned, that excess pump flow, which cannot pass through the flow control valve, is pushed over the system relief valve.

Meter-in flow control

This speed control method is also highly accurate, and is used where the load on the actuator resists the stroke at all times (no "runaway" condition).

The flow control valve is located in the feed-line on the actuator and where only one stroke is to be speed controlled, a reverse free-flow check valve would be required to provide rapid retraction. If both actuator strokes are to be speed controlled, the flow control valve may be installed between the pump and the directional-control valve. However, stroke speeds could then not be regulated individually.

Here too, excess pump flow is pushed over the system relief valve.

Bleed-off flow control

This speed control method has a power-saving advantage, as the pump operates always at the pressure required by the workload, and the excess pump flow returns via the flow control valve to tank, without being pushed over the relief valve.

The method is not as accurate as meter-in, since the measured flow goes to tank and the remaining flow into the actuator. This makes the actuator speed subject to varying pump delivery.

Bleed-off flow control does not require a reverse free-flow check valve. It must be noted, that bleed-off control is not suitable for "runaway" load conditions.

By-pass flow control valve

This valve controls flow to an actuator and diverts any excess (surplus) flow to tank (figs. 126 and 127). The in-built pressure relief valve is an additional feature of the by-pass valve and provides overload protection from excessive workload pressure build-up.

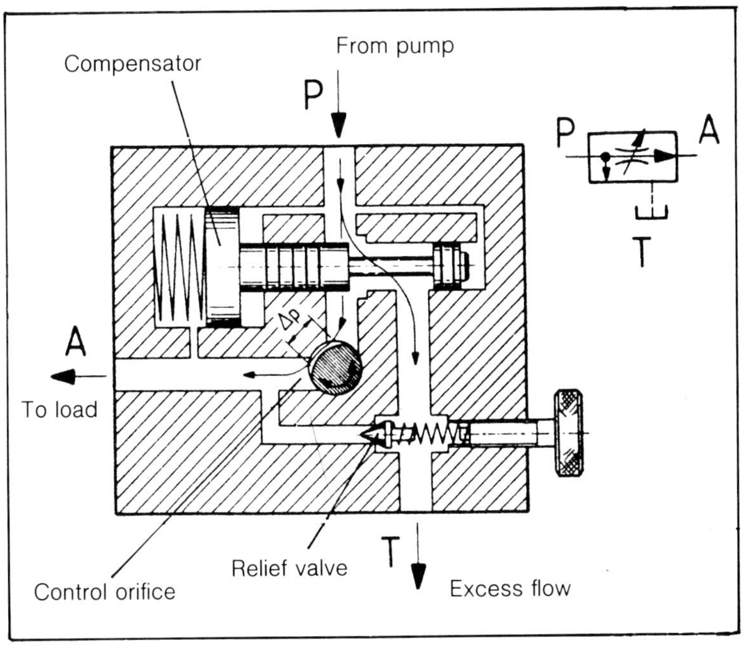

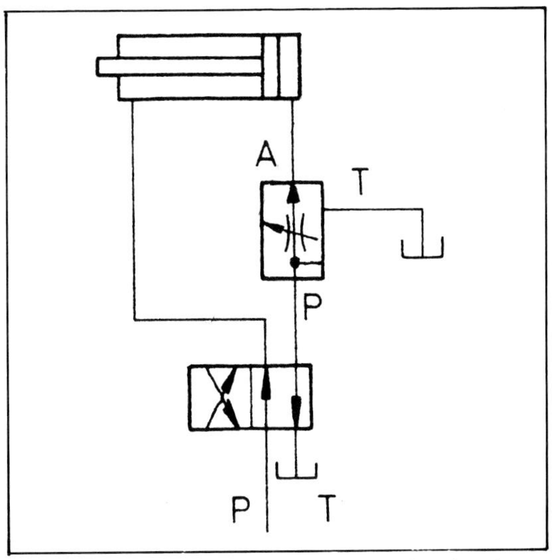

Fig. 127 *above:* By-pass flow control valve in a meter-in control application.

Fig. 126 *left:* By-pass flow control valve.

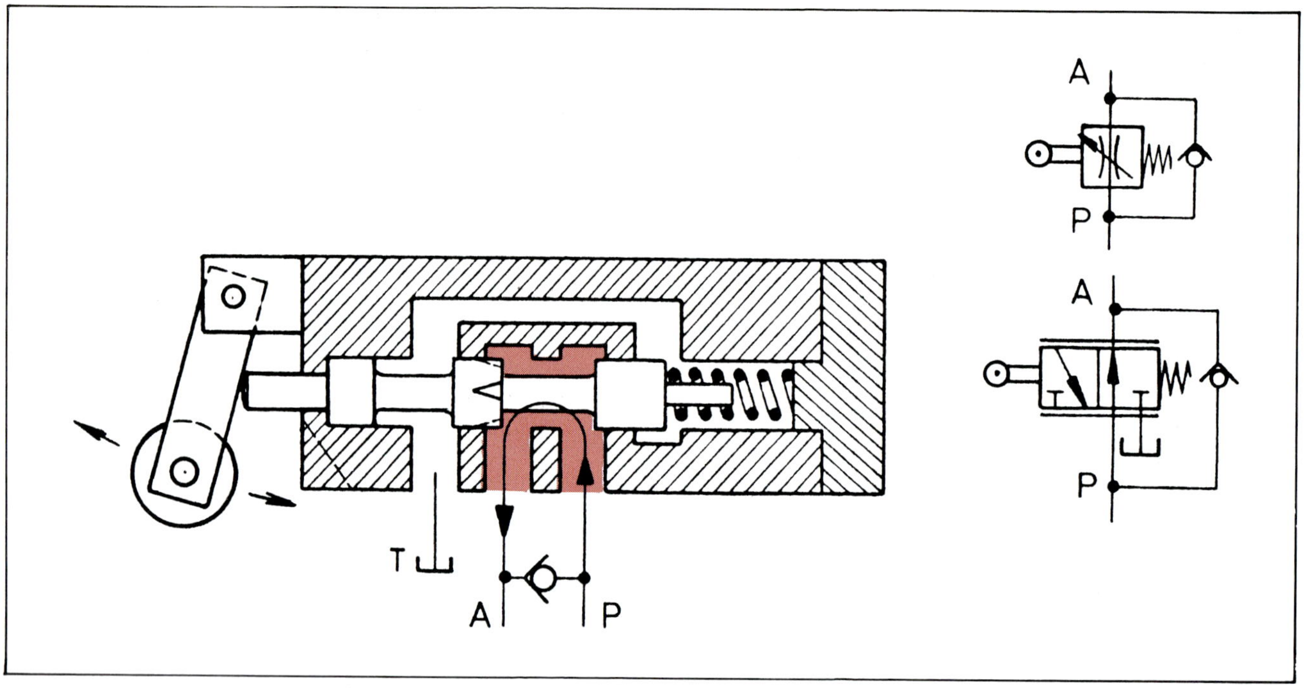

Fig. 128A Deceleration valve (normally open).

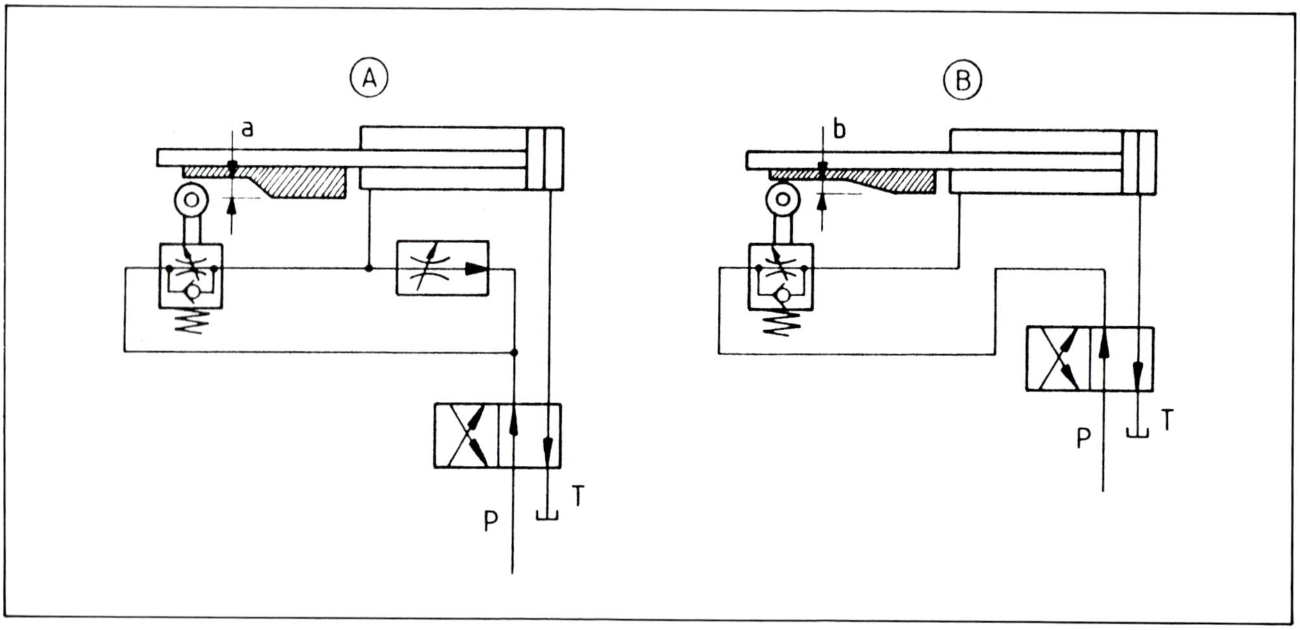

Fig. 128B Applications for deceleration valves.

Valve operation

The valve operates on similar principles to the pressure compensated valve depicted in fig. 124, but the compensator holds the tank line (T) closed when the pressure differential (Δp) across the control orifice is less than the bias-spring force.

When pressure on port P rises — due to the surplus flow from the pump — the compensator will move towards the left. Thus, any excess flow from the pump is diverted to tank. In this way, the valve provides a controlled pressure differential (Δp), and hence an accurately controlled flow from P to A.

Overload protection for the priority system is provided by the simple pressure relief valve, which also limits the pressure above the compensator. Thus, the valve functions in much the same way as a compound relief valve.

Valve application

The by-pass flow control valve must only be used in a meter-in flow control application. If used as meter-out, any excess exhaust flow from the actuator would flow unrestricted to tank. This would permit the load to "run-away".

The tank port must not be subjected to possible pressure surges, as this would disturb the compensator spool balance, thus resulting in a disturbance of the pressure differential and the flow to port A.

The by-pass flow control valve conserves pump input power, since excess flow is by-passed to tank at work-load pressure, and not pushed over the relief valve.

Deceleration valve (cam-operated flow control)

Deceleration valves are used to control piston speed during some parts of the actuator stroke. The valve is controlled by means of a cam, and depending on the cam shape, it can either increase or decrease (step-up and step-down) the actuator speed. A normally-open valve reduces flow when the plunger is depressed by the cam, whereas for the normally closed valve, a depression of the plunger increases the flow through the valve (figs. 128A & B).

Most manufacturers of deceleration valves build a reverse free-flow check valve into the valve body. This provides rapid piston retraction (fig. 128B). In circuit A of fig. 128B, the cam step (a) closes the flow control orifice completely. Exhaust flow for the final stroke length is then metered through the restrictor-type flow control valve. The latter arrangement is often available as a single valve body, containing the cam-operated spool and one of two flow control valves, for primary and secondary speed reduction, together with a reverse free-flow check valve.

In circuit B of fig. 128B, the orifice reduction, and thus the piston speed for the final stroke length, depends on the step heights of the cam (step b). Where several speed levels are required, the cam can be shaped or stepped accordingly, thus providing varying speeds within the stroke length.

Flow divider valves

Flow divider valves are used to split (divide) the flow from a pump into two equal or, where applicable, non-equal parts. The flow division is kept constant, regardless of load fluctuations on the flow outlets.

Valve operation

Inlet flow from the pump passes through two precisely matched orifices (one each spool Fig. 129) to their respective outlets. If flow to one outlet tends to increase, the greater pressure drop (Δp) across that spool will cause the spool to shift, thus restricting that outlet and the flows to their intended flow rates.

Valve application

Flow dividers may also be used as flow combiners, where a meter out flow control application is preferred. The valve is then mounted on the exhaust side of the two cylinders or motors.

Important

The accuracy and flow range of flow dividing valves are both limited, and care must be taken to avoid their use for unsuitable applications.

When synchronizing two linear actuators, it should be noted that any error occurring with a flow dividing valve can be cumulative, unless the actuators are brought up against mechanical stops to realign them. Apart from that, only the use of servo valves can give proper synchronization.

Priority flow divider valve (Priority flow control valve)

The priority flow divider is used on power steering or power braking on mobile hydraulic equipment. The variable input flow, for example, from a pump being driven off an idling diesel engine, is split into a constant (priority) flow and an excess flow which may be utilised for other functions. The priority flow leads to the power steering, the excess flow to other services on the machine.

Some pump manufacturers integrate the priority flow divider into the pump housing.

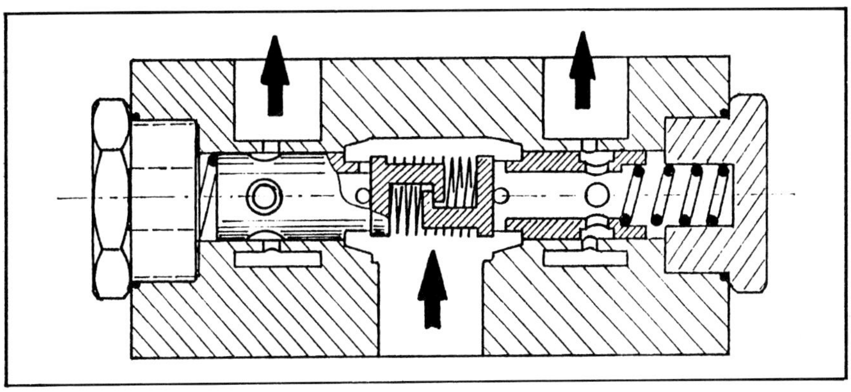

Fig. 129 Flow divider valve (flow combiner).

7 Rotary actuators (motors)

Hydraulic motors convert hydraulic energy into torque and consequently into power. Motors very closely resemble hydraulic pumps in their construction. In fact, many pumps can also be used as motors. Instead of pushing fluid into the system as pumps do, motors are being pushed by the fluid through which they develop torque and continuous rotary motion.

Hydraulic motors have several design features in common:

- Each design type must have a driving surface area (A) subject to a pressure differential (Δp). For vane and gear motors, this surface is rectangular. For radial and axial piston motors the surface is circular (fig. 130).
- In each design type this pressure exposed area (A) must be connected mechanically to the motor output shaft.
- Inlet and outlet fluid must have a timed porting arrangement to produce continuous rotation.

Maximum performance of the motor is determined by the ability of the pressure surfaces to withstand great force, by the internal leakage characteristics of the moving parts which seal the high pressure inlet from the low pressure outlet, and by the efficiency of the mechanism which links the moving pressure-exposed areas to the motor shaft (figs. 130 and 137). This maximum performance varies greatly between motor design types, and is expressed in terms of pressure, flow,

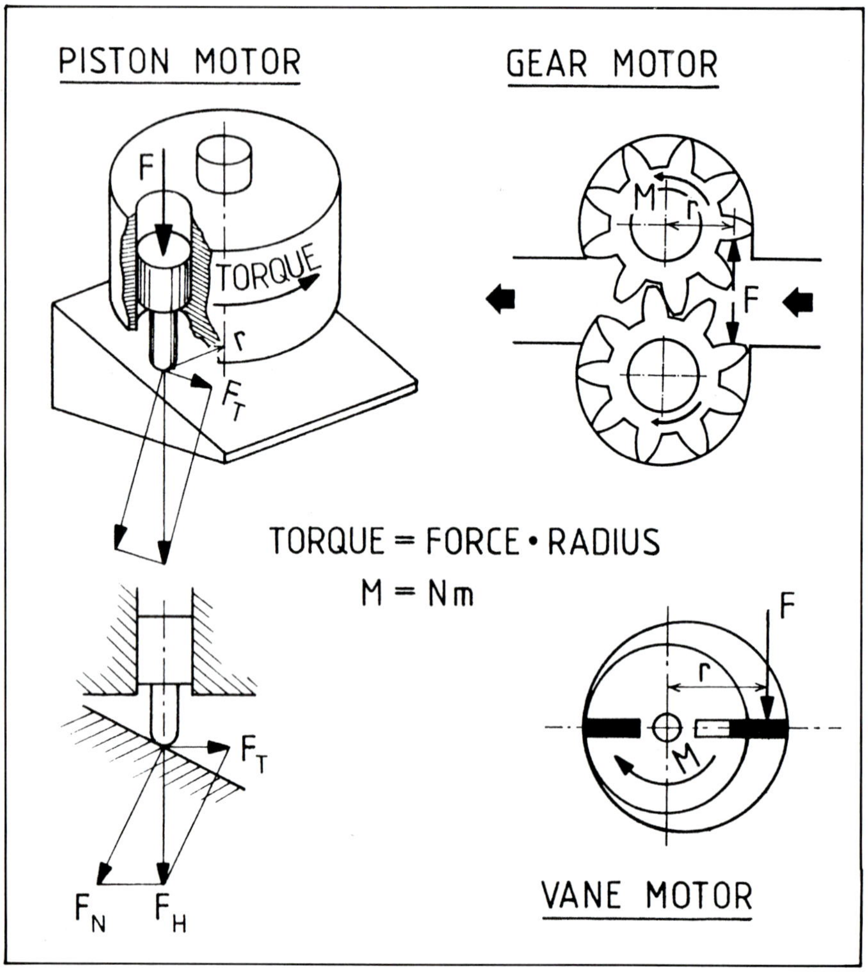

Fig. 130 Development of torque in hydraulic motors.

torque, output speed, volumetric and mechanical efficiency, life, and physical configuration.

Torque

Torque (M) may be defined as a twisting or turning moment and is expressed in Newton metres (Nm). Torque is a function of system pressure and leverage, whereby the leverage is measured from the centre of the drive shaft to the centre of the pressure exposed area. "Breakaway torque" is the turning moment required to accelerate a load from standstill into motion. This torque is generally much higher than "running torque". Running torque is the amount of torque required to keep a load in rotation after it has been started (fig. 130).

Motor displacement (geometric volume)

Motor displacement (V) refers to a quantity of fluid required to rotate the motor output shaft one revolution. Motor displacement (sometimes also called geometric volume) is measured in cm^3 (or mL) per revolution.

Hydraulic motors are being built with fixed or variable displacement. Fixed displacement motors provide constant torque and variable speed. Speed varies with the amount of input flow into the motor. Variable displacement motors provide variable torque and variable speed. With constant input flow and constant operating pressure, the ratio between torque and speed can be infinitely varied to meet load requirements.

Motor speed (rotations)

Motor speed (n) is a function of motor displacement and the input flow (flowrate) delivered to the motor. Motor speed is expressed in revolutions per minute (r.p.m.). Hydraulic motors should be operated at their most efficient speed range.

System failure can cause a motor to overspeed, thus causing accelerated wear, or in some cases total failure. Maximum motor speed is the speed at which the motor can safely operate without being damaged. Minimum motor speed is the speed at which the motor provides uninterrupted, reliable, and continuous shaft rotation. (For maximum and minimum r.p.m., see manufacturers' specifications.)

Vane motors

In vane motors torque is delivered by the pressurised fluid acting onto the rectangular vanes which slide in and out of their slots in a rotor splined to the drive shaft. As the motor turns, the vanes follow the contour of the cam ring, thus forming sealed cavities which transport the hydraulic fluid from the pressurised inlet side to the non-pressurised outlet side. Since the motor vanes must maintain cam ring contact at all times, and centrifugal force is absent during motor start, these vanes are usually fitted with springs to enable rotation to commence (fig. 131).

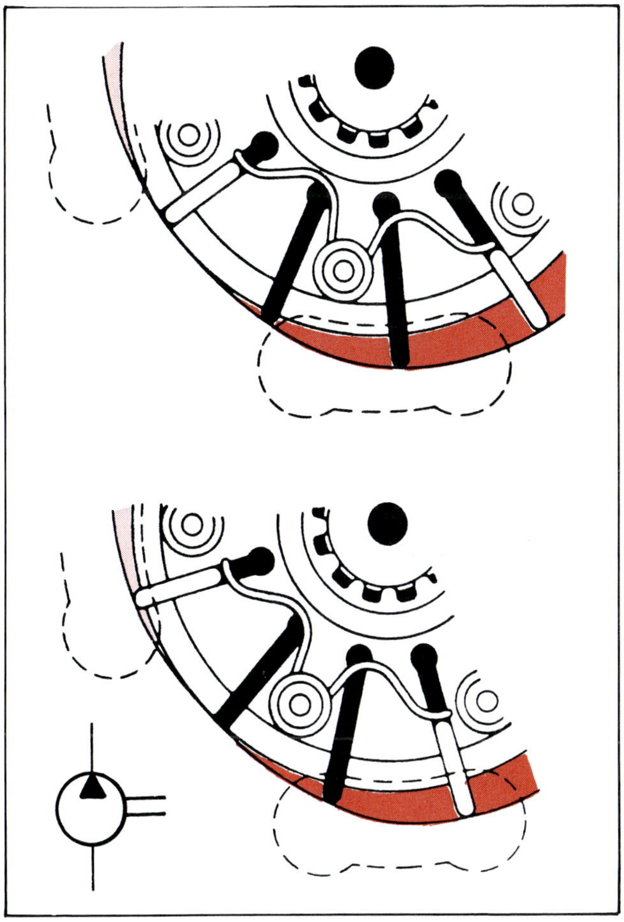

Fig. 131 Rocker springs hold the vanes against the cam ring.

Other designs use coil springs, or feeding of pressure to the underside of the vanes to force them firmly against the cam.

External gear motors

Some gear pumps can also be used as motors. The hydraulic fluid enters the gear chamber on the side where the gears mesh, and forces the gears to rotate. The fluid exits at low pressure on the opposite side to the inlet.

Reversible gear motors have an external case drain which must be connected to the reservoir. Non-reversible motors are internally drained to the exit port (fig. 132).

For motor construction, see external gear pumps.

Internal gear motors

Orbit motors, direct-drive gerotor motors, and crescent gear motors work in much the same way as internal gear pumps. The orbit motor and the direct-drive gerotor motor are designed and built for high torque and low

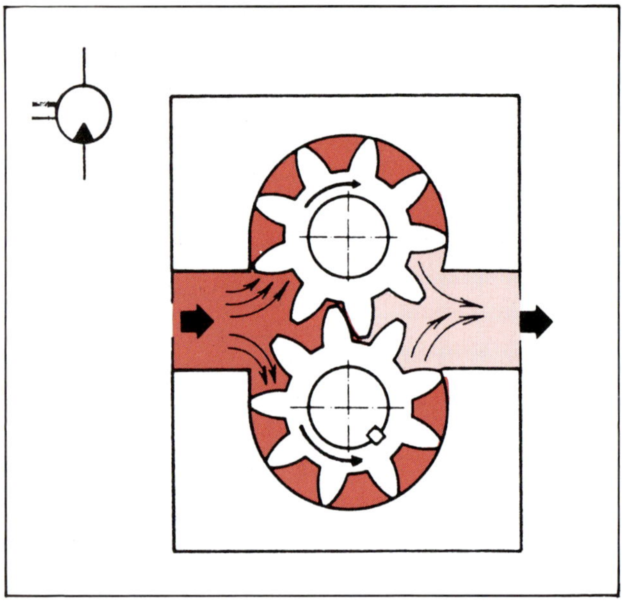

Fig. 132 Gear motor.

speed applications with volumetric efficiency ratings coming close to piston motors and piston pumps.

Most internal gear motors are flow reversible and thus require a case drain. For internal gear motor construction see internal gear pumps. All gear motors are fixed displacement machines which provide fixed torque, but speed can be varied provided the input flow is variable.

Piston motors

Piston motors are compact, provide extremely high torque and high acceleration, and have an excellent life expectancy. Axial piston motors are capable of operating at speeds from 0.5 r.p.m. to as high as 6000 r.p.m. with stable torque output for fixed or variable displacement.

Radial piston motors are up to 95% overall efficient, can operate at power outputs of up to 3000 kW, and reach peak speeds of 14,000 r.p.m. Some larger motors may require a motor input flowrate of 1600 L/min.

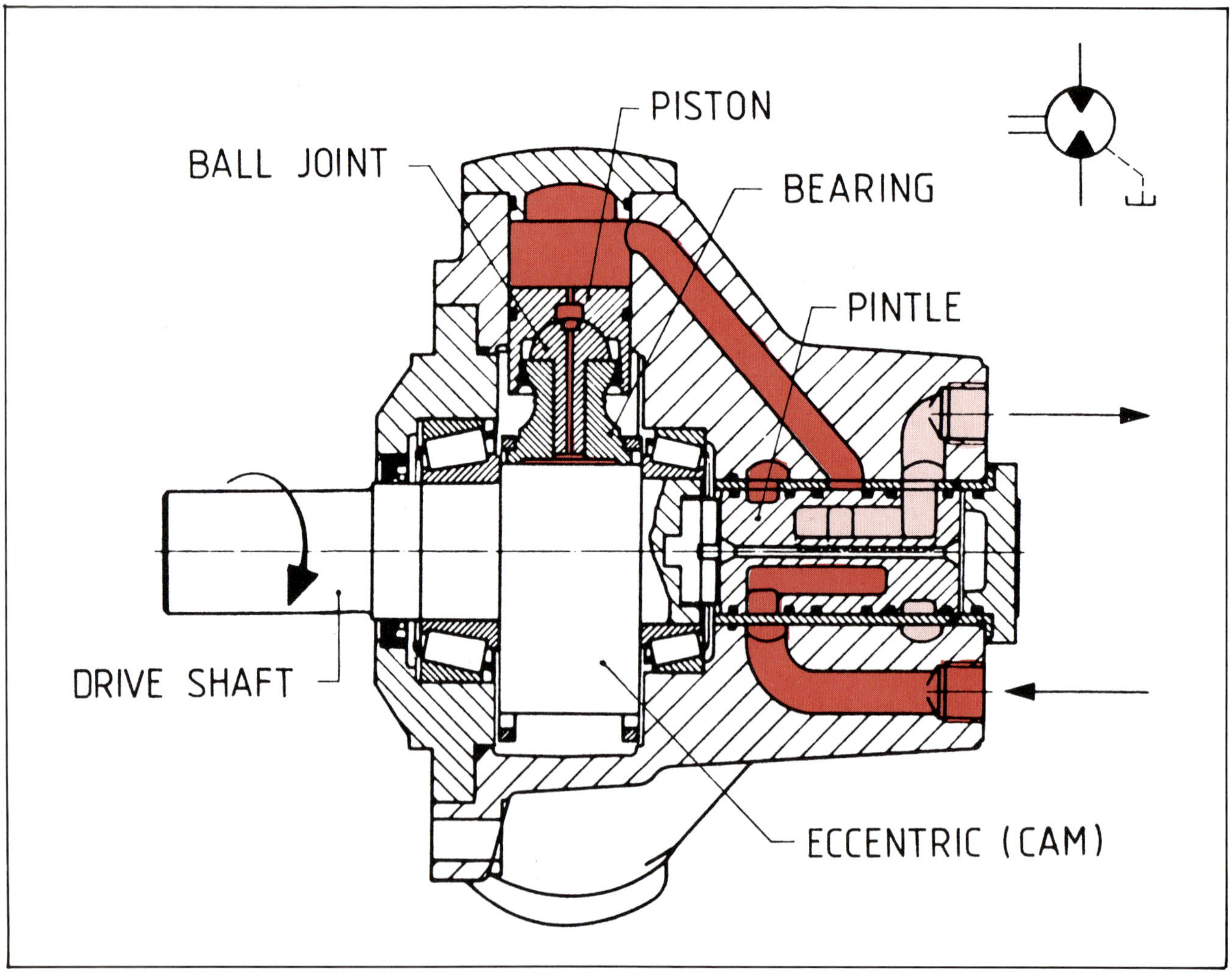

Fig. 133 Radial piston motor with stationary cylinder block and rotating shaft.

Radial piston motors

For the radial piston motor depicted in fig. 133, fluid is ported and distributed via the valve shaft (pintle) to each piston in turn. Force exerted by the pressure exposed pistons is transmitted by the connecting rods onto the crankshaft cam, thus causing the eccentric cam and its drive shaft to rotate. High pressure hydraulic fluid ported through the centre of the pistons and the connecting rods creates a static fluid cushion which reduces metal to metal contact and hydrostatically balances the connection-rod foot bearing.

This motor is reversible and develops identical torque in either direction. The motor casing is externally drained.

The motor depicted in fig. 134 has a stationary cylinder block in which the pistons reciprocate. The pistons guided by the cylinder block are connected to rollers which bear against the multi-lobed contour of the rotating cam ring. The rotating valve shaft (pintle) connected to the cam ring distributes pressure fluid from the inlet port to each piston in sequence, and drains the exhaust fluid from each retracting piston in sequence. Fixed side guides oppose reaction forces on the rollers, thus "unloading" the pistons of any tangential forces.

This type of motor generally has an even number of diametrically opposed pistons to assure hydraulic balance, which unloads the main bearing of radial forces.

The motor in fig. 134 is mounted on the device being

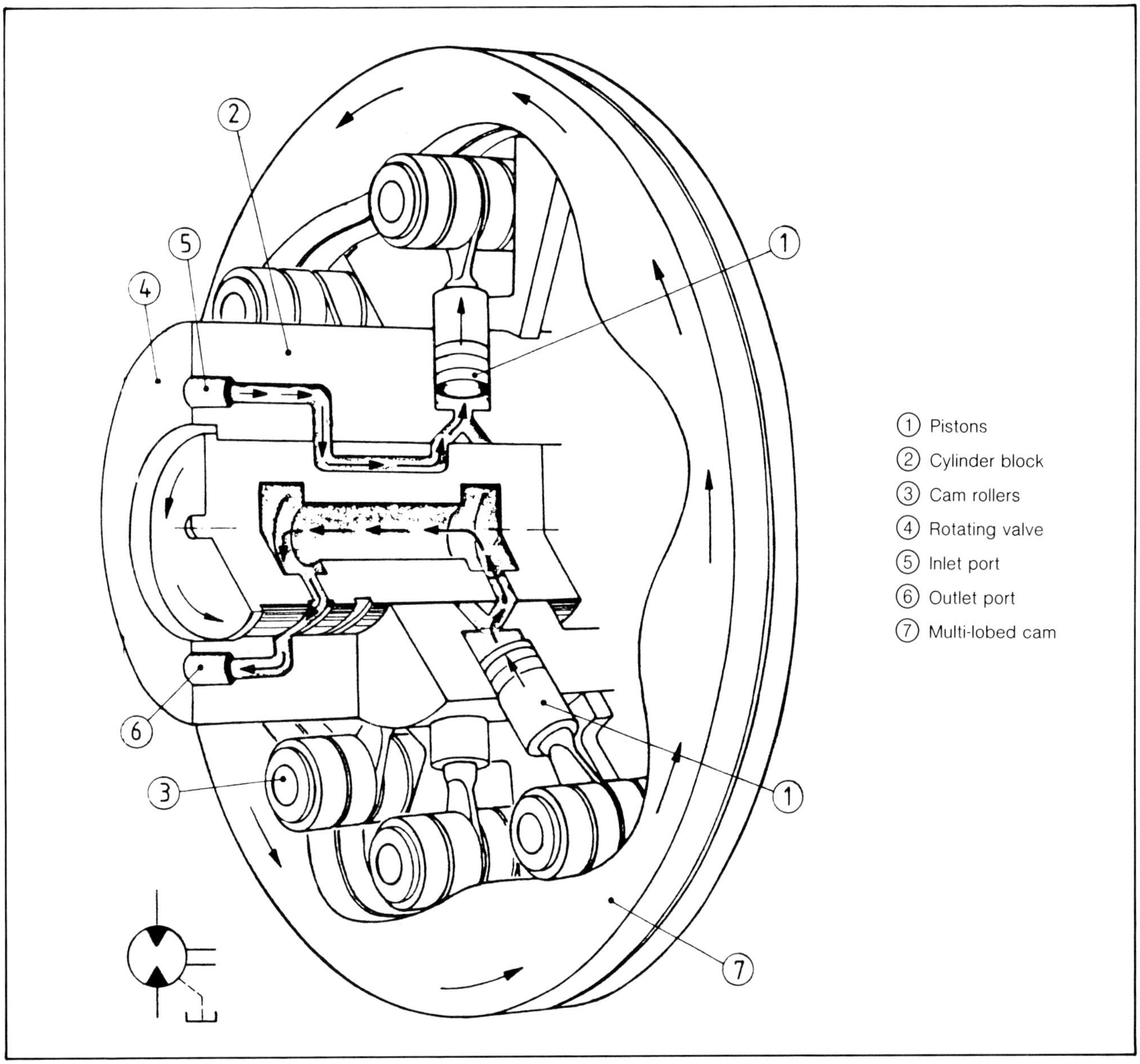

Fig. 134 Radial piston motor with stationary block and rotating cam ring.

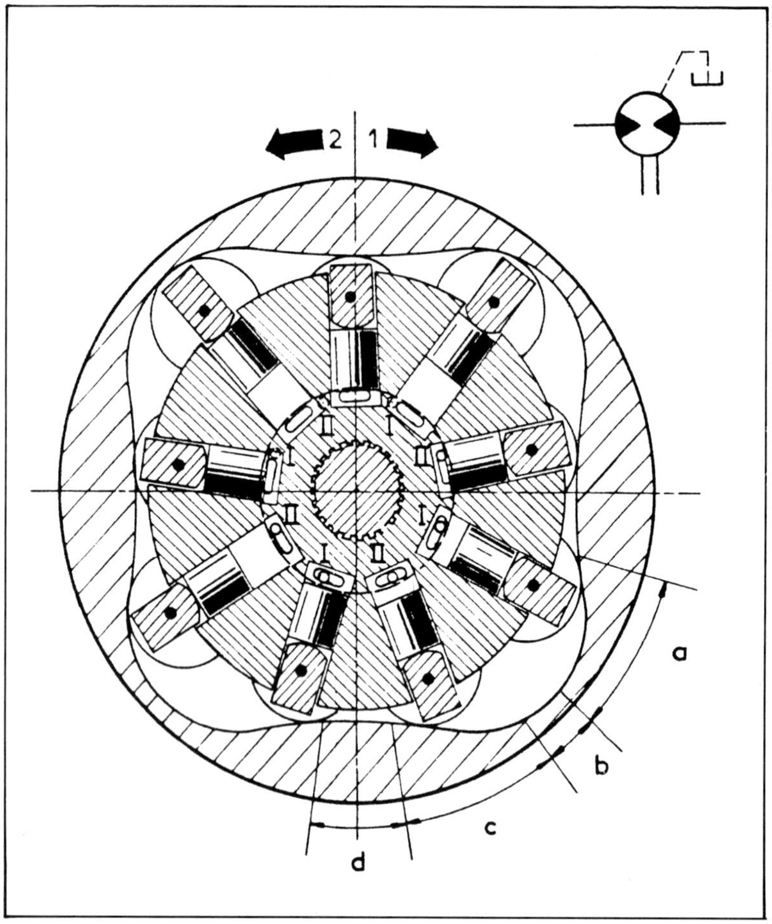

Fig. 135 Radial piston motor with power take-off from the drive shaft.

driven (e.g. a winch or a drum) with a torque arm holding the cylinder housing stationary. It has been designed for extremely high starting torque (breakaway torque) and extremely low speeds (0.5 r.p.m.). It can be free-wheeled and is reversible.

A similar motor with power take-off from a rotating shaft is shown in fig. 135.

Axial piston motors

The axial piston motor depicted in fig. 136 is a low speed, high torque unit with power take-off from the shaft. Hydraulic fluid is supplied to the pistons via the valve plate. The rotors are splined to the drive shaft. The pressurised pistons are forced against the multi-lobed cams where their resulting tangential force converts hydraulic pressure into rotary motion.

This motor can be free-wheeled and is reversible. For open circuit applications a 300–400 kPa outlet port pressure must be maintained to hold the pistons in contact with the cams.

The axial piston motor in fig. 137 has a rotating piston drum driven by the reaction forces of the pistons against the swash wedge. The rotating slipper pads are hydrostatically balanced and are kept to the swash wedge with a rotating spring-loaded shaft. Kidney-shaped slots port the hydraulic fluid to and from the reciprocating pistons. These motors can also be used as pumps. The motor has a fixed displacement and must be externally drained at all times.

Other axial piston motors operate on the principle explained for swash plate and bent axis pumps (chapter 3, pumps). These motors can be either of fixed or variable displacement, and some of the control mechanisms used on variable displacement pumps can also be used on variable displacement motors.

A fixed displacement bent axis motor is shown in fig. 138, and a variable displacement swash plate axial piston motor in fig. 139.

Output rotation is reversed by reversing the oil flow to and from the motor. It is not practical to reverse a motor by swinging the yoke or the swash plate over to centre, since the torque would go to zero and its speed infinitely high (if it did not stall before reaching centre).

Case drain connection

Reversible gear motors require external case drain, whereas non-reversible motors may be internally drained to the low pressure return line which is connected to the reservoir. The purpose of the drain is to

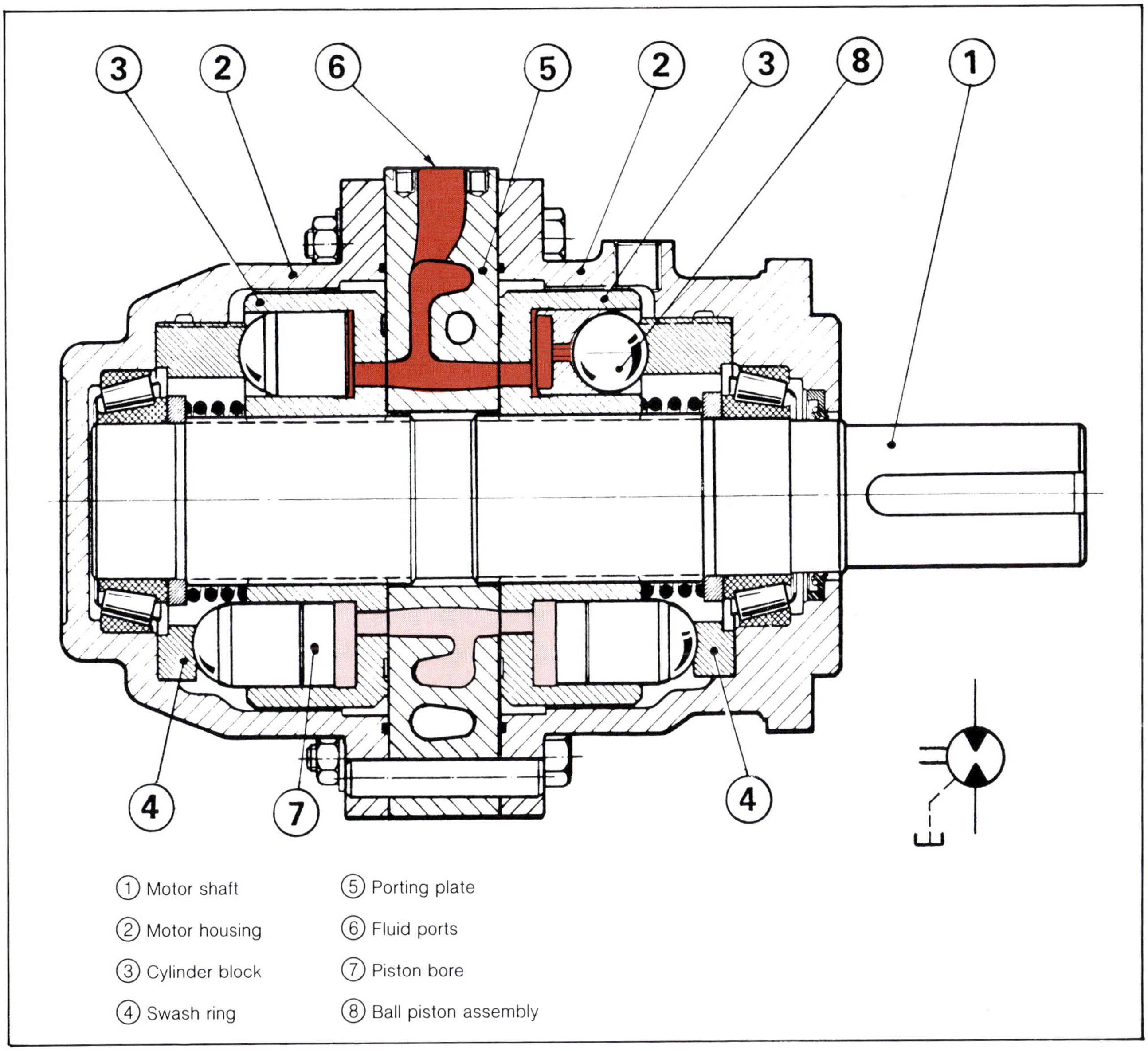

Fig. 136 "Carron" ball piston motor.

carry off the accumulated leakage fluid which seeps past the sealing surfaces. If not drained, this fluid would soon pressurise the motor case and blow out the shaft seal. Some reversible gear motors are internally drained by means of check valves (fig. 140), which direct the leakage fluid to the low pressure port.

Motor drain lines must not be connected to collection return lines of directional control valves, if such return lines are subject to pressure shocks. Gear motors connected in series may be drained as shown in fig. 141.

Freewheeling control

In some applications of hydraulic motors it is necessary for the motor to "freewheel" during part of the operating cycle. In some, freewheeling is achieved by means of suitable external valving (fig. 142) to permit flow circulation from port to port.

In piston motors freewheeling is sometimes achieved by connecting both inlet and outlet ports to tank, simultaneously ensuring a positive case pressure of approximately 50–100 kPa. This pressure moves the pistons clear of the cam ring. In this way, freewheeling for speeds of up to 700 r.p.m. may be achieved.

The type of motor depicted in fig. 133 is also available with dual displacement, one of which can be zero. This permits freewheeling up to a speed of 2000 r.p.m.

Deceleration control

Regulating the pressure on the motor outlet port controls the motor deceleration. In some applications inlet pressure is maintained while outlet pressure is

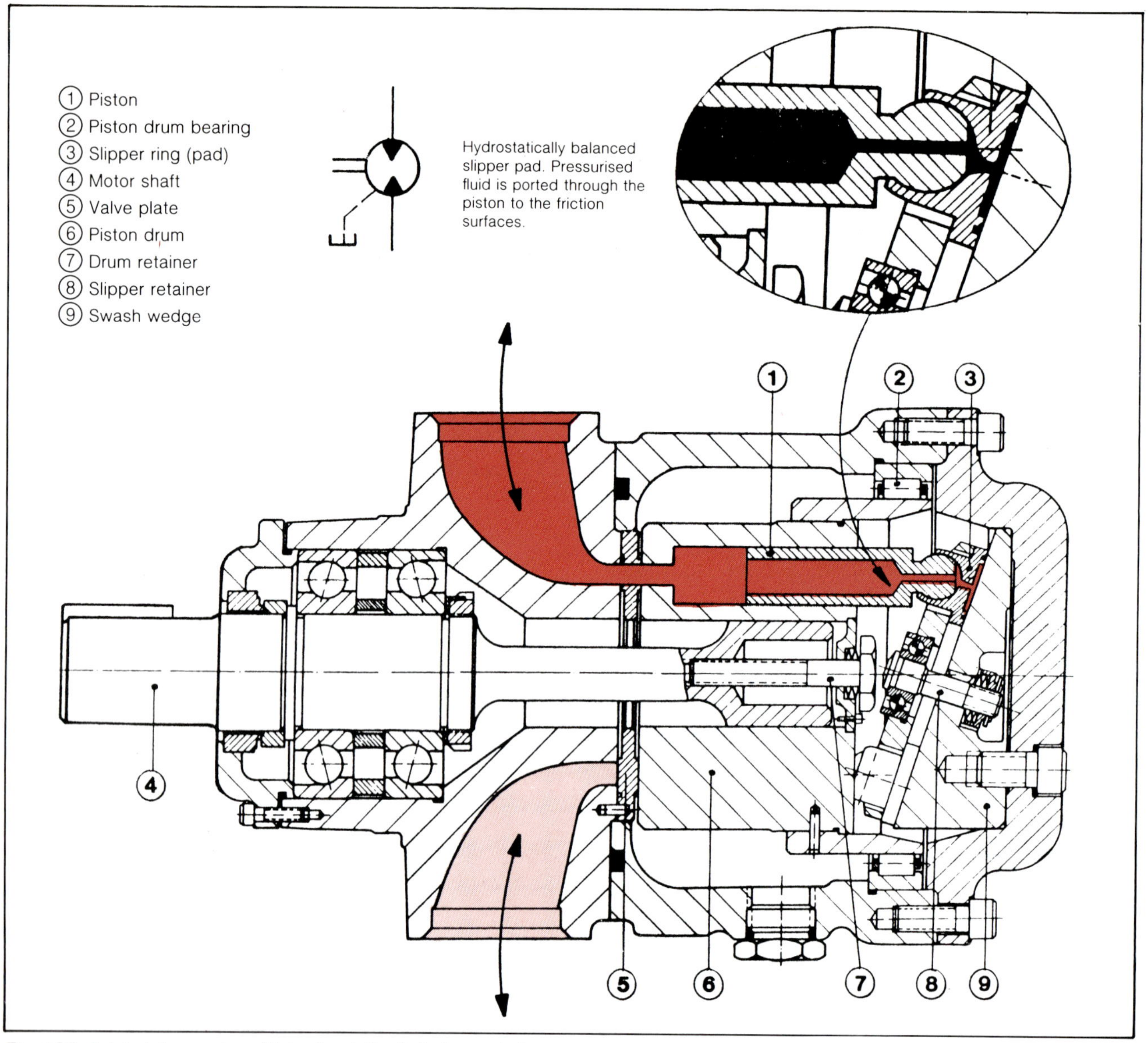

Fig. 137 Axial piston motor with hydrostatically balanced slipper pad. Pressurised fluid is ported through the piston to the friction surfaces (see enlargement).

Fig. 138 Bent axis piston motor (fixed displacement).

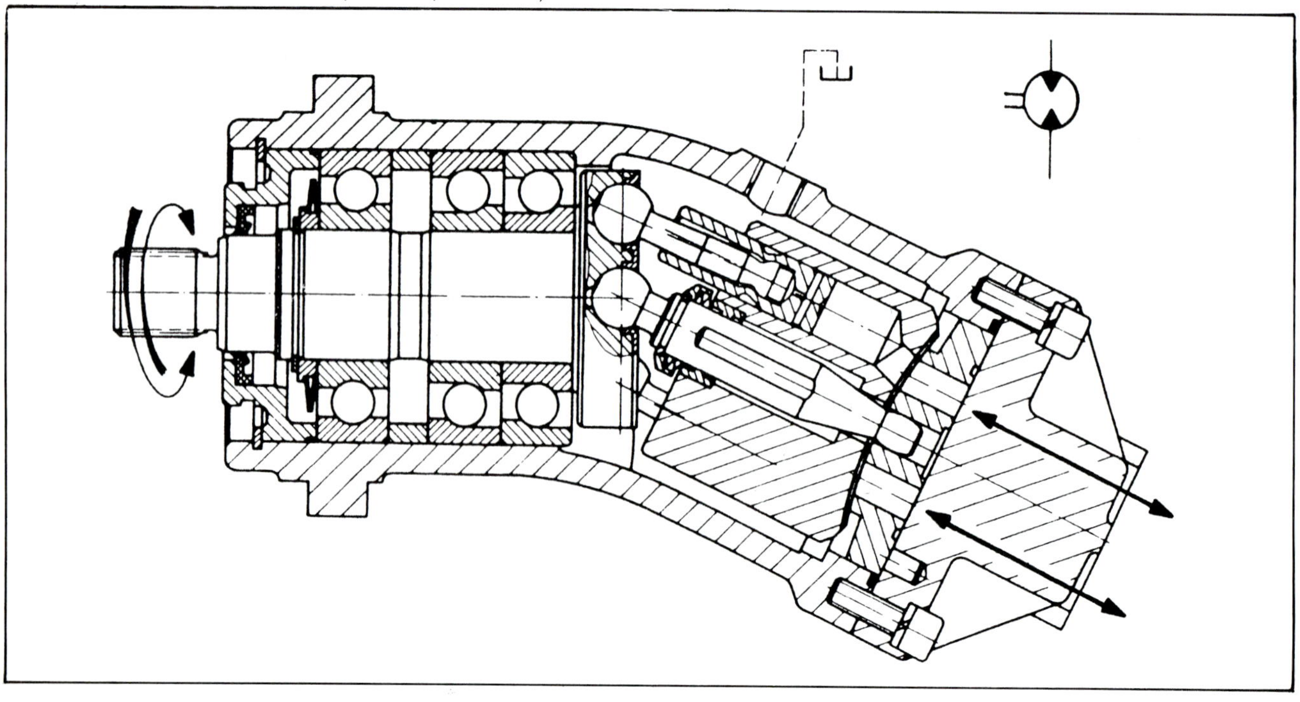

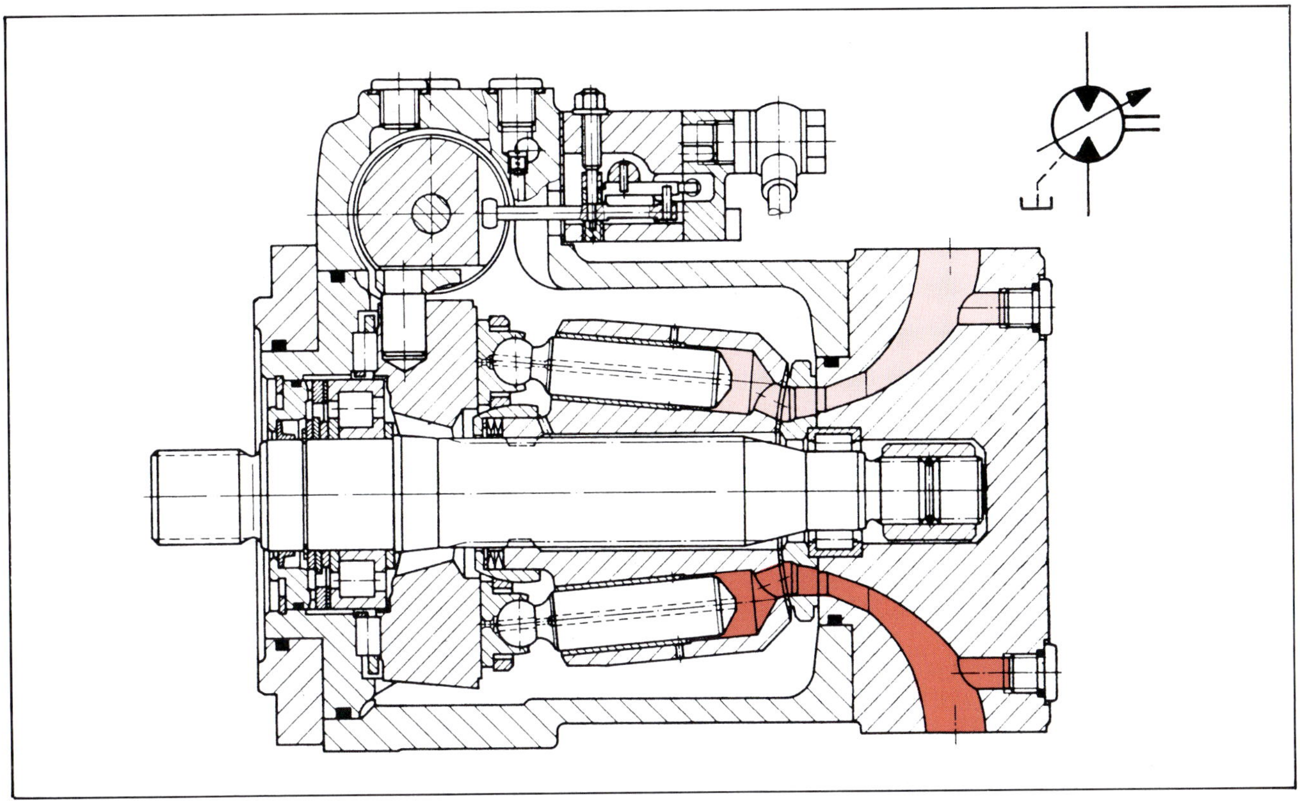

Fig. 139 Swash plate piston motor (variable displacement).

gradually increased until the motor stops. This method gives accurate deceleration control.

In other applications the directional control valve is shifted to neutral, while at the same time a brake valve imposes a restriction on the motor outlet flow. This restriction absorbs the energy of deceleration. The motor inlet port must not be starved of supply flow during deceleration. The pressure setting on the brake valve controls the rate of deceleration (figs. 143 and 144).

The operation of the brake valve is explained in Chapter 5, pressure controls.

Motor reversal control

Most hydraulic motors are built for bi-rotational operation (fig. 137). However, some are not, and immediate or early failure could occur if such motors would be pressurised on both sides, such as in meter-out speed control circuits.

Where a motor, driving a high inertia load, is suddenly reversed, there is a period of deceleration during which the motor acts as a pump, adding its pushed-out fluid to that of the supply pump at system relief pressure. During this period an adequate supply of hydraulic fluid must be ported to the low pressure port of the motor to prevent cavitation (fig. 148).

Cross-line pressure relief valves eliminate pressure shocks caused by rapid reversing or braking. These valves will also assist with the deceleration of high inertia loads (figs. 146 and 147), and prevent cavitation.

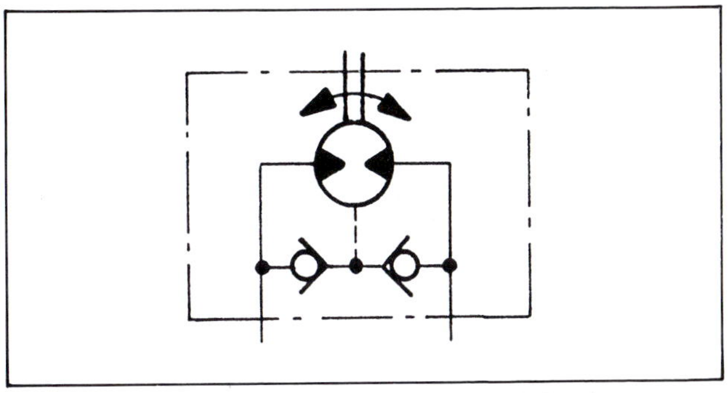

Fig. 140 Reversible gear motor internally drained.

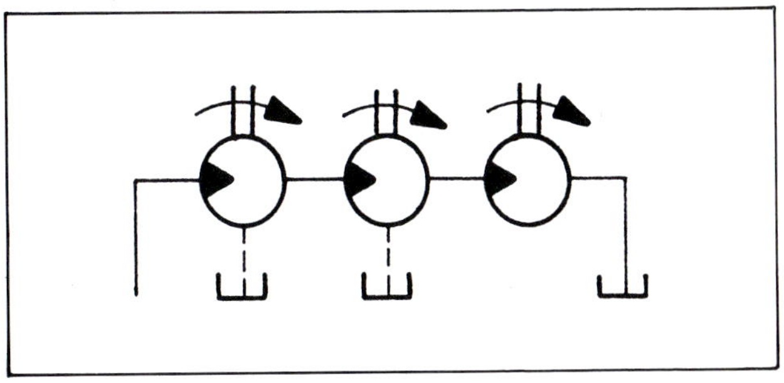

Fig. 141 Gear motors in series.

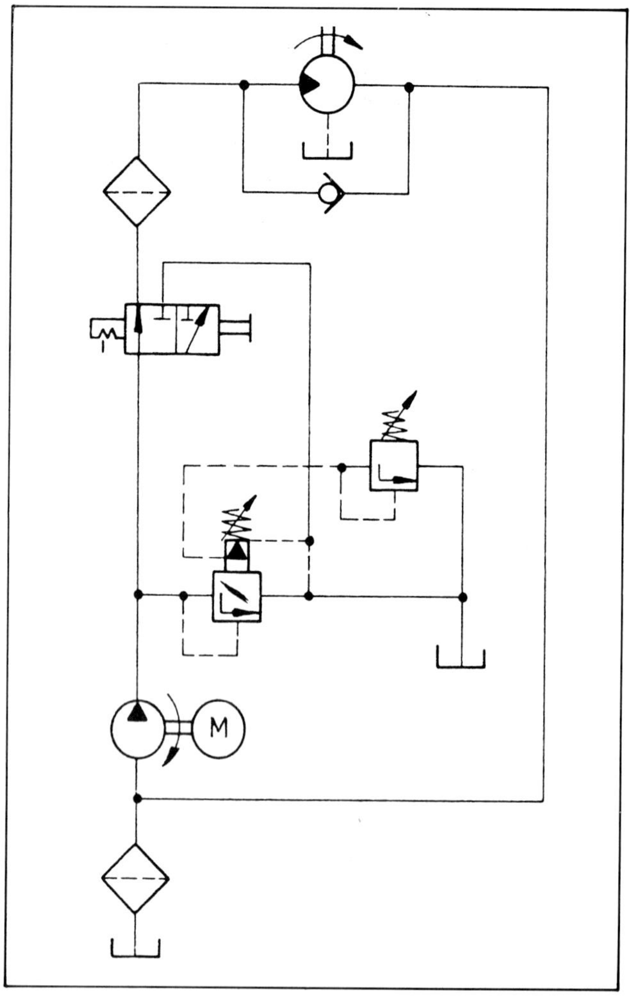

Fig. 142 Motor with port to port connection for 'freewheeling'.

Overhauling load control

Special attention must be given to motor circuits where during some part of the cycle the inertia of the load tends to drive the motor as a pump, or where the motor must hold the load while it is stopped. For such load-hold conditions several circuit solutions are possible. Some use pilot operated check valves others use counterbalance valves or friction brakes to hold the load (fig. 142).

For prolonged load holding conditions, the internal motor leakage to the drain port allows a small amount of rotational creep. Thus, the mechanical friction brake is essential if the load must be positively held.

For controlled lowering of loads it is essential to use a brake valve ("overcentre valve").

Motor speed control

The principles applicable to the control of linear actuator speed also apply to speed control of rotary actuators (motors). Meter-out speed control should be avoided for single direction drives, where the pump casing (shaft seal cavity) is drained to the low pressure

Fig. 143 *below:* Open circuit drive with reversible motor and brake valve for controlled deceleration. The low pressure port is replenished with hydraulic fluid drawn in through the open centre D.C.V.

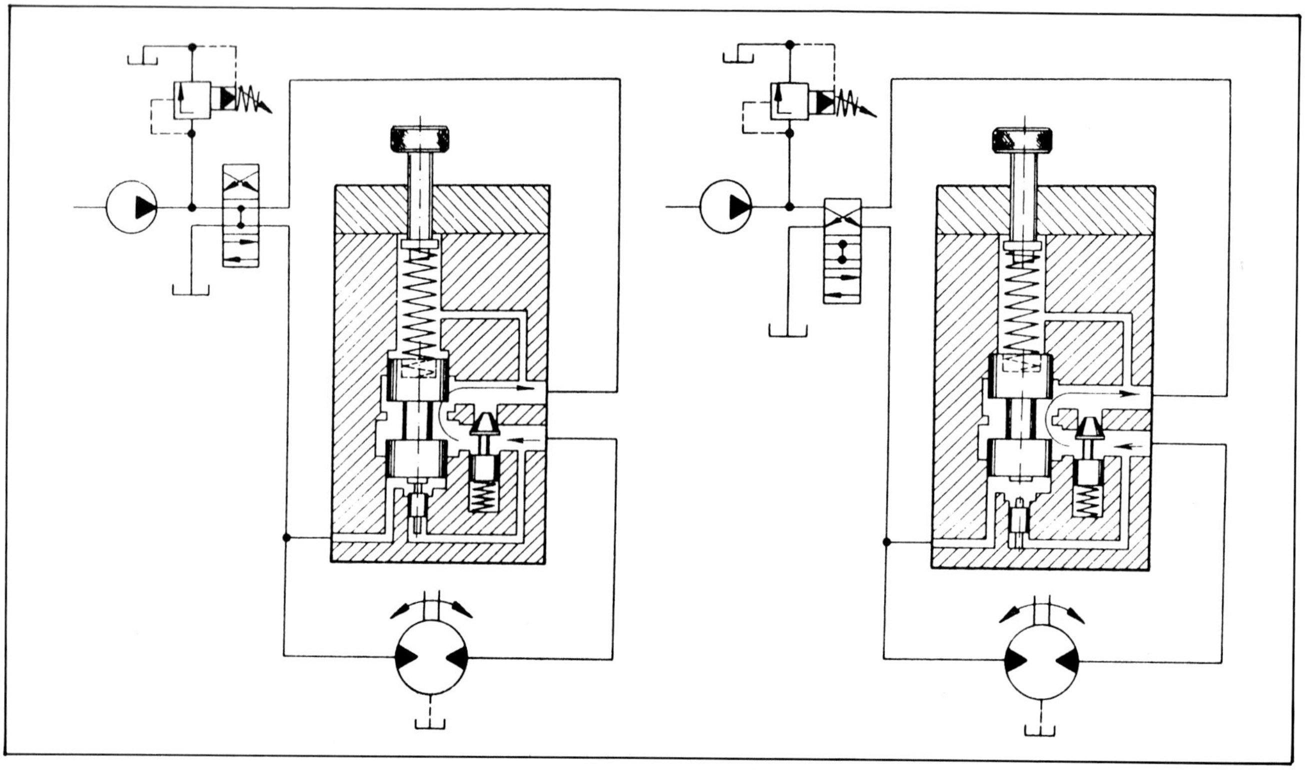

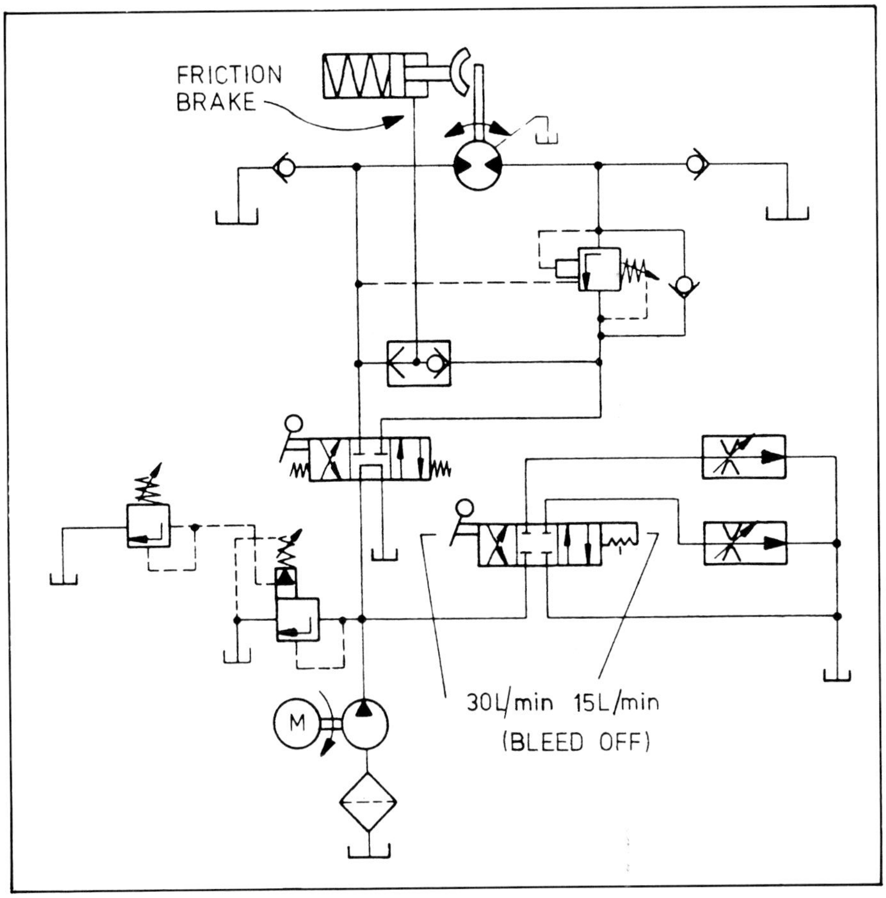

Fig. 144 Motor circuit with three speeds, remote torque control, brake control, and load hold with friction brake.

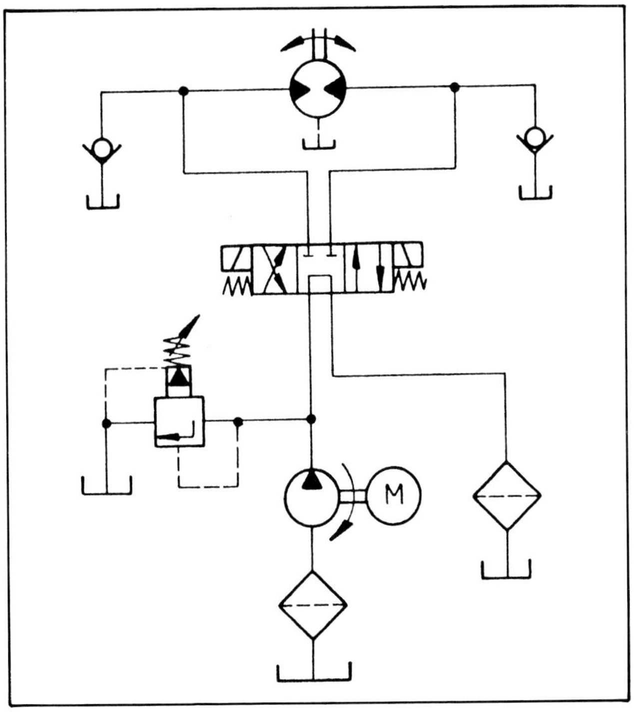

Fig. 145 *above:* Reversible motor drive with fluid replenishing lines.

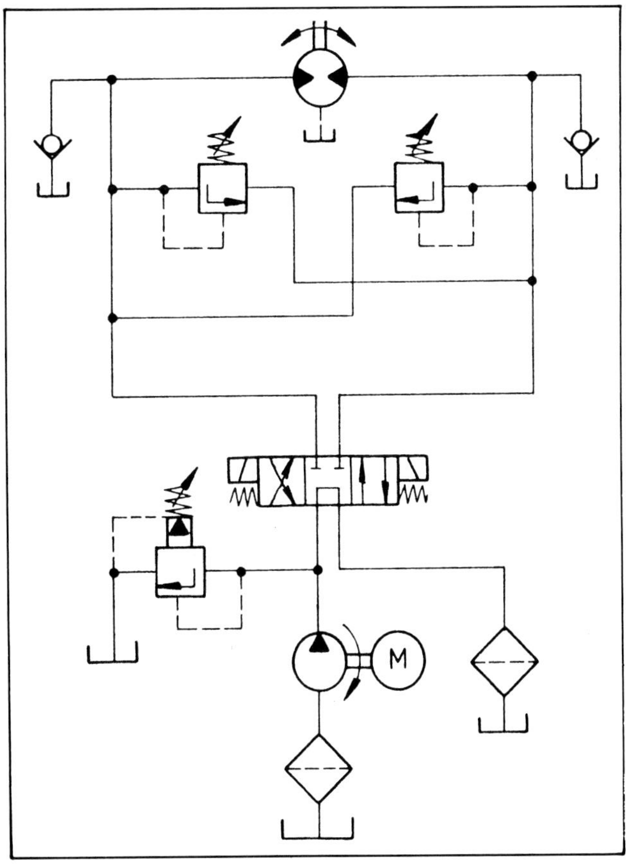

Fig. 146 *right:* Reversible motor drive with cross line relief valves.

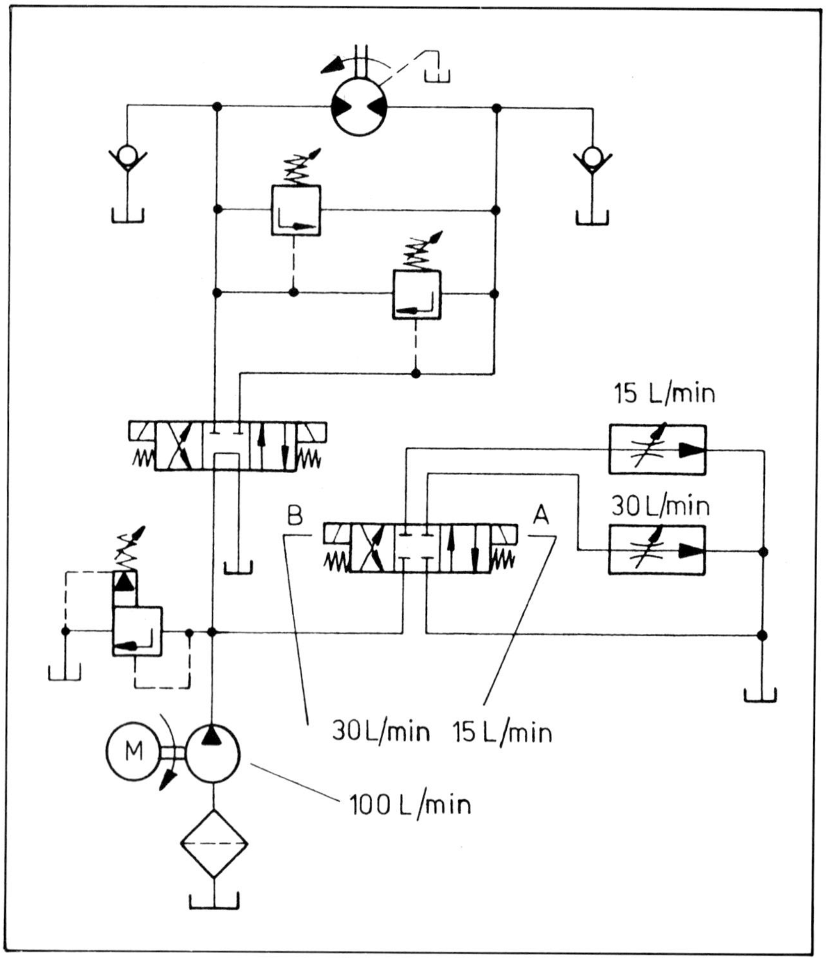

Fig. 147 Motor circuit with cross-line relief valves to prevent pressure shocks during motor reversal.

port. For reversible motors or motors with external drain, meter-out, meter-in, and bleed-off speed control is applicable.

A typical bleed-off speed control circuit with three selectable speeds is shown in fig. 147. Solenoid "B" selects speed no. 3 (85 L/min), solenoid "A" selects speed no. 2 (70 L/min), and with no solenoid actuation the full pump flow of 100 L/min goes to the motor, giving speed no. 1. The speed can thus be controlled for both directions of rotation.

The cross-line relief valves prevent pressure shocks during motor reversal and assist in the dynamic braking of loads. The port to port connection via the two cross-line relief valves provides fluid circulation to the low pressure port, and thus avoids cavitation. A typical meter-in speed control circuit is shown in fig. 148.

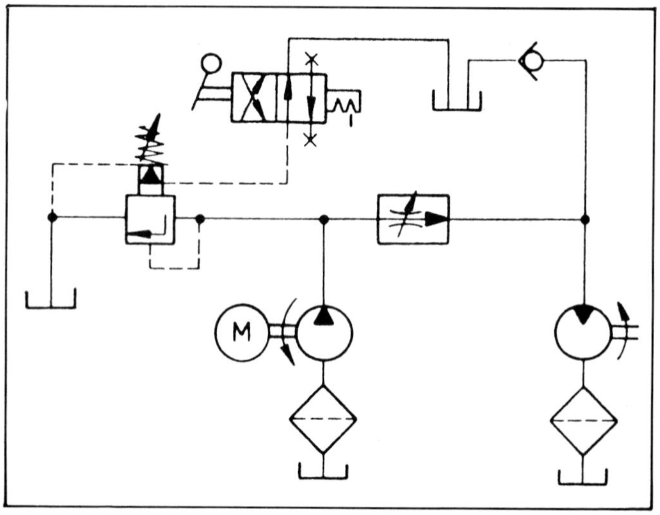

Fig. 148 Single direction motor drive with meter-in speed control and vented compound relief valve.

Motor torque control

Motor torque can be controlled and varied if the pressure in the motor circuit is variable. Figure 149 shows an open circuit drive with two selectable maximum torque levels. Maximum system pressure is limited by setting the system relief valve. The pressure reducing valves in the two alternative sub-pressure branch lines limit the pressure and thus the torque to the selected levels. The replenishing line maintains flow to the inlet port during braking.

Motor sizing

The application of the motor generally dictates minimum required power output and motor speed range, al-

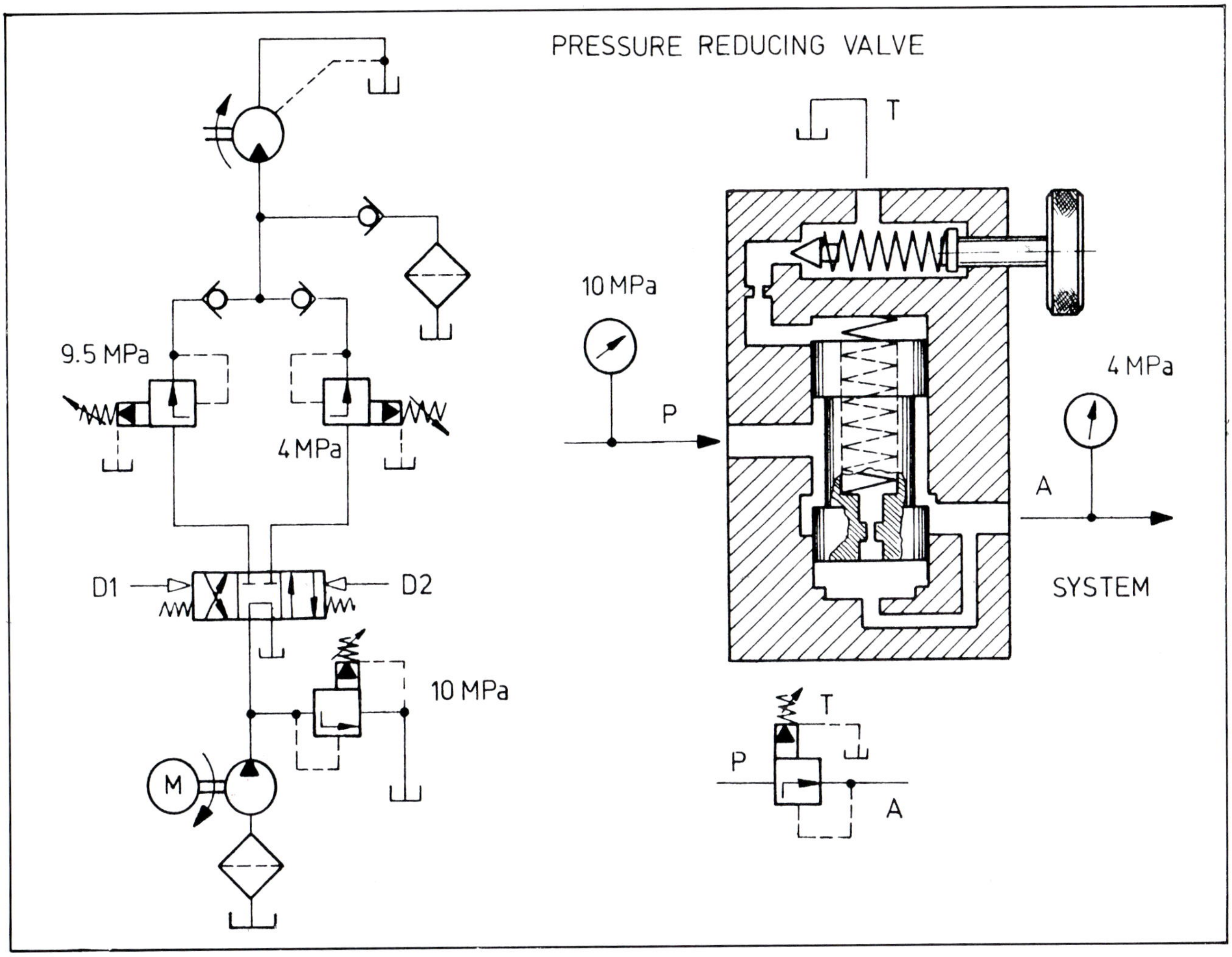

Fig. 149 Motor circuit with selectable torque control.

though for some applications both motor speed and torque may be varied to match prevailing conditions, while the power output is maintained. For such applications, the possible range of speed and torque, along with the desired maximum operating pressure, indicate that a wider range of motor sizes could be used.

Some factors for the sizing of a motor are:

- maximum torque required (Nm);
- maximum r.p.m. output (r.p.m.);
- maximum operating pressure (kPa);
- displacement volume per revolution (cm^3/rev. or mL/rev.);
- most efficient r.p.m.

A motor operating below its maximum rated capacity will provide a service life gain more than proportional to the loss in operating capacity.

Motor performance is separated into volumetric efficiency, mechanical efficiency, and overall efficiency, whereby the efficiencies are expressed as a percentage.

Volumetric efficiency is determined by the internal slippage of the motor, which means that some fluid "gets through" the motor without doing work. Volumetric losses increase with an increase in system pressure. But variations in the pump's revolutions have practically no effect on the internal slippage.

The formula for volumetric efficiency (η_v) is:

$$\eta_v = \frac{\text{Theoretical Flow Rate} \times 100}{\text{Actual Flow Rate} \times 1}$$

Mechanical efficiency (or hydro-mechanical efficiency) is determined by the pressure and the r.p.m. at which the motor is running. Most motors show high mechanical losses at high pressure and at low r.p.m.

The formula for mechanical efficiency (η_{hm}) is:

$$\eta_{hm} = \frac{\text{Actual Torque Output} \times 100}{\text{Theoretical Torque Output} \times 1}$$

Overall efficiency is used to calculate the power output of a hydraulic motor. It is expressed as the product of volumetric and mechanical efficiency:

$$\eta_o = \frac{\eta_v \times \eta_{hm}}{100}$$

General torque calculations

Torque may be defined as a twisting or turning moment (M) which is expressed in "Newton metres" (Nm), whereas rotary power is torque per unit of time expressed in Watts (W).

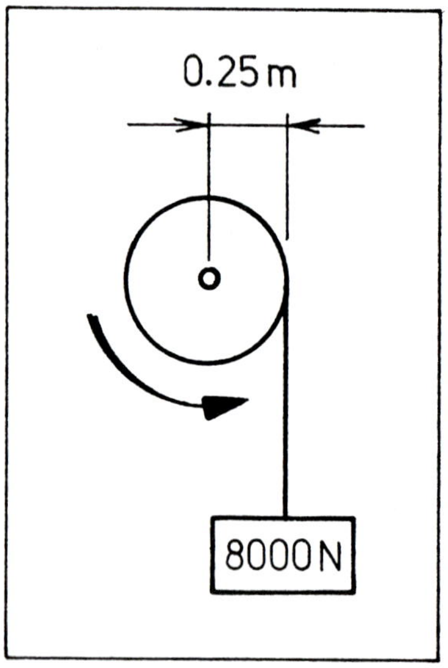

Fig. 150

Torque (M) = Force × Radial Distance

Torque = N × m

$$\text{Rotary Power (P)} = \frac{\text{Torque} \times \text{rev.}}{\text{sec.}}$$

$$= \frac{2 \times \pi \times n \times M}{60}$$

Example

Calculate the required motor torque to drive a cable drum with the following specifications:

Load mass = 800 kg → ≈ 8000 N
Cable drum dia. = 500 mm
1 kg × g = 9.81 N

$$\text{g} = \text{gravitational acceleration} = \frac{9.81 \text{ m}}{\text{s}^2}$$

Assume 100% efficiency

Solution

$$\frac{8000 \times 0.5}{10^3 \times 2} = 2 \text{ kNm}$$

Hydraulic motor calculations

Assuming that a motor has no elastic members and works without losses (efficiency = 100%), its displacement volume (V) per shaft revolution (n) is calculated as:

$$V = \frac{Q \times 60}{n}, \text{ or } V = A \times \pi \times d$$

where A is the total pressure exposed area per revolution and d is the mean diameter of the centre of this area.

The required flowrate (Q) per n revolutions per unit of time can be calculated as:

$$Q = \frac{V \times n}{60}, \text{ or } Q = V \times n, \text{ if revolutions given per second.}$$

The output torque (M) of the motor is proportional to its displacement volume (V) and the load induced pressure (Δp). This relationship can be developed from:

$$M = A \times \Delta p \times \frac{d}{2} \rightarrow \text{Torque} = \text{Force} \times \text{Radial Distance}$$

If the formula for displacement volume (V) is solved for area (A) and integrated into the formula for torque (M), then motor torque can be expressed as:

$$M = \frac{V \times \Delta p}{2 \times \pi} \rightarrow M = \frac{V \times \Delta p \times \cancel{d}}{\pi \times \cancel{d} \times 2}$$

The formula for motor revolutions (n) is derived by transposition from the formula of required flowrate (Q):

$$n = \frac{Q}{V} \text{ (gives n per second)}$$

Output power for hydraulic motors is calculated from the pressure and flowrate required to drive the load. Formulae given to calculate output power (P) (with n given in minutes) are:

$$P = \Delta p \times Q, \text{ or } P = \frac{2 \times \pi \times n \times M}{60}$$

In order to obtain precise and useful motor calculations, motor losses as described in "motor sizing" must be included in the calculations. The formula for required motor flowrate, including volumetric efficiency (η_v) is:

$$Q = \frac{V \times n \times 100}{60 \times \eta_v} \quad (\eta_v \text{ is expressed as a percentage})$$

The formula for produced output torque, including mechanical efficiency (η_{hm}) is:

$$M = \frac{V \times \Delta p \times \eta_{hm}}{2 \times \pi \times 100} \quad (\eta_{hm} \text{ is expressed as a percentage})$$

The formula for motor revolutions (speed), including volumetric efficiency (η_v) is:

$$n = \frac{Q \times \eta_v}{V \times 100} \quad (\eta_v \text{ is expressed as a percentage})$$

The formula for hydraulic motor output power, including overall efficiency (η_o) is:

$$P = \frac{2 \times \pi \times n \times M \times \eta_o}{60 \times 100} \quad (\eta_o \text{ is expressed as a percentage})$$

Nomenclature and corresponding units

A = area (m^2)
d = diameter (m)
F = force (N)
M = torque (Nm)
m = metre (m)
N = Newton (N)
n = revolutions per minute (r.p.m.)
P = power (Watt)
p = pressure (Pa) (bar)
Δp = pressure differential (Pa)
Q = flowrate (m^3/s)
r = radius (m)
s = second (s)
V = volume (m^3) (cm^3) (mL)
W = Watt (W)
η_v = volumetric efficiency (%)
η_{hm} = hydro mechanical efficiency (%)
η_o = overall total efficiency (%)
π = circle constant (3.1416)

Example

The winch depicted in fig. 150 is driven by a hydraulic motor with a direct drive (no gear box). Calculate the output power in kW, the required r.p.m. of the motor, and the flow requirement in L/min.

The load is to be lifted a distance of 157 metres in 30 seconds. The motor has a swept volume of 0.74 litres and operates at a pressure of 20 MPa. Efficiencies are: overall $\eta_o = 85\%$, volumetric $\eta_v = 90\%$, and winch = 100%.

$$\text{Required r.p.m.} = \frac{157 \times 60}{0.5 \times \pi \times 30} = 199.89 \text{ (use 200) r.p.m.}$$

$$\text{Flowrate (Q)} = \frac{V \times n \times 100}{60 \times \eta_v}$$

$$= \frac{0.74 \times 200 \times 100 \times 10^3 \times 60}{10^3 \times 60 \times 90}$$

$$= 164.4 \text{ L/min}$$

$$\text{Power output (P)} = \frac{2 \times \pi \times n \times M \times \eta_o}{60 \times 100}$$

$$= \frac{2 \times 3.1416 \times 200 \times 2 \times 10^3 \times 85}{60 \times 10^3 \times 100}$$

$$= 35.60 \text{ kW}$$

8 Accumulators

In hydraulic systems it is sometimes desirable to store hydraulic fluid under pressure, for release during peak demand. Unlike gases, such liquids cannot be compressed sufficiently to result in self-propelling release. The hydraulic accumulator solves this problem by storing the non-compressible fluid under external pressure.

Various means are applied to pressurise and drive the fluid from the accumulator into the hydraulic system and finally to the actuators. Weight loaded accumulators make use of gravitation, spring loaded accumulators use the elasticity of steel springs and gas charged accumulators use the compressibility of nitrogen gas to exert a force onto the hydraulic fluid (fig. 151).

Gas charged accumulators are more commonly used than spring or weight loaded accumulators, but the weight loaded accumulator has the advantage that the force exerted onto the fluid is always constant, no matter how full the fluid chamber is.

Many current hydraulic systems are equipped with one or more accumulators. In hydraulic systems, the storage of hydraulic fluid under pressure serves a number of purposes. The more common of these are:

- supplement pump delivery
- maintain system pressure
- emergency power source
- shock absorption
- noise elimination
- absorption of thermal expansion

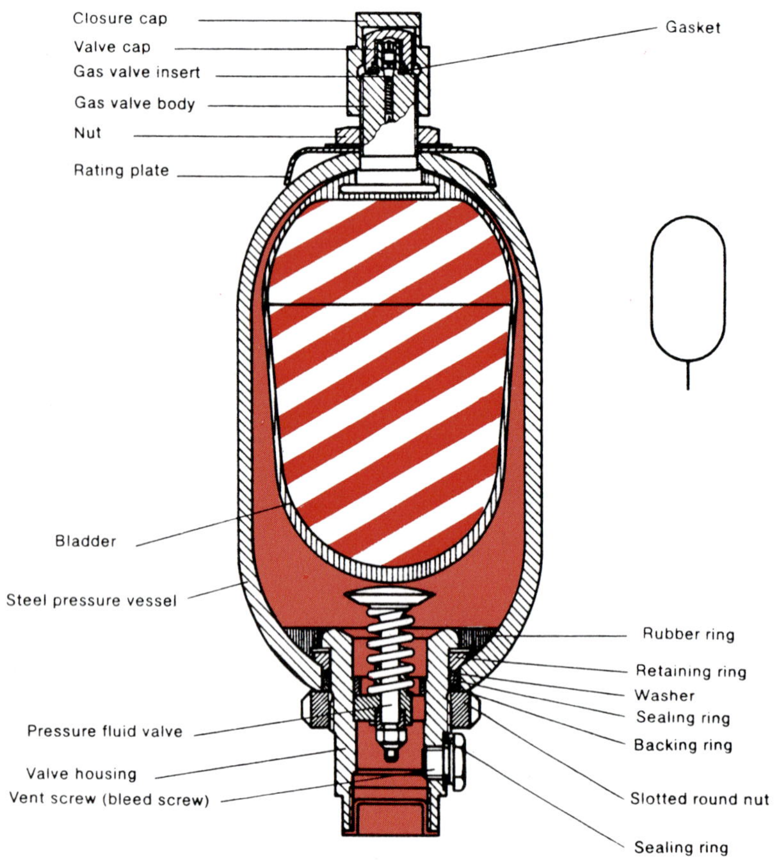

Bladder type accumulator (sectional drawing)

Fig. 151 Bladder type accumulator.

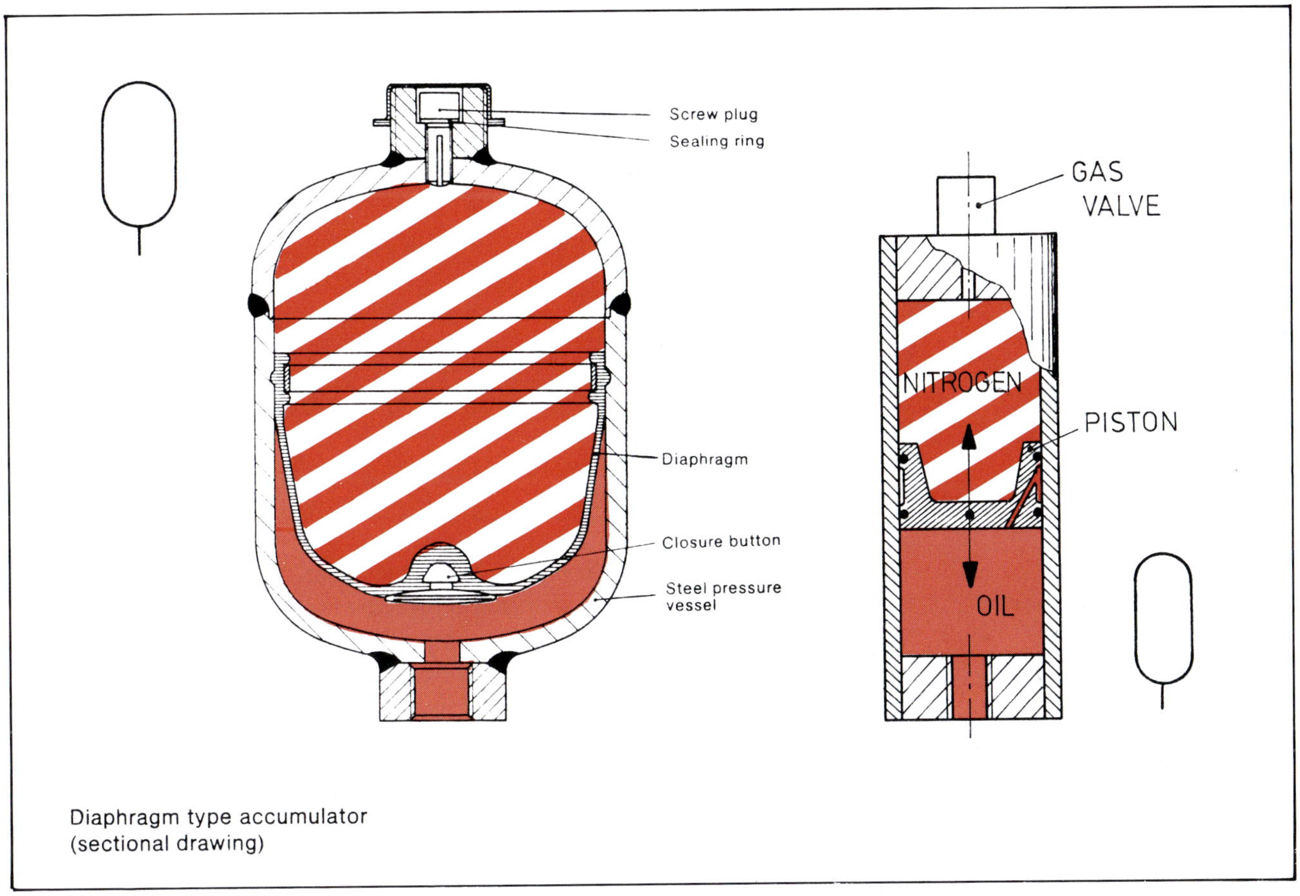

Fig. 152 Diaphragm type accumulator.

Gas charged accumulators

Bladder type accumulators are pre-charged with dry nitrogen gas. The pre-charge pressure will cause the bladder (bag) to fill the inside of the steel shell completely, and close the pressure fluid valve. As soon as the rising pressure within the hydraulic system reaches the gas pre-charge pressure p1 of the bladder, the fluid valve will open and the rising system pressure starts to force fluid into the accumulator. Further increases in system pressure will cause the bladder to be further compressed and even more hydraulic fluid is being forced into the steel shell.

Remember, that the bladder is filled with gas which is compressible, while the hydraulic fluid is not compressible. As the gas is compressed, the gas pressure will increase, equalling the system pressure. Thus, flow of fluid into the accumulator takes place only when the system pressure exceeds the gas pressure. Conversely, flow of fluid out of the accumulator takes place only when the system pressure sinks below the gas pressure. When the system flow demand is low or static the accumulator will be charged or filled. When the system demand is high or dynamic, the accumulator will release stored fluid to help the pump meet system flow demand.

Other gas-charged accumulators use a piston or a diaphragm to separate the gas from the hydraulic fluid (fig. 152).

Accumulator sizing

The nominal size of an accumulator must be carefully selected and calculated to ensure that a given volume (working volume) of hydraulic fluid can be absorbed or discharged by the accumulator. This transfer of fluid must take place within pre-determined pressure levels. Pressure p1 is assumed to be the gas pre-charge pressure discussed previously. Pressure p2 is the minimum system pressure for the safe functioning of the hydraulic system. Pressure p3 is the maximum system pressure controlled and limited by the system relief valve. All three pressure levels are shown in fig. 153.

To ensure that a minimum volume of hydraulic fluid remains in the shell, and to protect the gas bladder, the pre-charge pressure is normally kept somewhat below the minimum system pressure ($p1<p2$). The gas pre-charge pressure varies with the system application, but some basic guidelines are:

- supplement pump delivery (storage) — p1 is 90% of p2;

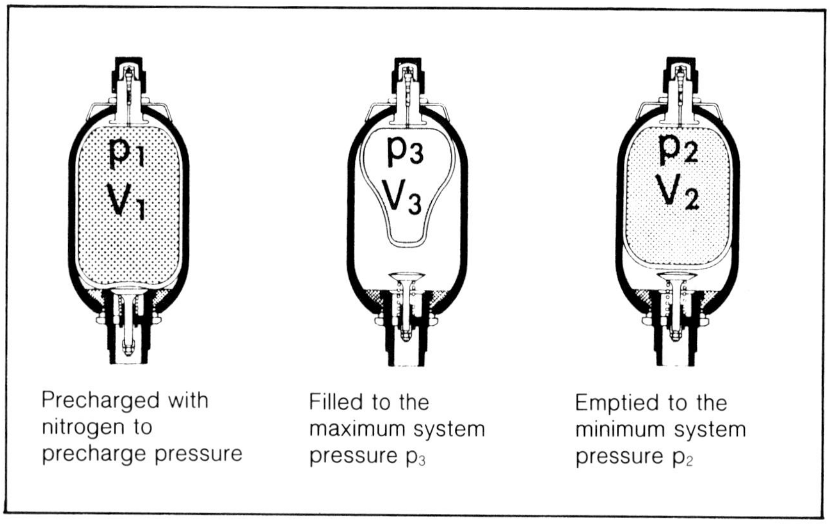

Fig. 153 Volume changes in a bag type accumulator.

- shock elimination — p1 is 60% of p2;
- emergency power source — p1 is equal p2;
- pulsation — p1 is 70% of p3.

Gas charged accumulators operate basically on the principle expressed by Boyle's Law (fig. 154):

$p1 \times V1 = p2 \times V2 = p3 \times V3$

This isothermal calculation, however, is only correct if the compression or expansion process of the gas within the bladder takes place slowly at constant gas temperature, allowing sufficient time for compression heat to be dissipated. The separator bag made of synthetic rubber with its high heat insulation characteristics inhibits rapid heat transfer. It is therefore advisable to

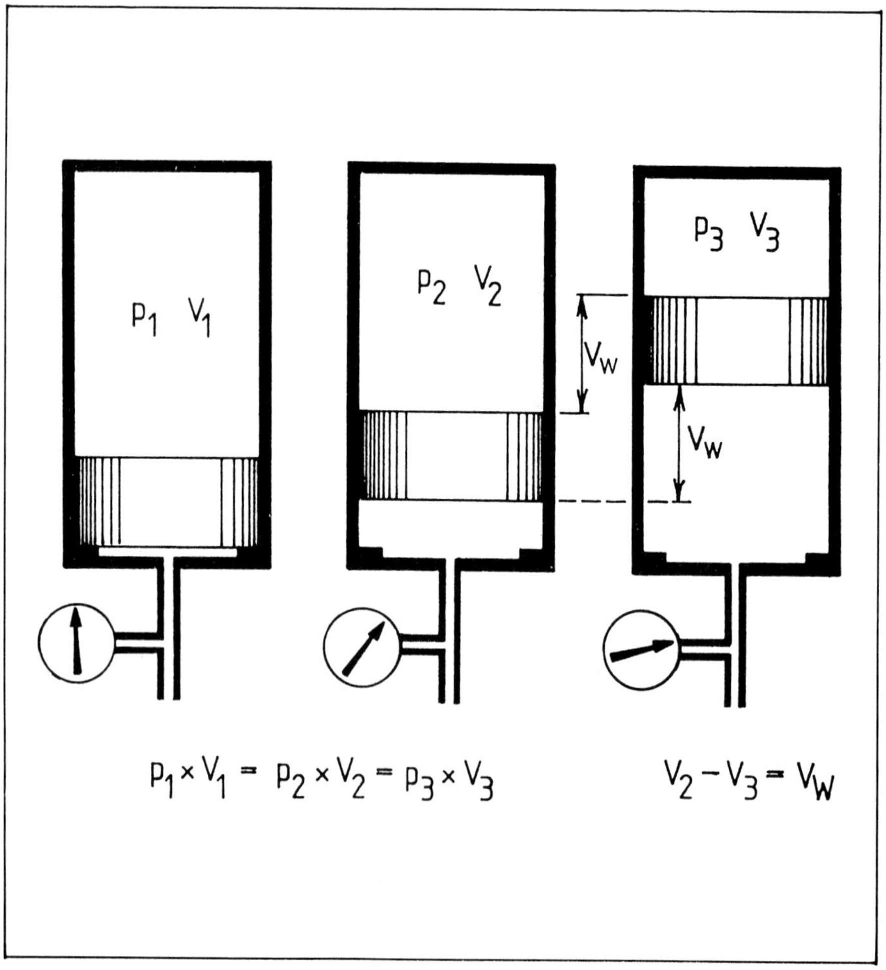

Fig. 154 Volume changes in a piston accumulator.

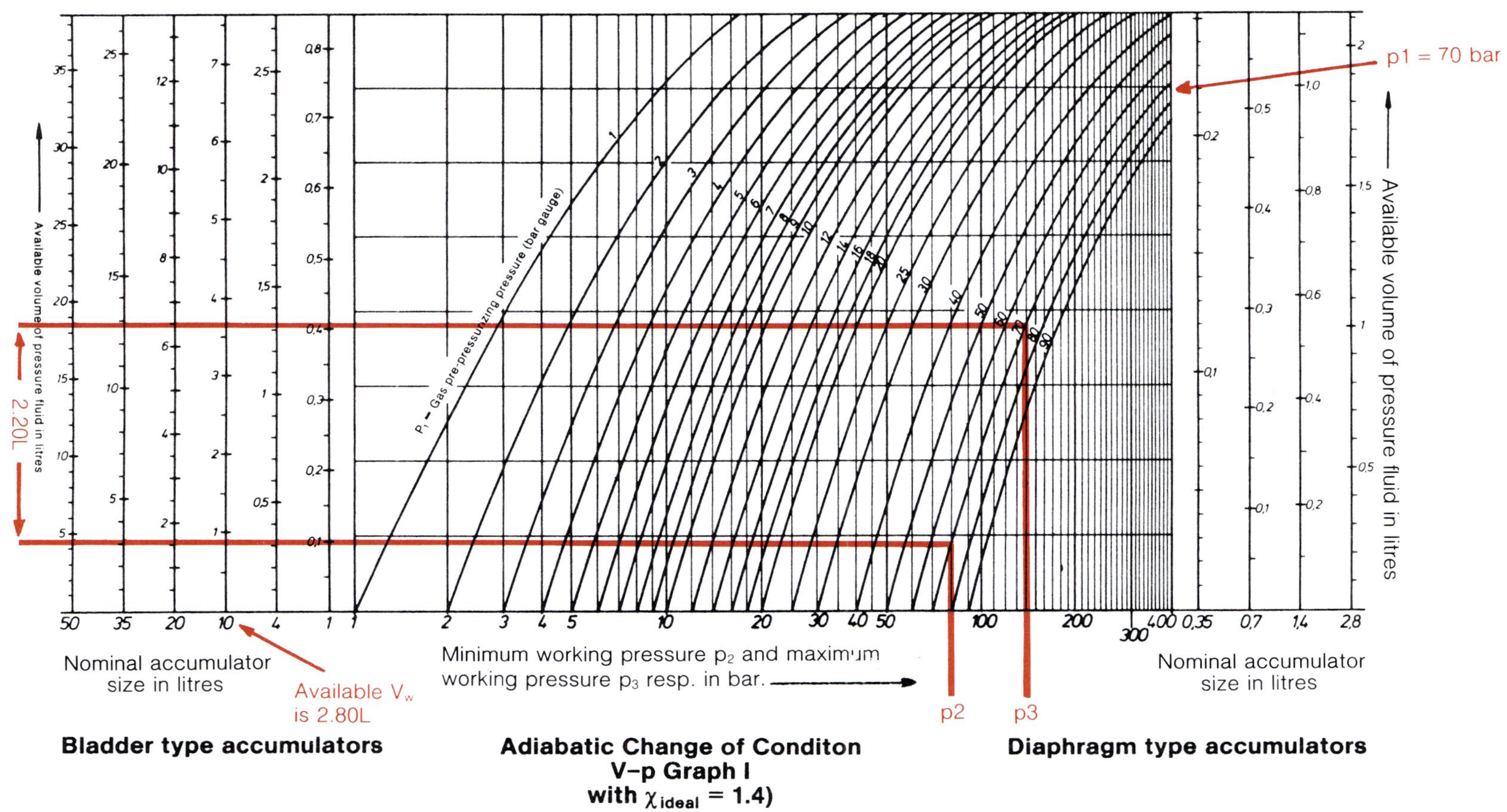

The values obtained from the graph represent average values, and they may deviate from the true value by ±4% in the case of bladder accumulators, and by ±1.5% in the case of diaphragm accumulators.

Fig. 155 Accumulator calculation, characteristic curves.

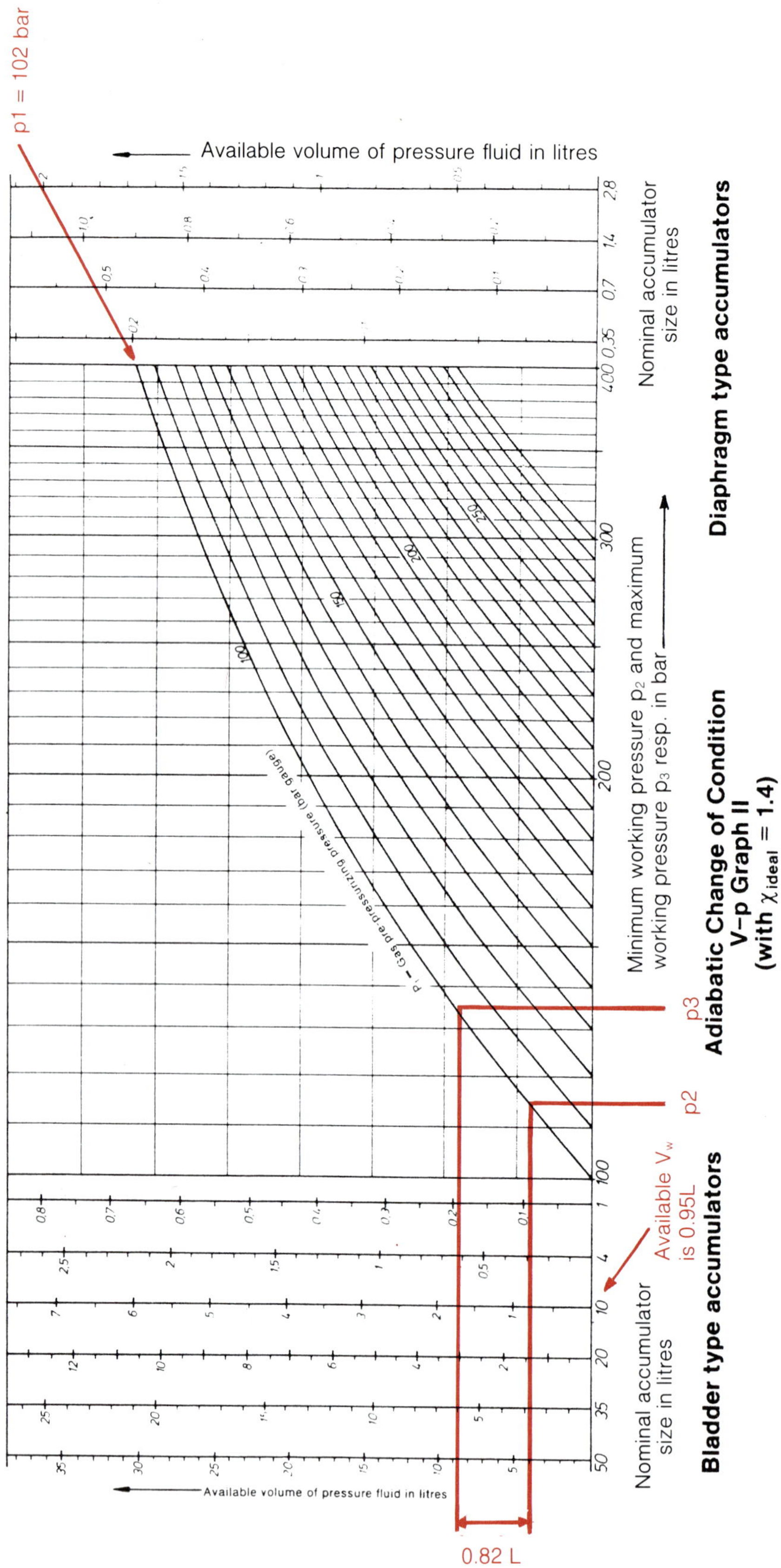

The values obtained from the graph represent average values, and they may deviate from the true value by ±4% in the case of bladder accumulators, and by ±1.5% in the case of diaphragm accumulators.

Fig. 156 Accumulator calculation, characteristic curves.

use adiabatic calculations or adiabatic power characteristic curves for the correct sizing of bladder type accumulators and piston type accumulators.

Accumulator calculations

Accumulator storage problems may be reduced to the simple question: what size accumulator will hold a predetermined quantity of fluid (V_w or working volume) between two selected pressure levels?

Example

Calculate the accumulator size (V1) if 4.1 litres of hydraulic fluid must be stored between a pressure level of 10.4 MPa (104 bar) and one of 20.7 MPa (207 bar). Use adiabatic, not isothermal pressure change for this calculation. Assuming that the conversion of gauge pressure to absolute pressure has a negligible effect on the calculated accumulator size, the following formula may be used:

$$V1=\frac{\left(\frac{p3}{p1}\right)^{\frac{1}{1.4}}\times V_w}{\left(\frac{p3}{p2}\right)^{\frac{1}{1.4}}-1}$$

This is derived from:

$$p1\times V_1^{1.4}=p2\times V_2^{1.4}=p3\times V_3^{1.4}$$

and

$$V_w=V_2-V_3$$

To ensure that a minimum of pressurised fluid remains in the shell the gas pre-charge pressure (p1) is set to 90% of the minimum system pressure (p2) (see guidelines, accumulator sizing).

$$p1=\frac{p2\times 90}{100}=9.36\text{ MPa}$$

p2 = 10.4 MPa

p3 = 20.7 MPa

$V_w=4.1$

$$V1=\frac{\left(\frac{20.7}{9.36}\right)^{\frac{1}{1.4}}\times 4.1}{\left(\frac{20.7}{10.4}\right)^{\frac{1}{1.4}}-1}=\frac{7.22760}{0.63503}=11.38\text{ L}$$

For ease of accumulator sizing most manufacturers provide sizing charts, graphs, or curves (figs. 155 and 156). These graphic aids are based on the adiabatic change of condition and are sufficiently accurate for most accumulator sizing problems.

Accumulator application

Two double acting hydraulic actuators and a hydraulic motor are required to operate in sequence as outlined in fig. 157. What size pump and accumulator are required?

Specifications

Maximum system pressure p3 = 14 MPa (140 bar)
Minimum system pressure p2 = 8 MPa (80 bar)
Gas pre-charge pressure p1 = 7 MPa (70 bar)
p3 = 85% of p4

Where unloading valves with an accumulator are used to control pump unloading and hold maximum system pressure (fig. 158), then the cut-in pressure of the unloading valve is taken for p3 and the cut out pressure is regarded as p4. The cut-in/cut-out difference varies from one manufacturer to another but is generally in the vicinity of 12 to 17%.

Solution

Let the pump supply the average flow demand and the accumulator the differential flow demand which exists during peak flow periods. The accumulator is being filled during low demand periods or at the end of the machine cycle.

1. With no accumulator in the system a pump with a flow delivery capacity of 5.40 L/sec. at approx. 8 MPa (80 bar) would be required (fig. 157).
2. With an accumulator in the system, and allowing 20% for system (internal) leakage, a pump of 1.46 L/sec. at approximately 16.5 MPa (165 bar) is required.

$$p4=\frac{14\text{ MPa}\times 100}{85}=16.5\text{ MPa}$$

It is evident from the flowrate diagram and the sequence-of-operation chart (fig. 157), that the average pump flowrate (allowing for 20% internal leakage) is:

$$\frac{\text{Total volume}\times 120\%}{\text{Total Time}}=\frac{7.95\times 1.20}{6.5}=1.467\text{ L/sec.}$$

3. This smaller pump will deliver the volumes listed in column 4 of the accumulator chart in fig. 157 for the corresponding time intervals of column 3.
4. If the accumulator stores X litres of hydraulic fluid at the beginning of sequence step 1 (see traverse-time diagram and flow rate diagram, fig. 157), then the accumulator will store the reduced cumulative volumes shown in column 6 at the end of each completed sequence step.
5. Therefore, if the stored volume X is greater than the minimum stored volume (X − 2.20 L), then the required differential volume (column 5) can be supplied by the accumulator to supplement pump flow.

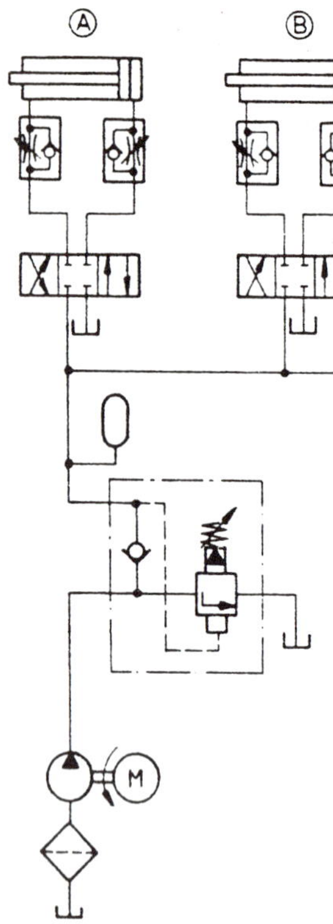

SEQUENCE OF OPERATION	VOLUME REQ'D	TIME	SUPPLY FROM PUMP	SUPPLY FROM ACCUMUL.	VOLUME LEFT IN ACCUMUL.	SUPPLY INTO ACCUMUL.
1. CYL. A (EXT)	1.96 L	1.0 s	1.46 L	0.50 L	x-0.50 L	0
2. CYL. A (RET)	1.19 L	1.0 s	1.46 L	0	x-0.23 L	0.27 L
3. CYL. B (EXT)	2.70 L	0.5 s	0.73 L	1.97 L	x-2.20 L	0
4. MOT. C (ON)	0.50 L	3.0 s	4.38 L	0	x+? L	3.88 L
5. CYL. B (RET)	1.60 L	1.0 s	1.46 L	0.14 L	x+? L	0
TOTAL VOL. →	7.95 L	6.5 s ← TOTAL TIME			V_W = 2.20 L	

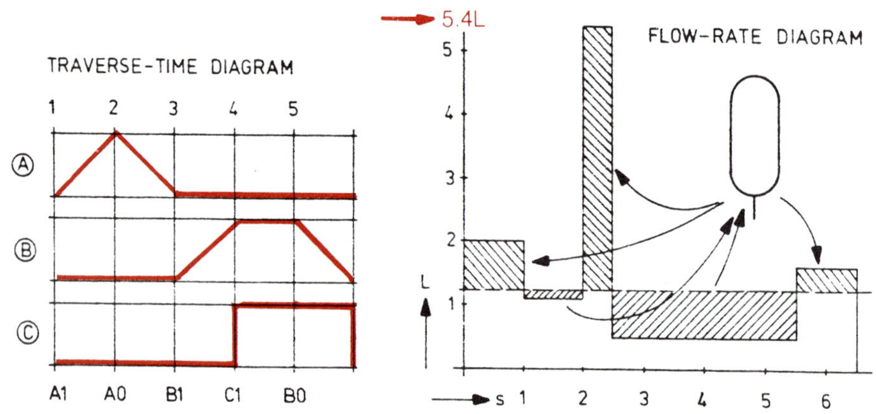

Fig. 157 Accumulator storage application.

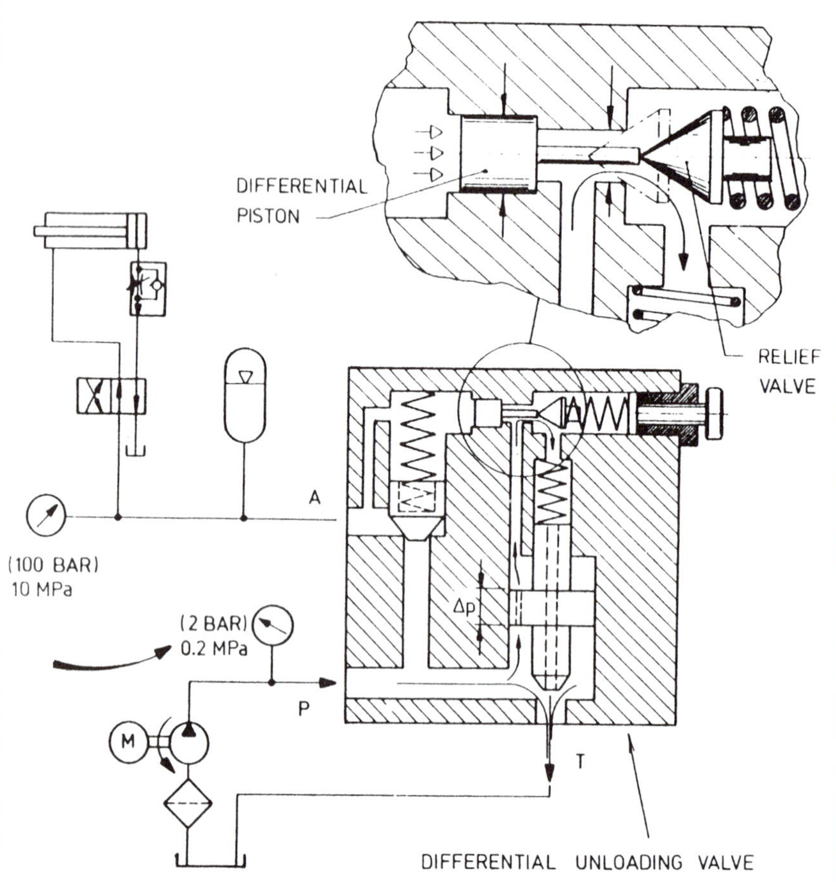

Fig. 158 Unloading valve in pump unloading position.

6. Since the pump delivers somewhat more flow than the average flowrate would require, excess flow will be absorbed by the accumulator and the unloading valve will unload (cut out) whenever pressure p4 is reached. Thus, excess pump flow during sequence step 4 will return to tank as soon as the accumulator is fully charged (pressure p4).
7. Because minimum guaranteed system pressure is 8 MPa (80 bar), an accumulator must be selected which will store at least 2.20 L (working volume) between the pressure levels of 14 MPa and 8 MPa.
8. During non-action or idle periods, the accumulator will maintain the maximum system pressure or cut-in pressure (p3) and will replace fluid lost by internal system leakage. For prolonged idle periods, the unloading valve may cut-in and cut-out several times to recharge the accumulator and replace fluid loss from leakage. Thus, maximum system pressure will fluctuate between p3 and p4.
9. From the accumulator sizing curves (figs. 155 and 156), the correct nominal size accumulator is selected. The pressure levels p2 and p3 are drawn into the graph to intersect with the gas pre-charge curve p1, which for this application is 70 bar (see specifications). The horizontal lines limit the required working volume (V_w) or storage volume between the pressures p2 and p3. Hence, a nominal size 10 L accumulator will provide at least 2.20 L (actual 2.80 L) working volume.

Measurement of gas pre-charge pressure

This may be accomplished by reading the system pressure off a pressure gauge.

Step 1: charge the system and the integrated accumulator with the pump to at least p2.

Step 2: switch off the pump and let the pressurised system fluid slowly drain out of the system back to tank.

Meanwhile, watch the pressure gauge, and when the system pressure has diminished to the pre-charge pressure (p1), the pressure fluid valve will close (fig. 151) and the pointer will drop sharply back to zero. The pressure reading on the gauge immediately prior to the sharp drop is the pre-charge pressure.

Gas pre-charge pressure can also be read off directly by means of a gauge permanently mounted in the gas filling port, or a permanently mounted gas pre-charging kit with a pressure gauge which is isolated during normal system operation.

Safety precautions

Gas charged accumulators must only be pressurised with dry nitrogen. Oxygen, for example, will cause a massive explosion if it comes in contact with hydraulic oil or grease!

Any work on hydraulic systems incorporating accumulators must only be carried out after release of the pressurised hydraulic fluid.

Gas bladder accumulators should normally be mounted vertically with the gas filling valve at the top.

Accumulators should be securely fastened to prevent them from being torn from their supports by recoil in the event of fracture in the piping system.

For maintenance work on accumulators, adhere strictly to the instructions provided by the manufacturer.

Accumulator circuits

An accumulator maintains system pressure while the unloading valve diverts pump flow back to tank. As soon as the system pressure diminishes below the cut-in pressure, the unloading valve will close and pump flow is directed into the system (see also figs. 104 and 105).

A	B	C	
0	0	0	PUMP FLOW TO TANK
0	0	1	ACCUMULATOR TO SYSTEM
0	1	0	PUMP TO SYSTEM
0	1	1	PUMP AND ACCUMULATOR
1	0	0	CHARGE ACCUMULATOR
1	0	1	———
1	1	0	———
1	1	1	———

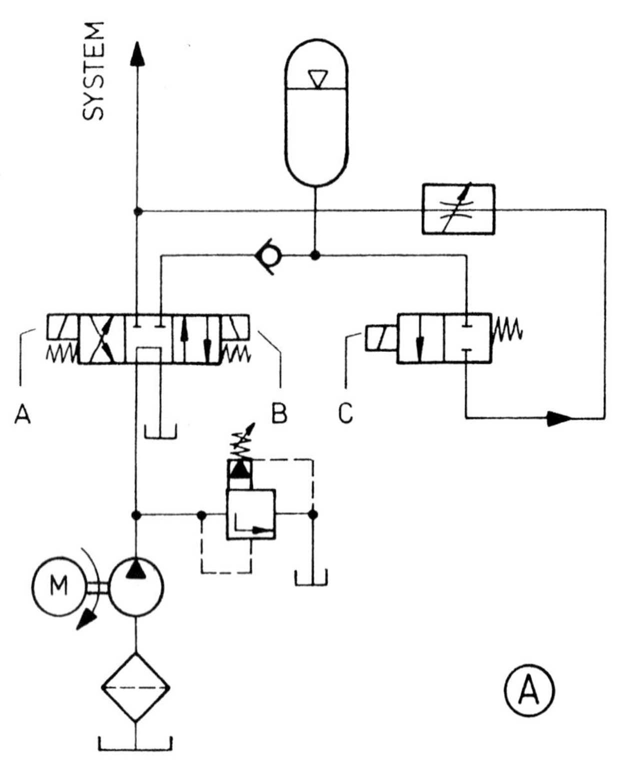

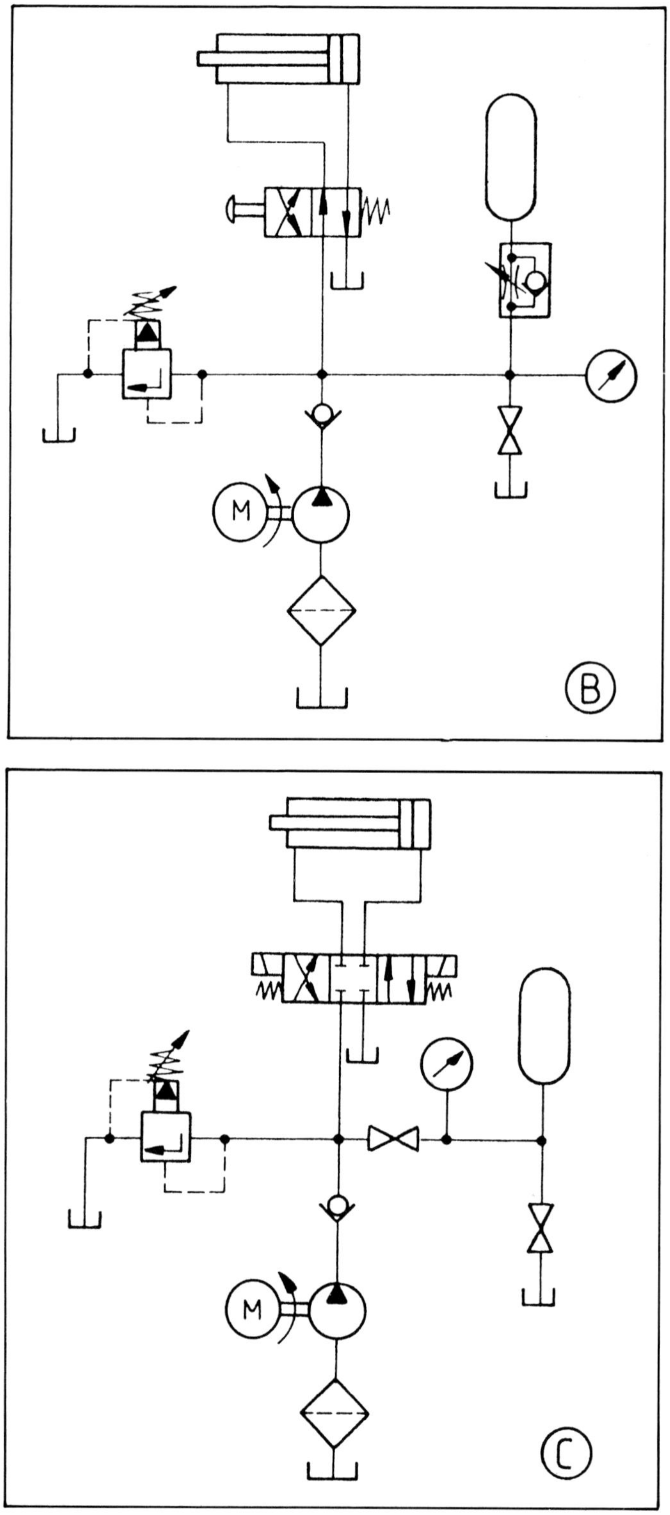

Fig. 159 (A) Typical accumulator control circuit with tandem centre pump offloading. (B) Simple pump flow storage circuit. (C) Typical pulsation elimination circuit.

9 Reservoirs

The hydraulic fluid reservoir, when correctly designed and constructed, has a considerable effect on the function and economical performance of the hydraulic system. The reservoir (also called tank) serves a number of important functions:

- It stores the fluid as it returns from the hydraulic system, and acts as a buffer for fluid fluctuations resulting from unequal flow displacement in the actuators.
- It dissipates fluid heat generated by power losses in the actuators and the control valves.
- It allows undispersed air (foam or bubbles) to separate out of the hydraulic fluid.
- It permits fluid contaminants to sink, and settle on the bottom of the reservoir, out of the fluid.

Reservoir construction

To perform these functions, certain design features are common to most reservoirs for industrial (stationary) applications. The reservoir is constructed of welded steel plate, with legs to raise the tank above ground level. This permits cooling by air circulation around all the walls and the bottom of the tank, to give optimum heat transfer.

For stationary applications, the reservoir is usually also designed to serve as a mounting platform for the pump, the motor, and related pressure and directional control valves. This demands that the reservoir construction be quite rigid and strong with a flat top.

The bottom of the reservoir is sloped or dished to drain the fluid to the drain plug (figs. 160 and 161). Large clean-out plates or lids are fitted on the side of the tank to permit access for cleaning and maintenance.

Some smaller reservoirs are made of aluminium castings with cooling fins. Such reservoirs are marketed as complete power units including the prime mover, the pump, and a system relief valve. Most modern small power units (up to 150 L) have the pump mounted vertically on the end of a bell-housing, immersed in the hydraulic fluid. A coupling permits the removal of the electric motor without having to disturb the pump. Also mounted on the top plate, for easy access and cleaning, is the return line filter (fig. 162).

Overhead reservoir (header tank)

Some press circuits may require a considerable quantity of fluid stored high in the machine structure. For example, with high-low circuits the high speed approach stroke of the main ram is gravity supplied from the overhead reservoir via a pre-fill valve, while the kicker cylinders move the platen into the work contact position (fig. 212). This type of reservoir, with its associated fluid lines, requires careful design to avoid undesired turbulence from the high fluid velocities during fluid transfer to and from the reservoir.

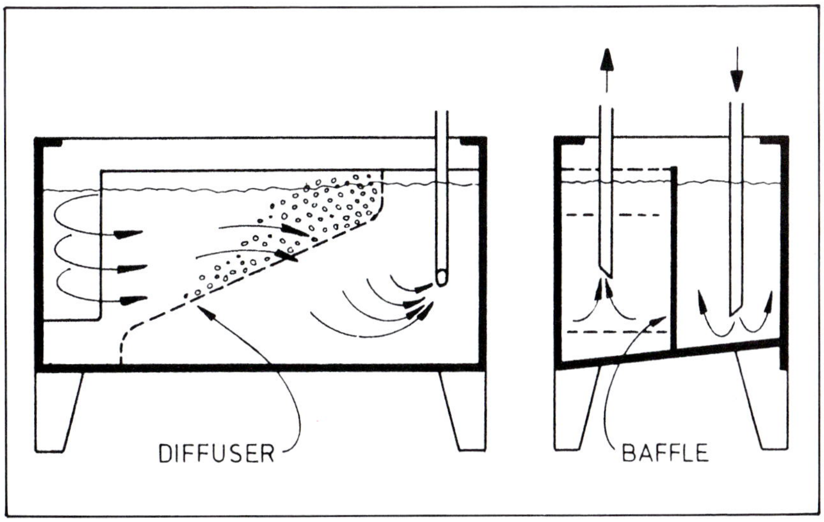

Fig. 160 Reservoir with baffle and air diffuser. Return line and suction line are separated by the baffle.

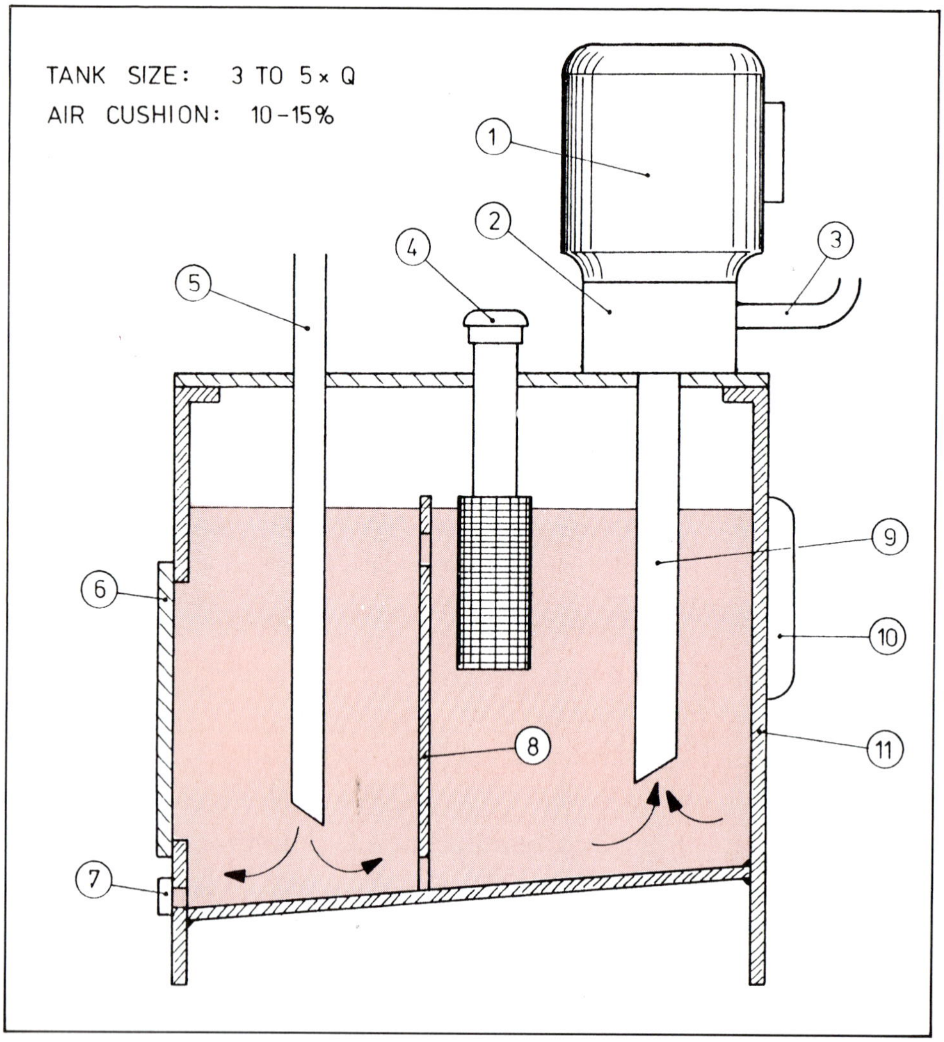

Fig. 161 Hydraulic power unit.

(1) Electric motor
(2) Pump
(3) Line to system
(4) Breather-filler
(5) Return line
(6) Clean out lid
(7) Drain plug
(8) Baffle plate
(9) Suction line
(10) Fluid level indicator
(11) Tank (container)

Elimination of contaminants

Even with adequate filtration, dirt particles such as fibres, abrasions, oxide scale, plastic elastomers, air-borne dirt, and silica sand can accumulate in the system. The particles which are not filtered out should be allowed to settle on the reservoir bottom. A baffle plate, which is a metal divider extending lengthwise through the middle of the reservoir (fig. 160), is used to circulate the fluid returning from the system, and to separate the pump suction line from the return line. The slow motion of the circulating fluid allows the heavier dirt particles to settle out.

Elimination of air

Air bubbles, sometimes created by entrained air or by the intermittent flow of low pressure return lines and drains, will have a chance to work their way to the fluid surface in the reservoir. This process is further stimulated by an air diffuser (a perforated plate or a sieve, fig. 160), and by the fluid circulation caused by the baffle.

Breather-filler

The fluctuating fluid volume in the reservoir (caused by variations in the flow demand and flow return by the actuators) produces a constant airflow in and out of the reservoir. This airflow passes through a specially designed breather-filler, which is built for a threefold purpose. It filters the air flowing into the reservoir, it is used as a strainer when the reservoir is replenished, and it provides a passage for constant air exchange to and from the reservoir. It is important to use a breather large enough to cope with the airflow caused by the fluid fluctuations, as the air volume in the reservoir must always be at atmospheric pressure to prime the pump.

In order to ensure that the reservoir is filled only with absolutely clean fluid, an increasing number of systems are fitted with a separate filling point in the fluid return line, and the breather is sealed. Thus, fluid is pumped (by hand, or power pump) through the return line filter into the reservoir. A check valve above the filter prevents fluid from going up the return line (fig. 163).

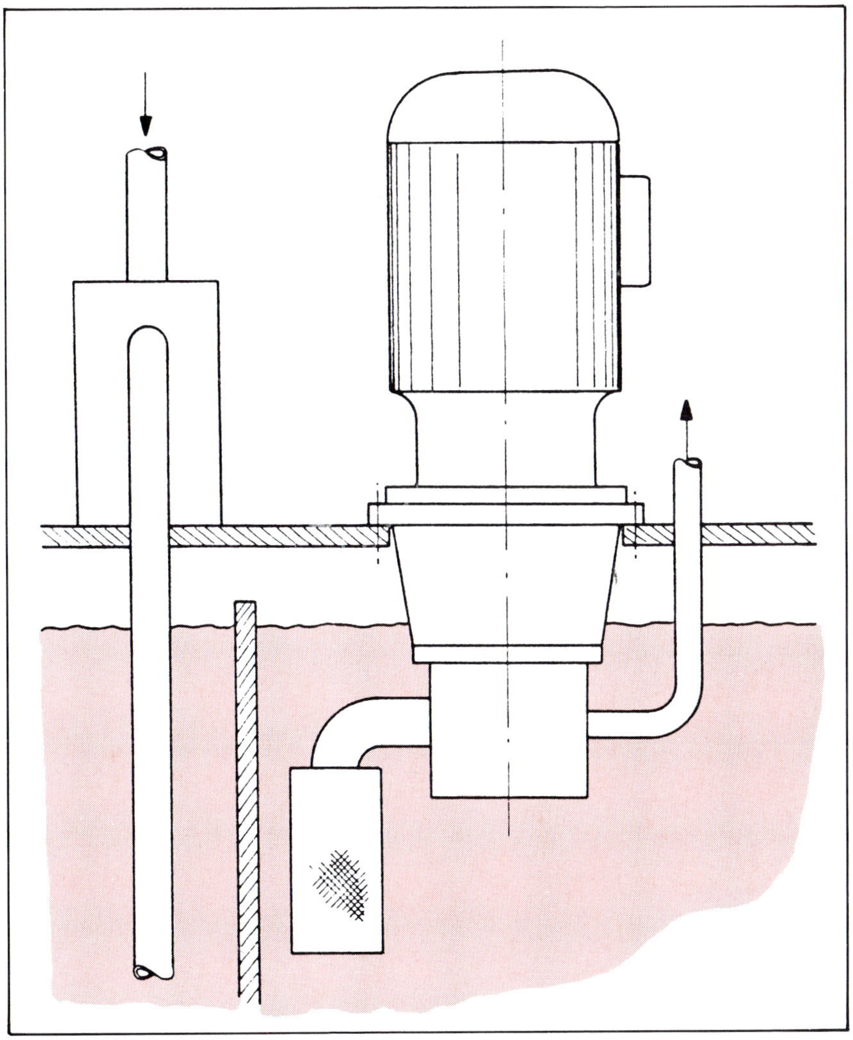

Fig. 162 Immersed pump with suction strainer.

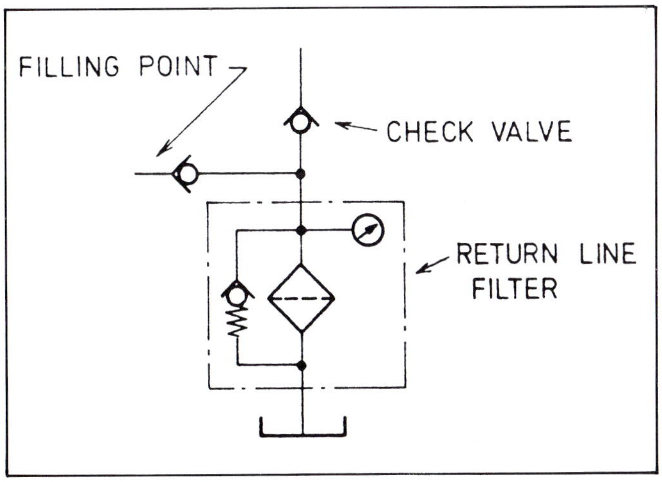

Fig. 163 Reservoir with filling point and filler line through the return line filter.

Elimination of fluid heat

Power losses in the hydraulic system are converted into heat. The bulk of this heat is absorbed by the hydraulic fluid, the control elements, the actuators, and the reservoir. The heat level in the system will therefore increase, until heat creation and heat dissipation (or heat transfer) are in balance. The temperature at this point is called the inertia temperature.

To operate the hydraulic system without separate cooling, this inertia temperature must be equal, or preferably less, than the maximum tolerable system temperature. The amount of heat that can be dissipated by the reservoir depends directly on:

- the size of the outside surface of the tank through which heat transfer can take place;
- the amount of hydraulic fluid in the reservoir and therefore its flowrate through the tank;
- the difference between hydraulic fluid temperature and the ambient air temperature around the reservoir;
- the air circulation around the hot reservoir.

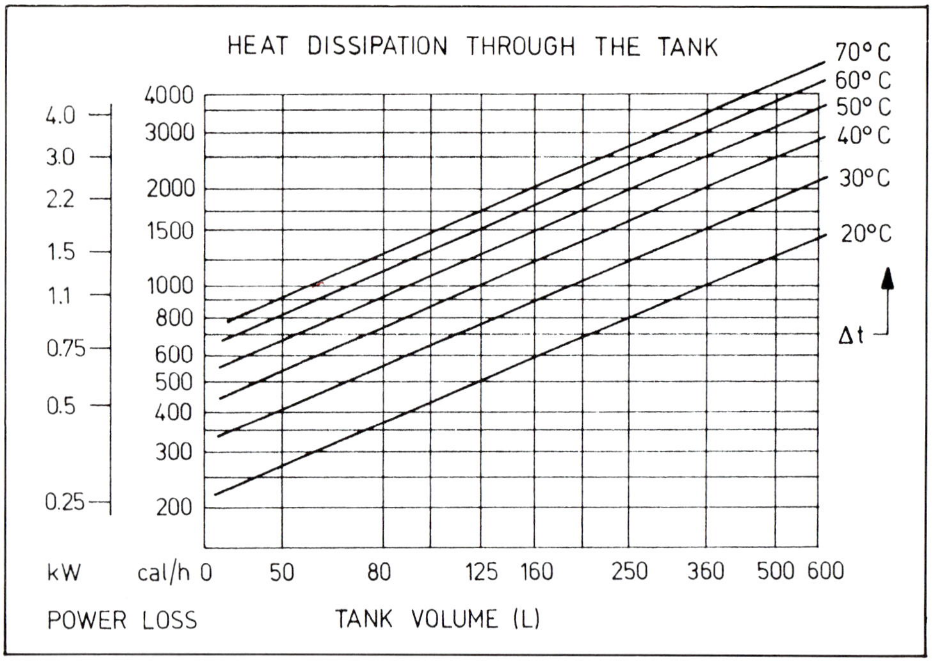

Fig. 164 Reservoir sizing chart.

Total power loss in a hydraulic system determines the amount of cooling required. These power losses (PL) may be calculated as:

$$PL_{TOTAL} = PL_{PUMP} + PL_{VALVES} + PL_{LINES} + PL_{ACTUATORS} \text{ (kW)}$$

For an average hydraulic system, these power losses may range from 20–30% of the output power of the prime mover driving the pump. For ease of calculation, a tank sizing chart is given in fig. 164. The power loss (PL) is given in kW, the temperature difference Δt in °C, and an average cooling capacity in a well-ventilated installation is assumed. The Δt should be taken as given for the hottest summer conditions.

In industrial (stationary) systems it is customary to use a reservoir which holds approximately 3 to 5 times the flowrate of the pump in L/min. Thus, a flowrate of 60 L/min would demand a reservoir of 180 to 300 litres. The biggest size would be necessary if the Δt is small, and the air circulation around the reservoir is minimal. The air space (cushion) above the fluid should be about 10–15% of the total reservoir capacity (fig. 161).

The sizing chart shows that with an improved Δt of 10°C (by natural or artificial air circulation) the tank volume for a power loss of 2.3 kW and an initial Δt of 40°C could be reduced by approximately 100 litres. Two important considerations in reservoir design are:

(1) that the fluid level never drops to the vicinity of the pump inlet pipe port (which could cause whirlpool action and draw air into the pump, see "entrained air"), and
(2) that the air space above the fluid level is adequate to allow air dissipation from the fluid.

Sealed reservoir

In some industrial applications the reservoir is completely sealed, and air can only enter through a stem into a separator bag. This is to ensure that the air is at no time in direct contact with the hydraulic fluid. If no moisture, corrosive gases, or contaminant particles enter the reservoir, the result is cleaner fluid, longer fluid life, reduced corrosion, and longer component life (fig. 165).

Instead of having an open passage from the bag to the outside of the reservoir, the bag can also be sealed and pressurised. Thus, the pre-pressurised fluid will charge the pump. This avoids cavitation and entrained air with its adverse side effects (see pumps, Chapter 3).

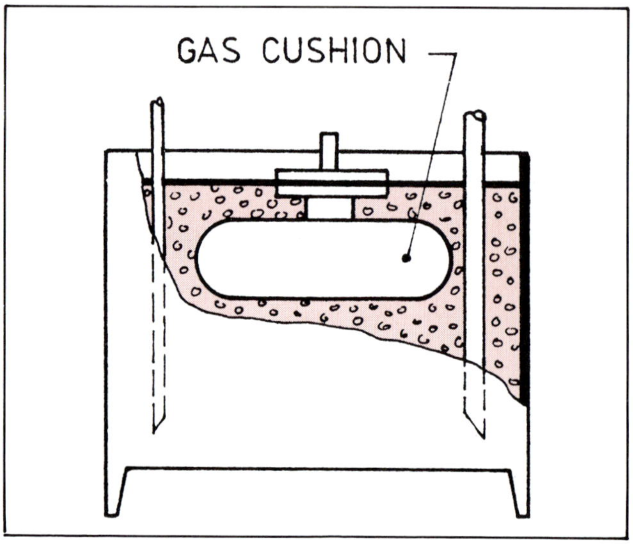

Fig. 165 Sealed reservoir with separator bag.

10 Fluid conductors and plumbing

Fluid tubes, pipes, and hoses interconnect the various hydraulic components and conduct the fluid around the system. These conductor lines, including the fittings, must be capable of withstanding not only the calculated maximum system pressure, but also the pressure shocks created within the system. The choice of conductor (tube, pipe, or hose) or fitting depends primarily on the following factors:

- static and dynamic pressure
- flowrate
- compatibility with the fluid
- serviceability
- vibration
- leak tightness
- environment
- application
- cost

Fluid conductors must have a cross-sectional area large enough to handle the required flowrate without producing excessive pressure losses. Seamless steel tubing is primarily used for rigid and semi-rigid conductors. Flexible hoses are used where the fluid is to be connected to moving machine parts (machine tools, cranes, mobile, and agricultural applications), or where vibration could cause leaking in the plumbing system.

Flow velocity in conductors

The quantity of fluid that must pass through the conductor in a given period of time (flowrate), is an important factor in selecting the inside diameter of a pipe, a tube, or a flexible hose. The required cross-sectional area is calculated with the formula:

$$A = \frac{Q}{v}, \text{ or } d = \sqrt{\frac{\text{Flow Rate}}{\text{Flow Velocity} \times \frac{\pi}{4}}}$$

$$\text{or } d = \sqrt{\frac{Q}{v \times 0.7854}}$$

where
Q = flowrate
A = cross sectional area
v = fluid velocity

Basically, line size depends on two factors:

1. The wall thickness of the pipe, which must be strong enough to convey the pressurised fluid even under extreme surge pressure peaks (shocks) (fig. 166).
2. The cross-sectional area, which must be large enough to prevent undue pressure drops. Such pressure losses would reduce transmitted energy and produce excessive fluid heat.

Outside diameter (mm)	Wall thickness (mm)	Calculated max. pressure (bar)
4	1	600
5	1	400
6	1	300
6	2	1200
8	1	228
8	2	686
10	1	172
10	2	458
12	1	137
12	2	343
14	1	128
14	2	309
15	1,5	192
15	2,5	365
16	1,5	177
16	2,5	331
18	1,5	154
18	3	365
20	2	193
20	3	313
22	1,5	122
22	3	273
25	2	147
25	3	230
28	1,5	92
28	3	199
30	2,5	119
30	4	265
35	2	100
35	4	216
38	3	136
38	5	261
42	2	81

Fig. 166

Example

Assume a required flow velocity of 2 m/s and a required flowrate of 3 L/s. Calculate the internal pipe diameter.

$$d = \sqrt{\frac{Q}{v \times 0.7854}} = \sqrt{\frac{3}{2 \times 0.7854 \times 10^3}}$$

$$= 0.044 \text{ m, or } 44 \text{ mm}$$

The flowrate nomogram (fig. 167) may also be used to find the flowrate (Q), the pipe inside diameter (mm), and the fluid velocity (m/s). To prevent turbulent flow in pressure lines, high back pressure in return or tank lines, and cavitation caused by too low a pressure at the pump inlet (fig. 168), the following figures for flow velocity should not be exceeded:

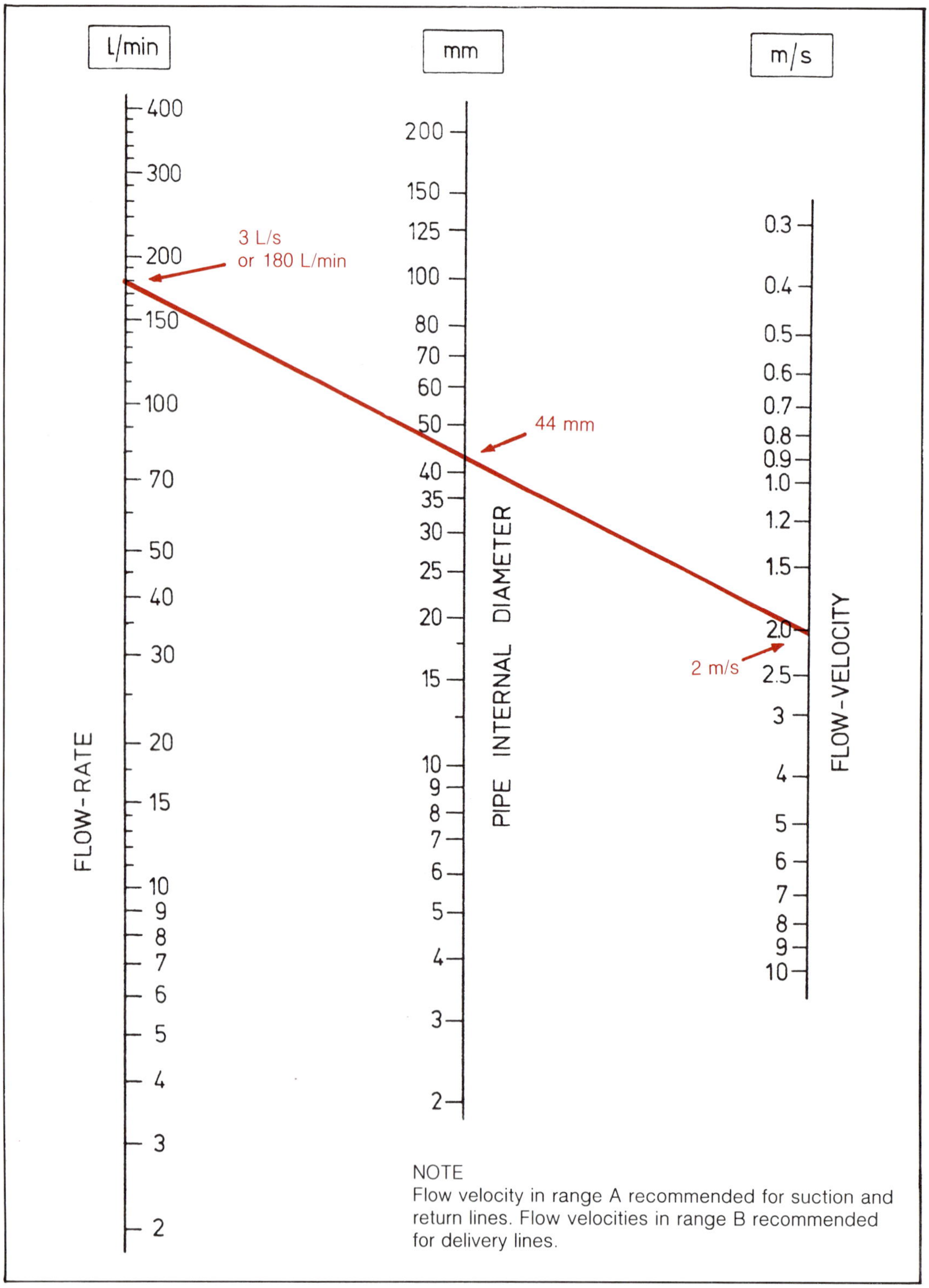

Fig. 167 Flowrate nomogram.

Fig. 168 Line designations.

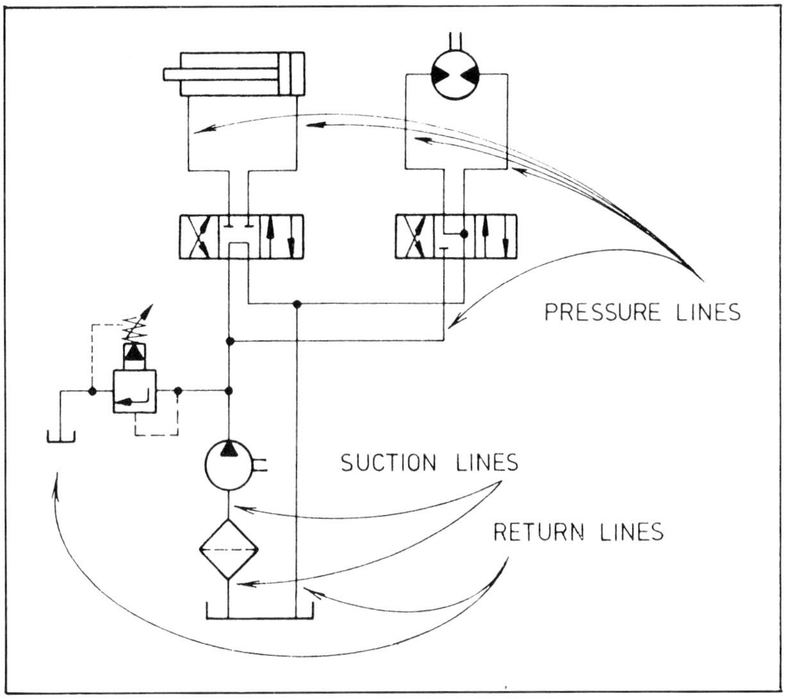

Suction lines (pump inlet):			0.5–1.5 m/s
Return lines (to tank):			2.0–3.0 m/s
Pressure lines:	p	< 5 MPa	4.0 m/s
	p=	5–10 MPa	4.0–5.0 m/s
	p=	10–20 MPa	5.0–6.0 m/s
	p	>20 MPa	7.0 m/s

Flexible hose

Flexible hose as used for hydraulic systems, is manufactured in layers of elastomeric materials, fibres, and braided wire or fabric. Hoses are available in various sizes and pressure ratings, and the innermost layer of the hose must be compatible with the hydraulic fluid in use.

Hydraulic hose is widely used, because of its ease in installation and its characteristics of absorbing pressure shocks and machine vibration. Hose installation requires less skill than pipe plumbing. The cost of the hose however, is far higher than that of rigid pipe installations.

Hydraulic hose for industrial and mobile use is manufactured to conform to S.A.E. specifications. The two types of hydraulic hose most commonly used are single wire braid (S.A.E. 100 R1) and double wire braid (S.A.E. 100 R2), which differ in their pressure ratings. Each type is made in two versions with different thicknesses of the outer elastomeric layer. They are:

- skive type, with a thick outer layer which must be skived to expose the wire braid prior to attaching the end fitting, e.g. S.A.E. 100 R2A; and
- non-skive type, with a thin outer layer which is not removed prior to attaching the end fitting, e.g. S.A.E. 100 R1AT.

End fittings are manufactured to suit the standard thread forms and flanges used throughout the world. They are designed for a particular hose designation and are not interchangeable for different types of hose. For example, a fitting for a ½" S.A.E. 100 R2A hose cannot be used on a ½" S.A.E. 100 R2AT hose.

They are further classified as:

- re-usable (subject to serviceability); and
- permanently attached, which is further divided into
  - — crimp type, where a tubular sheet steel ferrule fixed to the insert is deformed to crimp the hose between the ferrule and the insert; and
  - — pallet-swage type, where a separate, machined ferrule is swaged onto the hose end to grip the hose between the ferrule and the insert.

Provided that the fittings are correctly attached, it is generally accepted and proven by test results that re-usable as well as permanently attached hose end fittings will outlast the service life of the hose.

Hose selection and rating

Since peak or surge pressure spikes often reach 200 per cent or more of the calculated and adjusted maximum system pressure, hoses and pipes should be selected to withstand such pressure peaks. Three pressure ratings are usually given by the manufacturer:

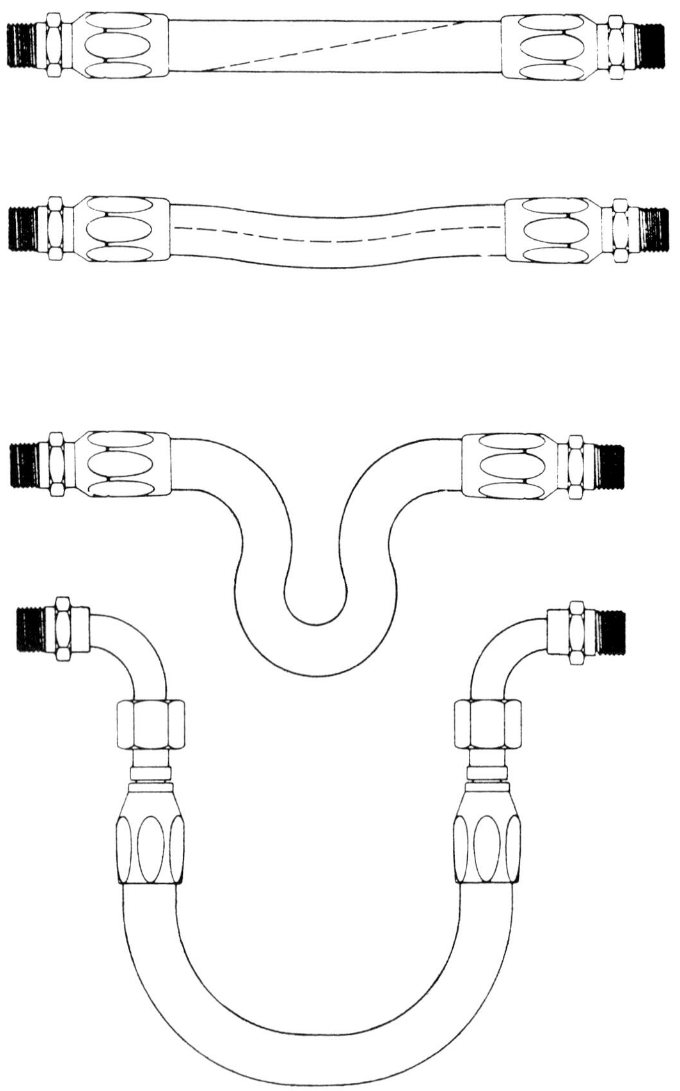

Fig. 169 Important hose installation guidelines.

- The recommended maximum system pressure (working pressure) at which the hose can be operated continually.
- The test pressure at which a hose is guaranteed to withstand pressure peaks.
- The burst pressure at which the hose will rupture.

The S.A.E. recommended working pressure for hydraulic hose is 25 per cent (or one quarter) of the minimum specified burst pressure. This safety margin allows for the absorption of pressure surges. For internal diameter calculations, use the nomogam (fig. 167) or the formula given for pipe sizing.

Hose and pipe installation

Tubing must be properly supported and securely clamped, to avoid vibration and movement, threading swarf and cutting burrs must be carefully removed to stop them from breaking loose and being flushed into the hydraulic system.

Although hoses are covered with a tough outer layer to reinforce and protect the inner layers, this outer layer should not be abraded or damaged. Hose must bend in the same plane as the connection to which it is attached, to prevent twisting. The hose appearance does not show whether the recommended minimum bend radius has been exceeded; the hose reinforcement determines this radius. It is best to check the manufacturer's specifications prior to installation (fig. 169).

Fittings

Hydraulic systems are seldom fitted entirely with lines of the same diameter. Often, pipes or hoses must step up or down from one diameter to the other, and port threads may differ between components which are to

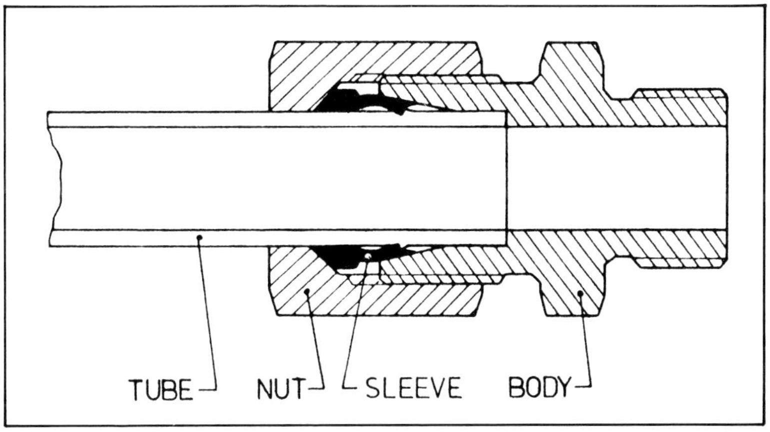

Fig. 170 Sleeve or ferrule type compression fitting.

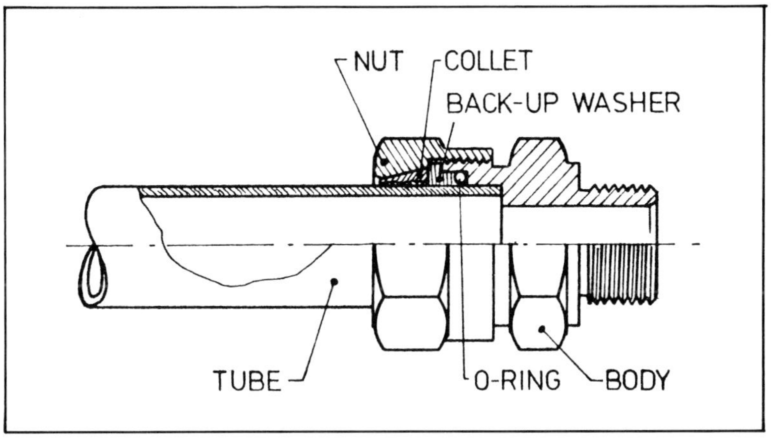

Fig. 171 O-ring seal fitting.

be connected to each other. Therefore, line size, thread type, and fitting material are only a few of the many variables encountered in system plumbing.

Fittings consist of two essentially separate components that provide mechanical and hydraulic linkage for a particular line size and component port. Hence, several basic requirements must be considered when selecting a hydraulic fitting:

- The mechanical bond between the fitting and the component body must remain firm, even under vibration, outside abuse, and temperature change.
- The hydraulic linkage between the pipe or hose, the fitting, and finally the component body (pump, valve, or actuator) must remain leak tight, even under vibration and pressure shocks.
- The ease with which lines can be connected to the components and the degree of fitting re-usability for maintenance and system re-design.

Flare fittings are still widely used, but have lost some ground to compression fittings and O-ring seal fittings. In flare fittings leak tightness is achieved by squeezing the flared end of the pipe against a sealing cone as the compression ring or nut is tightened.

Compression (sleeve) fittings seal by wedging the sleeve onto the pipe and forcing it partly into the pipe surface, so that a slight deformation occurs (fig. 170). Squareness of the pipe end is not as critical as with flare fittings, and absolute tightness can be guaranteed if tightening procedures and specifications are adhered to.

O-ring seal fittings hold the tube by means of a sleeve wedged onto the tube by the nut (fig. 171). The O-ring provides an absolute seal against leakage. Special O-rings must be used for non-flammable synthetic hydraulic fluids.

Welded connections may be used for high pressure systems. A flange is usually welded onto the pipe and mounted to the component. The mating surfaces must be absolutely flat, and an O-ring is used to assist in sealing. In the case of flexible hose, the flange is clamp connected or fixed permanently to the hose fitting (fig. 172).

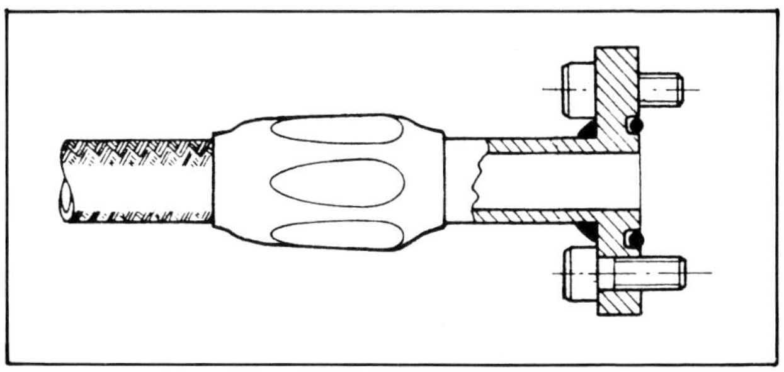

Fig. 172 Welded flange fitting for flexible hose.

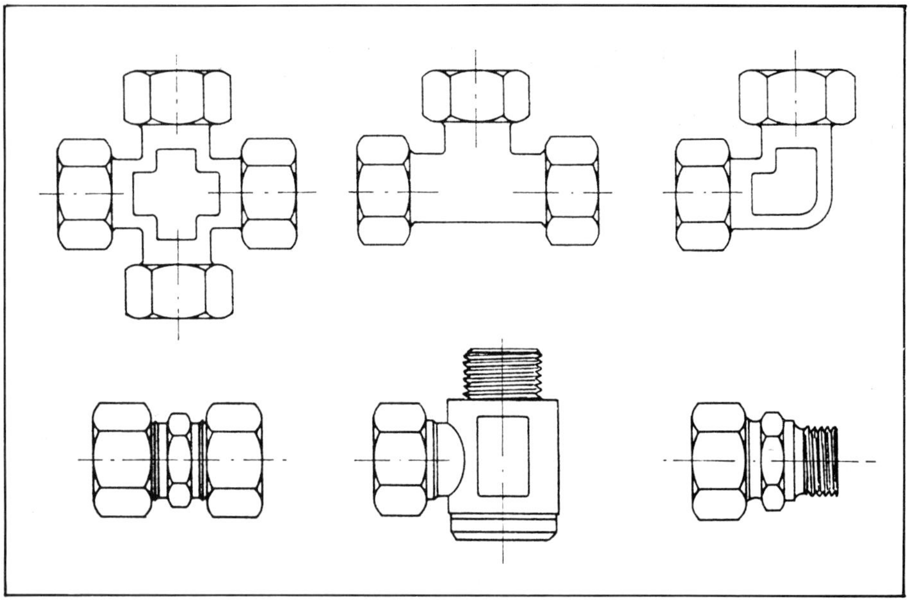

Fig. 173 Various types of fittings.

For various types of fittings and tube connectors, see fig. 173.

Port and fitting threads

Port and pipe fitting threads are designed to match standard pipes. Fittings usually come with British standard pipe thread. (B.S.P.), cylindrical or tapered, metric fine thread, also cylindrical or tapered, or standard American pipe thread (N.P.T.F.). Some common threads and matching pipe diameters are shown in fig. 174.

Pipe outside dia. (mm)	British standard pipe thread (BSP)	Metric fine thread
6	R¼″	M 22 × 1,5
8	R¼″	M 14 × 1,5
10	R⅜″	M 16 × 1,5
12	R⅜″	M 18 × 1,5
14	R½″	M 20 × 1,5
16	R½″	M 22 × 1,5
20	R¾″	M 27 × 2
25	R 1″	M 33 × 2
30	R 1¼″	M 42 × 2
38	R 1½″	M 48 × 2

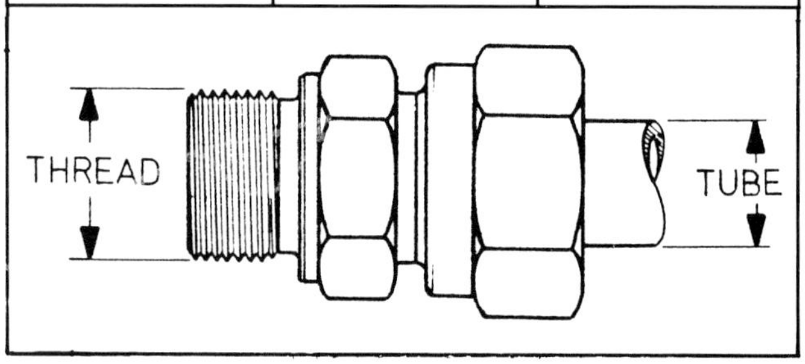

Fig. 174 Pipe diameter and matching fitting threads.

11 Filters, pressure switches, and pressure gauges

Filtration of the hydraulic fluid is of the greatest importance for maintaining the function and reliability of the hydraulic system. Fluid contamination occurs through a variety of sources:

- Contaminants left in the system during original assembly or subsequent maintenance work like welding scale and welding beads, silicone tape shreds, bits of pipe threads and seal materials, tubing burrs and grinding chips.
- Contaminants generated when running the system may include wear particles, sludge and varnish due to fluid oxidation, and rust and water due to condensation on the reservoir's interior.
- Contaminants introduced into the system from outside. These include using the wrong fluid when topping up, and dirt particles introduced by contaminated tools or repaired components.

The large quantities of highly pressurised fluid carry these contaminants endlessly through the system or deposit them in the small clearances of pumps, valves, actuators, and motors. Thus, if they are not filtered out, the system will soon either cease to function altogether, or the lapping action of the metal particles may prematurely wear pumps, motors, and valves, causing high internal leakage and power losses.

Quantity and size of contaminants

Filters are rated in micrometres (μm) which are equal to one-millionth of a metre. Tests on hydraulic fluids have shown a close connection between the grade of contamination and the size and number of dirt particles. According to S.A.E. standards, the degree of fluid contamination is divided into 7 classes, where class 0 is the best (fig. 175).

Filter ratings

Absolute filtration rating

This is a filter rating which states that 98% of particles in a fluid, passing through the filter above a specified micron size, will be retained. For example, a filter with a 40 micron rating should retain at least 98% of dirt particles on or over the size of 40 microns. Since there is no single consistent hole, pore, or mesh size in depth-type filters, they are given a nominal rating, whereas surface type filters may be rated in absolute values.

Nominal filtration rating

This rating is based on a retention rate of 50–95% of particles passing through the filter above the specified micron size. It is assumed that the fluid, when passing through the filter several times, will eventually be rid of all particles above the stated nominal size.

Beta rating

The beta ratio (rating) is an international standard defining the efficiency of filter elements. It is based on continuous contamination of the hydraulic fluid in the system reservoir, whilst taking samples of fluid both before and after the filter element. The ratio of upstream to downstream particle counts at a particular micrometre size (μ) is the beta rating (fig. 176). This is also called the multi-pass method for evaluating filtration performance (I.S.O. 4572).

A typical filter series is shown in fig. 177. Its beta ratio for 5 micrometre particles is 2.4, which means that a 58.33% efficiency can be expected. The same filter will retain 15 micrometre particles (β_{15}) with a ratio of 17.4. This means that 94.25% of particles with a size of 15 micrometre and over will be retained.

Size of particle µm	No. of particles in 100 cm³/classes						
	0	1	2	3	4	5	6
5–10	2700	4600	9700	24000	32000	87000	128000
10–25	670	1340	2680	5360	10700	21400	42000
25–50	93	210	380	780	1510	3130	6500
50–100	16	28	56	110	225	430	1000
100–	1	3	5	11	22	41	92

Fig. 175 Classification of fluid contamination.

BETA RATIO (β): DEFINITION

Number of particles greater than a given size (μ) in the influent fluid in relation to the number or particles greater than the same size in the effluent fluid. It accurately reflects the particle separation capability of the filter. The higher the BETA ratio, the greater the filter's capability to capture particles larger than the indicated BETA size.

Example: $\beta_{10} = 5.8$

BETA RATIO VALUE DERIVATION:

$$\frac{\text{particle count upstream}}{\text{particle count downstream}} = \textbf{Beta Ratio Value} \qquad \textbf{Example:} \quad \frac{36815}{6347} = 5.8$$

BETA EFFICIENCY DERIVATION:

$$\frac{\text{particle count upstream MINUS particle count downstream}}{\text{particle count upstream}} = \text{\% EFFICIENCY}$$

Example: $\frac{36815\ (-)\ 6347}{36815} = 82.76\%$ EFFICIENT Therefore, BETAμ = 5.8 is 82.76% EFFICIENT

TABLE OF EFFICIENCIES OF SELECTED BETA RATIO VALUES:

BETA μ =	1.0 =	0	% Efficient	BETA μ =	3.0	= 66.66	% Efficient	BETA μ =	32.0	= 96.875	% Efficient
BETA μ =	1.14 =	12.28	% Efficient	BETA μ =	4.0	= 75.00	% Efficient	BETA μ =	52.2	= 98.084	% Efficient
BETA μ =	1.5	= 33.33	% Efficient	BETA μ =	5.8	= 82.76	% Efficient	BETA μ =	100.0	= 99.0	% Efficient
BETA μ =	2.0	= 50.00	% Efficient	BETA μ =	16.0	= 93.75	% Efficient	BETA μ =	173.0	= 99.42	% Efficient
BETA μ =	2.4	= 58.33	% Efficient	BETA μ =	17.4	= 94.25	% Efficient				

Fig. 176 General "beta" element filtration information.

ELEMENT "SERIES"	β_5	β_{10}	β_{15}	β_{20}	β_{30}
MB	2.4	5.8	17.4	52.2	173.0

Fig. 177 Typical filtration performance.

Filter types and materials

In "depth type" filters the hydraulic fluid is forced through multiple layers of material. The contaminants are retained and embedded in the filter material because of the tortuous path the fluid must take (fig. 178). These filters are also called absorbent filters. Filter materials used for depth type filters are:

- porous and permeable paper (treated and resin coated)
- synthetic fibres in long strands, matted and pressed
- metal fibres woven, or matted and pressed
- glass fibres woven, or matted and pressed
- sintered granular metals (cartridge and disc-type elements)

In "surface type" filters the hydraulic fluid flows straight through a layer of woven mesh, and deposits the dirt particles on the surface of the mesh (fig. 179). To increase the filter surface, the mesh is often star-shape folded. Filter materials used for surface type filters are:

- steel wire cloth (woven)
- nylon monofilament cloth (woven)
- cellulose fibre cloth (woven)
- series of metal discs deparated by thin spacers

Paper filters cannot be cleaned, and must be discarded when filled with contaminants. Metal, glass, and synthetic fibre filters can be dismantled and cleaned. Some filters may have a provision for an electrical or

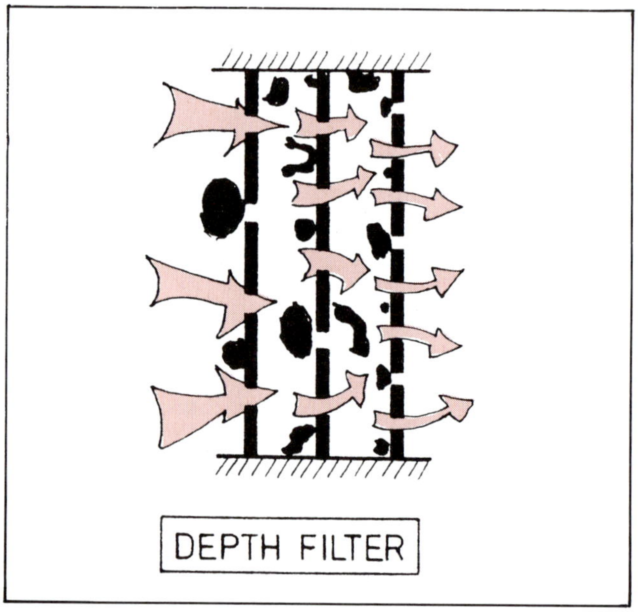

Fig. 178 Depth filter.

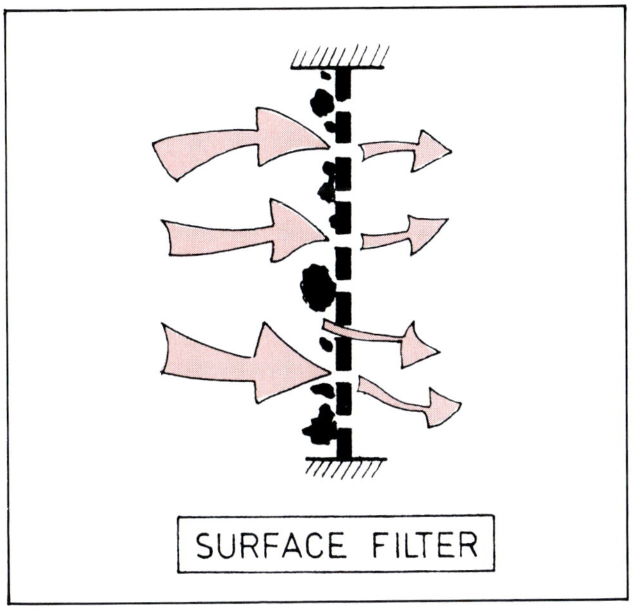

Fig. 179 Surface filter.

mechanical indicator which signals dirt saturation. As a rule a filter should never be used without an indicator, and pressure line and return line filters should always have a by-pass (fig. 180).

Return line filters

Return line filters are low pressure elements which filter the returning fluid prior to its entry into the reservoir. Return line filters are inexpensive, cause few problems during usage and maintenance, and filter the entire fluid volume. However, they have the disadvantage that only the returning fluid, and not the fluid entering into the system is cleaned. Thus, contaminants can enter the pump, the valves, and the actuators (fig. 181).

Pressure line filters

Pressure line filters must withstand maximum system pressure and are therefore strong and expensive. Pressure line filters are used upstream of motors and valves, to protect them from dirt particles. They are usually mounted directly into the component they must protect (fig. 181). Provided a high-strength element is used, these filters can be used without a bypass in order to shut down the system if clogged. They are used in this way to protect expensive servo valves against particle damage.

Suction line filters and strainers

Suction line filters are found in most industrial systems. It is a low grade filter which protects the pump. Its disadvantage is, that it is not easily accessible since it is

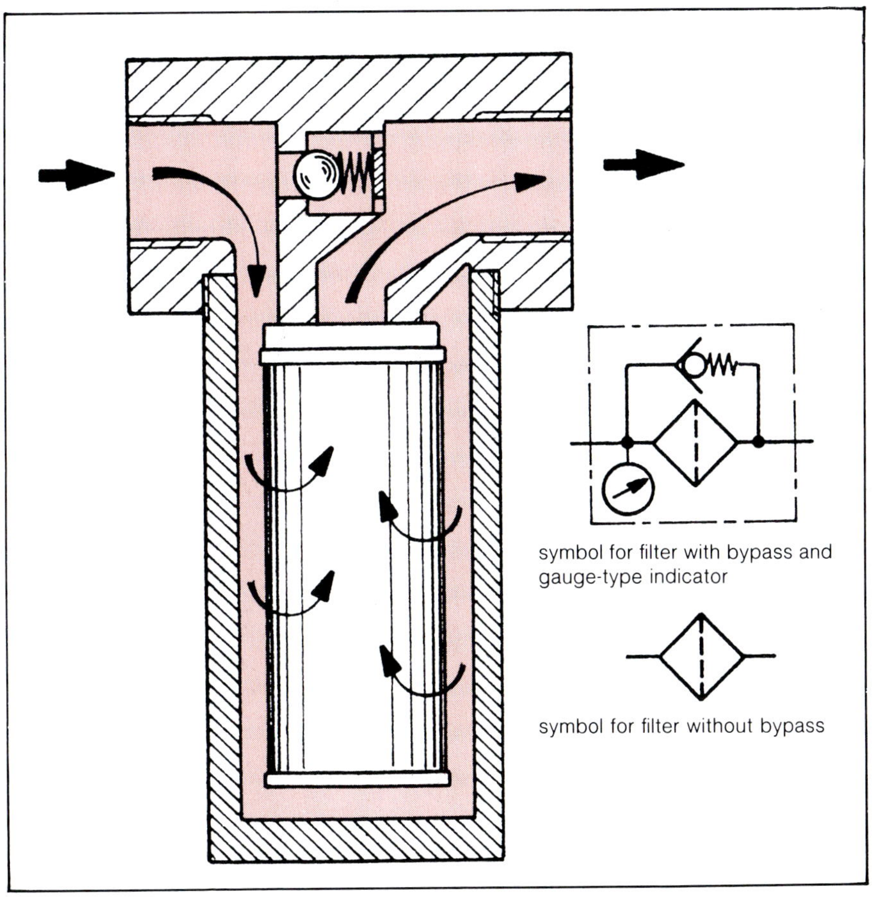

Fig. 180 Filter with by-pass or without by-pass check valve.

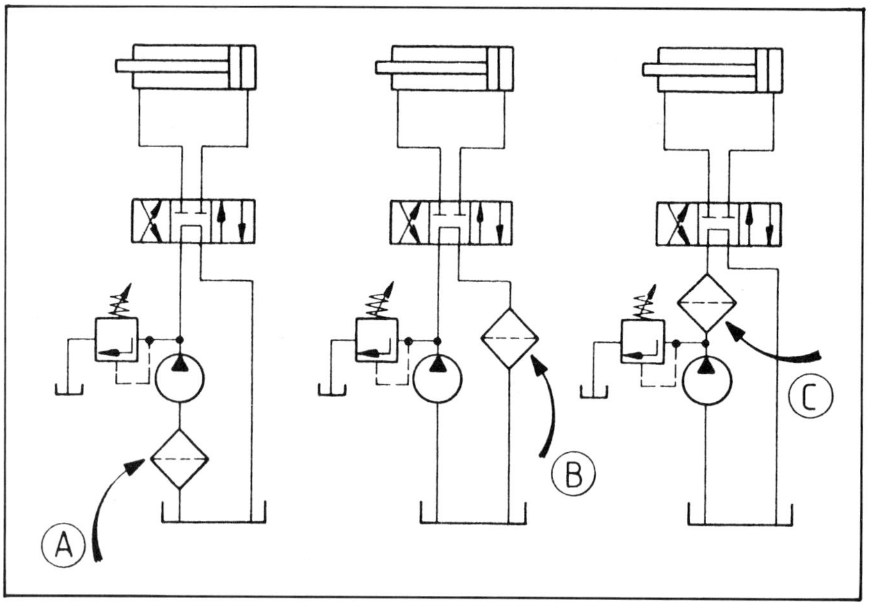

Fig. 181 Suction filter (A) return line filter (B) pressure line filter (C).

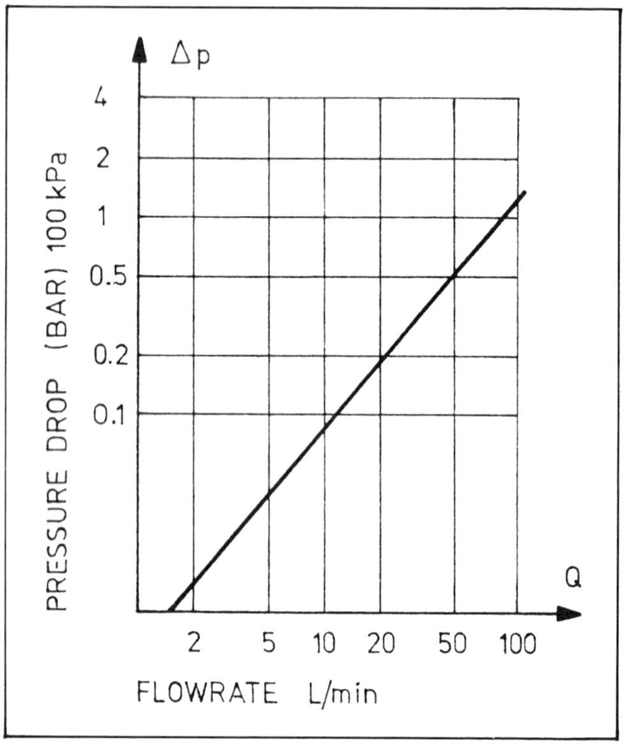

Fig. 182 Pressure drop curve for a pressure line filter.

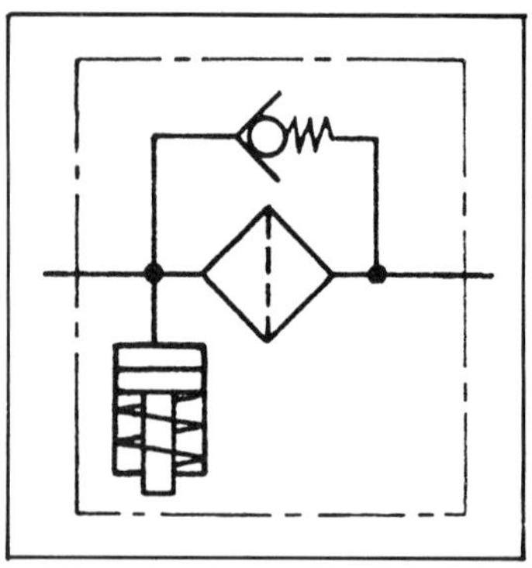

Fig. 183 Symbol for a return line filter with by-pass and mechanical clogging indicator.

mounted inside the reservoir (fig. 181). Some manufacturers offer an "in-tank" type suction filter which can be serviced without draining the tank. It is possible to have suction filters with ratings as low as 5 or 10 micrometre, but their size makes them expensive.

Suction filters, if underrated, can cause high pressure drop and cavitation. One way to provide suction line filtration without fear of cavitation is with a magnetic separator. This device will at least eliminate ferrous metal particles, which is valuable filtration at the pump inlet line.

Pressure drop

An important factor in the choice of filter is its pressure drop. Most manufacturers supply curves or charts which relate the pressure drop (Δp) for a given fluid viscosity to the relevant flowrate (Q) across the filter. Any increase in filter contamination or viscosity will automatically increase the pressure drop. Therefore, most filters have an inbuilt flow by-pass, which opens when the pressure drop increases beyond the safe flowrate of the filter (fig. 182).

Hydro-electric pressure switches

Hydro-electric pressure switches are interface elements which convert a fluid pressure signal into an electric signal by closing contacts in an electric circuit. Pressure switches are either used to switch solenoid-operated hydraulic valves in pressure-sequenced hydraulic functions, or to monitor pressure levels in the hydraulic circuit. These signals may then be used to operate visual or acoustic indicators (lamps and bells).

A typical piston pressure switch is shown in fig. 184, where hydraulic pressure is balanced against an adjust-

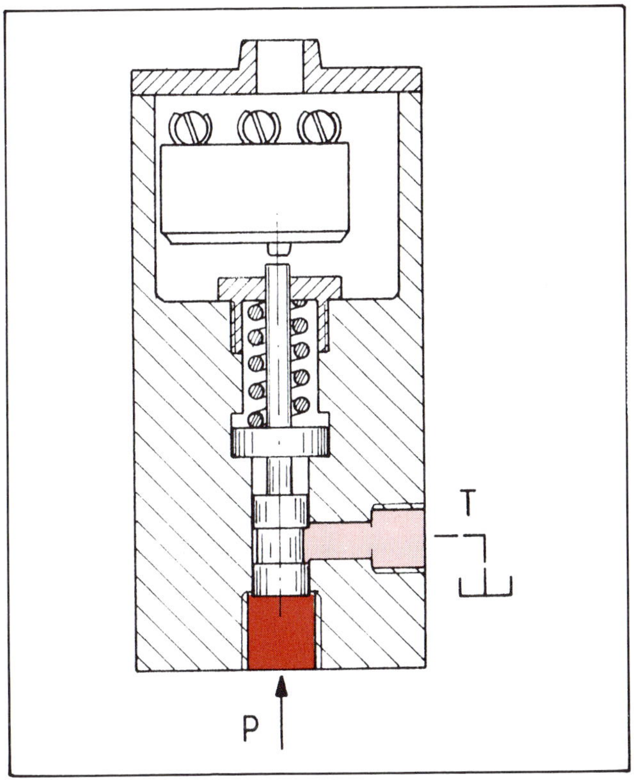

Fig. 184 Piston type pressure switch.

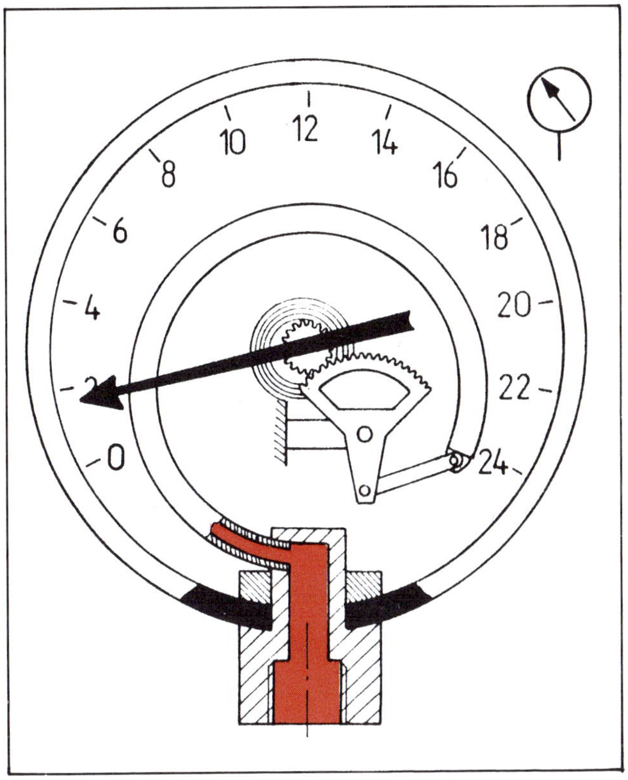

Fig. 186 Bourdon tube pressure gauge.

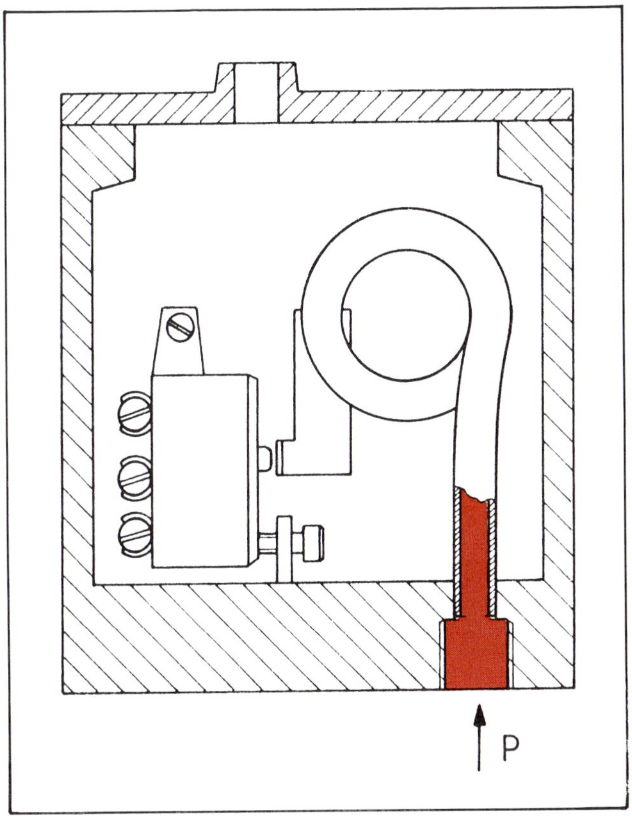

Fig. 185 Bourdon tube pressure switch.

able spring. When the fluid pressure exceeds the threshold set on the spring, the piston moves up and actuates the electric switch.

A Bourdon-tube pressure switch is shown in fig. 185. The hydraulic pressure makes the Bourdon-tube recoil and press against the plunger of the electric switch. Bourdon-tube pressure switches can also be used for gas and compressed air (switching accuracy is $< \pm 1\%$ of the setting range).

Pressure gauges

Pressure gauges are used in hydraulic systems for a number of reasons. They serve to adjust pressure settings of pressure control valves, to test components and fault find circuits, to charge accumulators to the required pressure, and to determine torque and the force exerted by rotary and linear actuators.

Bourdon-tube pressure gauges are the most widely used type (fig. 186). The recoil of the tube under pressure operates the pointer via a gear mechanism. Pressure gauges should be calibrated regularly, and tested to ensure precision. Master gauges or dead-weight testers are used for calibration.

Where pressure gauges are permanently installed in the system, a gauge isolator valve should be used to unload the gauge when it is not in use (fig. 187). Where the gauge must remain under pressure, a snubber (which is a porous flow restrictor element) should be installed. The snubber prevents the gauge from oscillating, and it dampens pressure shocks, which prolongs the service life of the pressure gauge (fig. 188).

Pressure gauges are available which have casings filled with a clear viscous liquid, usually glycerine. This too is done to increase gauge life and reliability, and more than justifies the greater cost. Their use is highly recommended.

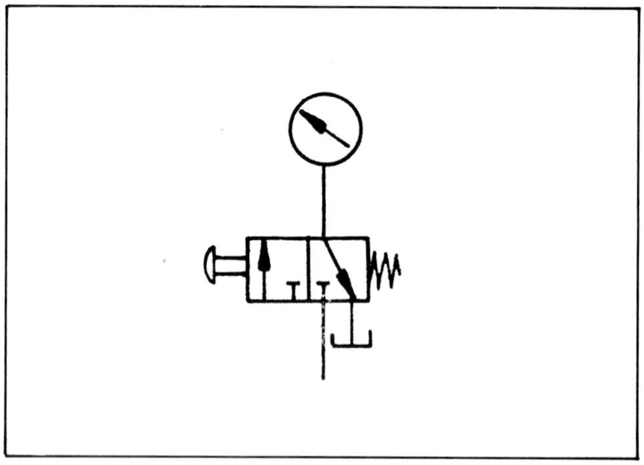

Fig. 187 Gauge with isolator valve.

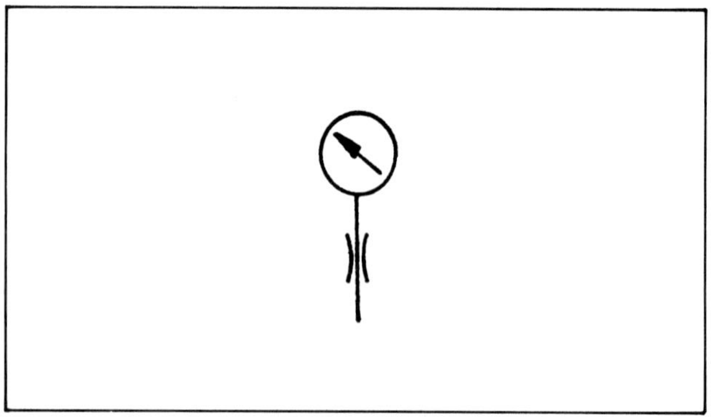

Fig. 188 Gauge with snubber.

12 Hydraulic fluids and decompression control

Pascal's Law states that, in a static condition, pressure exerted on a confined fluid is transmitted undiminished in all directions, and acts with equal force on equal areas. This characteristic makes hydraulic fluid an ideal energy transmitting medium. Some of the functions of a fluid in a hydraulic system are:

- Fluid must transmit an applied force from one part of the system to another.
- Fluid must reproduce immediately any change in direction or magnitude of the force transmitted.
- Fluid must provide adequate lubrication in bearings and sliding surfaces in pumps, valves and actuators.
- Fluid must have adequate film strength to seal clearances between moving parts against leakage.
- Fluid must transport system heat and contaminants to the filters and the reservoir.
- Fluid must prevent corrosion of the component parts of the system.
- Fluid must not react chemically with any materials with which it is normally in contact.

Hydraulic fluids must perform these functions and possess these properties, to suit the varying operating requirements of modern hydraulic systems.

Viscosity

Viscosity is generally regarded as one of the most important physical properties of a hydraulic fluid, since it affects both the ability to flow and to lubricate moving parts. Viscosity defines the internal resistance of the hydraulic fluid to flow. The viscosity of a fluid is low if it flows easily. The fluid is then said to be thin or to have a low body. The viscosity of a fluid is high if it flows with difficulty. It is then said to be thick or to have a high body.

In practical applications, the choice of viscosity is a compromise, often governed by the pump manufacturer's specifications. A high viscosity fluid provides good sealing between the clearance gaps of pumps, valves, and motors. But if the viscosity is too high, this could result in:

- high resistance to flow, causing sluggish actuator movement and pump cavitation;
- increased power consumption due to frictional losses;
- increased pressure drop through lines and valves.

Conversely, if the viscosity is too low, this could result in:

- excessive internal leakage losses;
- excessive wear due to insufficient lubrication in pumps and motors;
- decrease in pump and motor efficiency;
- increase in fluid temperature due to internal leakage losses.

Absolute viscosity

Absolute viscosity (μ) is defined as the force required to move a flat plate of unit area at unit distance from a fixed plate with unit relative velocity, when the space between the plates is filled with the fluid in question.

Kinematic viscosity

For the selection of hydraulic fluids, kinematic viscosity is used. Kinematic viscosity (ν) is obtained by dividing the absolute viscosity of the fluid by its mass density:

$$\text{Kinematic Viscosity } (\nu) = \frac{\text{Absolute Viscosity } (\mu)}{\text{Mass Density } (\rho)}$$

Expressed in traditional metric units, kinematic viscosity is:

$$\nu = \frac{\text{dyn} \cdot \text{s/cm}^2}{\text{dyn} \cdot \text{s}^2/\text{cm}^4} = \text{cm}^2/\text{s}, \text{ called Stokes (St) after Sir}$$

Gabriel Stokes (1819–1903), who made many fundamental contributions to fluid dynamics.

The Stoke is too large a unit for the types of fluid used in industrial hydraulics, and kinematic viscosity expressed in SI units is therefore given in centi Stokes:

$\nu = \text{mm}^2/\text{s}$, called centi Stokes (cSt)

Viscometers for viscosity measurement

Many types of viscometer are used to measure the viscosity of hydraulic fluids. Two common types are the falling-sphere type and the capillary viscometer.

In the falling-sphere type the fluid is located in a thermally controlled glass tube, which has two calibration lines. The sphere is allowed to sink through the fluid, and the time required for the sphere to drop the distance between the calibration lines is measured and related to viscosity (fig. 189).

In the capillary viscometer the hydraulic fluid is poured through inlet A into the container E where it is

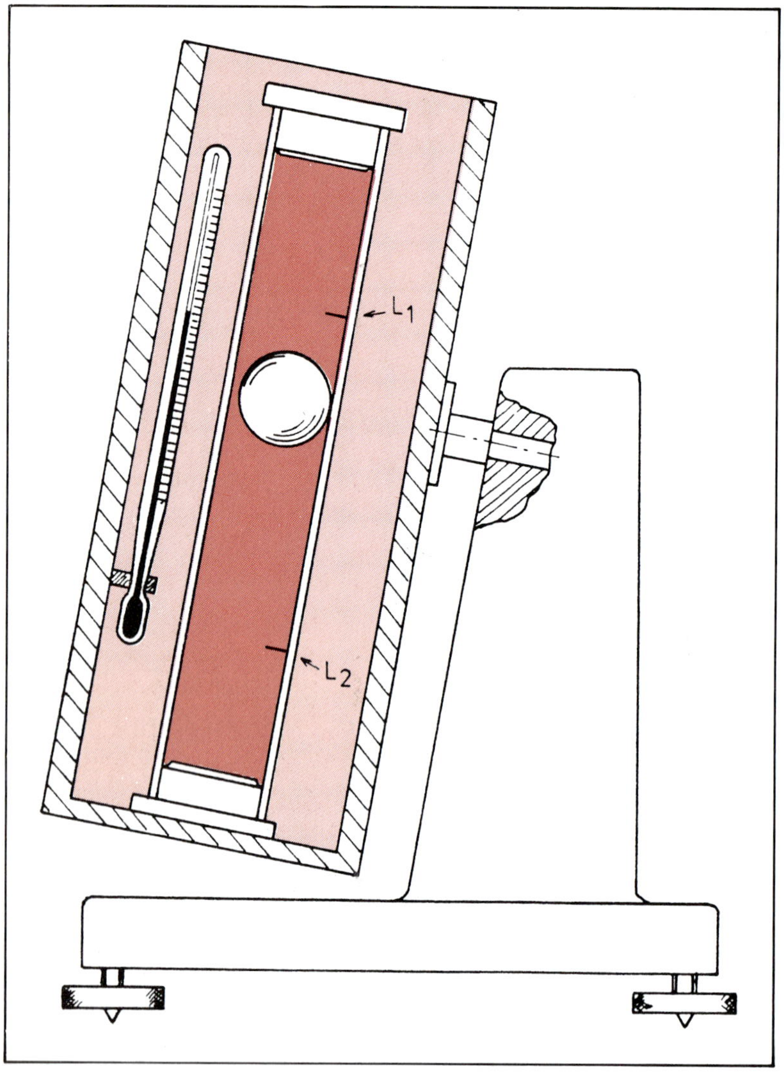

Fig. 189 Falling-sphere viscometer.

thermally controlled. It is then sucked through the capillary tube to the upper edge of bulb D, with port C closed. The ports A, B, and C are then opened, and a measurement is taken of the time required for the fluid level to drop from L_1 to L_2. This time is then related to viscosity (fig. 190).

Viscosity index

The viscosity index is an arbitrary indication of a given fluid's rate of change of viscosity corresponding with a certain change in temperature. A fluid that has a relatively stable viscosity at widely different temperatures has a high viscosity index. Hydraulic oils should have a viscosity index around the 100 mark. Most hydraulic oils have additives called "VI improvers" to achieve a viscosity index higher than 100.

In areas where extreme variations in ambient temperature are encountered, it is a widespread practice to use both heating and cooling to keep oil temperature variations to a minimum in order to maintain uniform viscosity. For viscosity index and temperature variation

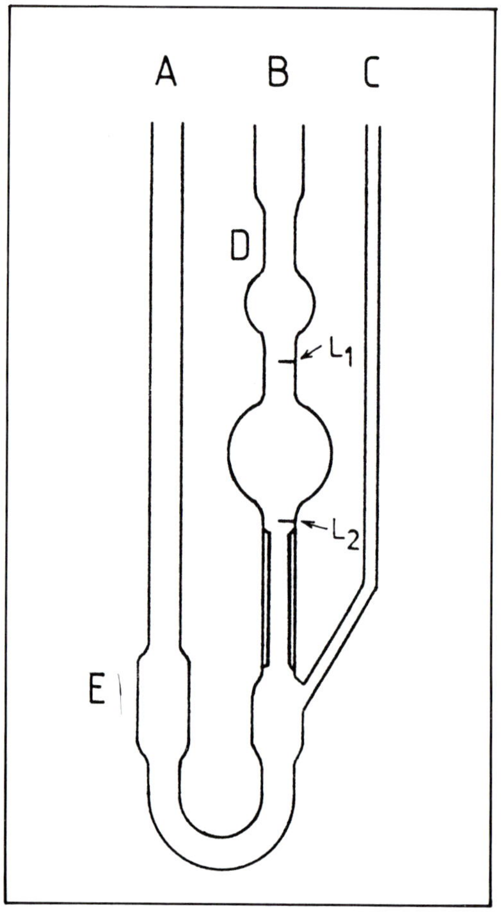

Fig. 190 Capillary viscometer.

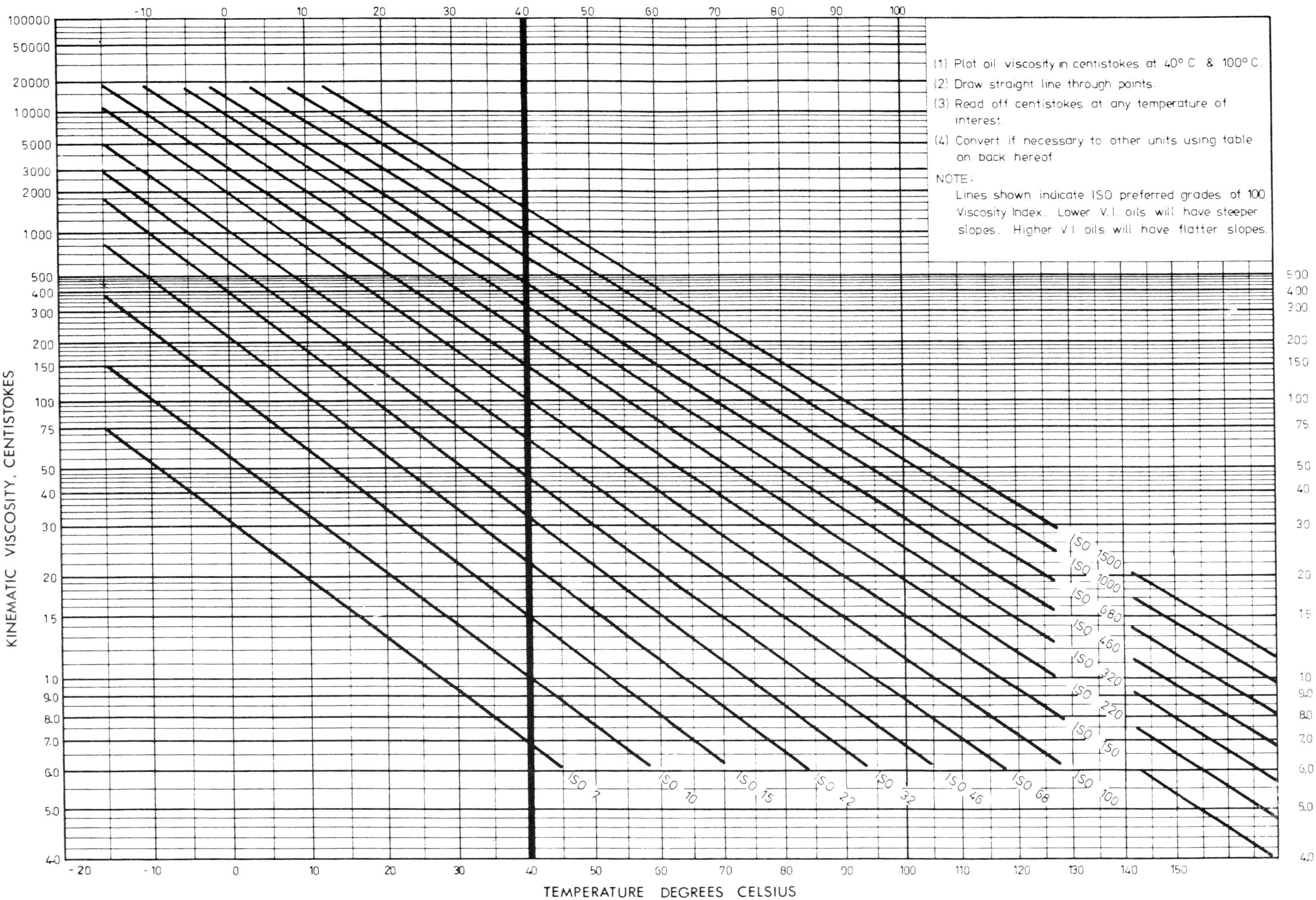

Fig. 191 Viscosity/temperature chart.

characteristics, see the chart of fig. 191. For the standard I.S.O. viscosity classification, see fig. 192, and for viscosity conversions, see fig. 193.

The ISO viscosity classification uses centiStoke (cSt) units and relates to the viscosity at 40°C. It consists of a series of 18 viscosity brackets between 1.98 cSt and 1650.0 cSt each of which is defined by a number. The numbers indicate, to the nearest whole number, the mid-points of their corresponding viscosity brackets.

ISO Viscosity Grade	Mid-Point Viscosity cSt at 40.0°C	Kinematic Viscosity Limits cSt at 40.0°C Min.	Max.
ISO VG 2	2.2	1.98	2.42
ISO VG 3	3.2	2.88	3.52
ISO VG 5	4.6	4.14	5.06
ISO-VG 7	6.8	6.12	7.48
ISO VG 10	10	9.00	11.00
ISO VG 15	15	13.50	16.50
ISO VG 22	22	19.80	24.20
ISO VG 32	32	28.80	35.20
ISO VG 46	46	41.40	50.60
ISO VG 68	68	61.20	74.80
ISO VG 100	100	90.00	110.00
ISO VG 150	150	135.00	165.00
ISO VG 220	220	198.00	242.00
ISO VG 320	320	288.00	352.00
ISO VG 460	460	414.00	506.00
ISO VG 680	680	612.00	748.00
ISO VG 1000	1000	900.00	1100.00
ISO VG 1500	1500	1350.00	1650.00

A few SHELL grades do not conform to the standard ISO classifications. For example the numbers 37, 78 and 800 are SHELL 'ISO type' numbers that have been allocated to meet certain important viscosity requirements that are not met by standard ISO numbers.

Fig. 192 ISO viscosity numbers.

Pour point

The temperature at which a fluid congeals is called the pour point. Practically all petroleum-base hydraulic fluids contain some waxy components. At low temperatures, these components tend to crystallise and hold the fluid portion of the oil immobile. For hydraulic machinery to operate in cold temperatures, the pour point should be 10 to 15°C below the lowest anticipated start-up temperature; this ensures that the oil will flow and supply the inlet side of the pump.

Oxidation stability

This property determines the oil's resistance to chemical deterioration and measures its resistance to forming insoluble varnishes and sludges. Varnish and sludge can interfere with the performance of the hydraulic system by plugging lines, filters, orifices, and screens. To remove such contaminants can be costly and time consuming.

Oxidation results from the reaction of the oil with the oxygen in the air of the reservoir. High fluid temeratures and excessive aeration and foaming accelerate this process, as does the presence of certain metals and water. High quality hydraulic fluids contain oxidation inhibitors which decrease oxidation considerably. But frequent fluid changes, periodical draining of water from the bottom of the reservoir, and the avoidance of excessive fluid heat, are still the best measures to keep oxidation and its serious side effects at bay.

Lubricity

High quality hydraulic fluid must have good lubricity to prevent undue friction between moving parts in pumps, motors, valves, and linear actuators. Increasing pressures and speeds, coupled with smaller clearances, cause the film of fluid to be squeezed very thin, and a condition called boundary lubrication occurs. When this happens, the oil no longer lubricates properly, the oil film breaks down, and metal to metal contact of moving parts occurs. Therefore, most hydraulic fluids contain anti-wear additives which reduce wear and provide adequate lubrication even under extreme conditions. (For example, zinc-dithiosphate was introduced to allow increased working pressures from high performance vane pumps.)

Demulsibility (water separation)

In a reservoir, moisture condensed from the atmosphere can mix with the hydraulic fluid. However, if the oil has a good demulsibility characteristic, that is if it resists mixing with water, entrained water will separate out easily and sink to the bottom of the reservoir. There it can be drained out periodically.

Foaming and aeration resistance

Foaming is essentially a system design problem and should be treated as such. Some of the common causes of foaming are too low a level of oil in the reservoir, permitting air to be drawn into the pump, excessive turbulence of the return oil due to improper design, and air leakage into the pump suction line. For added protection hydraulic oils contain additives that control foaming and encourage separation of air bubbles from the oil (see also Chapter 9).

Viscosities of lubricating oils have generally been quoted in one or other of the following terms depending on the instrument used for the viscosity determination.

Kinematic	Viscosity in centiStokes	(VK. cSt)
Redwood 1	Viscosity in seconds	(RI'')
Saybolt Universal	Viscosity in seconds	(SU'')
Engler	Viscosity in degrees	(°E)

A general move to the universal use of centiStokes is expected but, for many years, it will be necessary to convert back and forth between the various units.

For most application purposes in industry where appreciable tolerances in viscosity limits are normal, the following tables will serve for direct conversion of viscosities expressed in one form of measurement to viscosities in another form, provided always such conversions or comparisons have reference to identical temperature conditions.

Kinematic centiStokes	Redwood 1 Seconds	Saybolt Universal Seconds	Engler Degrees	Kinematic centiStokes	Redwood 1 Seconds	Saybolt Universal Seconds	Engler Degrees	Kinematic centiStokes	Redwood 1 Seconds	Saybolt Universal Seconds	Engler Degrees
2.0	31	32.6	1.12	33	137	155.2	4.46	104	426	484	13.73
2.5	32	34.4	1.17	34	141	159.7	4.58	106	435	493	13.99
3.0	33	36.0	1.22	35	145	164.3	4.71	108	443	502	14.26
3.5	35	37.6	1.26	36	149	168.8	4.84	110	451	511	14.52
4.0	36	39.1	1.31	37	153	173.3	4.96	112	459	521	14.78
4.5	37	40.7	1.35	38	157	178.0	5.10	114	467	530	15.05
5.0	39	42.3	1.39	39	161	182.4	5.22	116	476	540	15.31
5.5	40	44.0	1.44	40	165	187.0	5.35	118	484	549	15.58
6.0	41	45.6	1.48	41	169	191.5	5.48	120	492	558	15.84
6.5	43	47.2	1.52	42	173	196.0	5.61	122	500	567	16.10
7.0	44	48.8	1.56	43	177	200.5	5.74	124	508	577	16.37
7.5	45	50.4	1.61	44	181	205.0	5.87	126	517	586	16.63
8.0	46	52.1	1.65	45	185	209.8	6.00	128	525	595	16.90
8.5	48	53.8	1.71	46	189	214.5	6.13	130	533	605	17.16
9.0	49	55.5	1.75	47	193	219.0	6.26	132	541	614	17.42
9.5	51	57.2	1.80	48	197	223.7	6.38	134	549	623	17.69
10.0	52	58.9	1.84	49	201	228.3	6.51	136	558	632	17.95
10.5	54	60.7	1.89	50	205	233.0	6.64	138	566	642	18.22
11.0	55	62.4	1.94	51	209	237.5	6.77	140	574	651	18.48
11.5	57	64.2	1.98	52	213	242.2	6.90	142	582	658	18.74
12.0	58	66.0	2.03	53	218	246.8	7.04	144	590	667	19.01
12.5	60	67.9	2.08	54	222	251.5	7.17	146	599	677	19.27
13.0	62	69.8	2.13	55	226	256.0	7.30	148	607	686	19.54
13.5	64	71.7	2.18	56	230	260.7	7.43	150	615	695	19.80
14.0	65	73.6	2.23	57	234	265.3	7.56	152	623	705	20.06
14.5	67	75.5	2.28	58	238	270.0	7.69	154	631	714	20.33
15.0	68	77.4	2.33	59	242	274.7	7.82	156	640	723	20.59
15.5	70	79.3	2.39	60	246	279.2	7.95	158	648	732	20.86
16.0	72	81.3	2.44	61	250	284.0	8.04	160	656	742	21.12
16.5	74	83.3	2.50	62	254	288.5	8.18	164	672	760	21.65
17.0	75	85.3	2.55	63	258	293.6	8.31	168	689	779	22.18
17.5	77	87.4	2.60	64	262	297.7	8.45	172	705	797	22.70
18.0	79	89.4	2.65	65	266	302.4	8.58	176	722	816	23.23
18.5	81	91.5	2.71	66	271	307.0	8.72	180	738	834	23.76
19.0	82	93.6	2.77	67	275	311.7	8.85	184	754	853	24.49
19.5	84	95.7	2.83	68	279	316.3	8.98	188	771	871	24.82
20.0	86	97.8	2.88	69	283	321.0	9.11	192	787	890	25.34
20.5	88	99.9	2.94	70	287	325.5	9.24	196	804	908	25.87
21.0	90	102.0	3.00	72	295	335	9.51	200	820	927	26.40
21.5	92	104.2	3.06	74	303	344	9.77	204	836	946	26.93
22.0	94	106.4	3.11	76	311	353	10.03	208	853	964	27.46
22.5	96	108.5	3.17	78	319	363	10.30	212	869	983	27.98
23.0	97	110.7	3.23	80	328	372	10.56	216	886	1,001	28.51
23.5	99	112.8	3.29	82	336	381	10.82	220	902	1,020	29.04
24.0	101	115.0	3.35	84	344	391	11.09	224	918	1,038	29.57
24.5	103	117.1	3.41	86	352	400	11.35	228	935	1,057	30.10
25.0	105	119.3	3.47	88	360	410	11.62	232	951	1,075	30.62
26	109	124.0	3.59	90	369	419	11.88	236	968	1,094	31.15
27	113	128.5	3.71	92	377	428	12.14				
28	117	133.0	3.83	94	385	438	12.41				
29	121	137.5	3.96	96	393	447	12.67				
30	125	141.7	4.08	98	401	456	12.94				
31	129	146.0	4.21	100	410	465	13.20				
32	133	150.7	4.33	102	418	475	13.46				

For higher viscosities use the following factors:

RI = 4.10 VK.
SU = 4.635 VK.
E = .132 VK.

Fig. 193 Viscosity conversions.

Fire resistant hydraulic oils

A main disadvantage of petroleum-base hydraulic oil is its ability to burn. When a hydraulic system is located near high temperature equipment or other sources that could ignite the hydraulic fluid, the use of fire-resistant hydraulic fluid is imperative. There are three basic types of fire-resistant fluids:

- water-glycols
- water-oil emulsions
- synthetic fluids

Water-glycol fluids

Water-glycol fluids consist essentially of three main components. They are water (40%), glycol, and a high molecular-weight water-soluble poly-glycol. A fourth component is the additive package designed to impart corrosion resistance, metal passivation, and anti-wear and lubrication properties to the finished formulation. Water-glycol fluids give excellent fire resistance (fig. 194). When ignited, the water will boil off and the residue will burn. But even in this situation there is no possibility of a "flash" fire occurring, and there is always a significant time lag between contact with ignition and the development of a flame. When a flame occurs, the burning characteristics reveal a lazy low-output flame.

Water-in-oil emulsions

Water-in-oil emulsion type fluids are mixtures of water (35–40%), petroleum oil, and an emulsifying additive package. Each drop of water is encased in a skin of oil that breaks at high temperatures to release fire-smothering steam. Although not as fire resistant as water-glycol fluids, their flaming tendencies are significantly better than petroleum oils.

Oil-in-water emulsions

Oil-in-water emulsions are mixtures of 5% oil and 95% water (together with an emulsifying additive). Research is continuing into the application of these fluids, mainly in underground coalmining machinery. They are cheaper than synthetic fluids, but the hydraulic pumps and motors must be de-rated considerably for satisfactory wear-life. Their fire resistance is close to that of phosphate esters.

Phosphate esters (synthetic)

Often referred to as "synthetic" or "straight synthetic" fluids, the phosphate ester fluids are basically triaryl phosphate esters. Additives are incorporated to provide rust protection, metal passivation, anti-foam, and oxidation resistance. Although they do burn at very high temperatures, they do not propagate flame and are self-extinguishing when the ignition source is removed.

Oil-synthetic blends

These are blends of phosphate esters and refined petroleum stocks. They consist of 30–50% triaryl phosphate ester plus petroleum oil and a coupling agent to impart solution stability. Additives are included to improve corrosion and oxidation resistance. Their fire resistance is between that of phosphate esters and petroleum oil, and they are used where fire hazards are moderate.

System compatibility

The various fire resistant fluids, composed of many complex chemical components, display a variety of effects on the materials used in hydraulic systems. A convenient but generalised list of material behaviour to fire resistant hydraulic fluids is presented in fig. 195.

Fluid maintenance

Hydraulic fluid of any kind is expensive. Improper fluid treatment can be costly. Adherence to some basic rules can save downtime and system damage, and will guarantee long fluid and system life:

- Store hydraulic fluid drums on their side and keep them dry and cool (preferably under shelter).

	Type of Fluid				
	Petrol Oil	Water Glycol	Phosph. Ester	Oil in Water	Oil Synthetic
Fire resistance	P	E	G	F	F
Viscosity temp. properties	G	E	F	G	F–G
Seal compatibility	G	E	F	G	F
Lubricating quality	E	F–G	E	F–G	E
Temp. range (°C) above ideal	65	50	65	50	65
Relative cost comp. to oil	1	4	8	1.5	4

Key: E = Excellent, G = Good, F = Fair, P = Poor

Fig. 194 Properties of hydraulic fluids.

	Water-Glycol	Phos. Ester & Oil syn. blends	W/O Emulsion
Paints:			
Common Industrial	NC	NC	NC
Epoxy & Phenolic	C	C	C
Metals:			
Ferrous	C	C	C
Brass, Copper	C	C	C
Zinc	NC	C	C
Aluminium, Unanodized	NC	C	C
Aluminium, Anodized	C	C	C
Seals:			
Teflon	C	C	C
Viton	C	C	C
Neoprene	C	NC	C
Buna 'N'	C	NC	C
Butyl Rubber	C	C	NC
E.P. Rubber	C	C	NC
Silicone Rubber	C	C	C
Key: C = Compatible NC = Non-compatible			

Fig. 195 Fluid compatibility with hydraulic system materials.

Note Mineral Oil is similar to W/O (water/oil) Emulsion.

- Ensure the utmost cleanliness when topping up or renewing the fluid in the system. Use clean fluid transfer utensils.
- Pump the fluid through a pre-fill filter into the reservoir (fig. 163), and test the fluid frequently and regularly.
- Try to arrange it that the only reservoir filling point is a self-sealing coupling in the return line.
- Establish fluid change intervals so that oxidation and fluid breakdown never occur (check with fluid supplier).
- Prevent fluid contamination at all costs, and use proper air and fluid filtration (see reservoirs and filters, Chapters 9 and 11).
- Prevent excessive fluid heat build up; if required, use cooling or check for design faults (pump unloading, excessive resistance, etc.).
- Repair all leaks immediately, and use properly trained personnel for the maintenance of the system.
- Prior to fluid change (from oil to fire resistant fluids), make sure that components and seals are compatible with the new fluid (check with component and fluid suppliers), and ensure that the whole system is adequately flushed before it is re-commissioned.

Decompression control (bulk modulus)

Decompression is the slow release of potential energy stored in a hydraulic system. Potential energy may arise from the compressibility of the hydraulic fluid (bulk modulus), from stretched and widened machine members, or from the elasticity of the material being worked (fig. 197).

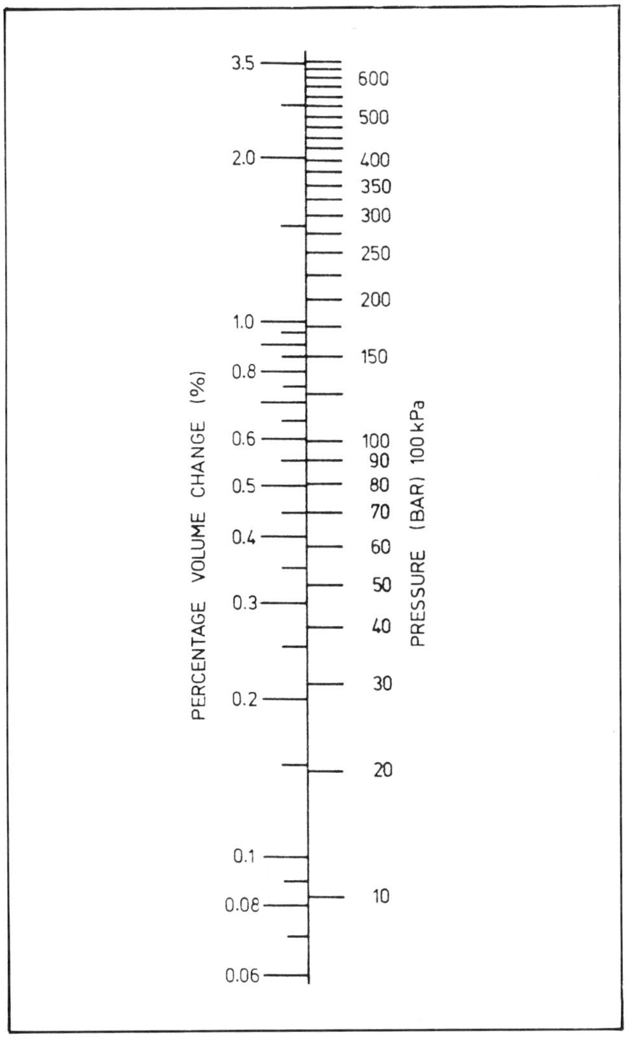

Fig. 196 Compressibility of hydraulic fluids.

Hydraulic oils are compressible to about 0.5% by volume for every 8 MPa (80 bar) (fig. 196). When the volume change of the fluid under pressure exceeds 160 mL, decompression (slow release) must be considered to avoid high pressure shocks and their resulting system damage. An automatic decompression control circuit is shown in fig. 197. Exhaust flow for actuator retraction closes the pressure reducing valve and decompression flow is metered out through the adjustable orifice. The pressure reducing valve will automatically open as soon as the pressure in the actuator matches the setting of the valve, and the bulk of the exhaust flow can then drain off through the pressure reducing valve. Two other decompression control circuits are depicted in figs. 198 and 199.

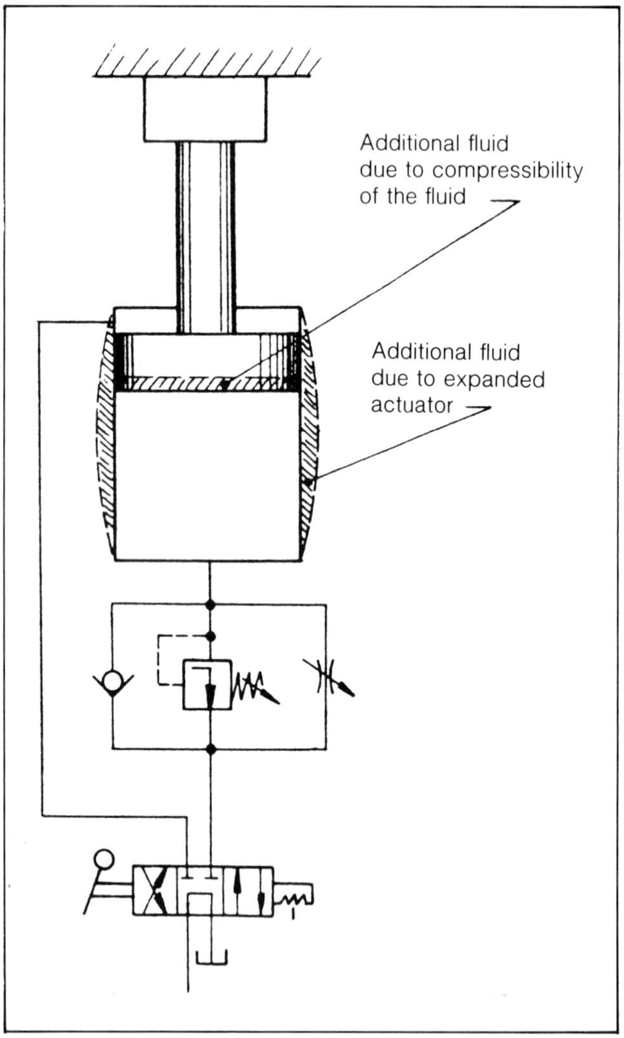

Fig. 197 Decompression control.

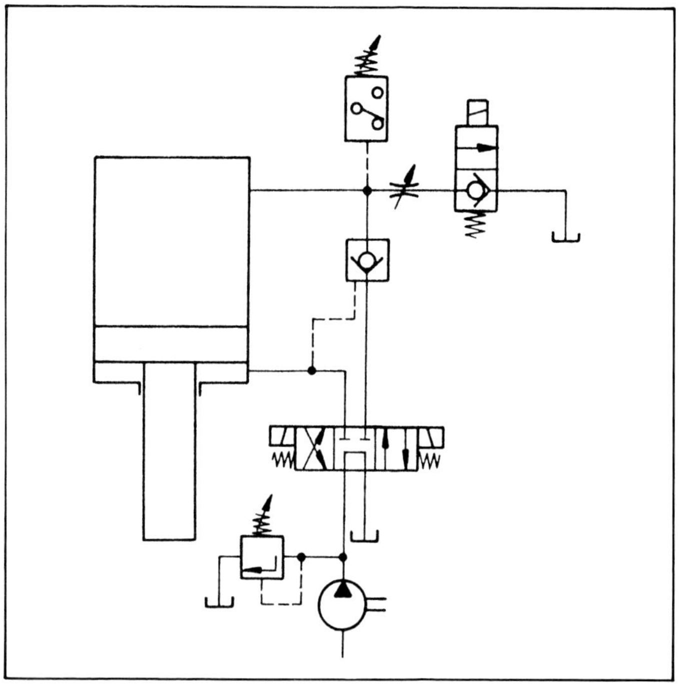

Fig. 198 Automatic decompression circuit electrically controlled (the electrical sequence and time controls are not shown).

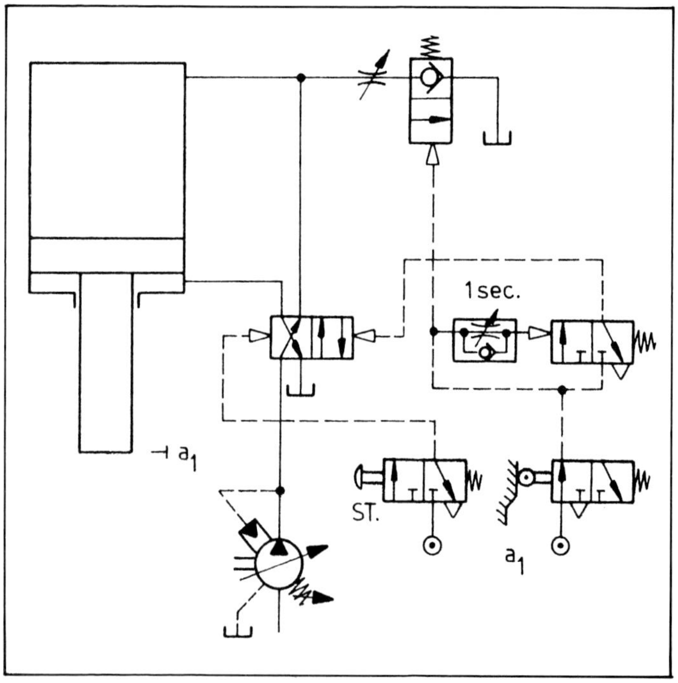

Fig. 199 Automatic decompression circuit with pneumatic sequence and time control.

13 Cartridge valve technology

Hydraulic cartridge valves, as a design idea, probably originated in the USA. They were primarily designed to provide a complete valve function in the form of a compact threaded barrel, which can be screwed into a circuit manifold block. Manifold blocks reduce, or often eliminate, the number of fittings and pipes used to interconnect the components of a hydraulic system. Hence, leakage points and their resulting fluid wastage are also eliminated or at least greatly reduced. Most major hydraulic valve manufacturers offer a wide range of screw-in cartridge valves for which there is practically no interchanging standard laid down; hence, once a particular brand of cartridge has been chosen, there is no flexibility for replacement by another brand at a later stage.

Furthermore, cartridge suppliers use different design concepts with regard to opening/closing area ratios for slip-in cartridge valves, cartridge popped ratios, pilot choke inserts and cover plate designs. Therefore, system design parameters vary greatly. For these reasons, changing to another brand of valves is often impossible or impractical.

Cartridge valves, as a general valve concept, may be regarded as a "bodiless" valve or a valve element without a housing. The cartridge is then inserted into a standardised cavity of a manifold block, and in spite of some preliminary international cavity standardisation, such as DIN standard 24342, cartridge valves are seldom interchangeable. Once the cartridge valve has been inserted into the cavity, it becomes a complete valve and operates like any conventional valve.

Screw-in cartridge valves

To understand the technology of cartridge valves, one must basically differentiate cartridge "screw-in" valves from cartridge "slip-in" valves. The screw-in cartridge is, when inserted into the manifold block, nothing more than a miniature version of a conventional hydraulic valve, providing a directional, a pressure or a flow control function (fig. 200). As its name reveals, the screw-in cartridge is screwed into a threaded cavity which secures it in its place. The cartridge then operates as a "stand-alone valve", requiring no companion valves or components to assist it in its basic function.

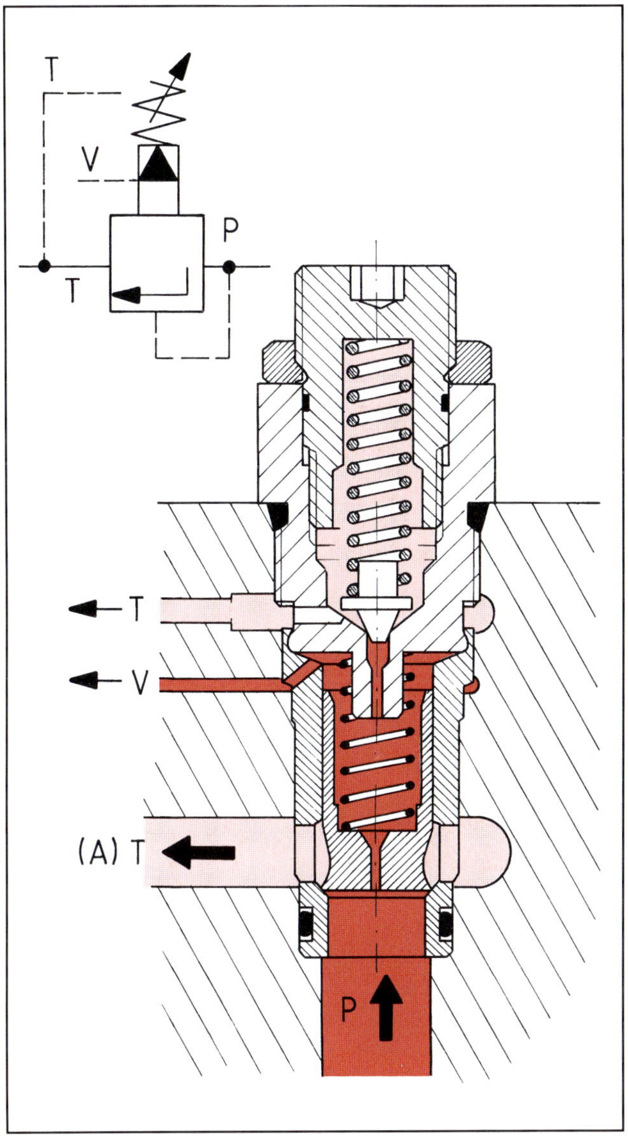

Fig. 200 Screw-in type cartridge valve for pressure relief function with venting port.

Screw-in cartridge valves offer a range of advantages over conventional valves. If the control valve cartridge becomes faulty for some reason, the cartridge is simply removed from its manifold cavity and replaced by a new element without disturbing any piping. Replacement time is short and thus machine down-time is greatly reduced. Leakage during operation is practically zero. Cartridge valves are space saving and are often not repaired when failure occurs, as their initial cost is much less than that of conventional valves.

For these reasons, screw-in cartridge valves have become increasingly popular with manufacturers of:

- injection moulding machines
- farm machinery
- utility services vehicles
- refuse equipment machinery
- marine and fishing applications
- forestry machinery
- mining and quarry equipment
- cherry pickers
- machine tools and robotics
- materials handling equipment.

For the following hydraulic circuitry functions, screw-in cartridge valves of all sizes are readily available:

- simple check valves
- pilot operated check valves
- flow restrictor valves
- flow divider valves
- priority flow control valves
- pressure relief valves
- counter balance valves
- sequence valves
- off-loading valves
- over centre valves
- accumulator charging valves
- pressure reducing valves
- anti-cavitation valves
- cross-over relief valves
- flushing valves
- motor break valves.

Since screw-in cartridge valves are no different, other than for size, from conventional hydraulic valves, there will be no need to deal with them any further, since their function, operation, circuit integration and pictorial appearance has been sufficiently described in the previous chapters.

Slip-in cartridge valves

Slip-in cartridge valves (or logic valves) are quite different in design and application from the screw-in cartridge. The slip-in cartridge uses a cover plate to secure the cartridge to the manifold block cavity. The valve usually consists of a valve sleeve, a poppet and a spring, and performs primarily an on–off or open–closed valve function (binary function) which is similar to a check valve (fig. 201).

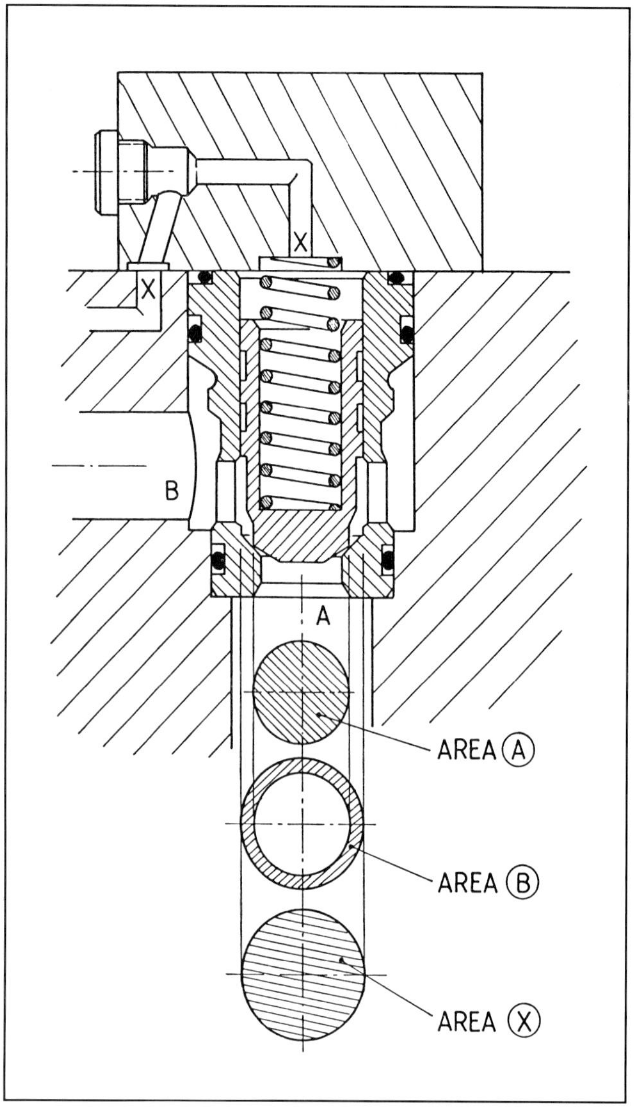

Fig. 201 Slip-in type cartridge valve (logic valve) with cover plate.

Slip-in cartridge valves offer a range of advantages when used correctly. They render:

- extremely high flowrate
- very high operating pressure capability
- zero internal leakage
- fast opening–closing response
- precise and smooth operating possibility
- low sensitivity to fluid contamination
- minimal wear — long life expectancy
- unlimited holding time
- compact and small installation dimensions
- minimal pressure peaks and gentle operation
- total design flexibility
- wide range of flowrate sizes available
- directional, flow and pressure control
- standardised cavity and cover plates.

With such an impressive range of advantages, slip-in cartridge valve technology has found its way into many fields of hydraulics. Just a few of these fields are: mobile hydraulics, press circuits, injection moulding machines, broaching machines and steel manufacturing plants.

Slip-in cartridge fundamentals

The slip-in cartridge is often referred to as a logic valve, since it primarily performs the logic "yes" valve function. (It opens when a pilot signal is applied.) It requires a completely different, and often new, line of design thinking strategy from the conventional valve system, and may aptly be considered the hydraulic equivalent to the electronic transistor. The slip-in cartridge is simply a hydraulic switch that opens or closes when a hydraulic pilot signal is applied to it or removed from it (see figs. 202 and 205).

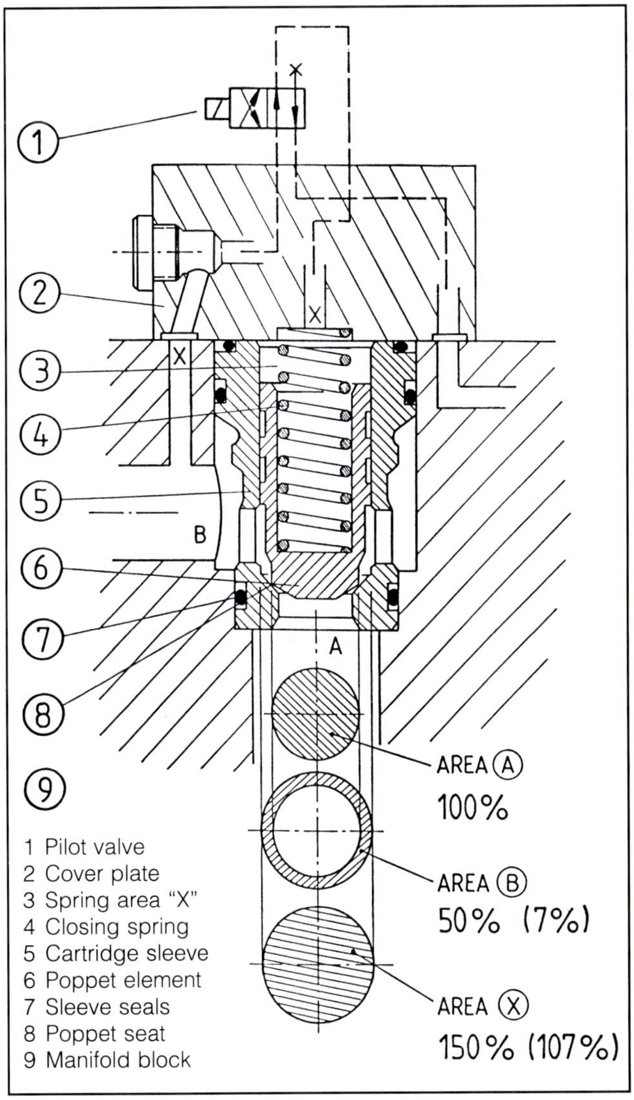

Fig. 202 Slip-in cartridge valve with cover plate and piloting valve.

There are numerous variations to the basic cartridge system, as is shown in figs. 218–221, but its essential components consist of:

- a sleeve which fits into the manifold block including its elastomeric static seals
- one only moving part, called the valve poppet element
- a closing spring
- and a cover plate.

With the absence of any dynamic seals to wear out and practically no frictional resistance to flow (particularly on large cartridges) it becomes obvious that this technology is often superior to conventional valving techniques, using sliding spool valves.

The manifold block becomes the valve housing with flow ports A and B, and the drilling for the pilot lines leading to the cover plate (see figs. 202, 211 and 219). The cover plate connects the pilot lines from the manifold block to the pilot valve and eventually to the spring area, labelled area X of the cartridge. It secures the cartridge sleeve to the manifold block.

Slip-in cartridge valves may also be used with adjustable stroke limiters fitted into the cover plate to limit the poppet opening aperture, so as to provide an additional flow control function to the on/off valve function. Furthermore, chokes (orifices), may be fitted into the signal lines of the cover plate to provide gradual valve poppet opening or closing, where required.

Slip-in cartridge operation

The slip-in cartridge can, generally speaking, be opened to allow flow in either way: A to B or B to A (see figs. 205 and 213). This opening action, however, is only possible if the pressure × area product force for opening exceeds the pressure × area product force for closing. To further clarify the valve's operation, it now becomes necessary to label the various areas and flow ports of the basic slip-in cartridge (see figs. 201 and 202). The area exposed to pressure in port A is called area A or A_A. The area exposed to pressure in port B, is called area B or A_B. The area exposed to pressure in pilot line X is area X or A_X. Area A is often also called the base area, whereas area B may be referred to as the annulus area. Area A added to area B equals area X. Thus, it may be said that:

$$A_A + A_B = A_X$$

Hence, whether a cartridge valve is open or closed depends entirely on the forces acting onto the poppet, and as a basic rule, we may say: "The operation of a cartridge logic valve is always purely pressure dependent!"

Manufacturers of cartridge valves today often follow somewhat differing valve design concepts and philosophies. This is particularly so when it comes to area ratios on cartridge poppets, cover plate designs and applying the cartridge to obtain compound valve functions (pressure and flow controls). The following table and illustration shows some of these valve design variations (see also fig. 202), as applicable for various valve manufacturers.

Cartridge area ratios are based on the area exposed to pressure in port A, which is called the base area (see fig. 201 and table below). It must also be noted that the spring force is based and calculated on the manufacturer's spring pressure rating (0.5–4.0 bar), with springs being available for every 0.5 bar increment. The area to which the spring is opposed is:

for base flow ⇒ area A
for annulus flow ⇒ area B

Manufacturer of cartridge	Type 1 $A_A : A_X$ (A_B)	Type 2 $A_A : A_X$ (A_B)
Bosch	1 : 1.60 (0.60)	——
Rexroth	1 : 1.50 (0.50)	1 : 1.07 (0.07)
Vickers	1 : 2.00 (1.00)	1 : 1.10 (0.10)
Duplomatic	1 : 2.00 (1.00)	1 : 1.10 (0.10)
Parker	1 : 1.67 (0.67)	——
Hägglund	1 : 1.50 (0.50)	1 : 1.07 (0.07)
Sun	1 : 1.80 (0.80)	——

Base flow and annulus flow concepts are depicted in fig. 213. The springs listed above determine the valve's "cracking pressure", which is the pressure required to crack the cartridge valve open.

Cartridge for directional controls

It is common practice to pilot slip-in cartridge valves from small directional control valves. Fig. 204 illustrates how a large double-acting linear actuator may effectively be controlled, using cartridge slip-in valve technology. The circuit represents the centre condition of the D.C.V. shown in fig. 203, whereby all four ports are blocked (closed centre position).

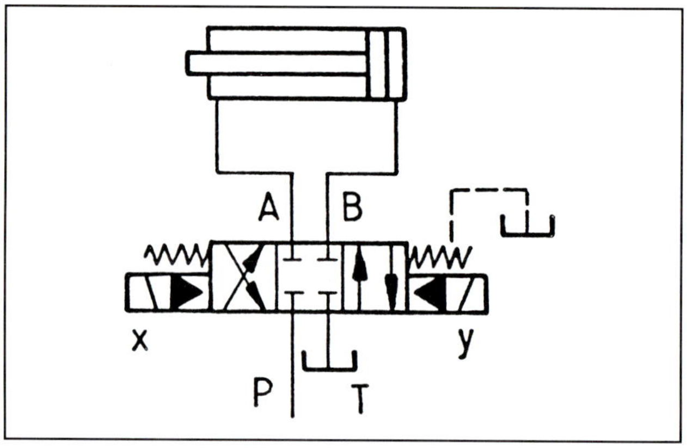

Fig. 203 Conventional solution to cartridge valve circuit shown in fig. 204.

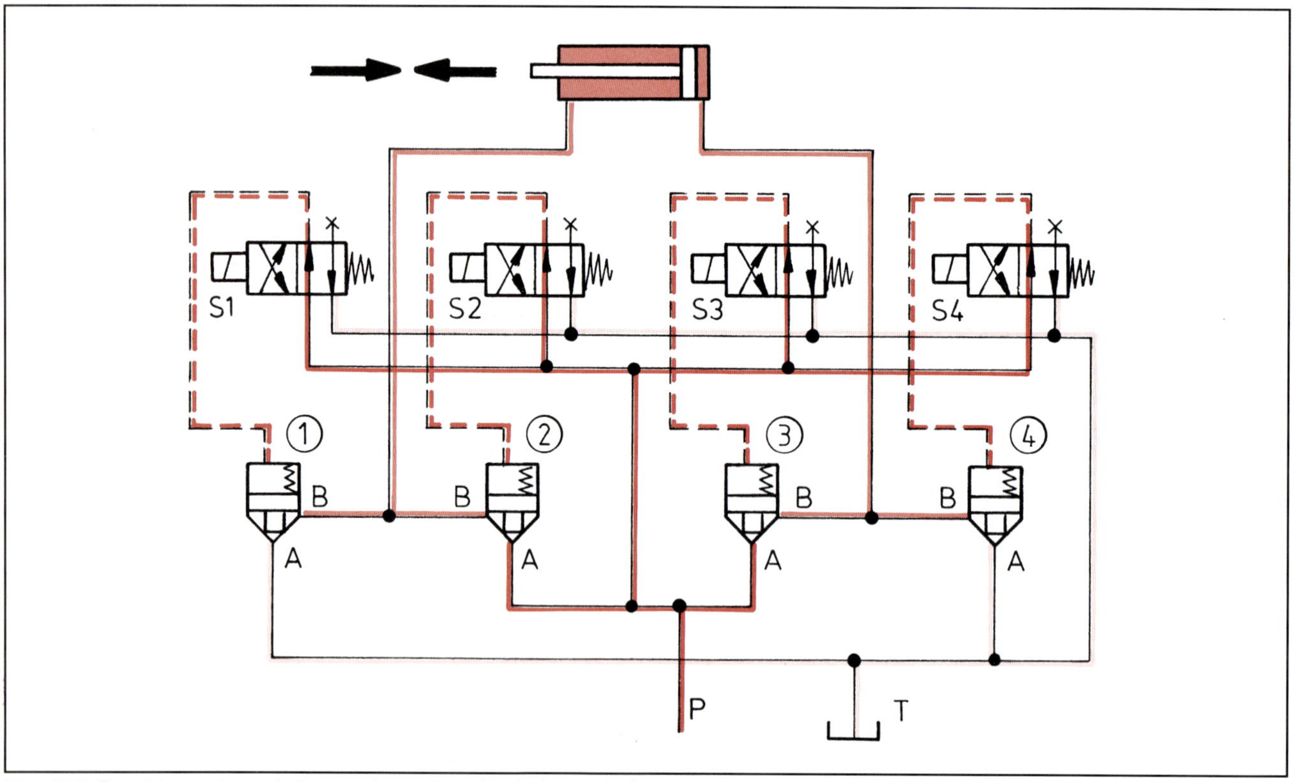

Fig. 204 Cartridge valve solution to alternative circuit shown in fig. 203.

Figure 203 aptly illustrates how this circuit would be achieved with a conventional solenoid pilot directional control valve. Illustrations 205 and 206 show the cartridge and pilot valve actuations for cylinder extension and subsequently cylinder retraction. The control uses four normally closed pilot valves.

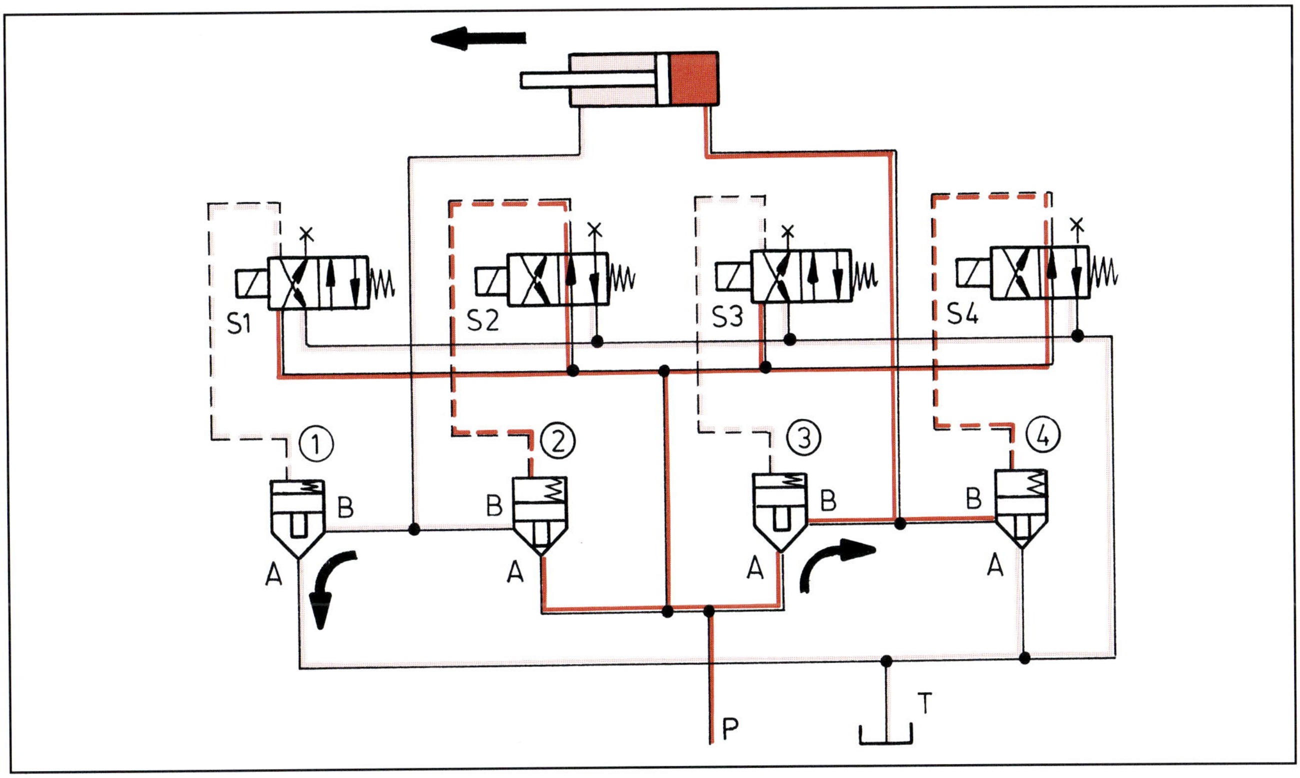

Fig. 205 Actuator extending with pilot valves S1 and S3 actuated.

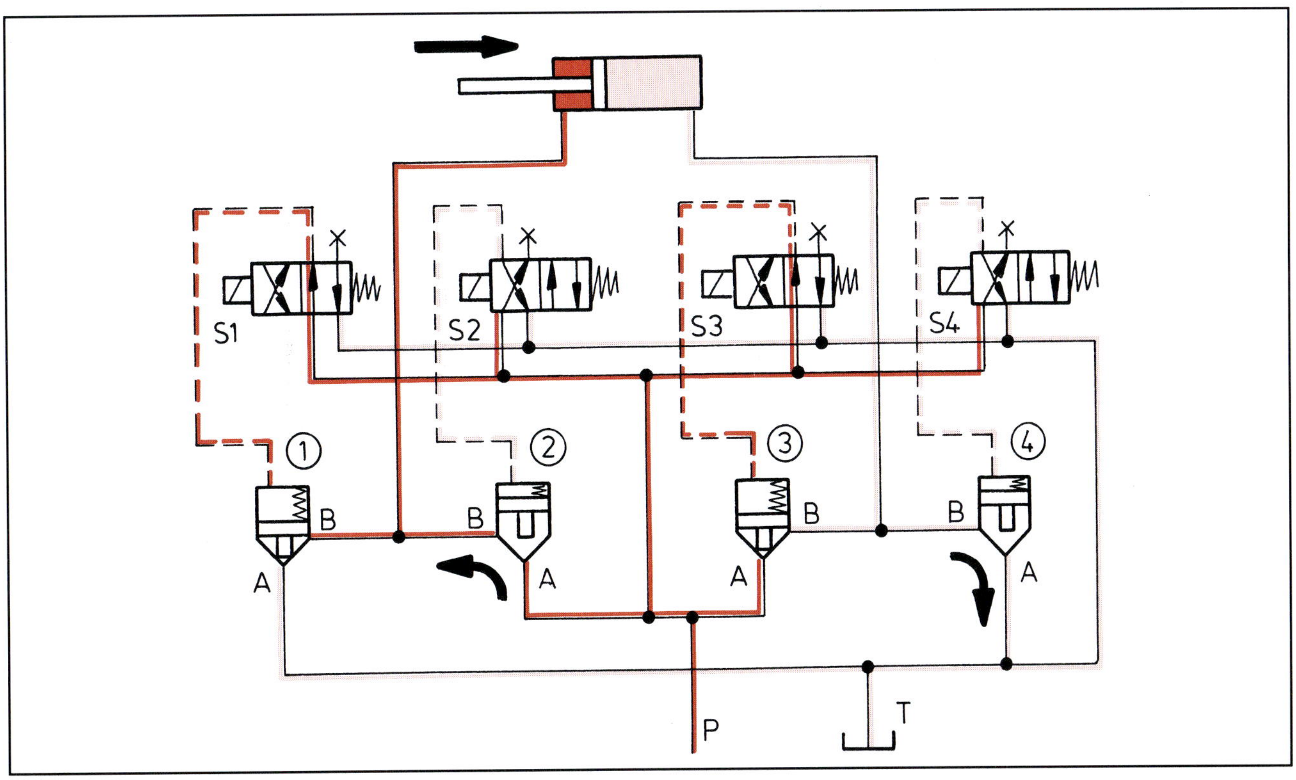

Fig. 206 Actuator retracting with pilot valves S2 and S4 actuated.

With the circuit solution of fig. 207 the cartridges are piloted with a three position, four port, directional control valve, using an A and B to P connection centre configuration. At first sight, this type of circuit appears more economical than the previous circuit concept, illustrated in figs. 204–206.

It must, however, be noted that using only one D.C.V. pilot valve instead of four piloting valves may ultimately prove more expensive, since this type of control concept requires time-consuming fine tuning on site during commissioning, using orifice inserts (pilot chokes). For pilot orifice inserts see figs. 214–216. The problem may be further compounded, because systems of this nature are also more dependent on fluid viscosity than multiple pilot valve systems.

If each of the cartridges is controlled by its own pilot valve (see fig. 208), the switching transitions can be timed and controlled at will by delaying (electronically) the actuation signals to the pilot valves. As a further benefit, by changing the solenoid signal actuation pattern, as shown in fig. 209, one may obtain the twelve flowpath configurations shown in that illustration called the "active valve" envelope. For example, actuating solenoids 2 and 3, one obtains the flowpath configuration A and P to tank, P blocked (no. 7 envelope in fig. 209).

For the flowpath configuration 4 and 13 in fig. 209 the pump is off-loaded back to tank and no system pressure is available to hold the cartridges closed. If firm closure is required, however, one may use pilot

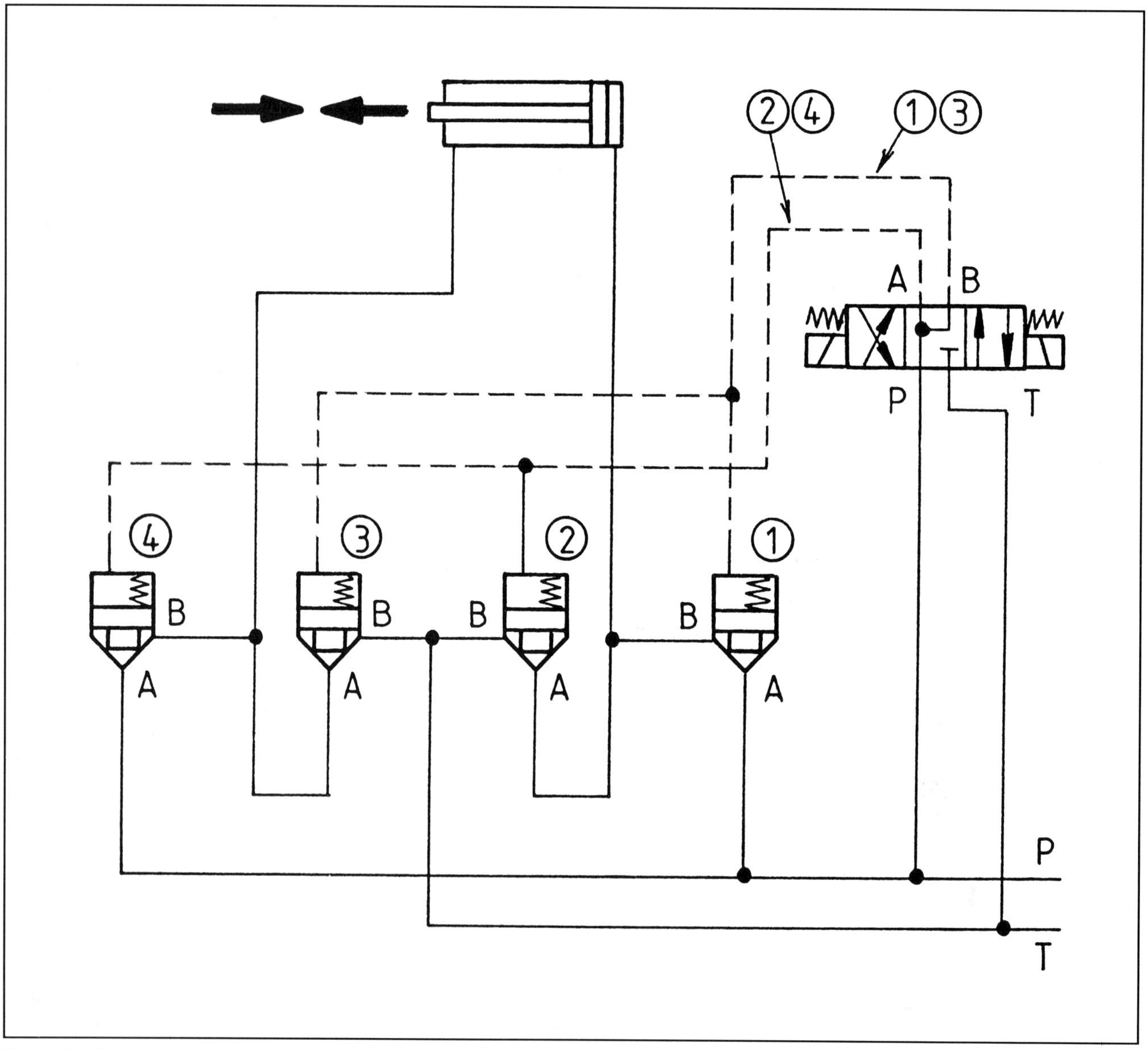

Fig. 207 Circuit using only one piloting valve instead of four, as shown in previous circuits.

pressure from an external source. With flow configuration 7 cavitation could take place if the actuator is moved due to external forces. By pressurising the return line, this problem can be alleviated.

To prove whether a slip-in cartridge exposed to certain pressure conditions would stay open or closed, one may calculate the effective forces on the valve as per the given example. The cartridge under investigation in this example is cartridge no. 1 of fig. 208. It is assumed that an external load force is trying to push the piston back, thus causing a pressure at port B on the cartridge annulus area of 70 bar.

The calculation example is based on a typical Rexroth size 25 cartridge, using a 4 bar spring, an annular area of 50% to the base area (ratio 1:1.5) and the following catalogue and circuit conditions:

Pressure at port A	$p_A = 240$ bar
Pressure at port B	$p_B = 70$ bar
Pressure at port X	$p_X = 240$ bar (as p_A)
Area A (base area)	$A_A = 3.30\ cm^2$
Area B (annulus)	$A_B = 1.61\ cm^2$
Area X (pilot)	$A_X = 4.91\ cm^2$

Opening forces: $(A_A \times p_A) + (A_B \times p_B)$

Closing forces: (spring force) + $(A_A \times p_A)$

$$F_{OPENING} = (0.000\,33 \times 24\,000\,000) + (0.000\,161 \times 7\,000\,000) = 9047\ N \text{ or } 9.047\ kN$$

$$F_{CLOSING} = (0.000\,33 \times 34500) + (0.000\,491 \times 24\,000\,000) = 11\,795\ N \text{ or } 11.795\ kN$$

(Spring force = effective area × required cracking pressure)

These calculations prove that the sum of closing forces is greater than the sum of opening forces; hence the slip-in cartridge will remain closed.

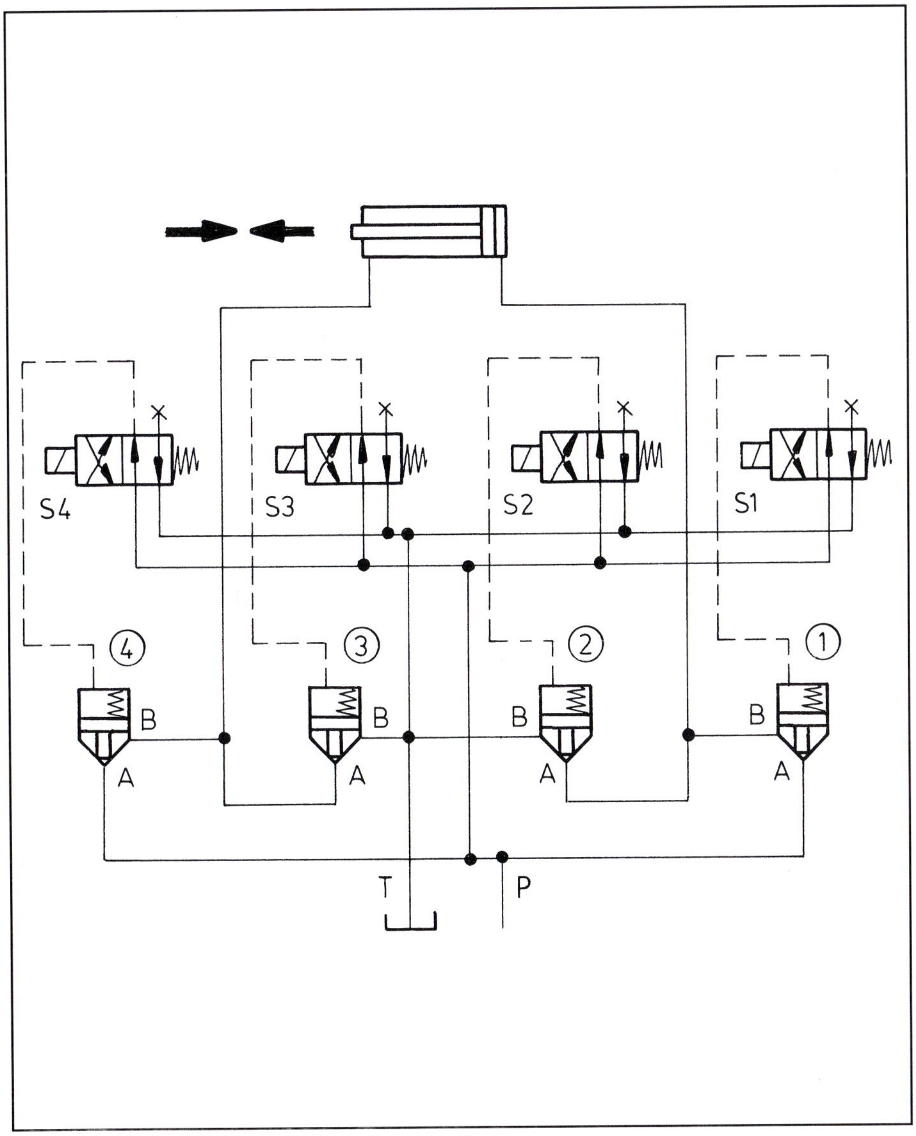

Fig. 208 Circuit using four pilot valves. This permits switching of all twelve valve envelopes as shown in fig. 209.

NO	ACTIVE VALVE	ENERG SOLEN. S4	S3	S2	S1
1	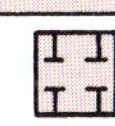				
2		A			
3	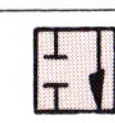		A		
4	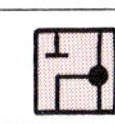	A	A		
5				A	
6		A		A	
7	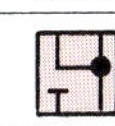		A	A	
8		A	A	A	
9	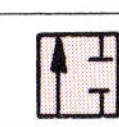				A
10	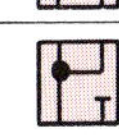	A			A
11	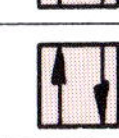		A		A
12		A	A		A
13	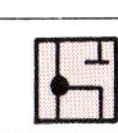			A	A
14		A		A	A
15			A	A	A
16	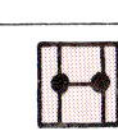	A	A	A	A

Fig. 209 Switching possibilities for circuit in fig. 208 (twelve switching possibilities).

Check valve function

The slip-in cartridge is basically a check valve, if it is used without a pilot valve and if piloting fluid is tapped from cavity B (see fig. 210). With piloting fluid taken from cavity A, the slip-in cartridge would also act as a check valve, but it would not be leak free (see valve leakage investigation and fig. 212). If the piloting fluid is channelled through a pilot valve, as is shown in figs. 211 and 213, the cartridge may become a pilot-operated check valve. But again, it must be noted that if leak free operation is required, the pilot fluid needs to be tapped from cavity B (annulus flow). An application for pilot-operated check control with cartridge valve is given in fig. 222.

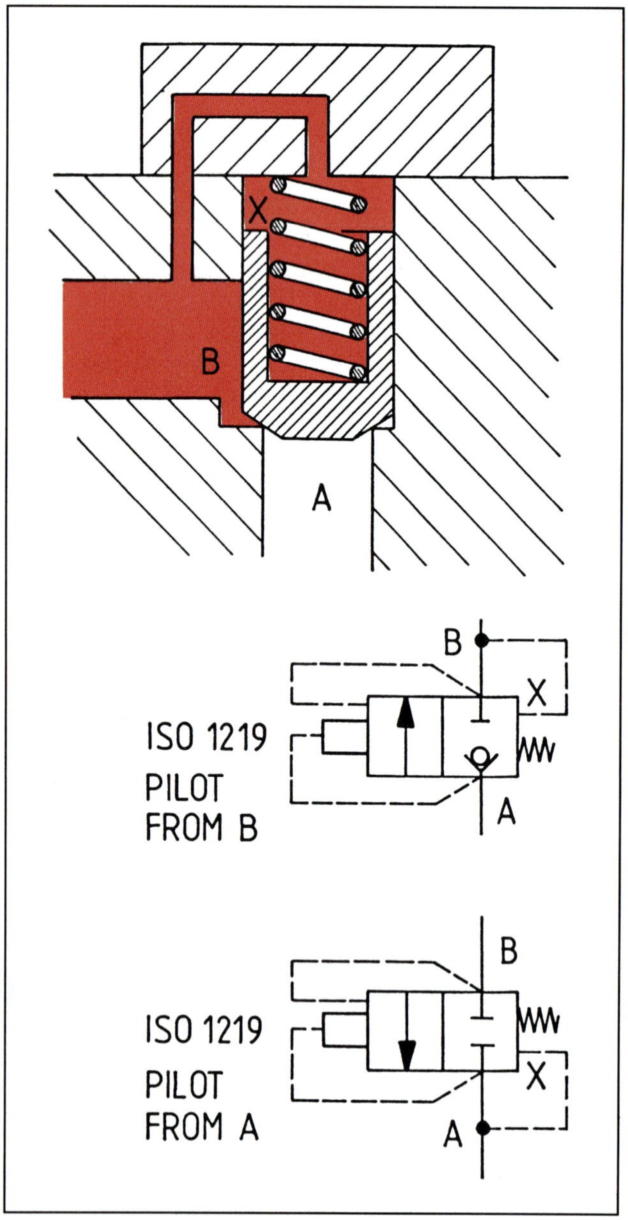

Fig. 210 Cartridge used for normal check valve function. ISO valve symbols for cartridge.

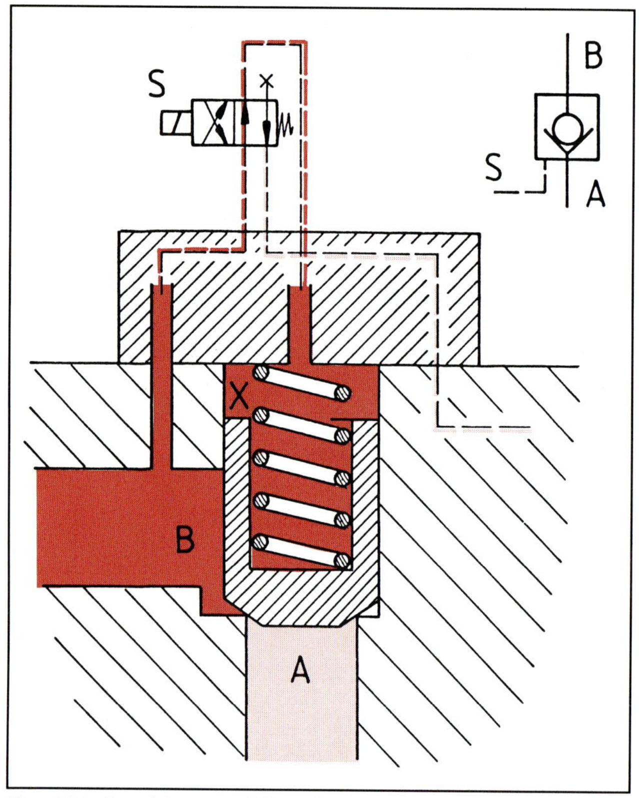

Fig. 211 Cartridge used for pilot-operated check valve function.

Flow direction

If the flow direction through the cartridge is from port A to port B this is then referred to as "base flow". If, however, the flow is from port B to port A, this is called "annulus flow" (see figs. 210 and 213). In most cartridge valve applications the flow may take place in either direction, depending on circuit application. This is particularly true if pilot valves are used to signal cartridge opening or closure (fig. 213). Pressurised flow should, however, preferably be annulus flow, if the pilot control fluid is taken from port B. This eliminates internal cartridge leakage and a more rapid cartridge closure (see fig. 212 and cartridge leakage explanation). Fig. 210 also shows the ISO 1219 symbols for cartridge valves with annulus and base flow.

Valve leakage investigation

To ascertain whether the slip-in cartridge valve is leak-free or not, one must consider the direction in which flow takes place, the location from which pilot flow is tapped (port A or port B), and the type of pilot valve being used to switch the pilot flow.

Generally speaking, spool valves as used for piloting operations do leak; hence, if a completely leak-free piloting circuit is required, one must use poppet type piloting valves instead.

As far as the valve cartridge leakage is concerned (see figs. 210 and 212), the following simple rule may be established:

Pilot oil taken from pressurised port A (base flow) will cause leakage from spring chamber X to tank pressure chamber B between the poppet outer diameter and sleeve inner diameter.

Thus, leakage fluid flows from port A to the tank port B. However, if pilot oil is taken from port B (annulus flow), then the pressure in the B cavity is equal to the pressure in the spring cavity (X) and thus, the valve is leak-free. Furthermore, system pressure forces the valve firmly onto its seat and no leakage can occur past the poppet seat aperture (fig. 213 B). Hence, annulus flow is to be preferred over base flow. To investigate the criteria of cartridge valve leakage, see figs. 212 and 213.

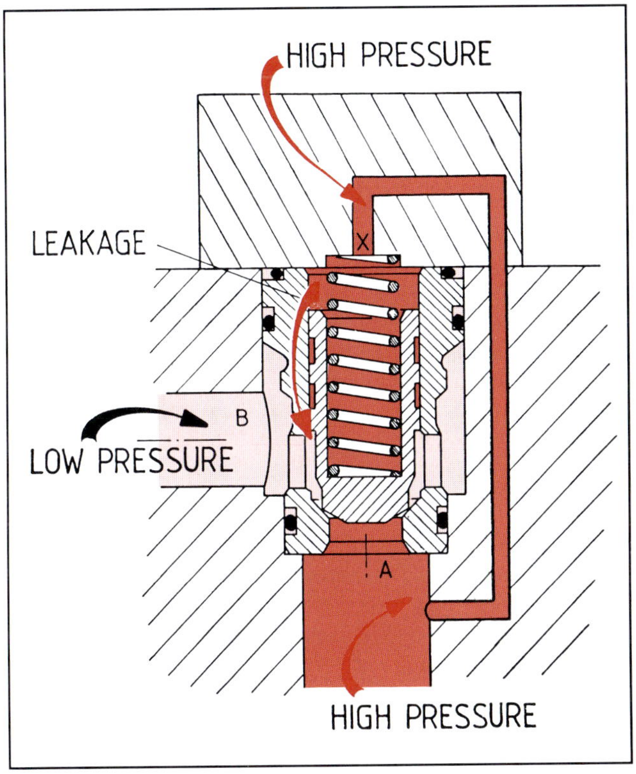

Fig. 212 Cartridge valve leakage occurs when pilot fluid is taken from port A.

Pilot control methods

To open a slip-in cartridge valve, pilot port X must be vented to tank (depressurised), and hence system pressure lifts the poppet from its seat (see fig. 213).

To close the valve, system pressure is connected (via the pilot valve) to port X. The poppet is now temporarily pressure balanced, since pressure acts onto area X as well as areas B and A (areas B and A equal area X).

The closing now takes place in two very rapid phases which may be regarded as two successive phases. In phase one, the closing spring starts to push the poppet, which is now pressure and force balanced, down in the closing direction, until a difference in pressure between the base cavity and annulus cavity occurs. In phase two, initiated by this pressure–force difference, the higher pressure in cavity B is conducted to area X and rapid final closure occurs, with full system pressure acting now onto area X, holding the valve poppet firmly closed.

With annulus flow and pilot pressure being taken from port B, the subsequent closing motion during phase two takes place more rapidly than with base flow (see fig. 213). The reason for this is that the pressure on the larger area A is relieved to tank and thus less force is opposing the closing process. In the case of annulus flow (see fig. 213), the pilot pressure required to ensure a firmly closed position is less than with base flow since system pressure trying to open the valve acts only onto the annular area (50% area) and not onto the 100% area, as is the case with base flow (compare fig. 213 A with fig. 213 B).

Fig. 213 shows the four most frequently used slip-in cartridge valve piloting methods. All four methods are applicable to cartridges with a poppet area ratio of 1:1.07 up to approximately 1:16.

Method A shows pilot fluid being tapped from cavity A and ported to the spring chamber of the cartridge poppet. This will always allow fluid flow from port B to port A, providing pressure $P_B > P_A$, regardless of the position of the pilot valve. Flow from port A to port B may be blocked or open, depending on the pilot valve position. Cavity A may, however, leak fluid to cavity B via the pilot valve, unless a leak-free poppet type pilot valve is applied. Method A shows base flow.

Method B shows pilot fluid being tapped from cavity B. This will always permit fluid flow from port A to port B, providing $P_A > P_B$. Flow from B to A, however, is either free or blocked depending on the pilot valve position. From cavity B fluid may leak via the pilot valve tank port T.

Method C shows pilot fluid tapping from cavity A or cavity B via the "or" function valve (shuttle valve). With the pilot valve not being actuated, flow from B to A as well as from A to B is blocked, which means the poppet does not permit any flow in either direction since the spring chamber of the poppet is pressurised regardless of flow direction. With the pilot valve being actuated, the pressure from the spring chamber is vented back to tank and the cartridge valve will permit flow in either direction (A to B, as well as B to A).

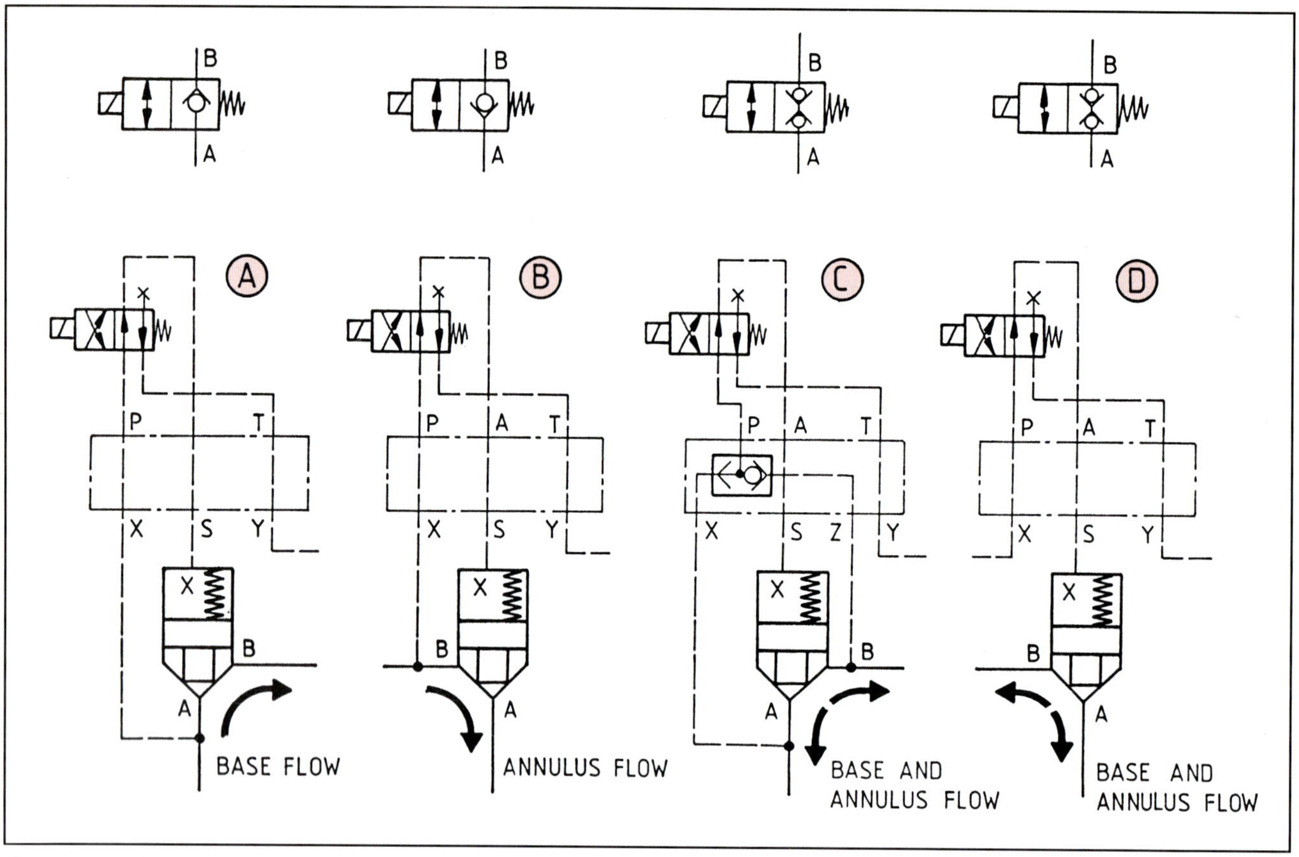

Fig. 213 The four most frequently used slip-in cartridge piloting methods.

Method D depicts pilot fluid being taken from an external source (accumulator or other part of the circuit). With the pilot valve not being actuated, the poppet blocks flow in either direction. With the pilot valve actuated, flow is possible in either direction. Leakage may occur from X to B or B to X, whichever pressure is the greater (see also valve leakage investigation, fig. 212).

Switching speed and control methods

Restricting the in-flowing or out-flowing fluid to and from the spring chamber of the cartridge (port X) will influence the closing or opening speed of the cartridge valve poppet. Illustration 214 depicts six basic methods of piloting speed control.

- In method A the orifice affects (reduces) the opening speed as well as the closing speed, as exit and in-flowing fluid needs to pass through the orifice.
- In method B the orifice reduces only the opening speed. Closing speed is governed by the flowrate through the pilot valve.
- In method C the orifice reduces only the closing speed. Opening speed is governed by the flowrate through the pilot valve.
- In method D orifice 1 reduces the closing speed and orifice 2 reduces the opening speed. Both opening and closure of the poppet valve may be controlled independently of each other.
- In method E poppet closure time is reduced by orifice 1, whereas opening is reduced by orifice 2. Opening speed is, however, also affected by orifice 1, since pilot fluid from the spring chamber plus the pilot fluid from cavity B must pass through orifice 2 to tank. With this piloting method, which is sometimes also called "passive control", pilot fluid will continuously flow via orifices 1 and 2 back to tank for as long as the pilot valve remains open. During that time, the cartridge poppet stays open and, as a negative effect, back pressure caused by orifice 2 will affect the poppet opening response.
- In Method F the negative effects of method E are eliminated by using a flow control with free return flow check. Passive control may be used in slip-in cartridge applications for pressure relief, pressure sequencing and pressure reducing. "Active control" is shown in methods A to D. With active control the spring chamber of the cartridge poppet is not subjected to back pressure and the rising poppet forces all the fluid completely out of the spring chamber.

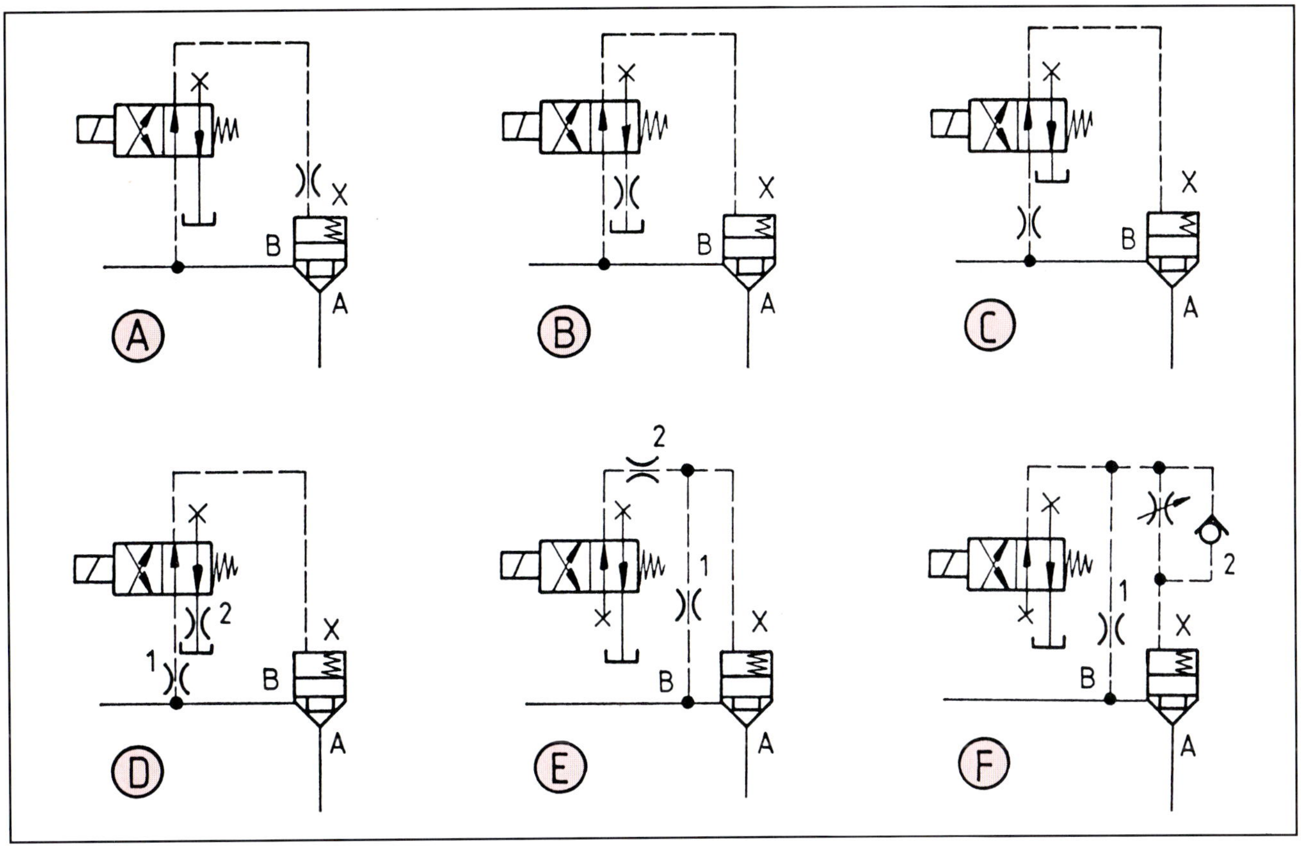

Fig. 214 Six basic methods of controlling cartridge operating speed.

Summary of pilot control methods

Methods A to D are regarded as "active control" methods, which operate without continuous pilot fluid loss, but one pilot valve can only control one cartridge. Methods E and F do operate with a continuous pilot fluid loss, but one pilot valve may control several cartridges, whereby each cartridge has its own orifice to give it independent and unique opening and closing speed control.

Cover plates, orifice inserts and cartridge cavities

Cover plates are an integral part for the use of slip-in cartridge valves. They secure the cartridge to the manifold block, provide connections (oil passages) from the manifold block to the cartridge and to the piloting valves, and house the orifice, or several orifices where required (see figs. 213–216). A plug port is usually also provided to connect a pressure gauge for system diagnostics and fault finding (see fig. 202). In addition, they may contain an "or" function valve (shuttle valve), if more than one pressure is sensed at port "X" of the cartridge (figs. 213 and 215). Cover plates are standardised and precise measurements are laid down in DIN 24342 standards. In the case of flow control functions with a cartridge, the cover plate may also be carrying the stroke limiter mechanism (see fig. 219). For cartridge cavities see data sheet fig. 234.

Orifices are also an integral part of slip-in cartridge valves. They are inserted (housed) in the cover plate (see fig. 216). The orifice determines the opening or closing time of the cartridge valve. Its purpose is to reduce the speed at which the valve poppet opens or closes. Various orifice diameters are available to determine (or tune) cartridge response in relation to the overall hydraulic control system. Orifice inserts are carefully selected by calculation or "trial and error", to obtain maximum system response to switching with minimal hydraulic shocks.

It must be noted that the minimum (shortest) possible valve opening response can never be less than the operating time of the pilot valve! This response will always be at least 25–30 ms. In addition to this, the cartridge itself has a response time of at least 10 ms. Hence, total minimum opening response for a pilot-operated size 16 cartridge is at least 35 ms.

Closing the cartridge depends on a number of factors such as spring size, pressure drop across closing poppet and size of cartridge valve. Its minimal response time is also about 20–25 ms plus the response time of the pilot valve.

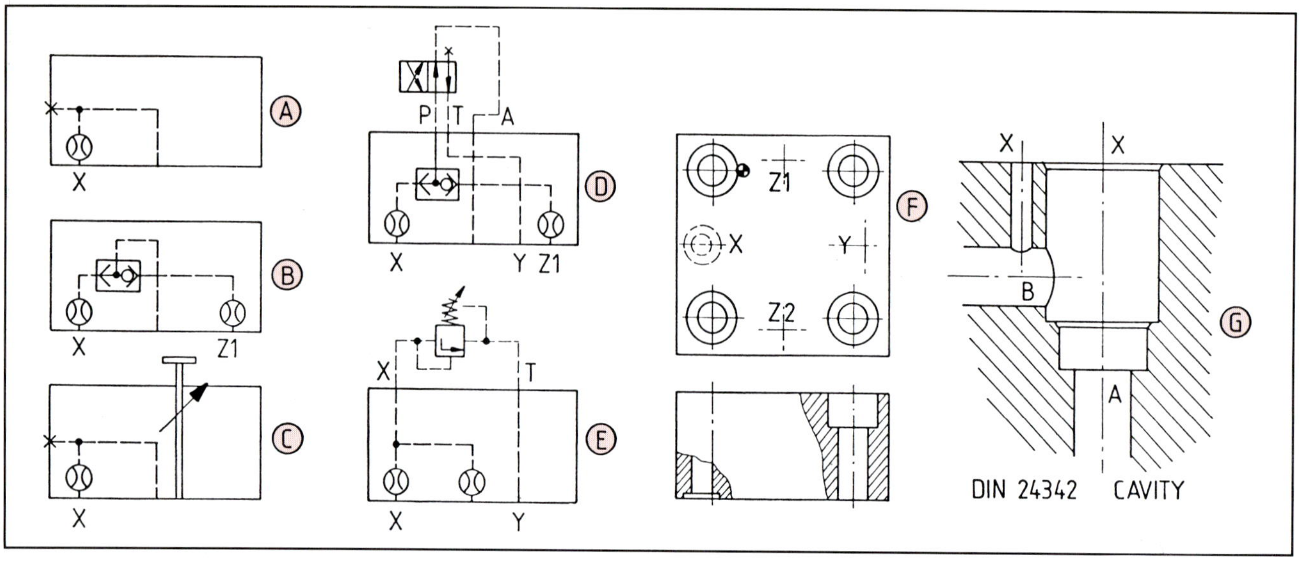

Fig. 215 Cover plates and cavity standard DIN 24342 (see also fig. 234).

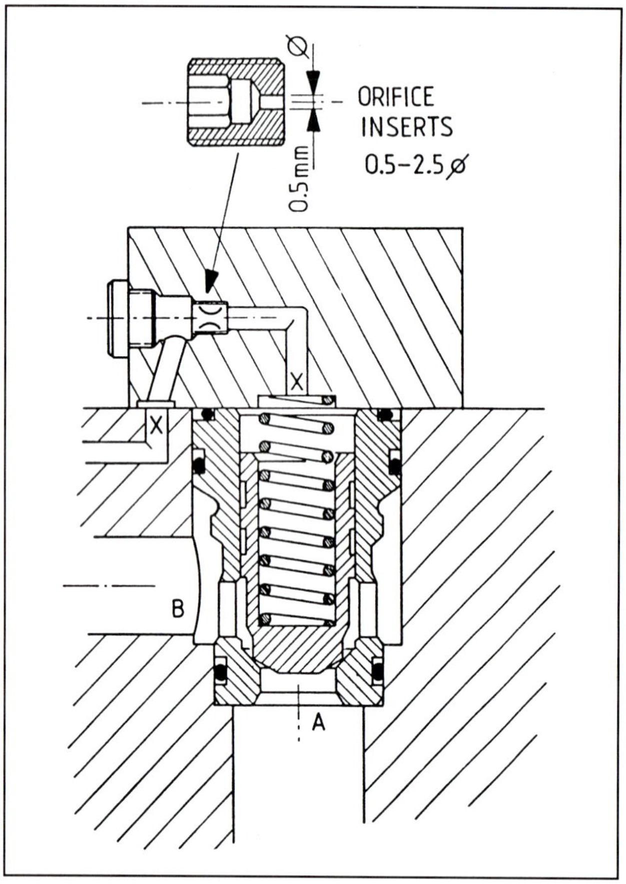

Fig. 216 Orifice embedded in cover plate and plug for pressure gauge connection.

Orifice sizing by calculation

Selection of orifice aperture by calculation must be based on the following three parameters:

- expected opening or closing time of the cartridge poppet (t)
- anticipated pressure drop (maximum system pressure to tank pressure) through the orifice (Δp)
- fluid volume passing through the orifice during poppet opening or closure (L/min).

Average expected opening or closing time of the poppet is supposed to be about 40 ms. Shorter times do make the system work faster, but as a trade-off, one has to expect pressure shocks.

The pressure drop through the orifice is usually between maximum system pressure upstream, and zero (tank) pressure downstream.

The piloting fluid volume passing through the orifice depends entirely on the geometrical volume of the cartridge spring cavity. This information is normally provided by the cartridge valve manufacturer (see fig. 233).

A close and precise observation of the valve opening process shows the following: the initial pressure drop is causing high exit flow velocity. As the poppet opens progressively, both the measured flow velocity and pressure drop show a corresponding fall. Hence the opening process is not a linear function. Experience has shown that in normal cases the accelerating actuator piston should have reached maximum stroking velocity in about 40 ms or 0.04 s, with the poppet at this point being 30% opened. This also means that 30% of the pilot fluid volume in the spring cavity must have passed through the orifice by that time (30% of the poppet stroke).

To give a realistic calculation example, the figures used here are based on a Rexroth size 40 slip-in cartridge without dampening nose and a geometrical spring cavity volume of 16.6 cm^3 (see data sheets).

$V = 16.6\ cm^3$, $t = 0.045\ s$, stroke = 30%

$$Q = \frac{V}{t} = \frac{30 \times 16.6 \times 60}{10^2 \times 0.04 \times 10^3} = 7.47\ L/min$$

Since the pressure drop during poppet opening is not linear, it is therefore suggested, for practical purposes, to use only two-thirds of the maximum pressure drop. Hence, with a maximum system pressure of 300 bar, two-thirds would be 200 bar. With these two quantities (flowrate 7.47 L/min and a pressure drop of 200 bar) transferred into the orifice selection nomogram in fig. 217, one will find a theoretical orifice of 1.0 mm or 1.2 mm. Rexroth suggest the following orifices:

Cartridge size	Orifice diameter
size 16	0.7 ∅
size 25	0.8 ∅
size 32	1.0 ∅
size 40	1.2 ∅
size 50	1.5 ∅
size 63	1.8 ∅
size 80	2.0 ∅
size 100	2.5 ∅

"Rexroth" standard orifices

Poppet with dampening nose

Poppets may also be selected with a dampening nose. This nose increases the opening or closing time (fig. 218) of the poppet and it is particularly used with base flow.

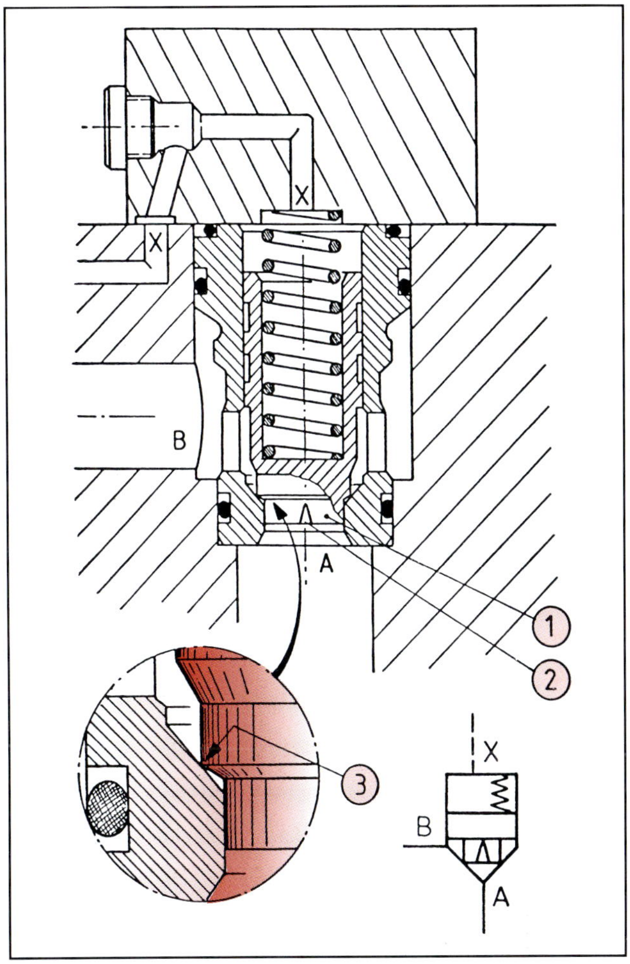

Fig. 218 Poppet with dampening nose ① and flow control notches ②. The sharp poppet edge ③ seals against the 45° sleeve face angle.

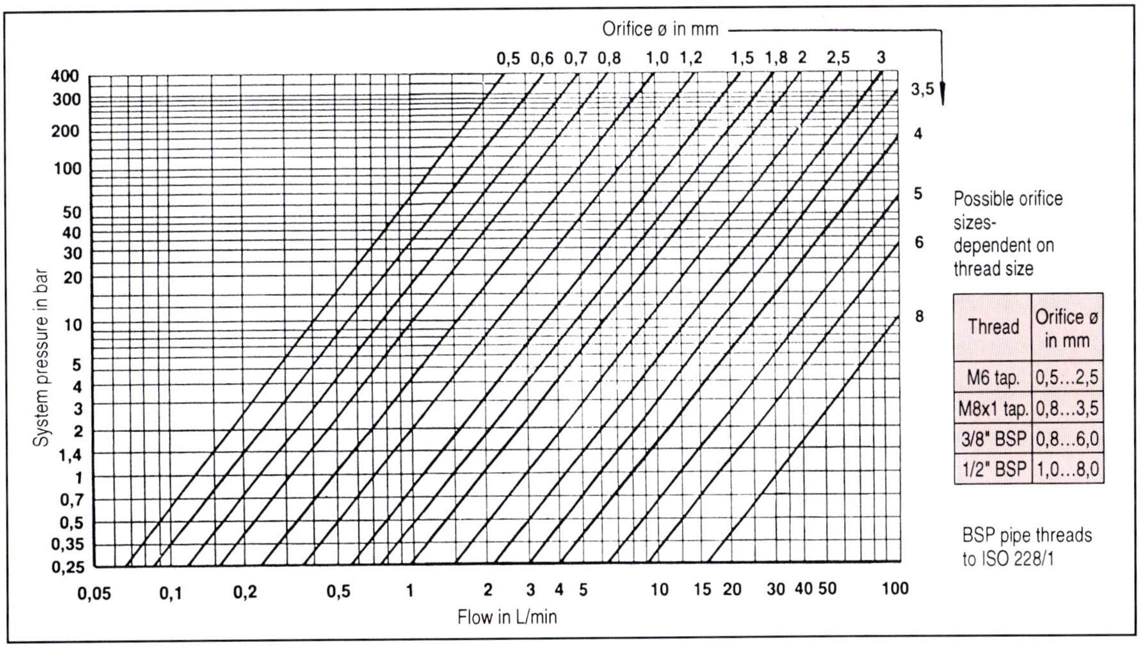

Fig. 217 Orifice selection nomogram.

If it is used with annulus flow, the final closure motion of the poppet is more rapid, due to an increased differential pressure. Hence, the valve tends to slam shut (metallic banging noise).

Poppets with dampening nose may be useful in decompression control circuits, deceleration circuits (fast to slow speed), flow control applications in conjunction with stroke limiter devices and shock-free operation of a linear actuator.

Slip-in cartridge for flow control

Although in itself not a flow control, a cartridge valve may in conjunction with a variable stroke limiting device become a flow control cartridge valve. A stroke limiter restricts the opening movement (stroke) of the valve poppet and thus causes a flow passage restriction (see fig. 218). By screwing the stroke limiter spindle in or out of the cover plate one can infinitely vary the flowrate through the cartridge valve; hence the cartridge becomes a directional control valve plus variable flow controller all in one component.

The poppet on such flow-controlled slip-in cartridge valves is usually furnished with a dampening nose to provide better control of the throttling function. An extension to this element is an electronically proportionally adjustable stroke limiter to achieve pressure-compensated flow control.

Slip-in cartridge for pressure control

In pressure limiting or pressure sequencing functions, the cartridge is used as the main flow-passing valve to cause the pump flow to return to tank or in a sequencing function to open a flow passage into the secondary system (see Chapter 5, "Pressure controls", figs. 99, 113 and 114). A direct-acting

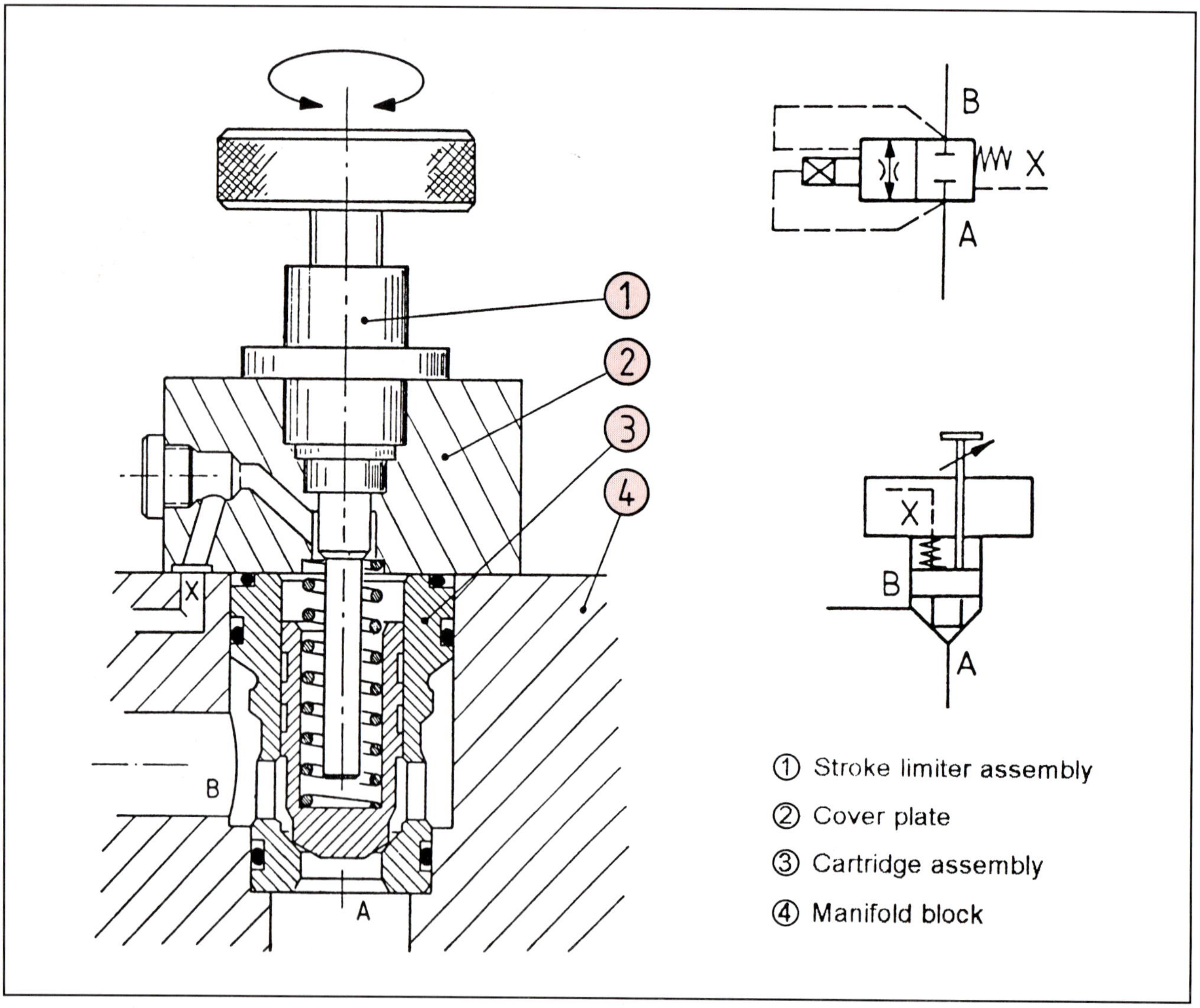

Fig. 219 Slip-in cartridge with stroke limiter for flow control.

pressure relief valve is used to pilot the cartridge poppet (see fig. 220). Together the two valves become a compound system relief valve, or if an external spring chamber drain is used on the pilot valve, the compound component functions as a pressure sequencing valve. Since the secondary system valve outlet opens into a pressurised section of the circuit, the spring chamber of the pilot valve needs a separate tank return line. For relief function, this spring chamber drain may be emptied into port B cavity of the manifold block.

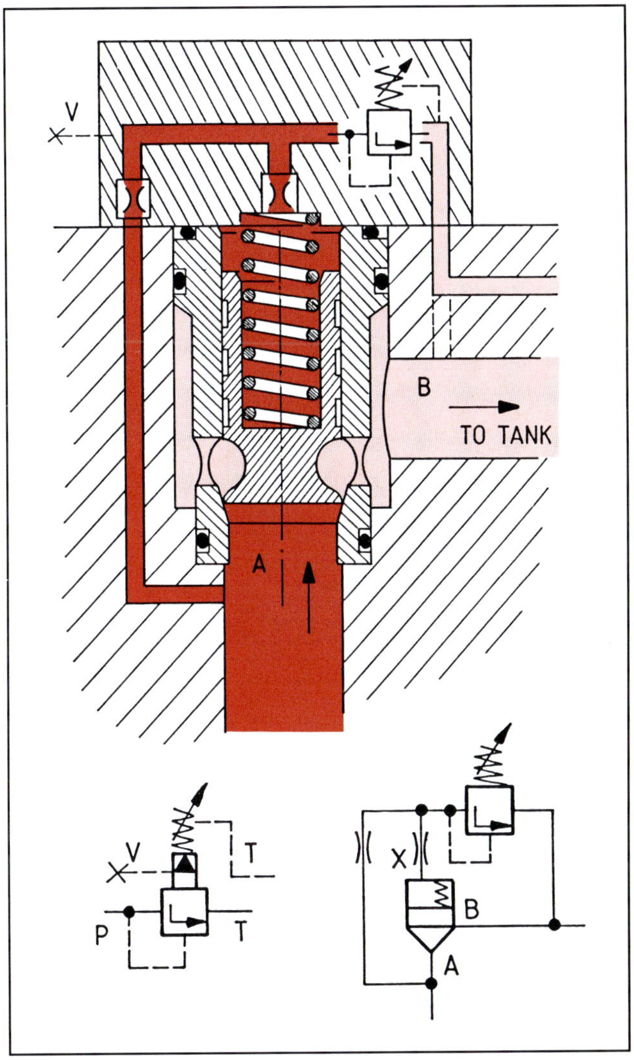

Fig. 220 Slip-in cartridge for pressure control (no B area). Normally closed pressure relief valve acts as piloting valve.

Cartridge valve for pressure reduction

The pressure reducing valve function in this compound element is accomplished by means of a "normally open" cartridge (see fig. 221) and a direct-acting relief valve (see also Chapter 5, figs. 114 and 115). This type of pressure-reducing valve is ideal for passing high flowrate with a minimum of power loss.

The flow direction is from B to A. The valve functions primarily the same way as the valve described in Chapter 5. As long as the inlet pressure at port B is below the calibration of the pilot valve, the cartridge spool remains open (held down by the spool spring). When the outlet pressure in cavity B reaches the calibrated pressure setting of the pilot valve, control fluid flows through the pilot valve to tank. This creates a pressure drop Δp across the metering orifice, which causes the cartridge spool to close the through passage partly off. Such partial closure diminishes the flowrate to port A or closes it off altogether, if no further flow is required at port A. Meanwhile, the pilot valve remains open and pilot flow passes back to tank.

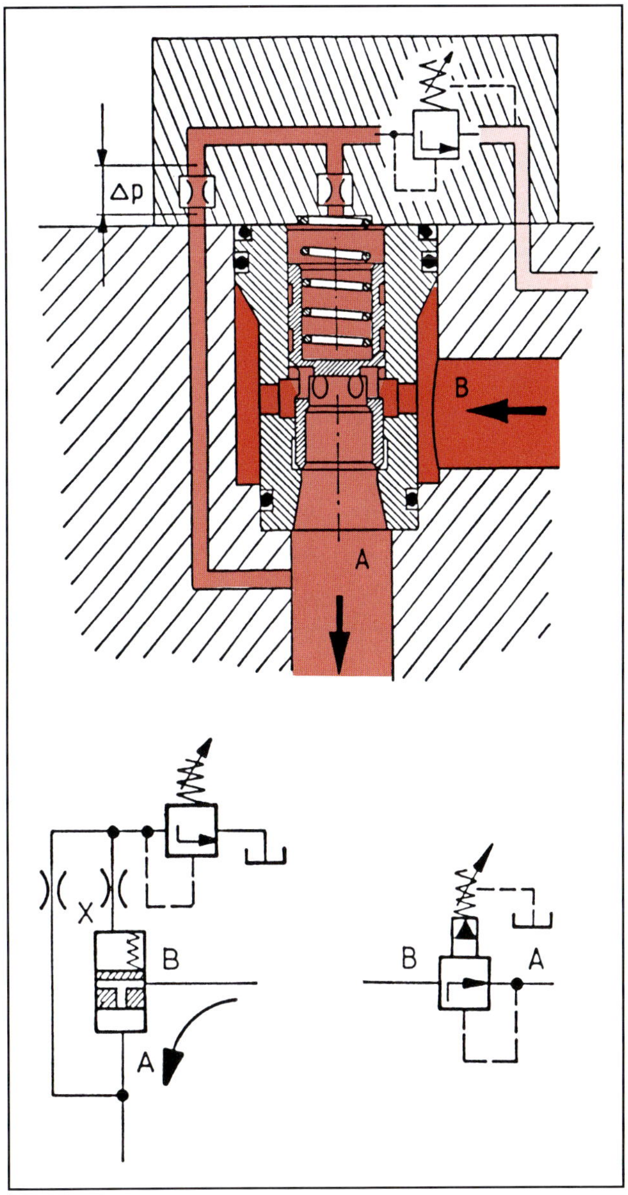

Fig. 221 Slip-in cartridge for pressure reduction. The cartridge is normally open and of spool type design. The valve is piloted by a simple pressure relief valve.

Cartridge poppet position monitoring

Some manufacturers, in their cartridge valve range, also offer a special cartridge cover with inbuilt electrical position sensor. The solid state proximity switch with integral amplifier switches when the poppet reaches the fully closed position. The proximity switch, its amplifier and the electrical cable connection plug are all part of the cover plate (see fig. 222, symbol A). The limit switch usually has no dynamic seals. It provides precise and fully reliable monitoring of the closed position of the cartridge poppet, and for covers with stroke limiter; it also has an exact external adjustment provision. The symbols for position monitoring in covers, with or without stroke limiter, or combined with a signalling valve, are shown in fig. 222.

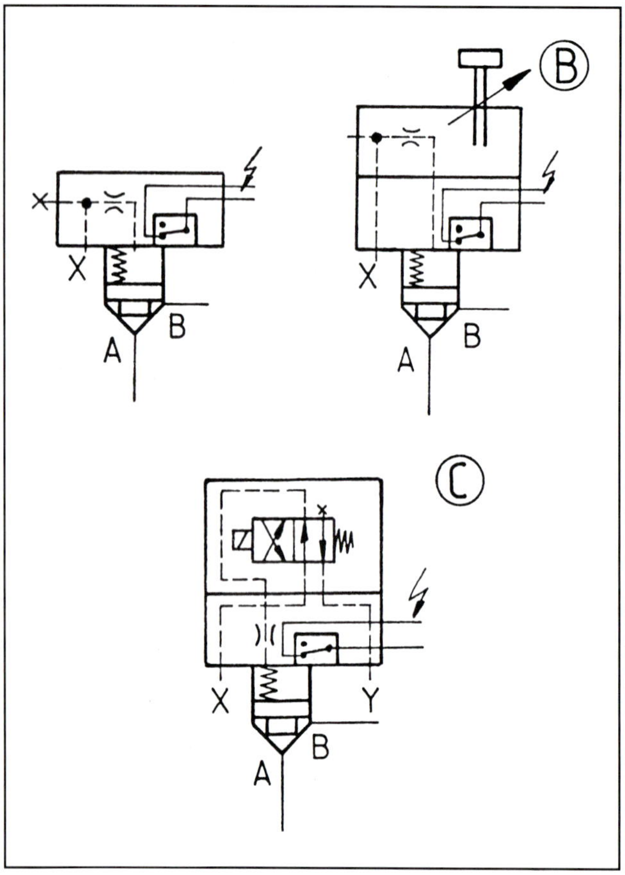

Fig. 222 Slip-in cartridge covers with (A) electronic poppet position monitoring, (B) position monitoring plus stroke limiter, and (C) position monitoring plus signalling valve.

Circuit design with slip-in cartridges

Control problem 1: Regenerative circuit

A conventional circuit for a flow regeneration application is shown in fig. 222. The left-hand part of the table indicates which solenoid signals are to be actuated (symbol "A"), to achieve the various control modes. Fast approach with flow regeneration is signalled if solenoid signals S2 and S4 are actuated, whereby flow from the rod-end cylinder chamber is regenerated to the cap end. For slow final stroking, only solenoid S4 is actuated and the rod end fluid flows back to tank. Retraction is signalled by solenoids S1 and S3 being actuated. The circuit in fig. 224 shows the same control, using slip-in cartridge valves. Cartridges 3 and 4 act as a pilot operated check valve. The control is used on a plastic forming and bonding press, where the actuator must remain extended and apply force for some time without creeping back. Slip-in cartridge valves allow large flowrates with extremely small pressure drops, thus making this circuit an ideal model for the use of slip-in cartridges instead of using extremely large conventional directional control valves with their inherent large pressure and power losses.

	CONVENTIONAL CIRCUIT				CARTRIDGE CIRCUIT			
	S1	S2	S3	S4	S1	S2	S3	S4
STOP								
FAST EXTEND		A		A		A		A
SLOW EXTEND				A	A			A
RETRACT	A		A			A	A	

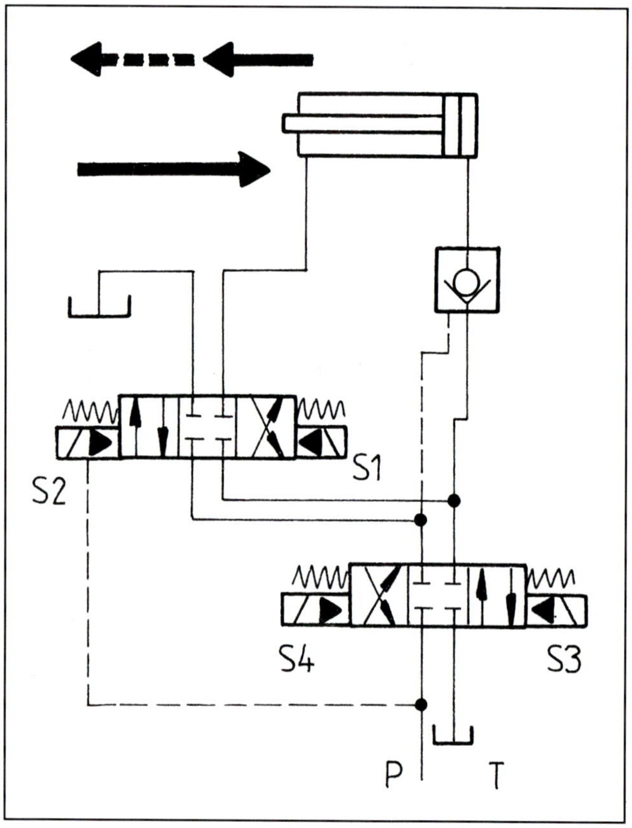

Fig. 223 Regenerative control using conventional valves.

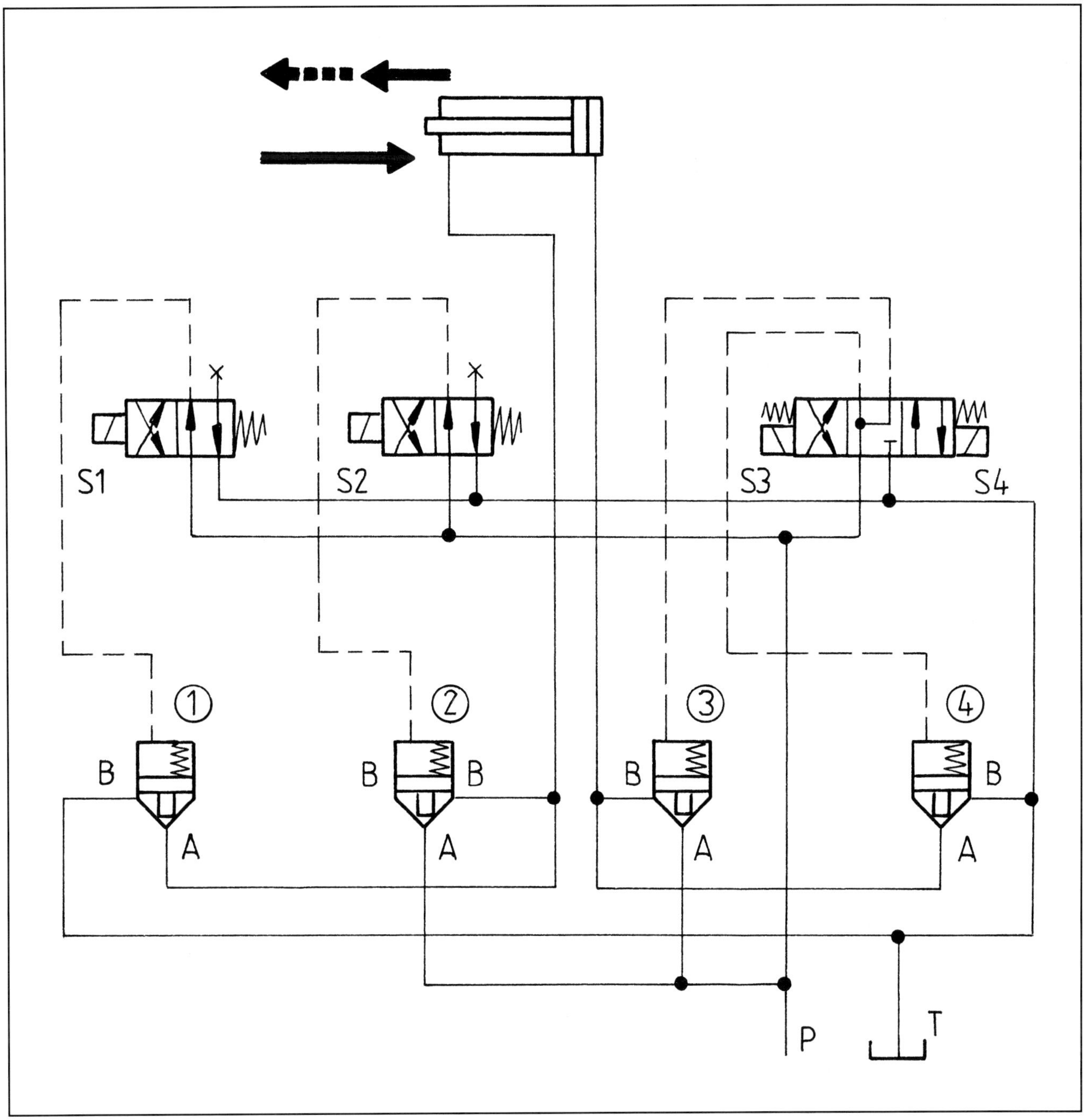

Fig. 224 Regenerative control as shown in fig. 223, but using slip-in cartridges instead.

Control problem 2: Pressure sequence valve

As explained before, with pressure sequence valves, the spring chamber of the pilot valve must not be drained into the pressurised B system, and for satisfactory system functioning, a free reverse flow check is required. If the cartridge in use has no B area (annulus area, see figs. 201, 220, 225–227). A pressure sequencing circuit is shown in fig. 111 (Chapter 5, "Pressure controls"). The cartridge circuit in fig. 226 uses passive control with pilot fluid continuously flowing to tank (see also fig. 214).

If the sequencing circuit is required to allow pump flow to stream into the B circuit when an electrical signal is given, or when the pressure threshold is reached, then the circuit in fig. 228 could be used. The cartridge opens when the pressure-reducing pilot valve opens, or when the directional pilot valve opens, thus permitting the cartridge spring chamber to be unloaded to tank.

If pressure sequencing is to be combined with position sequencing, then the circuit in fig. 229 may be used (see also fig. 112). In this application, the position sequencing directional control valve 1 is connected in series to the pilot valve 2. Their combined action vents the spring chamber of cartridge 3 to tank, and thus permits flow from the pump to stream into system B.

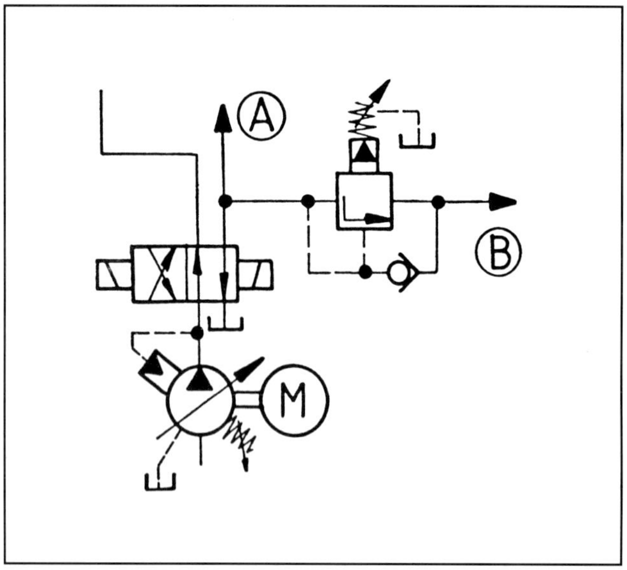

Fig. 225 Conventional pressure sequencing circuit.

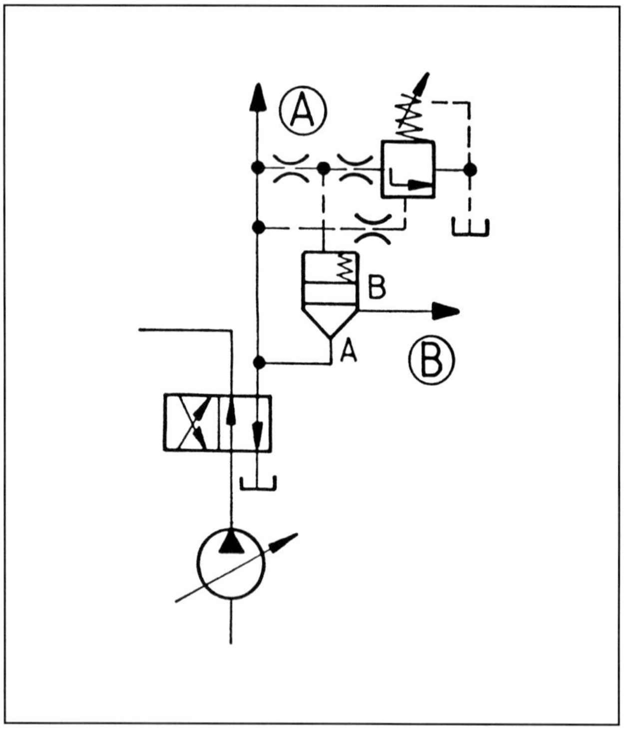

Fig. 226 Pressure sequencing circuit with cartridge valve for flow into circuit B. Cartridge sequencing is accomplished with a direct-acting relief valve.

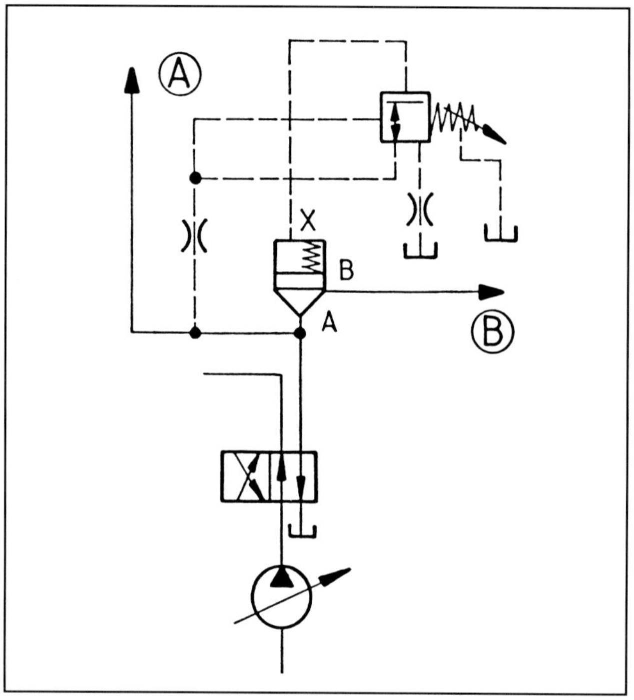

Fig. 227 Pressure sequencing circuit with cartridge valve for flow into circuit B. Cartridge sequencing is accomplished with a direct-operated pressure reducing valve.

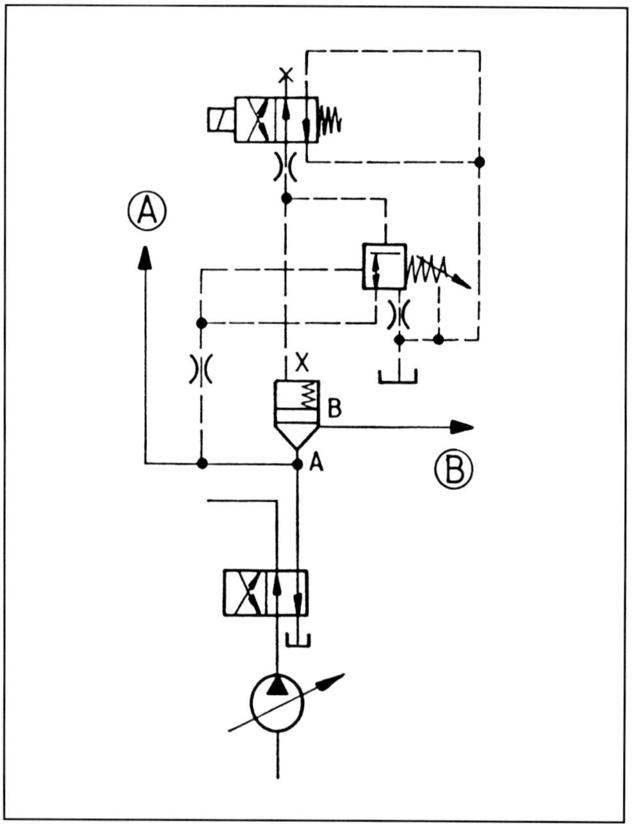

Fig. 228 Pressure sequencing circuit with cartridge valve for flow into circuit B. Cartridge sequencing is accomplished with an electrical solenoid valve or alternatively with a direct operated pressure reducing valve.

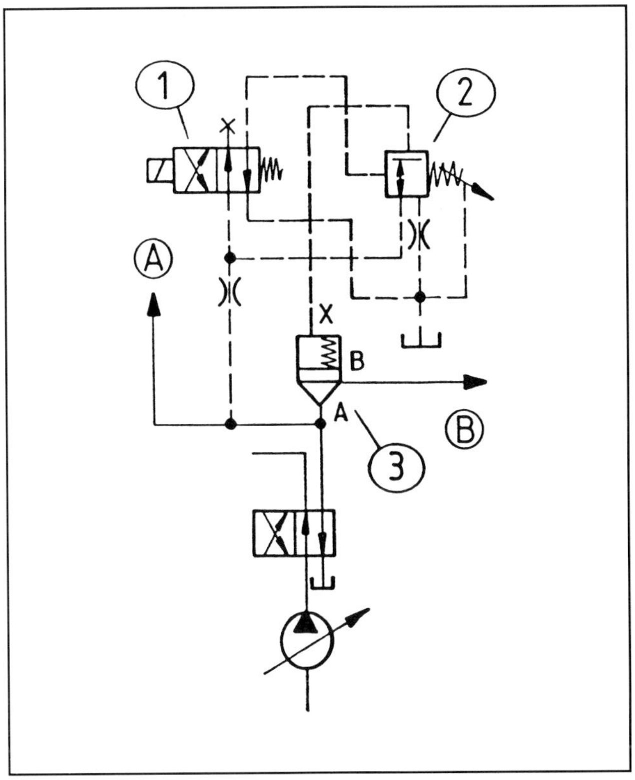

Fig. 229 Pressure sequencing circuit with cartridge valve for flow into circuit B. Cartridge sequencing is by means of an "and" function, consisting of position and pressure sequencing.

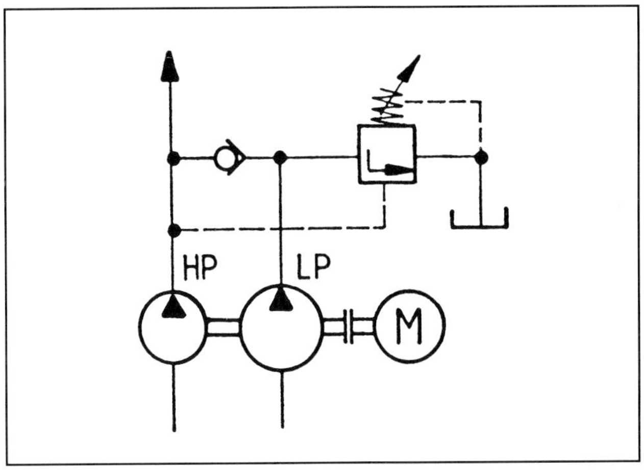

Fig. 230 Conventional low pressure pump off-loading circuit.

Control problem 3: Pump off-loading

Pump off-loading with a conventional off-loading valve system has been explained in Chapter 5 and is illustrated in fig. 107. The low pressure/large volume pump, with a conventional system, would necessitate a very large off-loading valve. The same may be achieved in a manifold block system with a much more compact slip-in cartridge arrangement (compare fig. 230 with fig. 232). Passive pilot control is again used in fig. 231 and the check valve preventing high pressure from exiting through the off-loading valve is also a slip-in cartridge.

Warning!

In some countries, poppet position monitoring is a requirement for press control circuits. Consult local industry safety authorities and their regulations before designing any hydraulic circuits with slip-in cartridge valves!

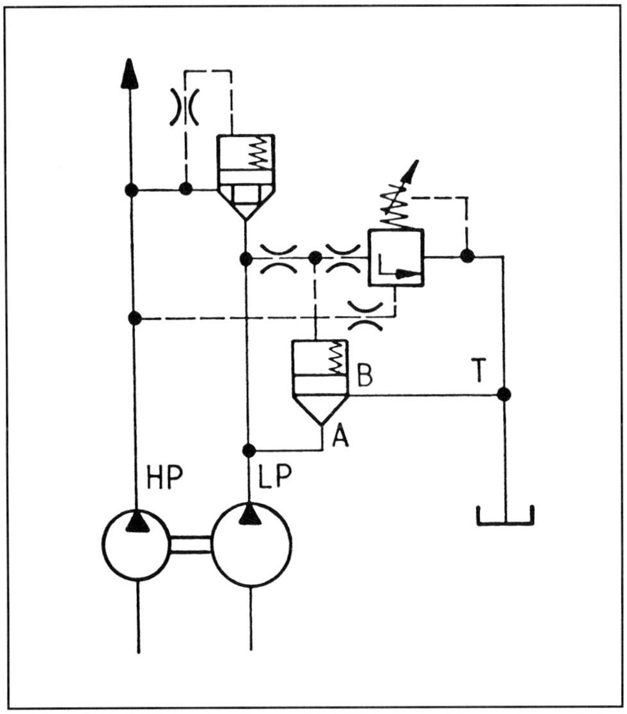

Fig. 231 Pump off-loading circuit with cartridge valve for pump off-loading and cartridge valve for check valve. The off-loading cartridge is controlled by a direct-acting relief valve.

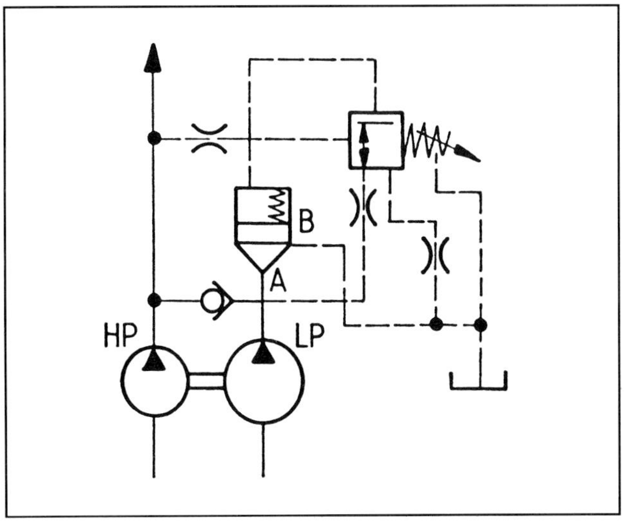

Fig. 232 Pump off-loading circuit with cartridge valve for pump off-loading, conventional check valve and direct operating pressure-reducing valve to control cartridge operation.

Technical data (For application outside these parameters, please consult us!)		
Fluid		Mineral oils to DIN 51524 (HL, HLP) Phosphate ester (HFD-R)
Fluid temperature range	°C	– 20 ... + 70
Viscosity range	mm^2/s	2,8 ... 380
Operating pressure at ports A, B, X, Z1 and Z2	bar	... 315 (Cover with built-on directional spool valve; Cover with built-on directional poppet valve [315 bar model])
	bar	... 420 (Cover without built-on directional spool valve; Cover with built-on directional poppet valve [630 bar model])
Operating pressure at port Y	bar	≙ Tankpressure for built-on valves

2 way cartridge valves – directional control function

Sizes		16	25	32	40	50	63	80	100	125	160
Area A1 cm^2	LC..A..	1,54	3,3	5,3	9,24	16,6	22,9	37,9	63,6	95	160,6
	LC..B..	2,14	4,6	7,55	12,95	22,9	32,2	52,8	89,1	133,7	224,8
Area A2 cm^2	LC..A..	0,73	1,61	2,74	4,61	8,03	11,3	18,84	31,4	48	79,9
	LC..B..	0,13	0,31	0,49	0,9	1,73	2,0	3,94	5,9	9,3	15,7
Area A3 cm^2	LC..A.. LC..B..	2,27	4,91	8,04	13,85	24,63	34,2	56,74	95	143	240,5
Stroke cm	LC..E..	0,7	0,78	0,92	1,2	1,6	1,9	2,4	3,0	3,8	5,0
	LC..D..	0,7	1,0	1,22	1,6	2,0	2,4	3,0	3,8	4,8	6,5
Pilot volume cm^3	LC... E	1,6	3,8	7,4	16,6	39,4	65	136	285	544	1203
	LC... D	1,6	4,9	9,8	22,2	49,3	82	170	361	687	1563
theo. pilot oil flow for a control time of 10 ms in L/min	LC... E	9,6	22,8	44	100	236	390	816	1710	3264	7218
	LC.. D	9,6	29,4	59	133	296	492	1020	2166	4122	9378
Weight	Cartridge kg	0,2	0,4	1,0	1,8	3,8	7,0	13,0	27,0	44,0	75,0
	Cover kg	1,2	2,3	4,0	7,4	10,5	21,0	27,0	42,0	80,0	150,0
Cracking pressure											
Direction of flow A → B	LC..A00..	0,02	0,025	0,05	0,05	0,05	0,07	0,07	0,1	0,15	0,15
	LC..A05..	0,43	0,45	0,46	0,43	0,45	0,42	0,44	0,43	0,43	0,45
	LC..A10..	0,86	0,88	0,91	0,87	0,87	0,85	0,88	0,88	0,88	–
	LC..A20..	1,76	1,77	1,85	1,73	1,74	1,7	1,75	1,75	1,76	1,94
	LC..A30..	–	–	–	–	–	–	–	–	2,05	–
	LC..A40..	3,4	3,45	3,34	3,49	3,35	3,32	3,13	3,04	–	–
	LC..B00..	0,014	0,02	0,035	0,035	0,035	0,05	0,05	0,07	0,1	0,1
	LC..B05..	0,31	0,32	0,32	0,31	0,32	0,3	0,31	0,31	0,31	0,32
	LC..B10..	0,62	0,63	0,64	0,62	0,63	0,61	0,63	0,63	0,62	–
	LC..B20..	1,27	1,27	1,3	1,24	1,26	1,21	1,26	1,25	1,25	1,4
	LC..B30..	–	–	–	–	–	–	–	–	1,45	–
	LC..B40..	2,45	2,47	2,35	2,5	2,43	2,36	2,25	2,17	–	–
Direction of flow B → A	LC..A00	0,04	0,05	0,1	0,1	0,1	0,14	0,14	0,2	0,30	0,33
	LC..A05	0,9	0,92	0,89	0,86	0,93	0,85	0,88	0,88	0,86	0,91
	LC..A10	1,8	1,8	1,77	1,74	1,8	1,73	1,77	1,78	1,73	–
	LC..A20	3,7	3,6	3,6	3,46	3,6	3,44	3,53	3,54	3,50	3,9
	LC..A30	–	–	–	–	–	–	–	–	4,00	–
	LC..A40	7,2	7,1	6,5	7,0	6,9	6,7	6,3	6,2	–	–
	LC..B00	0,24	0,25	0,5	0,5	0,5	0,8	0,7	1,0	1,5	1,5
	LC..B05	5,0	4,8	4,9	4,1	4,3	4,7	4,2	4,6	4,4	4,6
	LC..B10	10,0	9,4	9,8	8,2	8,4	9,6	8,4	9,4	8,9	–
	LC..B20	20,6	19,0	20,0	16,4	16,7	19,0	16,9	18,7	17,9	20
MANNESMANN REXROTH	LC..B30	–	–	–	–	–	–	–	–	20,7	–
	LC..B40	40,0	36,8	36,0	33,2	32,2	37,0	30,2	32,5		–

Fig. 233 Technical data for cartridge sizes 16–160 taken from "Rexroth" hydraulics data sheets. *Note*: spring chamber volumes, cracking pressures and theoretical pilot oil flow for 10 ms cartridge response in L/min.

Installation to DIN 24342 (except sizes 125 and 160) (dimensions in mm)

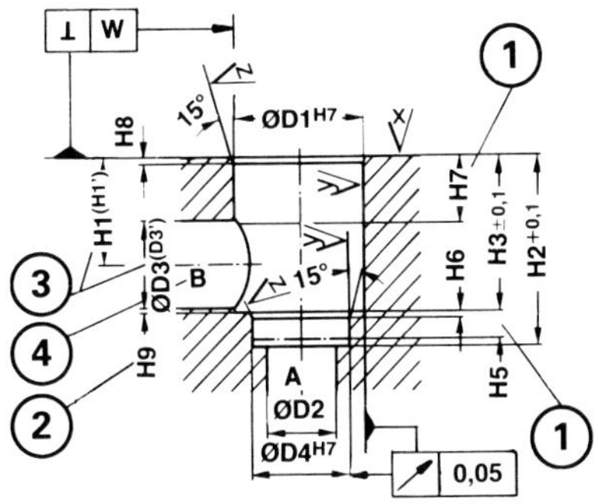

Size	16	25	32	40	50	63	80	100	125	160
ØD1	32'	45	60	75	90	120	145	180	225	300
ØD2	16	25	32	40	50	63	80	100	150*	200*
ØD3	16	25	32	40	50	63	80	100	125	200
(ØD3')	25	32	40	50	63	80	100	125	150	250*
ØD4	25	34	45	55	68	90	110	135	200	270
ØD5	M8	M12	M16	M20	M20	M30	M24	M30	–	–
ØD6*	4	6	8	10	10	12	16	20	–	–
ØD7	4	6	6	6	8	8	10	10	–	–
H1	34	44	52	64	72	95	130	155	192	268
(H1')	29,5	40,5	48	59	65,5	86,5	120	142	180	243
H2	56	72	85	105	122	155	205	245	300+0,15	425+0,15
H3	43	58	70	87	100	130	175±0,2	210±0,2	257±0,5	370±0,5
H4	20	25	35	45	45	65	50	63	–	–
H5	11	12	13	15	17	20	25	29	31	45
H6	2	2,5	2,5	3	3	4	5	5	7±0,5	8±0,5
H7	20	30	30	30	35	40	40	50	40	50
H8	2	2,5	2,5	3	4	4	5	5	5,5+0,2	5,5+0,2
H9	0,5	1	1,5	2,5	2,5	3	4,5	4,5	2	2
L1	65/80	85	102	125	140	180	250	300	–	–
L2	46	58	70	85	100	125	200	245	–	–
L3	23	29	35	42,5	50	62,5	–	–	–	–
L4	25	33	41	50	58	75	–	–	–	–
L5	10,5	16	17	23	30	38	–	–	–	–
W	0,05	0,05	0,1	0,1	0,1	0,2	0,2	0,2	0,2	0,2

* max

X/ = √R max 4; Y/ = √R max 8;
Z/ = √Rz10

Hole pattern for cover

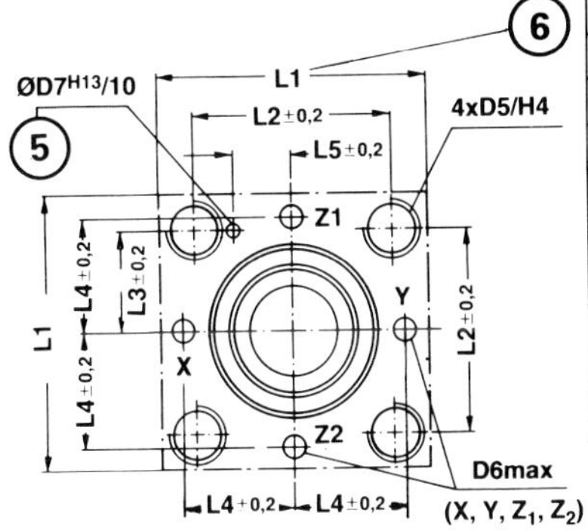

Sizes 16 ... 63

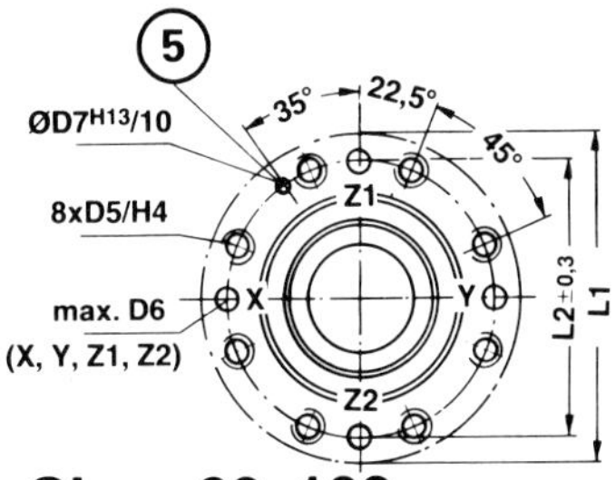

Sizes 80, 100

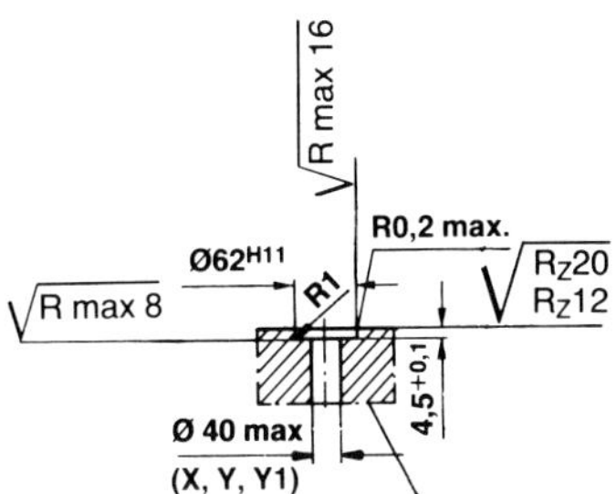

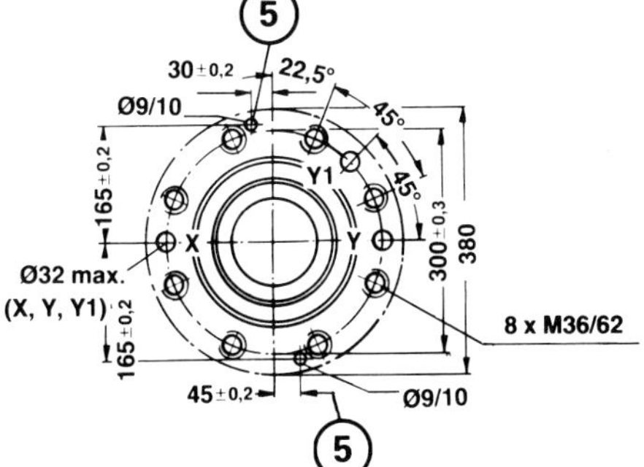

Size 125

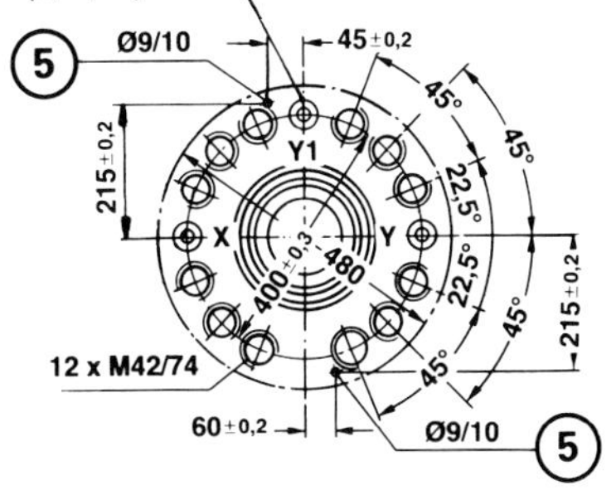

Size 160

Bore details

1 depth of fit
2 reference dimension
3 For other sizes of port B than ØD3 or (ØD3'), the distance from the cover face to the centre of the port must be calculated.
4 Port B can be move around the central axis of port A, but care must be taken to ensure that metal is retained around the mounting screw holes.
5 Drilling for locating pin
6 Note on Size 16: Length L1 (along the x - y axis) becomes 80 mm with a built on pilot directional valve.

MANNESMANN REXROTH

Fig. 234 Technical data for cartridge sizes 16–160 taken from "Rexroth" hydraulics data sheets, depicting engineering information and measurements for cavity machining and cover plate mountings including mounting hole sizes.

14 Power saving pump controls

The second half of the twentieth century has been plagued by global pollution and its consequence, the "greenhouse effect". For this and for other reasons, government authorities in Europe, Japan and numerous other countries no longer tolerate machinery that is excessively power wasting. Hence machine design engineers had to develop concepts that will make more economical use of power at all times; this, of course, applies also to fluid power drives. One of these engineering concepts that has been developed is hydraulic load sensing.

The concept of load sensing is simple. It ensures that the prime mover and the pump produce exactly enough hydraulic power to satisfy the demands of the actuators (motors, cylinders). Thus a well designed load-sensitive hydraulic system virtually eliminates the heat problem with its numerous negative effects on fluid power components and the running costs of the hydraulic system.

Load-sensing methods

Load-sensing methods vary from manufacturer to manufacturer, particularly in relation to variable displacement pump controllers. However, load sensing may be broadly grouped into pump control load sensing and relief valve load sensing. In order to ease the grasping of its concepts, the principle of relief valve spring chamber loading is explained first and illustrated. At this point it may also be helpful to consult the chapter "Pressure controls" (pages 62–66) in the textbook *Industrial Hydraulic Control* by Peter Rohner. In the illustration (fig. 235), both valve poppets, that of the main valve 4, and in series the poppet of the remote simple relief valve 3, resist the pilot oil flow to tank (through line 1). Hence, the pilot pressure at point 2 is the sum of both resistances, and it determines the opening pressure 5, at which the main relief valve 6 opens and pump flow is redirected to tank 7. Consequently, the lowest system pressure is set on the main relief valve 8, and higher system pressures may be adjusted by adding a back pressure 9 to the poppet 4 in the main relief valve pilot spring chamber 9.

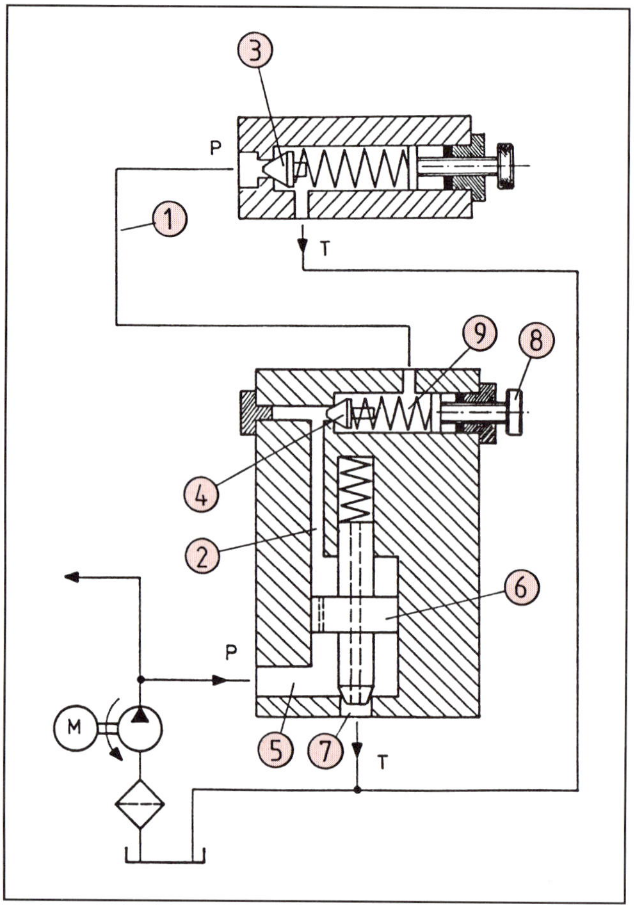

Fig. 235 Relief valve spring chamber loading.

Relief valve load sensing

Load sensing is much better understood if one investigates it with a case problem. The first case problem version has no load sensing applied to it and for the given operating conditions (fig. 236) is grossly power wasting. The second version has relief valve load sensing, which is again not the ultimate solution, but may be regarded as a form of load sensing (fig. 236). Load, pressure and flowrate data are given for both versions, and calculations plus a power graph illustrate power wastage or power saving.

For the first version of the case problems presented, the motor is assumed to operate but at a small percentage of its maximum possible load capacity (torque). Furthermore, a flow control valve reduces the motor input flowrate to 10 L/min. Its pump, however, produces a flowrate of 20 L/min to cater for other actuators that are presently not in use (figs. 236 and 237).

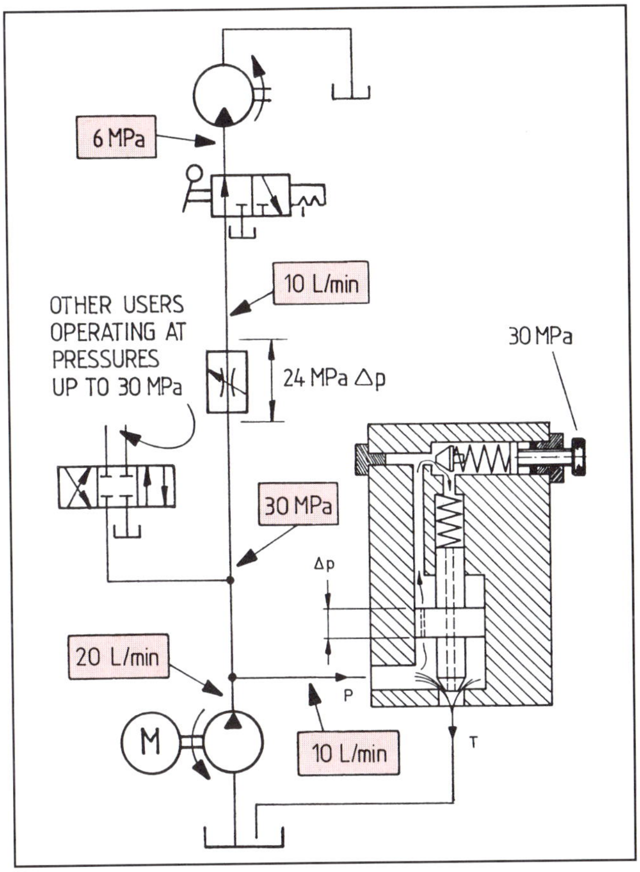

Fig. 236 Circuit without load sensing.

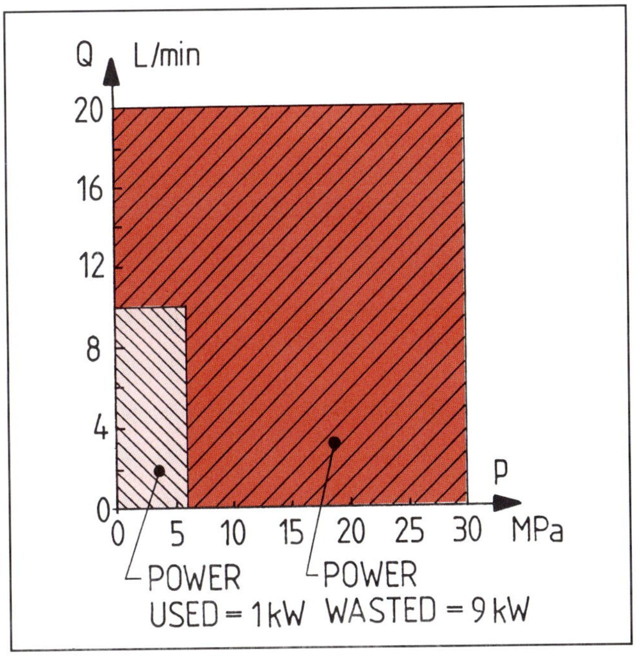

Fig. 237 Flow–pressure–power diagram for circuit without load sensing.

The system relief valve is set to 30 MPa, whereas motor load-induced pressure, measured at the motor inlet port, is at a mere 6 MPa.

Case problem version 1 calculation:

$$\text{Pump input power} = \frac{30 \times 10^6 \times 20}{10^3 \times 60 \times 10^3} = 10 \text{ kW}$$

$$\text{Motor output power} = \frac{6 \times 10^6 \times 10}{10^3 \times 60 \times 10^3} = 1 \text{ kW}$$

$$\text{Power wastage} = 10 \text{ kW} - 1 \text{ kW} = 9 \text{ kW}$$

For simplicity reasons, we assume 100% pump, motor, and system efficiency (see fig. 236).

For the second version of the case problem we assume operating conditions identical to the first version, except that the circuit is fitted with a relief valve load-sensing modification. This consists of a "load sensing pilot line" plus a remote simple relief valve for spring chamber loading of the main relief valve (see fig. 238).

When the directional control valve is closed to the motor, the main relief valve will open at a pump pressure of 1 MPa, and the fluid flowing past its pilot poppet will pass to tank via the directional control valve. This pilot poppet flow is then governed by the restrictions in the main valve and the orifice in the load sensing pilot line. The bulk of the pump flow (remainder) will flow through the main relief valve back to tank.

When the motor is running, and is loaded to induce a load pressure of say 6 MPa, this load induced pressure is then added to the closing force of the pilot valve spring in the main relief valve. The pressure on the pump is therefore 7 MPa and the pressure on the motor inlet is 6 MPa. Hence the main relief valve always opens at a pressure of 1 MPa above load sensing pressure and surplus pump flow can be diverted to tank via the open main relief valve at 1 MPa above load sensing pressure. For example: should the load induced pressure be say 9 MPa, the pump will then operate at 10 MPa and surplus pump flow will return to tank at 10 MPa.

The simple relief valve for remote pressure loading will not open until load induced pressure reaches 29 MPa. Hence maximum system pressure is limited to 30 MPa (29 MPa + 1 MPa). For the given case problem, the pump will therefore operate at 7 MPa (compare figs. 235 and 238). The size of the orifice limiting the pilot oil flow in the load sensing line is extremely important to the proper operation of the system. It should therefore be a variable orifice to permit fine tuning. Furthermore, the load sensing line should be kept as short as possible and must be rigid tubing (not hose). Relief valve load sensing, although a clever idea, is not as effective as variable pump load sensing.

Case problem version 2 calculation:

$$\text{Pump input power} = \frac{7 \times 10^6 \times 20}{10^3 \times 60 \times 10^3} = 2.333 \text{ kW}$$

$$\text{Motor output power} = \frac{6 \times 10^6 \times 10}{10^3 \times 60 \times 10^3} = 1 \text{ kW}$$

$$\text{Power wastage} = 2.333 \text{ kW} - 1 \text{ kW} = 1.333 \text{ kW}$$

Compared with the first example, there is a power saving of 7.666 kW. For verification, compare fig. 236 with fig. 238 and fig. 237 with fig. 239.

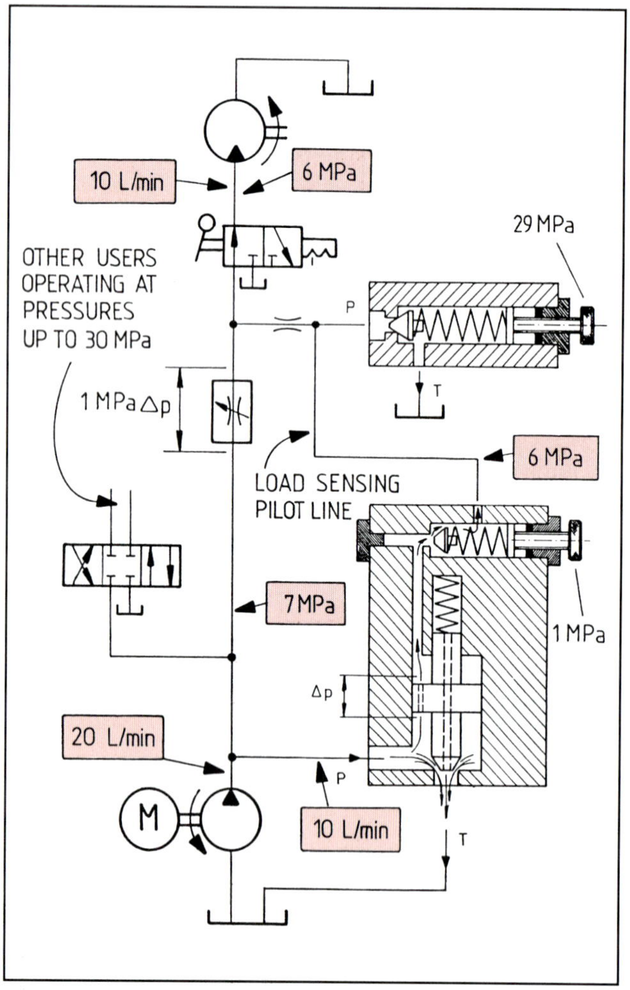

Fig. 238 Circuit with relief valve load sensing.

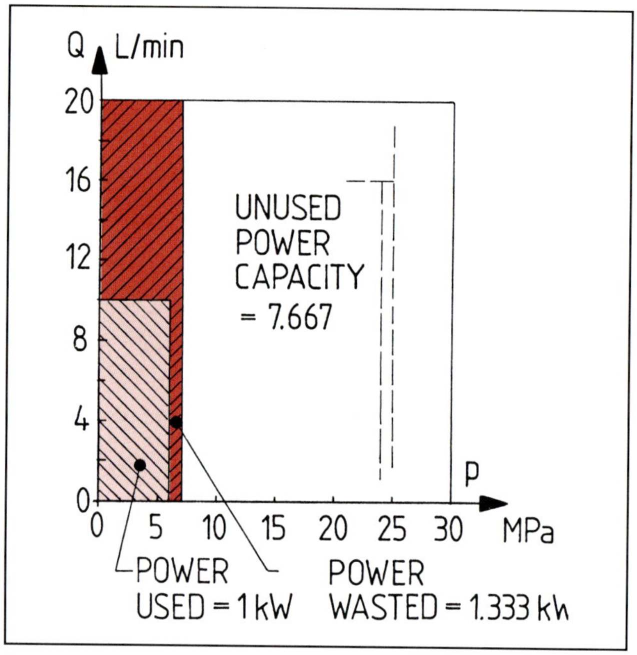

Fig. 239 Flow–pressure–power diagram for circuit with relief valve load sensing.

Load sensing controls with variable pump

Load sensing is nowadays widely applied in a multitude of hydraulic controls, but the way load sensing works, and its effect on the overall system, are often misunderstood or misapplied. How then does load sensing work and what does it achieve? To answer these pertinent questions, one must firstly establish and fully grasp the following basic hydraulic pressure–flow–power concepts:

1. Pressure in hydraulic system is caused whenever the actuators resist the pump flow or when the flow is forced through an orifice (valve, restrictor, etc.). This pressure is generally called "load induced pressure" or simply "load pressure".
2. The magnitude of this pressure depends entirely on the magnitude of the load or system resistance and is limited by the relief valve or pump pressure compensator in variable pumps.
3. Hydraulic pump input power is calculated by multiplying system or load pressure by pump flowrate:

$$W = Pa \times m^3/s, \quad P = p \times Q$$

   Therefore varying either the load pressure (p) or the pump delivery (Q) will have an effect on pump input power.
4. Flow across an orifice will cause a pressure drop or pressure differential often called delta p (Δp). If the orifice area is varied, the pressure drop will also vary accordingly and proportionally.

5. Variable displacement pumps deliver varying flow-rate (Q) if their geometrical volume is varied, or in mobile applications, if the prime mover rpm are varied (see fig. 239). In a swash plate piston pump, for example, by varying the swash plate angle, the flowrate will vary, since the pump's geometrical volume has been varied.
6. Variable displacement pumps without load sensing but with a pressure controller (pressure compensator) do not reduce flowrate until maximum system pressure, as adjusted at the compensator, is reached (see Chapter 3, figs. 76 and 77).
7. A load sensed pump works at approximately system pressure and system flow demand even if the load induced pressure is below relief valve or pump compensator pressure.

System efficiency and power loss

For fixed displacement pumps, as long as the actuators (cylinders and motors) use all of the flow produced, the power wastage is relatively small or practically zero (see fig. 240). The system pressure would then be determined by the actuator which is demanding the greatest pressure (Pascal's law). However, as soon as the actuators use only a portion of the produced pump flow, the surplus flow must be disposed of by the system relief valve (see figs. 97, 99 and 236), and this will raise the pressure in the entire system to the pressure at which the system relief valve opens (maximum possible system pressure). At that moment, power wastage can be calculated as:

Power = Disposed flowrate × Maximum system pressure

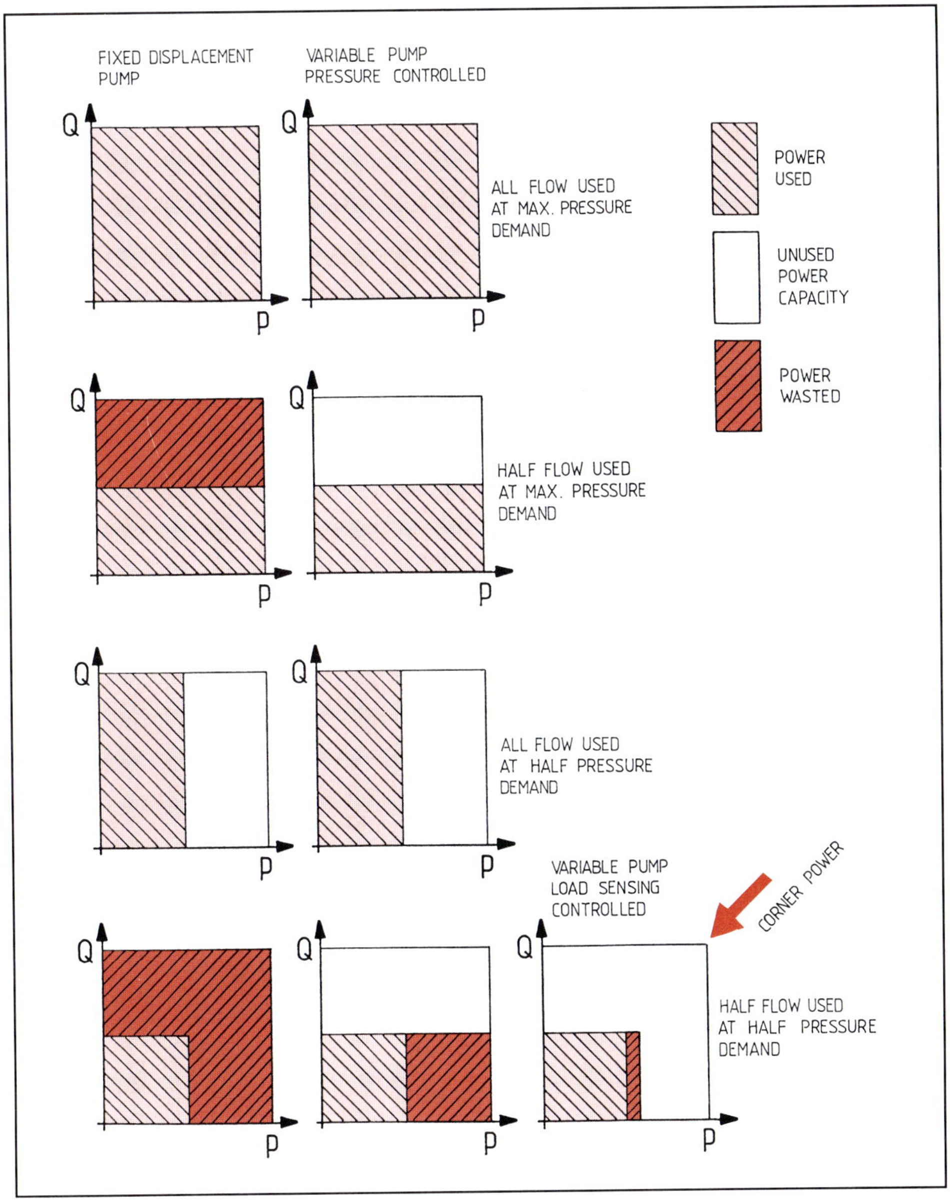

Fig. 240 Flow–pressure–power diagrams for fixed, variable and load-sensing controlled pumps.

Here, the disposed flowrate is the amount of flow passing over the system relief valve; the remaining flow passes to the actuators. Such potentially large power losses can be avoided if variable displacement pumps are used instead of fixed displacement pumps. A variable displacement pump adjusts its flow output automatically to zero when the actuators stop moving. At the same time it maintains system pressure at maximum, depending on the setting of the pump pressure compensator (controller; see fig. 241). If the system uses only a portion of the maximum available flow, and the variable pump delivers full flow, then the system pressure must rise to the pre-adjusted pump controller setting and, as a result of this, the controller then reduces the flow output to the flow demanded by the system; but this of course will be at maximum system pressure. This, however, will then also constitute a power loss, which may be rather significant if only a small portion of maximum pressure is required by the actuators. Separate pressure–flow–power diagrams are provided to show each of these cases (see fig. 240).

Load-sensing concept

A load-sensing controller built onto a variable displacement pump has a dual function:

1. It automatically adjusts the pump flow (Q) to that required by the hydraulic actuators. It is therefore often referred to also as a flow controller. It also adjusts its pressure output to the preset pressure differential necessary to force the flow through the directional control valves. It must be remembered at this point that any valve is an orifice (fixed or variable) and therefore creates a resistance to pump flow. This differential pressure multiplied by the pump flow results in the power loss shown in fig. 244, but this power loss, compared to that of a system without load sensing, is very small.
2. It limits the system pressure to that adjusted at the pressure compensator and reduces flow automatically to near zero, if adjusted pressure is reached. Thus, it maintains stand-by pressure and just enough flow to cater for internal system losses within pump and valves. It is, therefore, often not necessary to install a maximum pressure system relief valve (see fig. 244).

Summarising these features: the load-sensing pump will automatically adjust itself to any load demand within the pressure and flow boundaries shown by the pressure–flow diagram characteristics of that particular pump.

Where the hydraulic system is subject to wide fluctuations in load (that is, pressure and/or flow), the load-sensing system offers substantial energy savings. This is one of many reasons why load-sensing systems have become extremely popular for mobile, and of late also industrial, hydraulic systems.

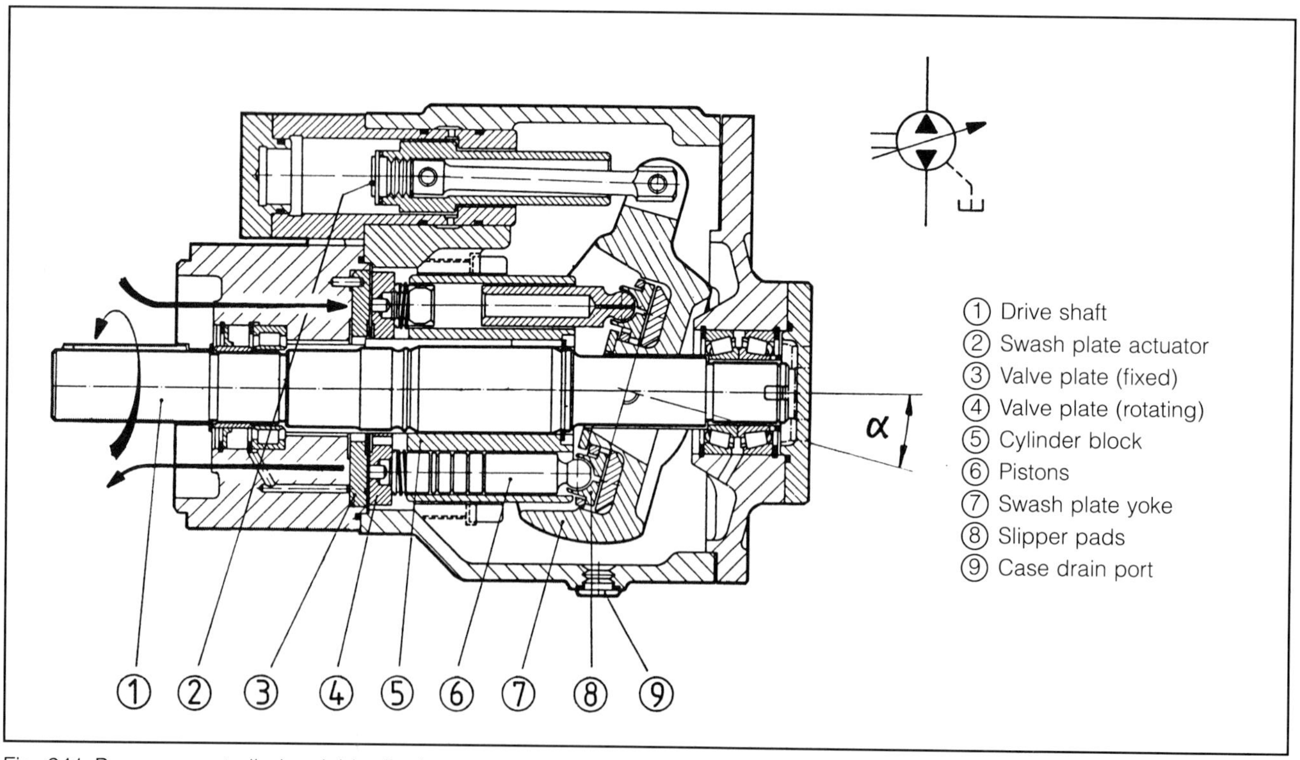

Fig. 241 Pressure-controlled variable displacement pump.

Load sensing — how it works

Although the design philosophies of pump manufacturers differ somewhat, the basic concept of how a load-sensing pump controller works is almost identical in all designs. The load-sensing pump controller always acts onto the pump-volume control mechanism, whether this be a swash plate, bent axis, radial piston or vane pump. However, it is easier to understand the load-sensing controller if one compares the load-sensing mechanism and its controller with a pressure-compensated flow control valve.

The pressure-compensated flow control valve consists of a compensator spool (hydrostat) with precisely equal areas at both ends, and downstream a control orifice connected in series. The ends of the compensator spool are hydraulically pilot connected to the inlet and outlet of the control orifice and therefore make the compensator subject to the pressures sensed at these two orifice extremities. The compensator spring gives the spool a bias to open the throttle land. In a flow condition, any pressure differential (Δp) greater than the bias spring will reduce the throttle opening until the pressure differential is again achieved. Thus, it can be said that the compensator adjusts flow to the control orifice to maintain a consistent pressure differential. The flow passing to the actuator is therefore maintained at a constant level regardless of actuator pressure fluctuations (see figs. 124 and 242).

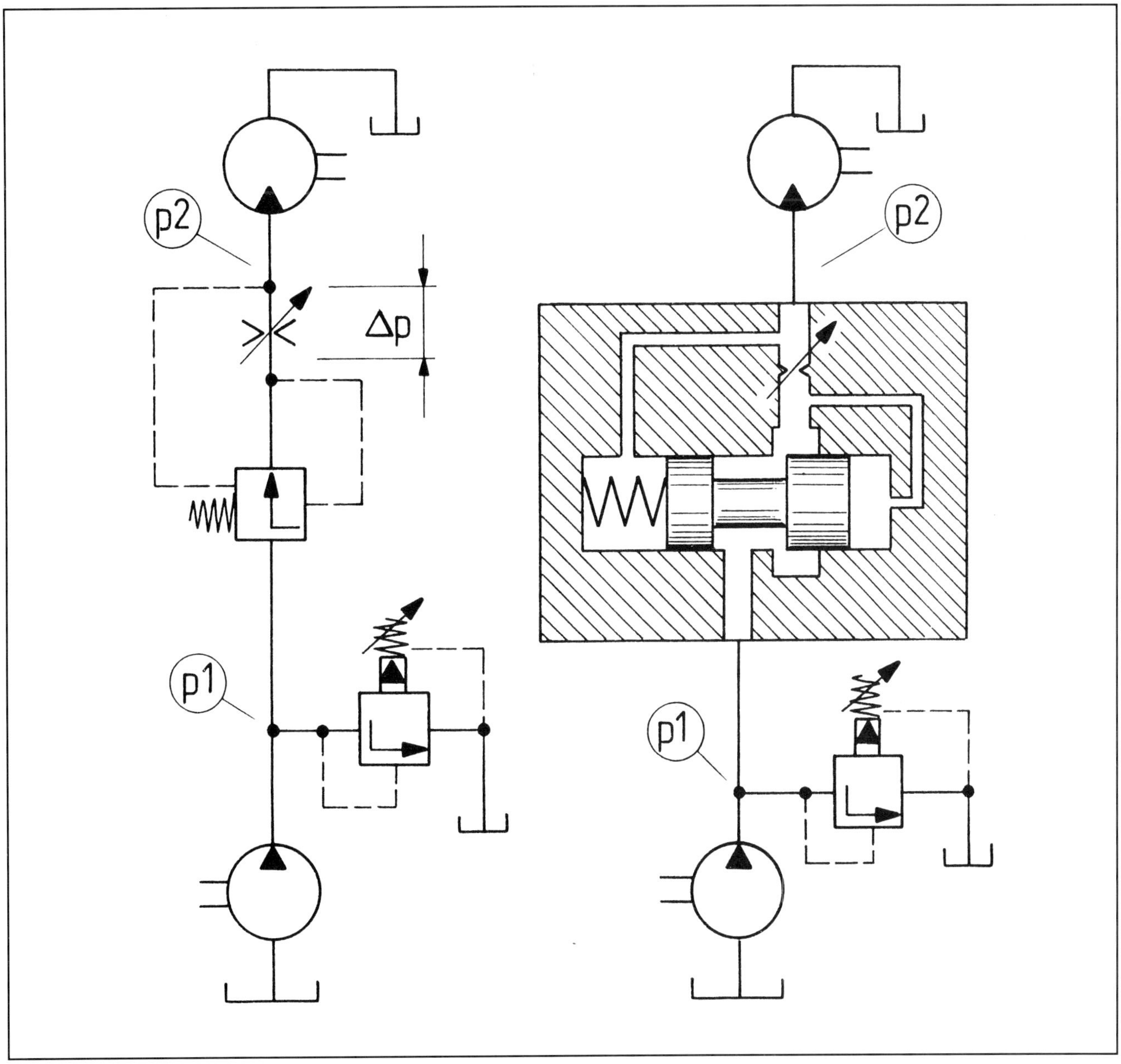

Fig. 242 Pressure-compensated flow control valve controlling the speed of a hydraulic motor.

The load-sensing controller on a variable pump works on precisely the same principle. The spring-biased compensator spool is hydro-mechanically connected to the pump volume control mechanism (see fig. 243). Any minute change in the pressure differential across the control orifice affects the angle of the swash plate in a piston pump, or the track ring eccentricity in a vane pump, or the bent axis angle in a bent axis piston pump. By way of illustration, the operation of a load-sensing controller for a swash plate piston pump is investigated.

Note: The pump under investigation is also equipped with a pressure-limiting controller which limits maximum system pressure. Both controllers are an integral part of the pump (see fig. 244).

The orifice across which Δp is sensed may be variable or fixed, or it may be a directional control valve that can be selected either by hand lever control or proportional control, and may assume an infinite number of positions (see fig. 245). The differential pressure spring holding the load-sensing spool to the left is preset to approximately 10 bar. A second spring holds the pressure-limiting control spool to the right-hand side. It is adjusted to limit maximum system pressure, similarly to a pressure relief valve.

It is now assumed that the directional control valve is in a position (D.C.V. spool opening) such that the pressure differential sensed from its outlet port (p2) to the prevailing pressure at the pump outlet (p1) is less than the value preset at the differential pressure spring. For that situation the load-sensing spool will be minutely left of the situation shown in fig. 244 and the pressure-limiting spool will be minutely to the right. The swash plate actuating piston is then vented to tank and the swash plate bias spring holds the pump at maximum displacement (see figs. 241 and 244). The pump now delivers maximum flow at 10 bar above load pressure.

As soon as the directional control valve becomes even partially closed, so that it restricts pump delivery, pump outlet pressure rises and this increased pressure is sensed at the left-hand face of the load-sensing spool. When this increased pressure exceeds

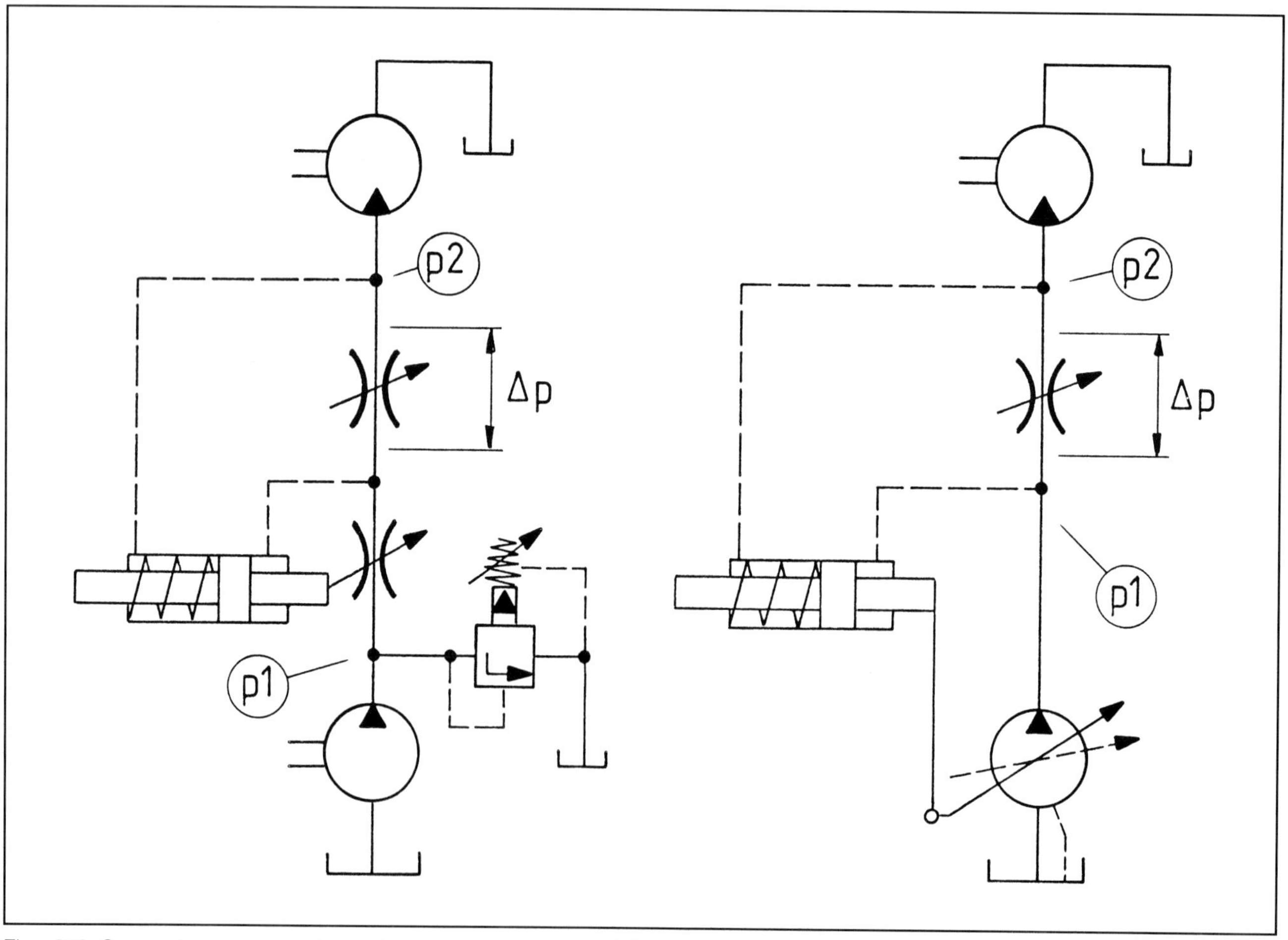

Fig. 243 Comparison and analogy of pressure-compensated flow control valve mechanism to pressure-compensated pump control mechanism.

the sum of load-pressure force in the load-sensing chamber plus load-sensing spring force, the spool becomes unbalanced and moves minutely to the right, thus passing a small amount of pump outlet fluid to the swash plate actuating piston. This reduces the pump delivery (Q) and keeps on reducing it until the pump output pressure is again equal to load pressure plus load-sensing spring force. At that time the load-sensing spool is again force balanced and closes the fluid passage to the swash plate actuating piston. The reduced swash plate angle now results in a reduced pump delivery (Q) until either the directional control valve orifice or the load induced pressure downstream of the D.C.V. changes (e.g. the Δp changes).

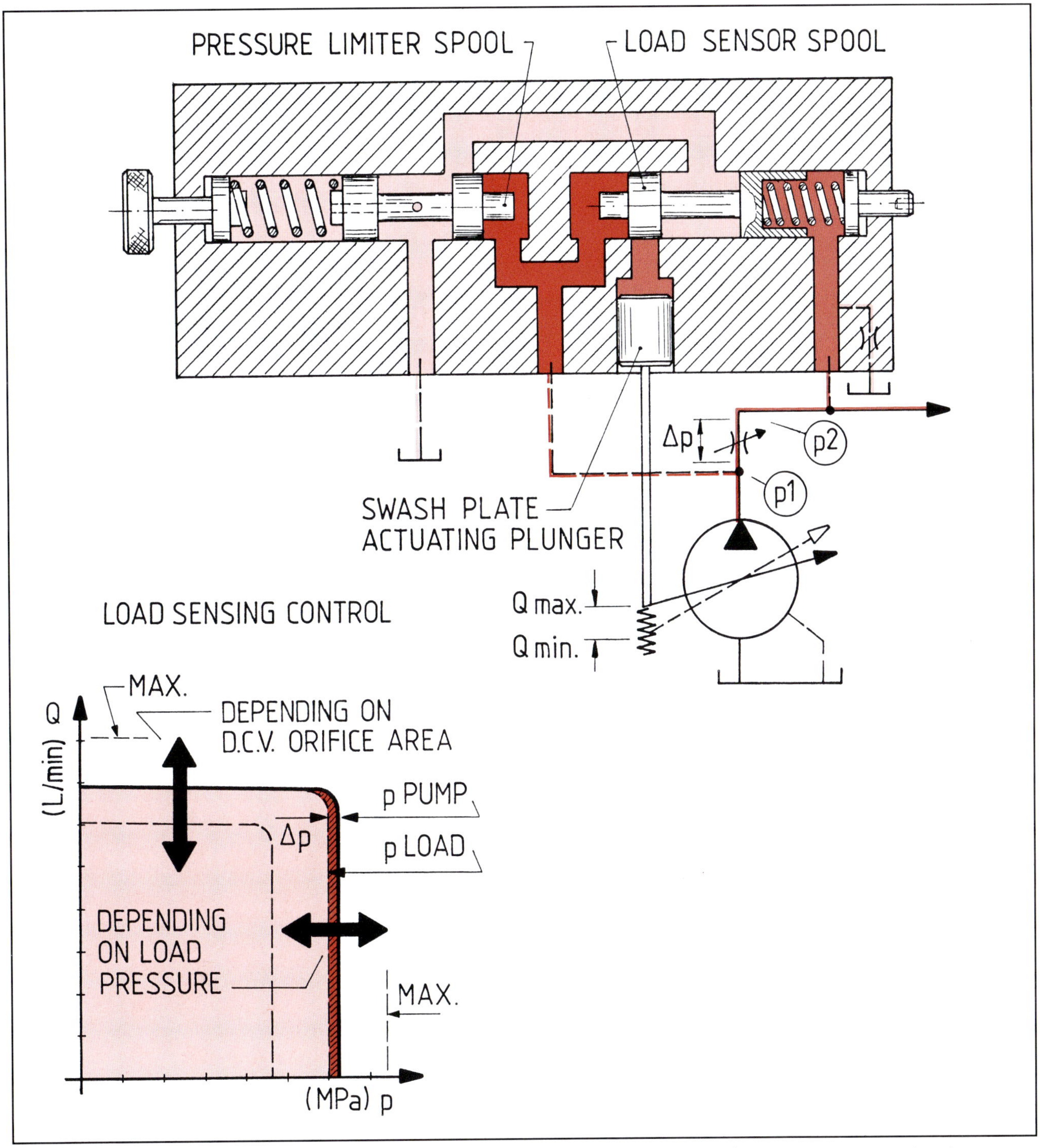

Fig. 244 Load-sensing controller mechanism including maximum pressure-limiting control (pressure compensator) and pressure–flow–power diagram for load-sensing pump.

If the system pressure reaches the setting of the pressure limiting controller, its spool is then forced to the far left. This permits pump outlet flow to pass to tank and at the same time forces the swash plate to near-zero pumping. This situation prevails until load resistance falls below maximum system pressure.

Figure 245 shows the graphic symbol circuit for a load-sensing controller plus maximum pressure limiter control (pressure compensator) for the control mechanisms depicted in fig. 244. The system shown is similar to a Vickers pump controller.

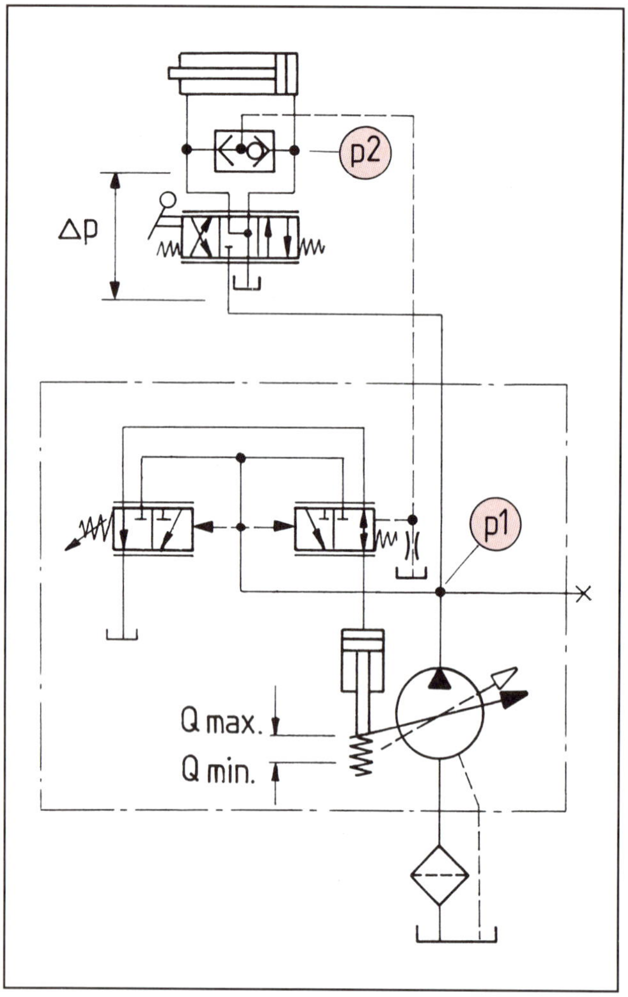

Fig. 245 Graphic symbol circuit for pump with load-sensing control and maximum pressure-limiting control.

Sensing the load pressure

A number of important points are illustrated in fig. 245. Sensing load pressure correctly and selecting a directional control valve to give satisfactory actuator and load motion is often not an easy task. In this illustration the load-sensed pump drives a reversible hydraulic linear actuator and a shuttle valve is therefore used to sense load pressure for either actuator direction. Note that a float centre type directional control valve has been selected. The float centre ensures that with the valve in the middle position the load-sensing line is completely vented to zero pressure. The pump will then "dead head" against the shut pressure port of the directional control valve. With the load-sensing spring chamber now vented to zero pressure (see figs. 244 and 245), the pump will force the load-sensing spool to the right at 10 bar pressure. Pump outlet pilot fluid, now being at 10 bar pressure, forces the swash plate actuating piston down. This de-strokes the swash plate against the bias spring and the pump operates now at near-zero flow, and at a mere 10 bar stand-by pressure. As soon as the directional control valve is selected to drive the actuator, the pump will come on-stroke again and it must then deliver flow to the actuator at selected D.C.V. orifice opening and at requested load pressure.

Load sensing for closed centre directional control valves

If closed centre directional control valves are used in conjunction with load sensing, one must take care that the following situations are avoided:

1. With the D.C.V. valve in the middle position, all ports are blocked and the load-sensing line is no longer vented by the valve spool, as was the case with the float centre valve previously discussed. The load-sensing pilot line is then vented to tank via one of several orifices (see figs. 245 and 246). This orifice permits a small amount of pilot fluid to flow to tank at all times, and thus constitutes a small power loss. This power loss, however, is insignificant compared to the overall system power wastage.
2. For accurate and fast responding load-sensing adjustments, it is important that load-sensing pilot lines are properly drained, and that the fluid in the pilot line is at system temperature and free of air bubbles.
3. To prevent suspended or hydraulically supported loads from drifting, pilot-operated check valves should be built into the flow lines. These valves also prevent spurious load-sensing pressure from being caused by external forces acting onto the actuator piston (see fig. 246).

Load sensing for several actuators

If in a hydraulic circuit several actuators are supplied from the same load-sensed pump, one must then use shuttle valves to sense load pressure at all directional control valve outlet ports (see figs. 245–247). The pump can only sense load pressure from one directional control valve. To make sure that the actuator with the highest load pressure need is also supplied, the pump must logically be working at that particular pressure and, therefore, all directional control valve outlet ports are connected to each other in a multiple input "or" logic function. The common output of this "or" logic function then becomes the load-sensing input to the load-sensing pump controller.

As already mentioned, fast and accurate pump response to varying load conditions can only be achieved if the load-sensing pilot line is properly vented and free of air pocket inclusions, and the pilot fluid is at system temperature. To achieve this, system designers recommend a continuous flushing of the load-sensing line. An orifice is built into the load-sensing line to keep the flow loss at a small, manageable level (see figs. 238, 244, 245 and 247).

Figure 247 shows a load-sensing control with conventional mobile hydraulic directional control valves. Isolation check valves are used at every directional control valve pressure port, so that the pump control can accurately sense the actuator demanding the highest working pressure.

If the actuators are also to be speed controlled, it is advisable to select non-pressure-compensated flow controllers. Standard pressure-compensated flow controllers generally cause a minimum pressure drop of 5–10 bar across the metering orifice. This can induce too much flow resistance to the measured pressure differential (Δp) on which the load-sensing pressure is based. The pump may, therefore, not come on-stroke or be curtailed in its flow delivery.

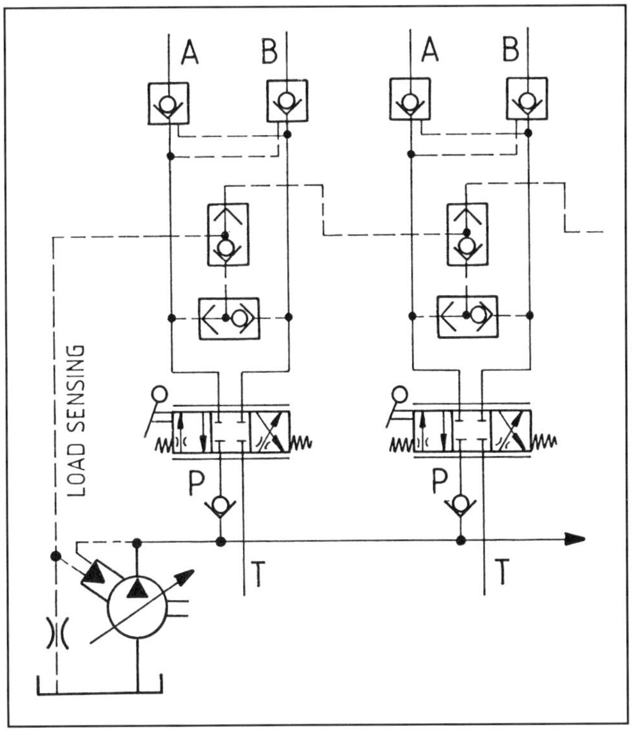

Fig. 246 Load sensing for closed centre directional control valves and supported loads.

Note: The pump manufacturer "Rexroth" also mentions that this can cause interference between the springs in the flow controller and maximum pressure controller, which may lead to control instability (see fig. 244).

Considering these negative effects on pump control performance, it is therefore advisable, wherever possible, to use non-pressure-compensated flow controls. It must also be noted that load sensing in itself acts as a compensator for varying load pressures across the metering orifice.

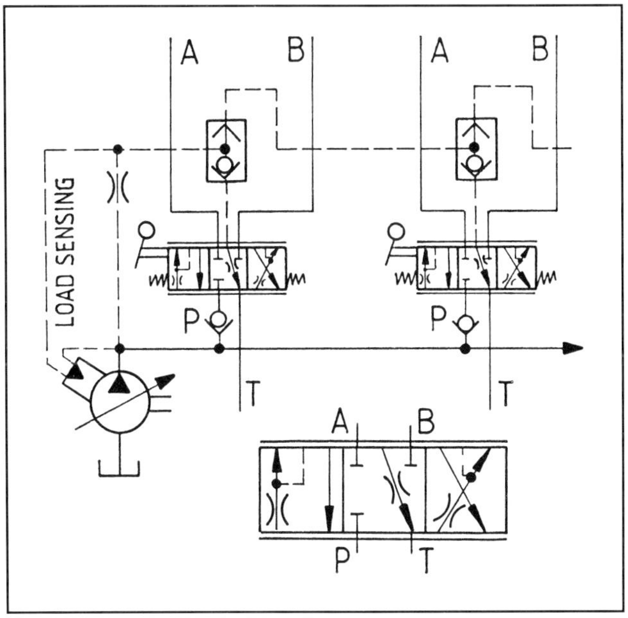

Fig. 247 Load sensing with conventional mobile hydraulic directional control valves and isolation check valves.

Sensitive load sensing for multiple actuator controls

As previously mentioned, when several actuators operating at differing load pressures are in motion, only the actuator with the highest load pressure is load sensed. All other actuators are then also working at the current pump pressure. Furthermore, the speed and forces of the other actuators may then be negatively influenced from pressure variations upstream or downstream of their respective directional control valves. In a typical mobile, hand lever operated D.C.V. actuation, the machine operator can correct these variations on the D.C.V. hand lever. But a better way is to use pressure compensators in front of every D.C.V. pressure port (see fig. 248). Some mobile hydraulic valves have these compensators built in as standard by the valve manufacturer. The pressure compensator, together with the D.C.V. spool-selected orifice opening, works much the same as a pressure-compensated flow controller. It maintains constant flowrate as selected by the spool, but independent of pressure variation of the pump or the actuator pressure (see also Chapter 6, fig. 124). This provides load-independent operation of all actuators, which is free of negative cross-influence among the various actuators and their directional control valves. The spring tension (e.g. pressure differential), adjusted at the pressure compensators, must be slightly higher than the differential set at the load-sensing controller of the pump. This ensures that the compensator spool is fully open when its D.C.V. load-sensing signal reports the highest load pressure to the pump controller. The pump controller will then adjust the pressure drop across the D.C.V.

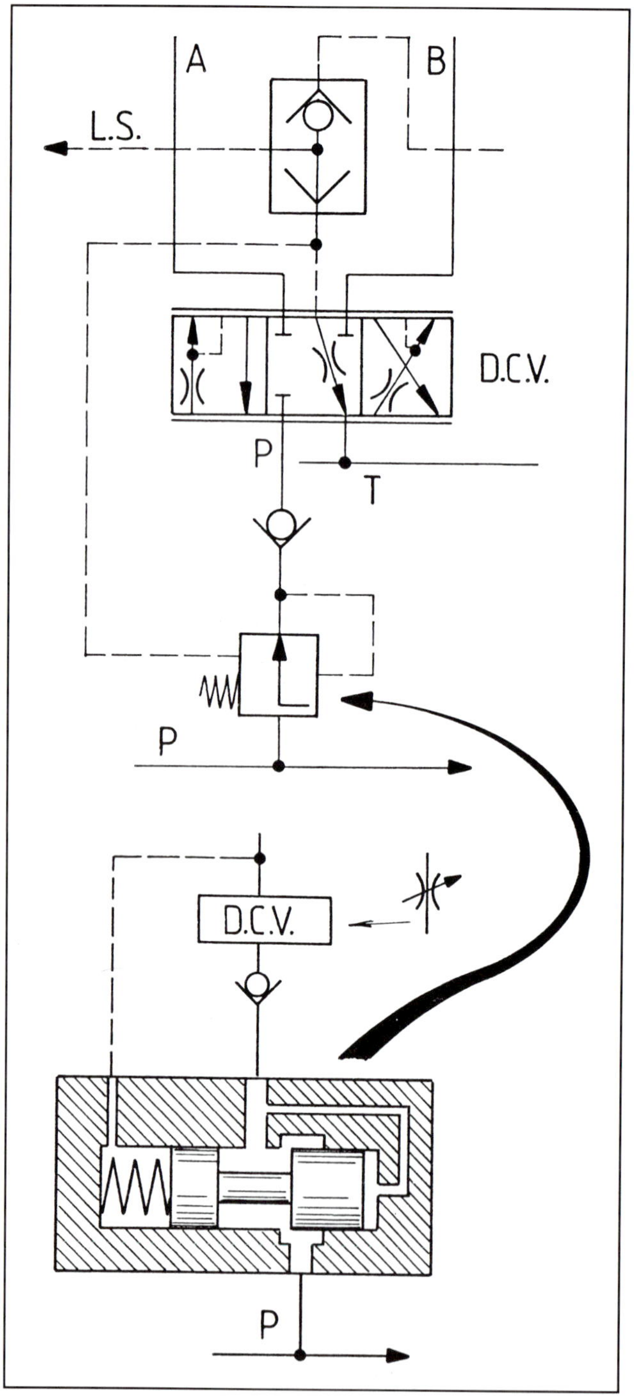

Fig. 248 Pressure compensators in the pressure line of each D.C.V. avoid negative pressure influences on actuators working at lower load pressure.

Conclusions on load sensing

Load-sensing controls find more and more acceptance in industrial hydraulics. However, the complexity of their application, particularly the way the load-sensing signal is obtained from the various directional control valves, demands that the system designer has sound knowledge of specific components used, and their integration and adjustments within the total system. Maintenance and service personnel also require excellent knowledge of product function and operation in order to maintain load-sensed machinery in good working order.

The majority of load-sensing controls are still used for mobile and agricultural applications, mainly because of their random methods of actuator operation and sequence (bucket, lift, turn, drive, mast, etc.). Many of the most recently designed and built garbage trucks and compactors are equipped with a load-sensing pump. This eliminates the necessity to disengage the pump when compacting is completed. The pump then rotates at near-zero flow in low pressure stand-by mode, when its hydraulic actuators are idle, thus using virtually no power from the prime mover. Even with the compactor system operating, the only power used is that of the compactor actuator pressure demand multiplied by the pump flowrate; thus power losses and oil heating are kept to a minimum.

Another aspect of load-sensing control is that it provides constant flow regardless of variations in pump input drive speed. If applied to a concrete mixer and delivery truck, where the diesel engine drives the vehicle as well as the hydraulic pump, then diesel engine rpm, affected by road speed conditions, will have no effect on the flowrate from the load-sensed pump. The concrete mixer drum will therefore rotate at constant rpm while the vehicle travels between loading base and delivery sites.

Concept of pump power control (torque limiting control)

Pump output power, expressed in watts or the derived units kilowatts (kW), or megawatts (MW), is the product of pump output pressure and pump flowrate:

$$P_{pump} = p \times Q$$

The pump pressure, of course, is developed by restrictions to its flow, and is caused by load resistance against the hydraulic actuators. Variations in either p or Q will have an effect on the resulting power.

Pump input power, naturally, is somewhat higher than its output power, since no mechanical apparatus works without losses (see Chapter 3). The pump being driven by a prime mover, therefore, causes resistance to that prime mover. In mobile, agricultural and construction machinery, the prime mover is usually an internal combustion engine (diesel or petrol). Combustion engines run most efficiently at a particular rpm range, or torque range. It is desirable, therefore, to draw only so much power from the pump coupled to its prime mover engine, as not to exceed this range. Prime mover output power may also be expressed as:

$$\text{Power} = \text{Torque (M)} \times \text{Angular velocity } (\omega)$$

$$P_{OUT} = \frac{\text{Nm} \times \text{rpm} \times 2 \times \pi}{60 \times 10^3} = \text{kW}$$

Prime mover output power is equal to pump input power. It can, therefore, also be expressed as:

$$\text{Power} = \text{Pressure} \times \text{Flowrate}$$

With these concepts in mind one can understand why pump manufacturers have developed controls to regulate maximum pump output power. Since it is often not necessary to have both of these quantities (pressure or flowrate) available at maximum level at the same time, controllers have been developed which, like a computer, constantly multiply system pressure demand and system flow demand and compare the result to a predetermined and set power value. If the result of this computation and comparison exceeds the set value on the power controller, the pump flowrate is reduced to the maximum available value. Thus, using a constant power controller, a variable displacement pump may be operated so that available input power from the prime mover can be fully utilised, but never exceeds the preset value adjusted and set on the power controller. In short, the product p × Q is maintained constant or below the maximum set value at all times.

Example

For a light load, high flow is available for rapid machine cycle times. This is possible while still retaining the essential ability to move much heavier loads when necessary, at a rate not exceeding the total adjusted power product of speed and force.

Power control — how it works

For illustrative purposes, the power control on a "Parker" axial piston pump is investigated (see figs. 240 and 241). Pump flow regulation is achieved by extension or retraction of a swash plate actuating piston. Pump flow is increased if this piston extends (is forced out), and decreased when it retracts. The combined effect of the hydraulic off-setting forces of the pumping pistons and the slipper pad retaining spring in the middle of the rotating cylinder block (barrel) tends to force the swash plate to zero pumping (see fig. 241). These combined forces are counteracted by the swash plate actuating piston.

Note: The centre line of the rotating cylinder block with its pumping pistons is offset from the pivot point of the swash plate. Therefore, the pumping pistons' effective summation force tends to destroke the swash plate to zero pumping. This destroking effect is, however, balanced by the swash plate actuating piston.

Since the power controller multiplies the demanded and prevailing system pressure by the pump flowrate at that time, both values need to be sensed. Pump flowrate may be sensed if one ascertains the position of the swash plate deviation angle from zero. Pressure is sensed (or limited) with a power-controlled relief valve. The power controller, therefore, requires a

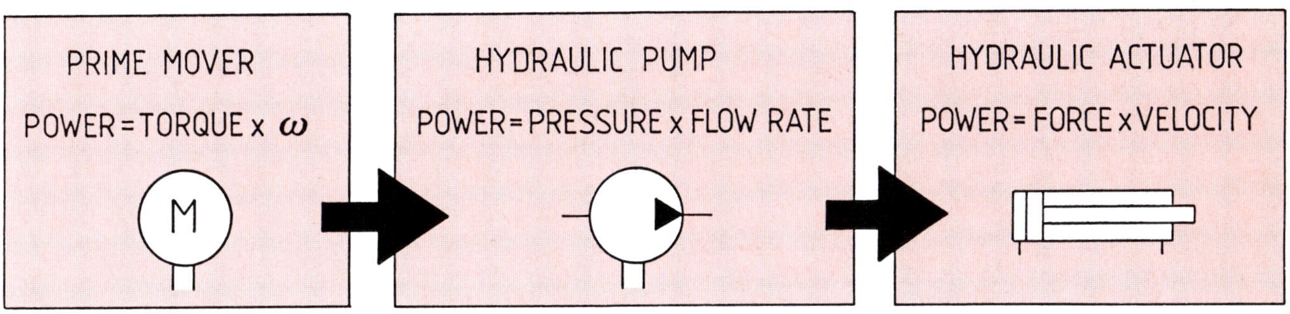

Fig. 249 Hydraulic power transmission and articulation from prime mover input to actuator output.

mechanical sensing linkage to the swash plate actuating piston (see fig. 250). When the swash plate actuating piston is retracted, the swash plate causes low flow and the power control piston develops maximum spring force onto the power control relief valve. This permits pump pressure to rise, if required, without exceeding maximum power. When the swash plate actuating piston is extended, causing maximum flow, the power control relief valve is set at its minimum. This permits it to open under increasing pressure sensed in the load-sensing chamber, thereby venting some of the pressure in the load-sensing chamber. As a result of this, the load-sensing spool becomes unbalanced and moves down, thus starting to vent the swash plate actuating piston cavity. The swash plate now swings back to reduce pumping, thereby maintaining maximum available power (increased pressure and decreased flow). When system pressure is again reduced, and therefore the power control relief valve shuts, and pressure in the load-sensing chamber again reaches prevailing system pressure, the load-sensing spool starts to move back because the bias spring forces it back. Pressure at both spool ends is now equal and system pressure sensed via the load-sensing spool holds the pump on stroke.

Some power controllers work exactly to the perfect power parabola (see fig. 250), whereas others achieve only an approximate parabola. Control always starts at maximum swivel angle (maximum flow) and then adjusts the swivel angle to the prevailing load pressure. Both maximum system pressure and maximum flow may be adjusted individually, but one has to be careful with these adjustments.

Too much flow setting may cause pump cavitation and hydraulic motor over-spinning whereas, if the

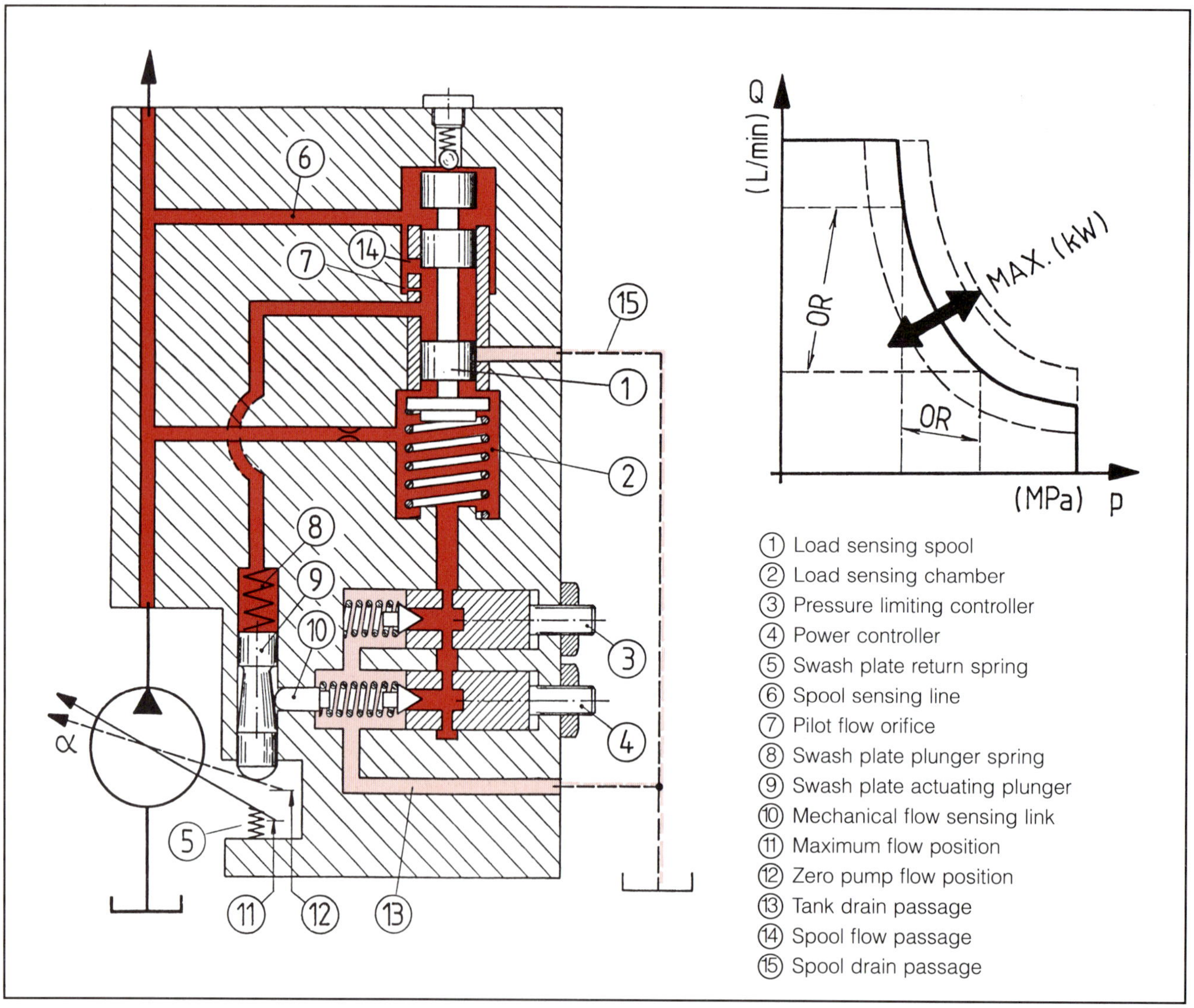

Fig. 250 Variable displacement pump with maximum power controller and maximum pressure limiter.

minimum set angle is too high, so that the minimum pump flow is higher than tolerable, the prime mover motor may become overloaded in the high pressure range (excessive torque).

The maximum available pressure in a power-controlled pump is usually limited by a pressure compensator control (see figs. 244 and 250). For maximum productivity and minimum input energy consumption, with the least amount of system power loss in circuits for frequently varying loads, a combination of power control and load-sensing control is used. Most pump manufacturers provide pumps with such a combined control option. This combination of pump controls gives the advantage of high productivity intrinsic to constant power operation, but at the same time provides full use of power over the entire range of load demand and giving maximum possible load velocity at all times.

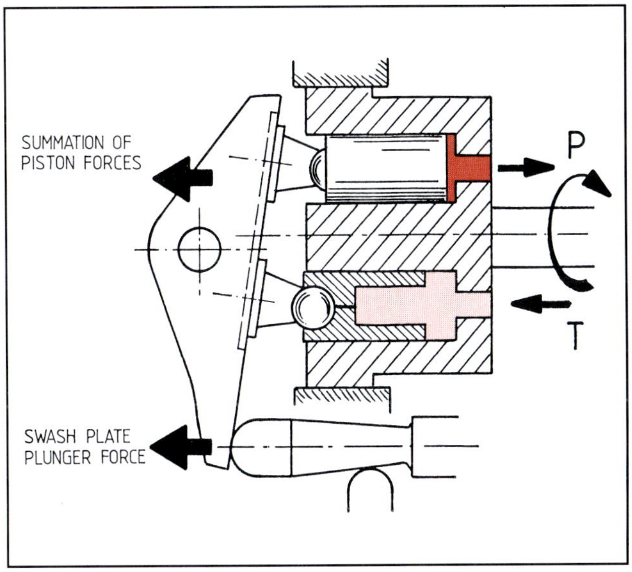

Fig. 251 Swash plate actuation method with offset swash plate pivot for "Parker" pump.

Pump input power and prime mover output power calculations

Some power information has been provided in Chapter 3, "Hydraulic pumps", where pump input power was defined as:

$$P = p \times Q$$

with the hydro-mechanical losses adding to the input power requirement (see page 32). With this calculation method, one needs to know the geometrical volume of the pump (V_g), the performance characteristics of the pump, the system pressure and the rpm of the prime mover.

If the input power calculation is based on pump torque, however, then a different formula is required. Pump input torque (M) is proportional to the pressure differential Δp (p_{in} to p_{out} on the pump ports) and the geometrical volume of the pump. The theoretical

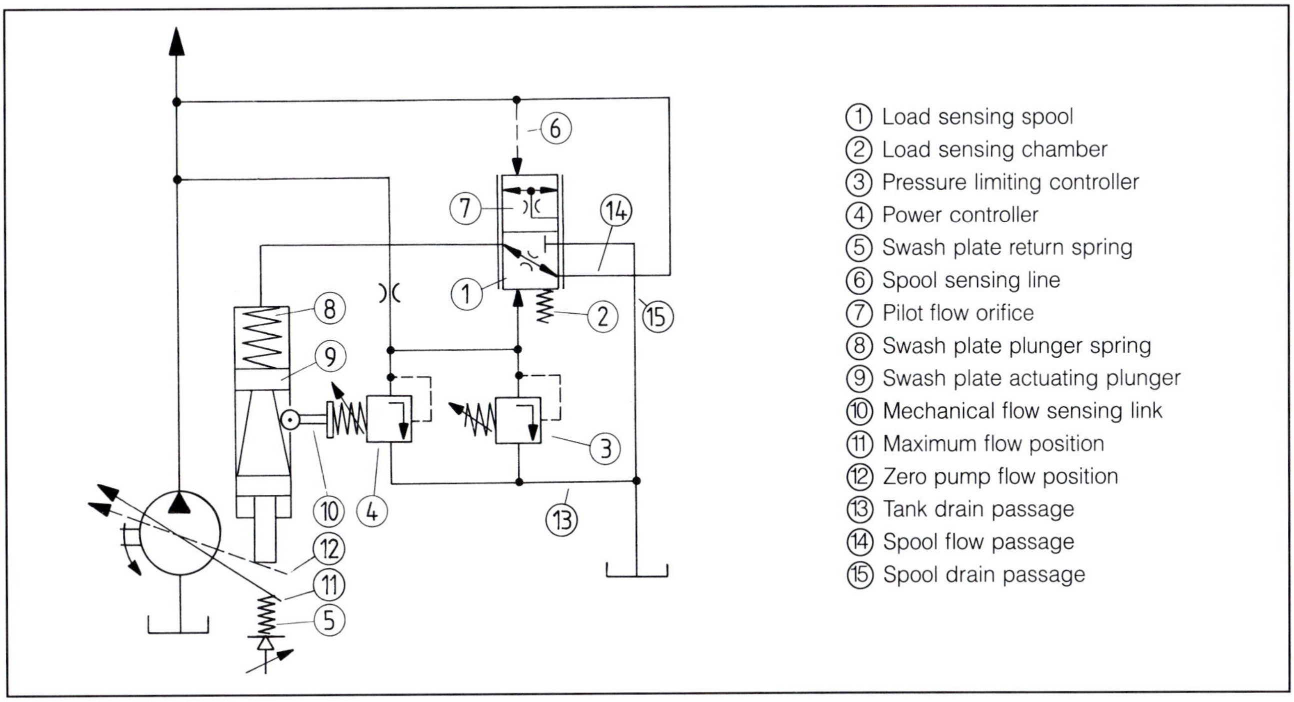

Fig. 252 Hydraulic circuit for the constant power pump control illustrated in fig. 250.

geometrical volume is equivalent to the amount of flow discharged per pump shaft rotation. The formula used for this calculation approach is:

$$P = M \times \omega$$

The above formula contains pump input torque and the necessary angular velocity (ω). Torque (M) then is calculated as:

$$M = \frac{V \times \Delta p}{2 \times \pi}$$

The angular velocity may be calculated as follows:

$$\omega = 2 \times \pi \times n$$

In order to obtain output power in watts, one needs to insert the pump revolutions (n) per unit time, e.g. rps (revolutions per second).

With load-sensing control, the pump input power fluctuates because pump pressure fluctuates. If, with a load-sensed pump, the rpm is fluctuating, as is the case with a mobile prime mover drive, then pump output flow is kept constant because the load-sensing controller adjusts the geometrical pumping volume (swash plate angle).

With constant power control, constant prime mover speed is mandatory and power is kept constant by the power controller (torque control). Therefore, pump input torque is also kept constant.

Most pump manufacturers offer variable displacement pumps (rotary vane, bent axis, axial or radial piston) with all or at least some of the following flowrate and pressure controls:

1. Maximum pressure control (pressure compensator) for ramp or step type control action (see also page 48).
2. Load-sensing control with or without superimposed pressure compensation control (see fig. 244).
3. Power control (torque control) with superimposed pressure control (see fig. 250).
4. Power control with superimposed pressure control and summation power control for multiple pump arrangements (fig. 253).
5. Load-sensing control with superimposed pressure control and summation control option for multiple pump arrangements.
6. Load-sensing control combined with power control plus superimposed pressure control.

Load sensing combined with power control (torque control)

For minimum input energy consumption and maximum productivity in a hydraulic system, designed to handle widely varying loads, a combination of torque limiting (constant power) control and load sensing is best. Such a combination of pump controls provides the obvious advantage of high productivity unique to constant torque control, while making full use of available power over the entire load range and giving a maximum load velocity in response to pressure demands. However, it also allows loads to be moved at less than the maximum possible, with minimum power losses. Last but not least, it places the pump in a condition of minimum power consumption during stand-by, when no flow is required.

Pump performance characteristics

Figure 253 shows the typical pump performance characteristics diagram for a "Kämper Hydraulik" fixed displacement axial piston pump with swash plate angle $\alpha = 17.5°$, geometrical volume $V = 65.0 \times 10^{-6}$ m^3 (65 mL), dynamic oil viscosity $\eta = 30 \times 10^{-3}$ Ns/m^2. The interrupted line, ascending from a pump drive speed of 25 s^{-1} (25 rps or 1500 rpm) in the direction of the arrows, intersects with the designated system pressure of 150 bar (15 MPa) at point BP. Moving along parallel to the fine lines to the left, leads to an input torque result of 166 Nm. From intersection point BP, following parallel to the diagonal power curve, one obtains a pump input power requirement of 26 kW. Descending from intersection point BP, parallel to the curved flowrate line, one obtains an effective flowrate of 1.6×10^{-3} m^3/s (equivalent to 1.6 L/s). From the near-circular, but partially oblong efficiency curves, the diagram shows a hydromechanical efficiency (η_{hm}) of approximately 91%.

Figure 254 shows a typical pump performance characteristics diagram for a "Kämper Hydraulik" variable displacement piston pump, operating at a momentary adjusted geometrical volume $V_{eff} = 53.5 \times 10^{-6}$m^3 and a nominal system pressure of 150 bar (see intersection point BP). Using similar diagram reading guidelines to those given for fig. 253, one obtains a power input torque of 143 Nm, an input power requirement of 21 kW and a hydro-mechanical efficiency of approx. 89.5%.

Pump power and prime mover calculations

The power formula given below is used to calculate pump input power as well as prime mover output power. It is also valid for power calculations as regards hydraulic motors (see Chapter 7, "Rotary actuators (motors)").

$$P_{OUT} = \frac{Nm \times rpm \times 2 \times \pi}{60 \times 10^3} = kW$$

Example

Calculate the required prime mover output power (electric motor) for the "Kämper Hydraulik" pump depicted in fig. 254. The pump operates at a nominal system pressure of 150 bar and in accordance with the pump performance diagram provided, develops an input torque of 143 Nm and requires a motor drive speed of 1500 rpm (25 s^{-1}).

$$\text{Angular velocity} = \omega = 2 \times \pi \times n$$
$$\text{Power} = P = M \times \omega$$

Solution

$$\omega = 2 \times 3.1416 \times 25 \text{ L/s} = 157 \text{ L/s}$$
$$P = 143 \text{ Nm} \times 157 \text{ L/s} = 22462 \text{ Nm/s} = 22462 \text{ W} = 22.46 \text{ kW}$$

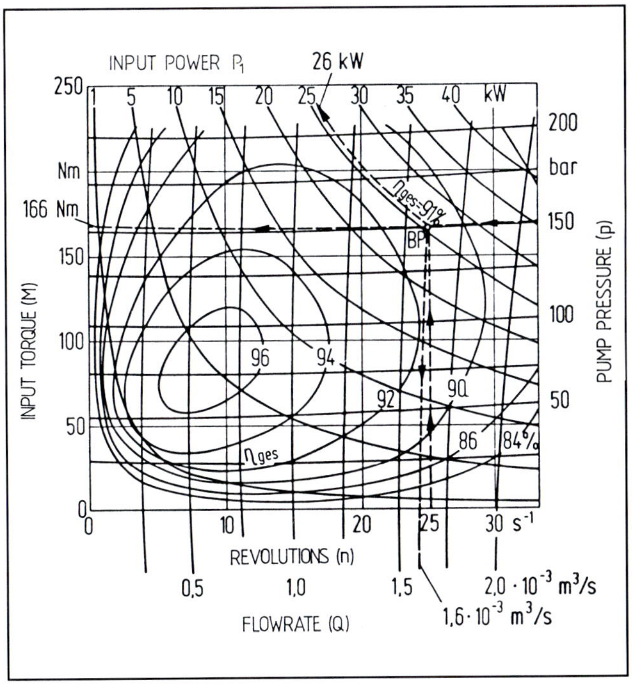

Fig. 253 Typical pump performance characteristics diagram for a fixed displacement axial piston pump, showing pump input power, pump input torque, pump flowrate and pump efficiency for a given pump drive speed, given system pressure and fixed geometrical volume.

Example

Using fig. 253, derive pump input power (P), input torque (M), output flowrate (Q) and overall efficiency (η), if the prime mover develops a drive speed (n) of 1800 rpm and the actuators cause a pump pressure of 17.5 MPa. Show in red all required diagram solution lines and the operating pressure point BP.

Solution

From the graph in fig. 253 we derive:

$$P = 38 \text{ kW} \quad M = 200 \text{ Nm}$$
$$Q = 1.9 \times 10^{-3} \text{ m}^3/\text{s} = 1.9 \text{ L/s}$$
$$\eta = 87.5\% \text{ approx.}$$

Example

For the example given above, derive pump input power (P), input torque (M), angular velocity (ω) and pump efficiency, if pump operating pressure is decreased to 125 bar.

Solution

From the graph in fig. 253 we derive:

$$P = 27 \text{ kW} \quad M = 140 \text{ Nm}$$
$$\omega = 30 \times 2 \times \pi = 188.5 \text{ L/s}$$
$$\eta = 89\% \text{ approx.}$$

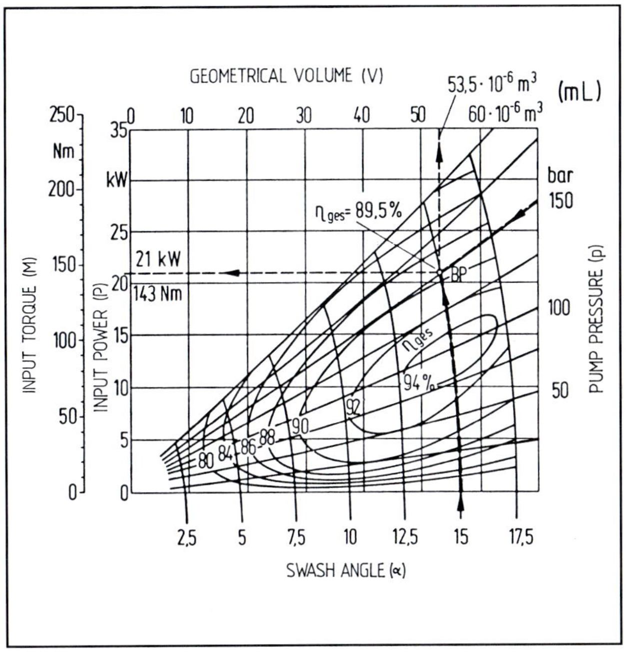

Fig. 254 Typical pump performance characteristics for a variable displacement axial piston pump, operating at a momentary adjusted geometrical volume and a nominal system pressure, showing pump input power, pump input torque, pump flowrate and pump efficiency.

Optimum operation – with special controls

Power-control Follows power hyperbola exactly unlike spring-controlled regulating systems which only approximate to the curve.	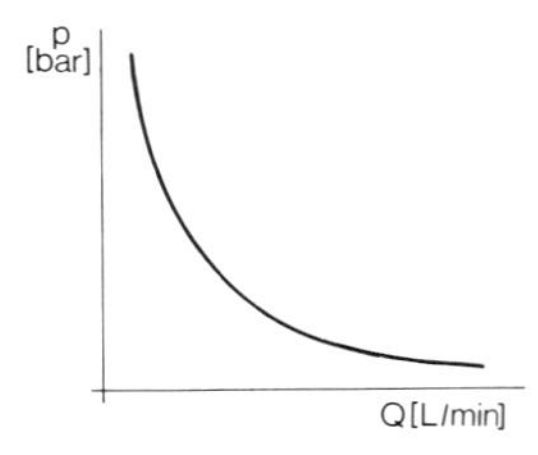	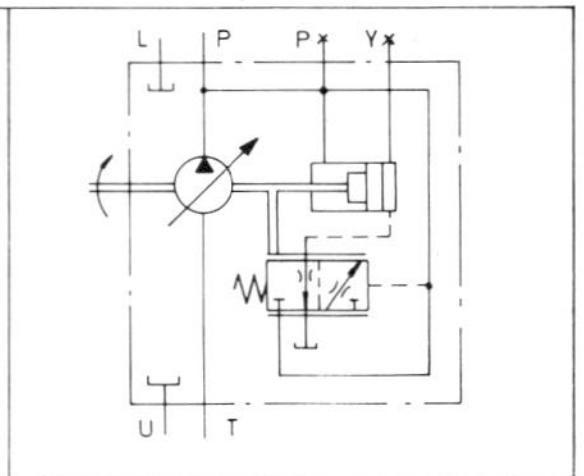
Pressure compensator System-pressure activated. At preset pressure the pump automatically returns to the position necessary to maintain the selected pressure.	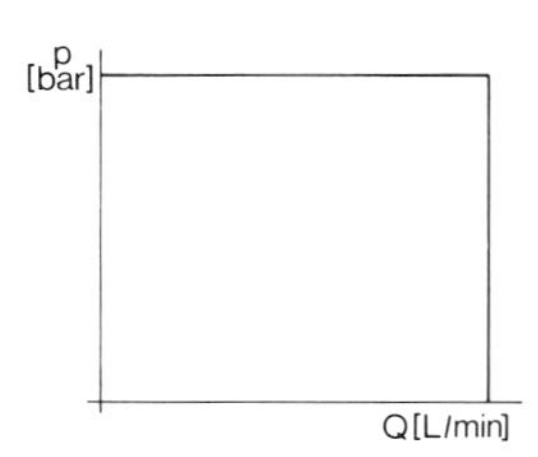	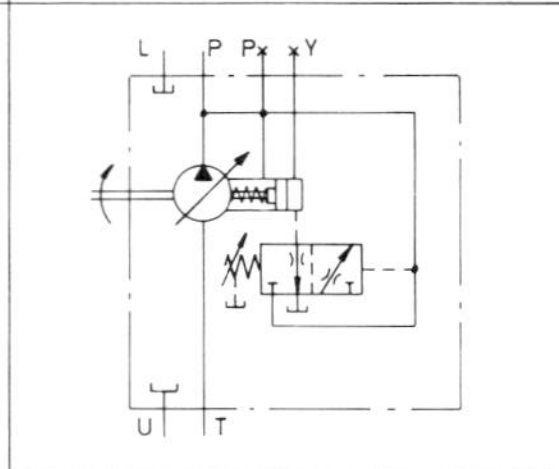
Power-control with hydraulic stroke limiter for summation power-control In a two-circuit system with summated power-control both pumps are regulated by the respective system pressure. With this system the sum of the powers remains constant.	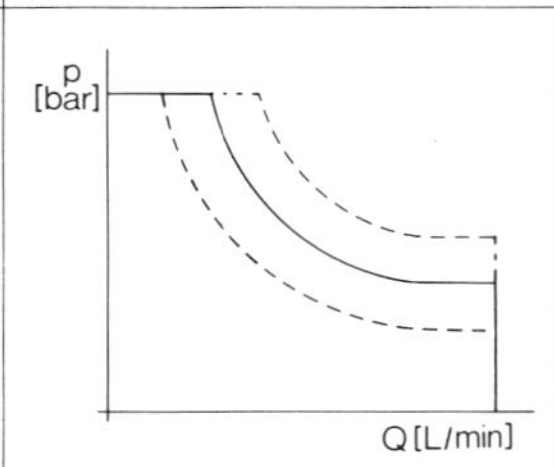	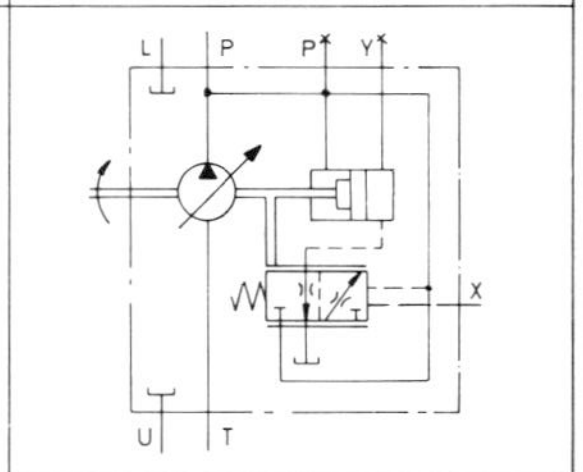
Pressure compensator and load-sensing The pump pressure is always 20 bar above system pressure, and the pump flow adjusts to the service requirements.	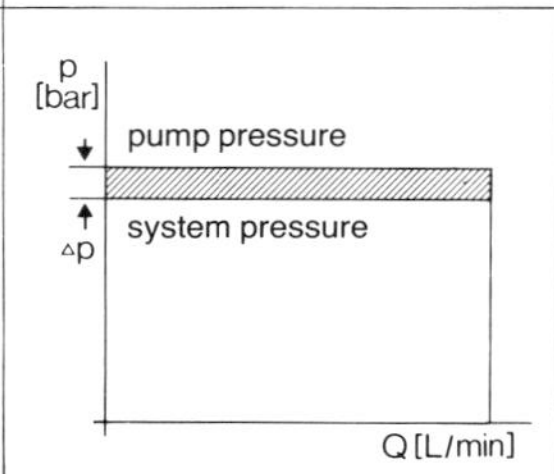	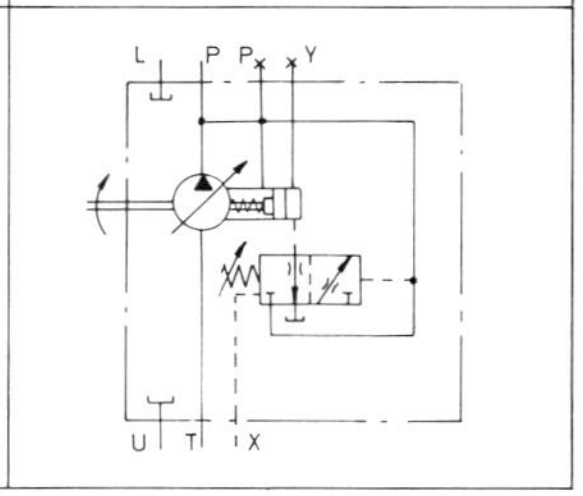

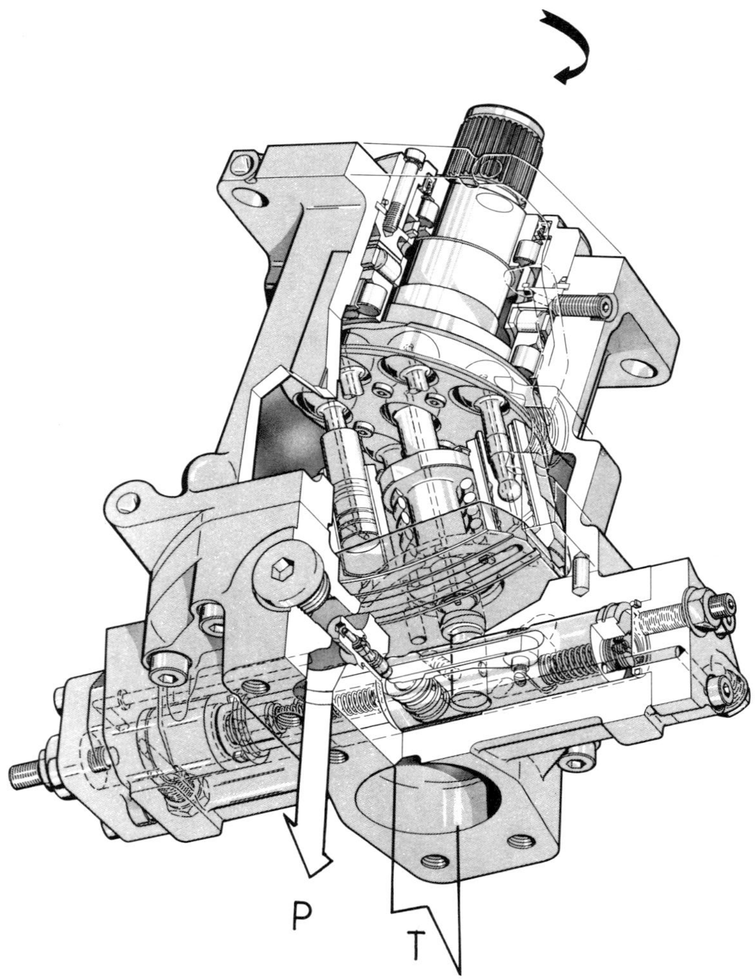

Fig. 255 "Linde" bent-axis variable displacement piston pump with four control options.

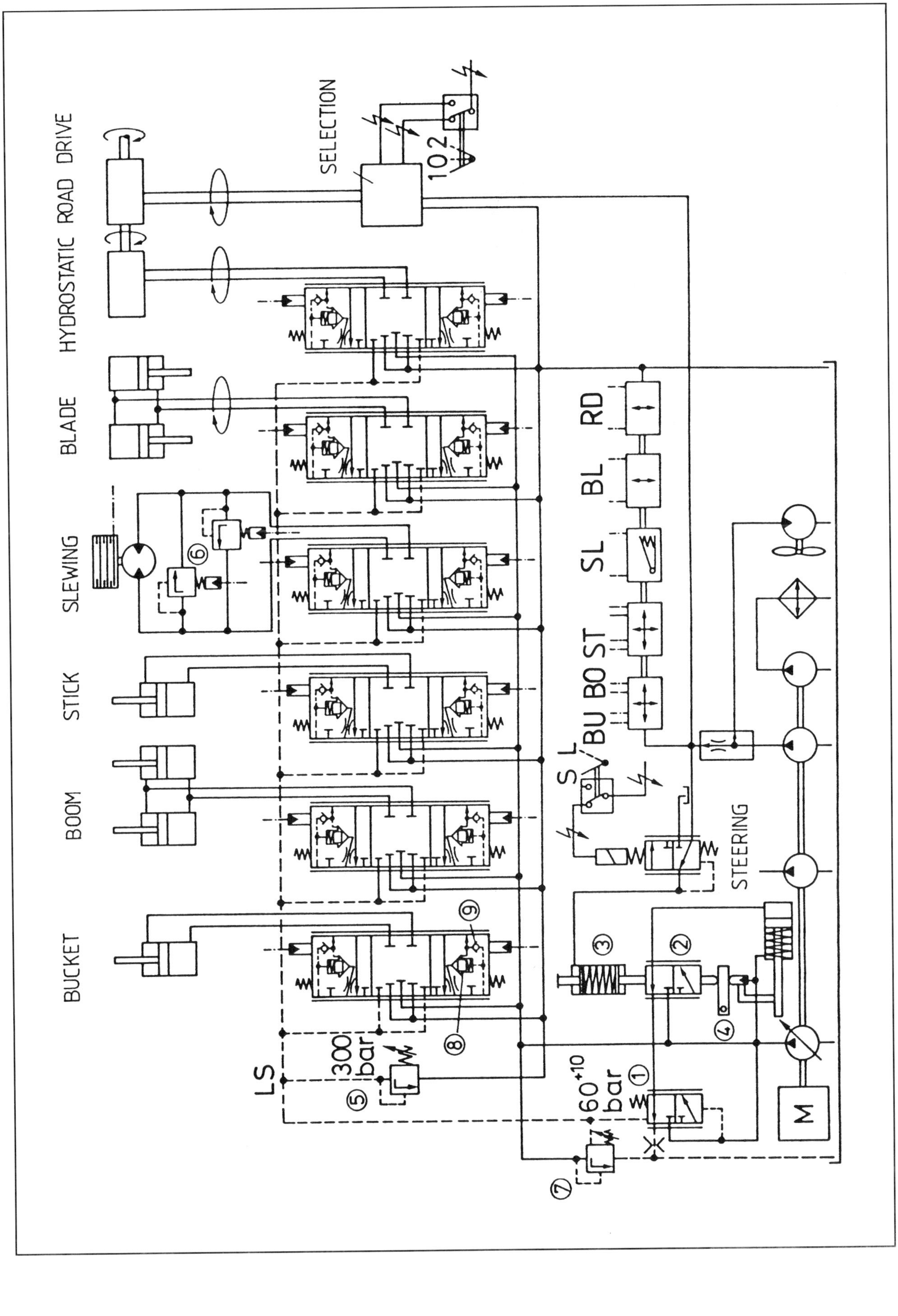

Fig. 256 Mobile dredger/digger with load-sensing control and superimposed power control.

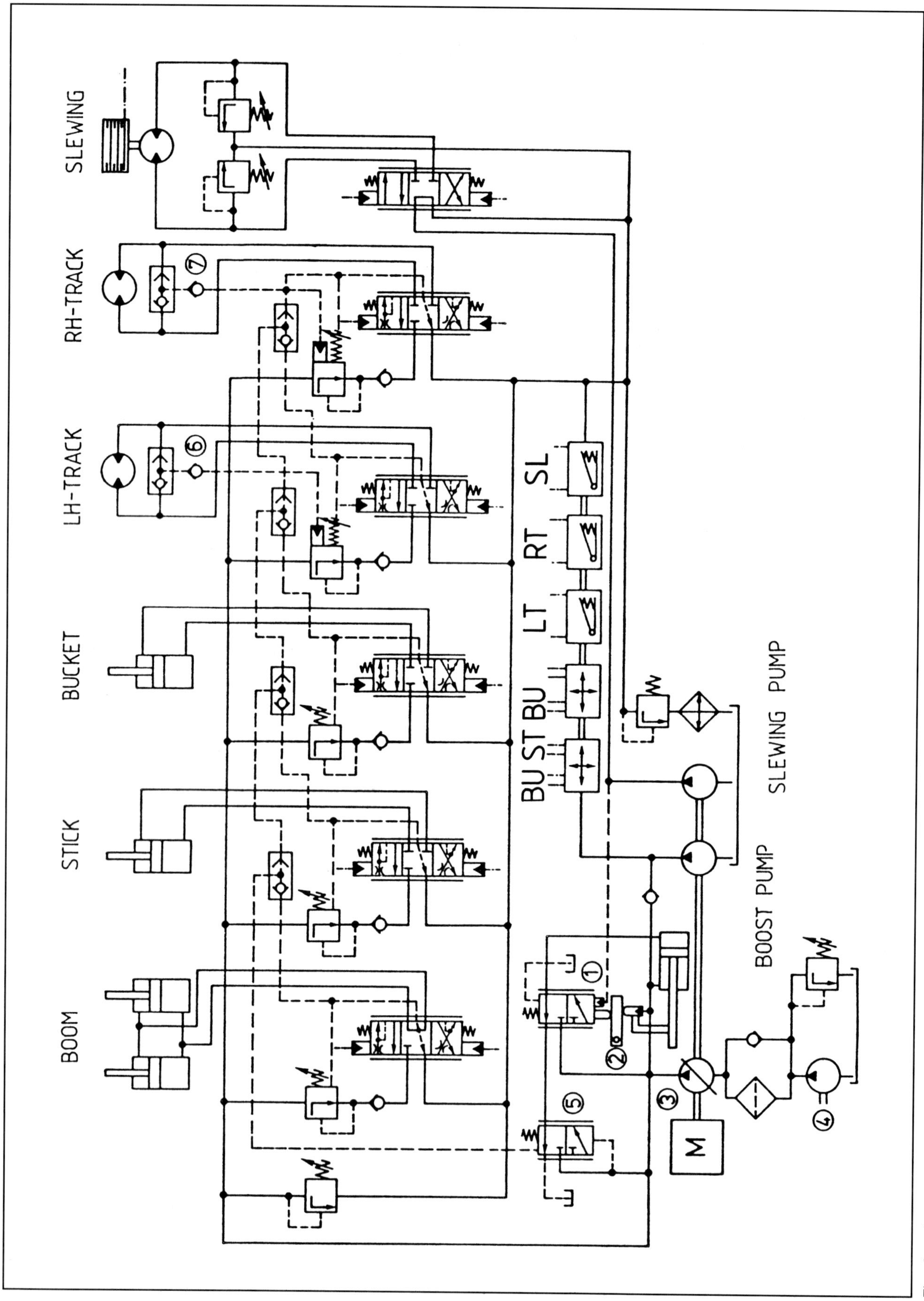

Fig. 257 Load-sensing and constant power controlled caterpillar-driven dredger with separate pump for tower rotation.

15 Industrial hydraulic pattern circuits

The previous chapters have dealt with basic principles and system components. In addition, there are numerous principles that can be more simply explained in a discussion of circuitry.

Designing hydraulic circuits is an intriguing responsibility. The circuit designer is limited only by his own ingenuity and knowledge of components and their correct and economical integration. A circuit is an arrangement of individual components, interconnected to provide a desired form of hydraulic power transfer (fig. 258). Many of these components have multiple uses by merely changing a cover plate, using an adapter, or connecting the valve to different ports. Fluid power overlaps with practically every branch of engineering, controlling or powering some part of it.

With all this versatility and flexibility, the circuit designer has a big responsibility towards the user of his circuits. He must therefore be aware that the final product of his design, the circuit, is:

- designed with the utmost care, to operate safely even under adverse or unexpected conditions;
- functional, and meets the required performance specifications for its purpose of control and power transfer;
- efficient and economical without wasting power, and is simple in design and hardware application.

Power saving circuits

Many industrial hydraulic circuits do not require a high force output over the total stroke of the actuator. (This applies to, e.g. injection moulding machines, scrap baling machines, presses, etc.) In such cases it is possible to incorporate valves and components which provide a means of increasing the rate of actuator movement with a low force output, until a predetermined pressure or position is reached. The control then provides automatic changeover to a slower actuator speed, with higher force output for the same power input. Thus, pump input power is conserved. The following four methods are frequently used to achieve rapid approach with low force, and automatic changeover to slow work stroke with high force:

- high-low double pump system
- kicker actuators with pre-fill valve and header tank
- sequenced regenerative system
- variable displacement pressure controlled pump

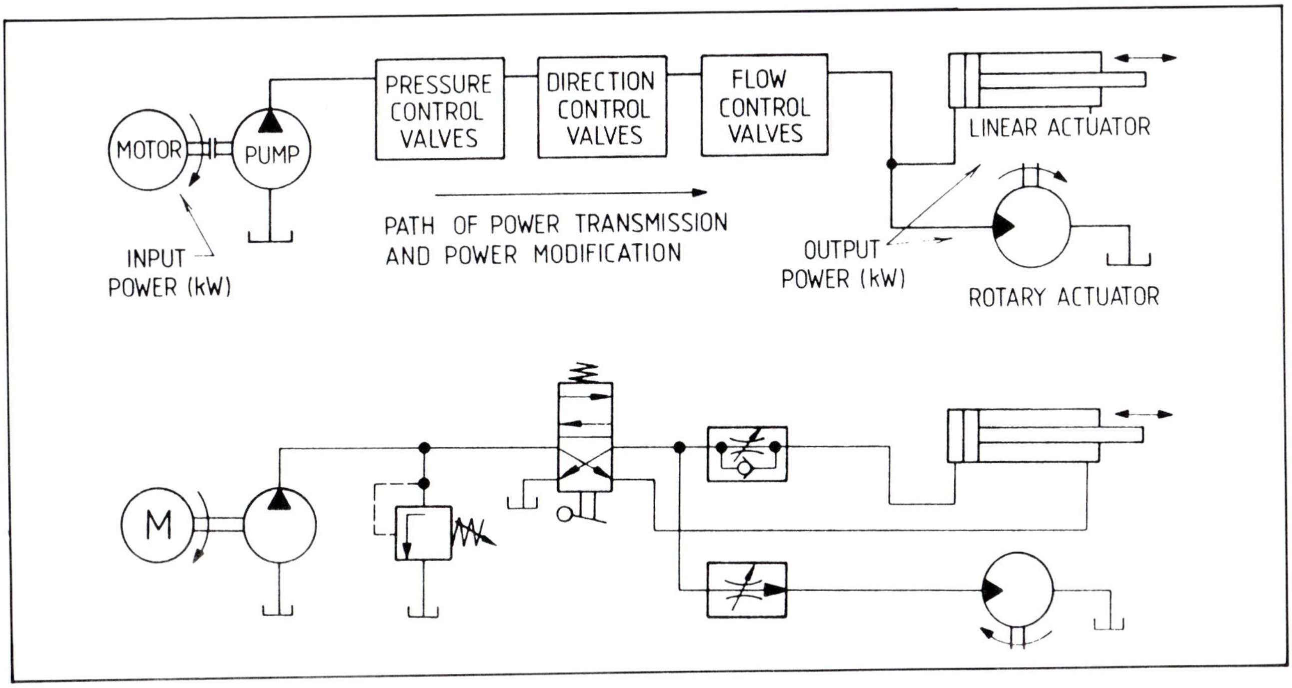

Fig. 258 Basic hydraulic system.

Power curves for power saving circuits can be plotted so that any two force/speed coordinates intersecting on the curve represent the input power for which the curve is designed.

Example

Power (W) = Force (N) × velocity (m/s)

Power (kW) = MN × mm/s, or $\text{mm/s} = \dfrac{\text{kW}}{\text{MN}}$

14 kW = 7 MN × 2 mm/s

This means that an actuator with a fast approach speed of 14 mm/s and a force output of 1 MN would require an input power of 14 kW. When switched to slow work speed of 2 mm/s, at the same power consumption, the actuator would produce a force output of 7 MN. (See applications, figs. 260–263).

High-low double pump system

Powered by both pumps, the actuator extends rapidly until the system pressure, due to an increase in the work load, exceeds the pressure setting of the off-loading valve. The high volume pump is then off-loaded to tank, and the actuator proceeds with high force output powered by the high pressure pump. The total power consumption must not exceed 14 kW (fig. 259). The rapid approach speed is 14 mm/s at a force output of 1 MN. The slow working speed is 1.84 mm/s at a force output of 7.6 MN. Maximum system pressure is limited by the system relief valve. During non-action periods, both pumps are unloaded to tank via the tandem centre D.C.V. (fig 260). The actuator retracts as soon as full extension is signalled.

Kicker actuators with pre-fill valve and header tank

Rapid approach with minimal force output is achieved by directing the whole pump flow to the two kicker actuators only. During rapid advance, the pressure sequence valve is closed, the main actuator is extended by the two kicker actuators and is supplied with hydraulic fluid from the header tank via the pre-fill valve (fig. 261).

Automatic changeover to slow work stroke with high force output takes place when the actuators meet the workload resistance, and high pressure builds up in the system. This pressure increase

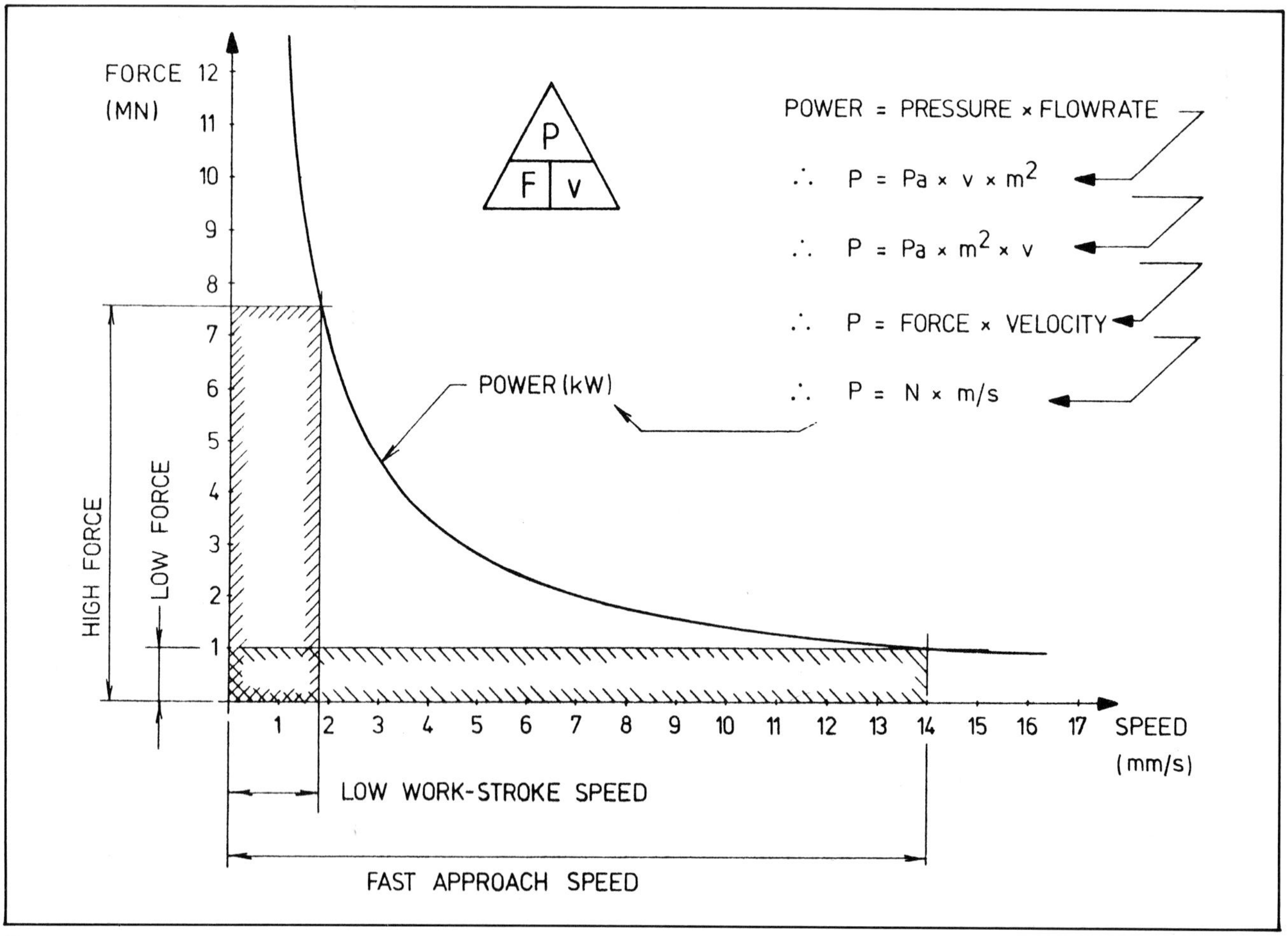

Fig. 259 Power curve for high-low control.

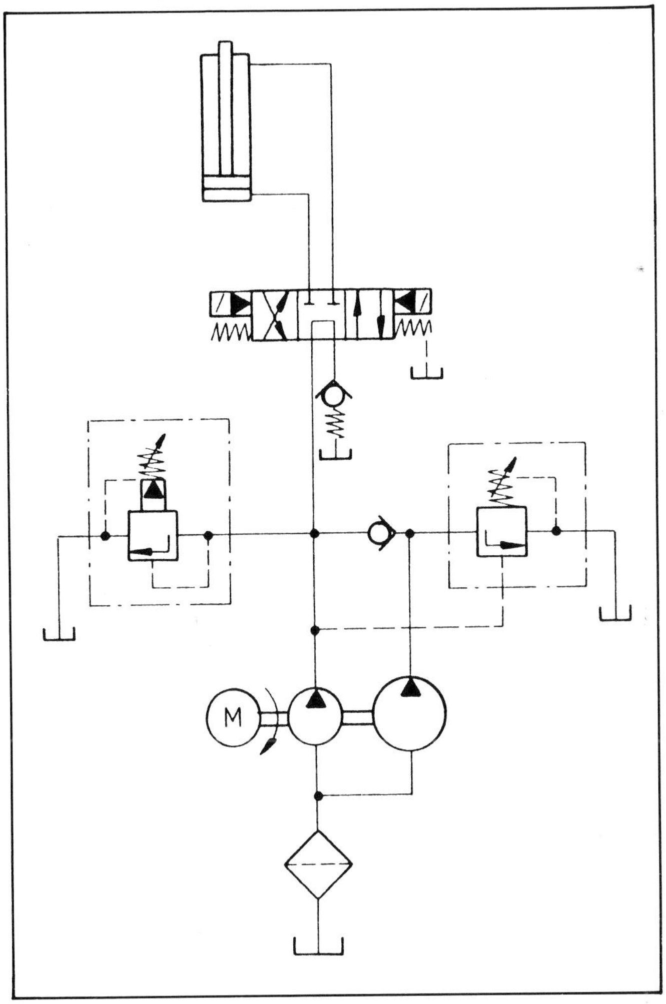

Fig. 260 *above:* High-low control with double pump system.

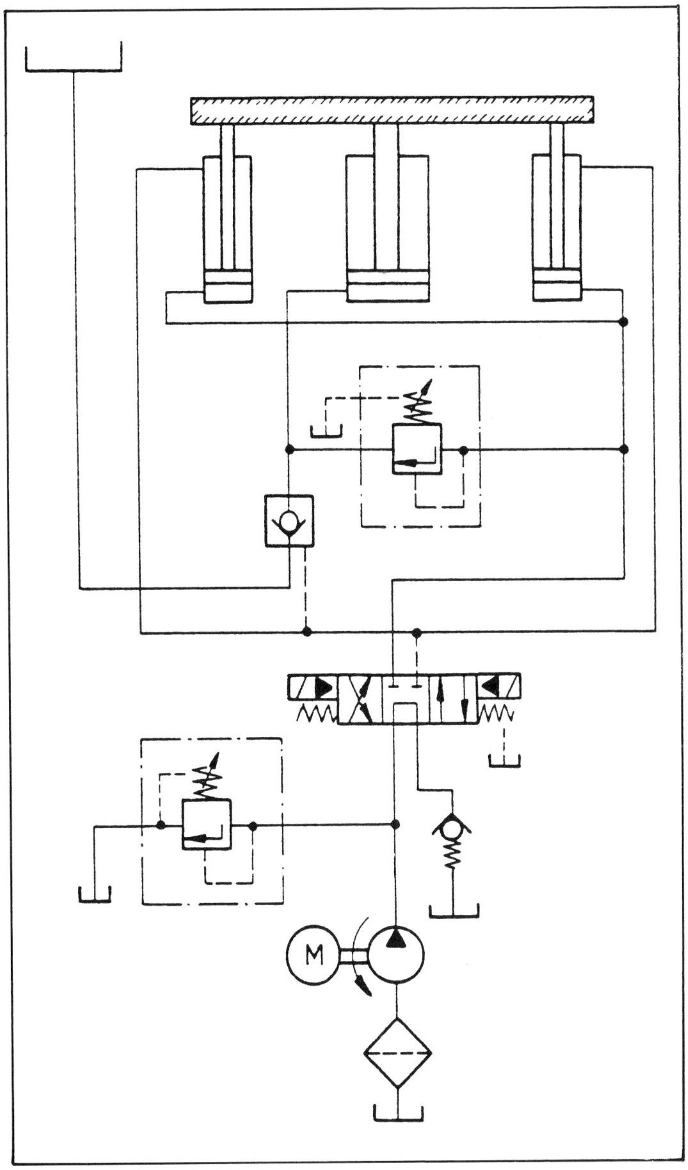

Fig. 261 *right:* Kicker actuator circuit with pressure sequence valve, prefill valve and header tank.

causes the sequence valve to open and pump flow is now directed to all three actuators. The increase in total piston area (three pistons) increases the force output and decreases the actuator velocity. (Actuator speed equals flowrate divided by the piston area.)

For the retraction stroke, the pre-fill valve is piloted open and the two kicker actuators pull the main actuator back, forcing its discharge fluid back into the header tank. During non-action periods, the pump is unloaded via the tandem centre directional control valve.

Sequenced regenerative system

Rapid approach with reduced force proportional to the rod area is achieved during the actuator extension, when the rod-end discharge flow is regenerated via line R and combines with the pump flow at merging point M. The piston speed is determined by the pump flow plus the rod-end discharge flow, divided by the full piston area (fig. 262).

Automatic changeover to slow work stroke takes place as soon as the actuator meets the workload resistance and high pressure builds up in the system. This increased pressure causes the sequence valve to open and the rod-end discharge flow is diverted via the directional control valve to the reservoir. Thus, the speed changes to slow with full force, proportional to the piston area.

Rapid return stroke is obtained when solenoid X is selected and pump flow is redirected via the integral check valve (in the sequence valve) to the rod end of the actuator. The discharge from the actuator flows via the directional control valve to the reservoir, and the piston speed is determined by the pump flow divided by the annulus area (see Chapter 4).

The check valve in line R permits the discharge flow from the rod end to combine with the pump flow during rapid advance, but prevents reverse flow during slow advance. If the full piston area has a ratio of 2:1 to the rod area, the actuator will extend and retract with the same speed. (Rapid approach speed and retraction speed are equal.)

Variable displacement, pressure controlled pump circuit

A variable displacement pump with a pressure controller will also achieve fast approach with low pressure, and slow work stroke with high pressure. In a swash plate type pump for example, the swash angle would be at its maximum during the approach stroke, giving full pump flow (fig. 263). When the pressure controller senses an increase in system pressure, the swash angle will automatically reduce and the pump will deliver whatever flow is required during the work stroke. (See pages 47–48)

Power saving circuits with additional functions

The kicker actuator circuit in fig. 266 maintains a holding pressure with the unloading valve and the accumulator, but the holding pressure will fluctuate between the cut-out and cut-in pressure levels. In order to maintain a constant (non-fluctuating) holding pressure, one can use either of the two circuit solutions shown in figs. 264 and 265. The circuit in fig. 265 maintains constant holding pressure with the differential unloading valve and the pressure reducing valve. The circuit in fig. 264 uses a small, variable displacement, high pressure pump to achieve the same result. A counterbalance valve prevents the load from falling.

Position sequenced pump off-loading

The changeover from high approach speed to slow work stroke speed may either be pressure sequenced, as shown in figs. 260–262, and 264–266, or position sequenced as shown in figs. 267 and 268.

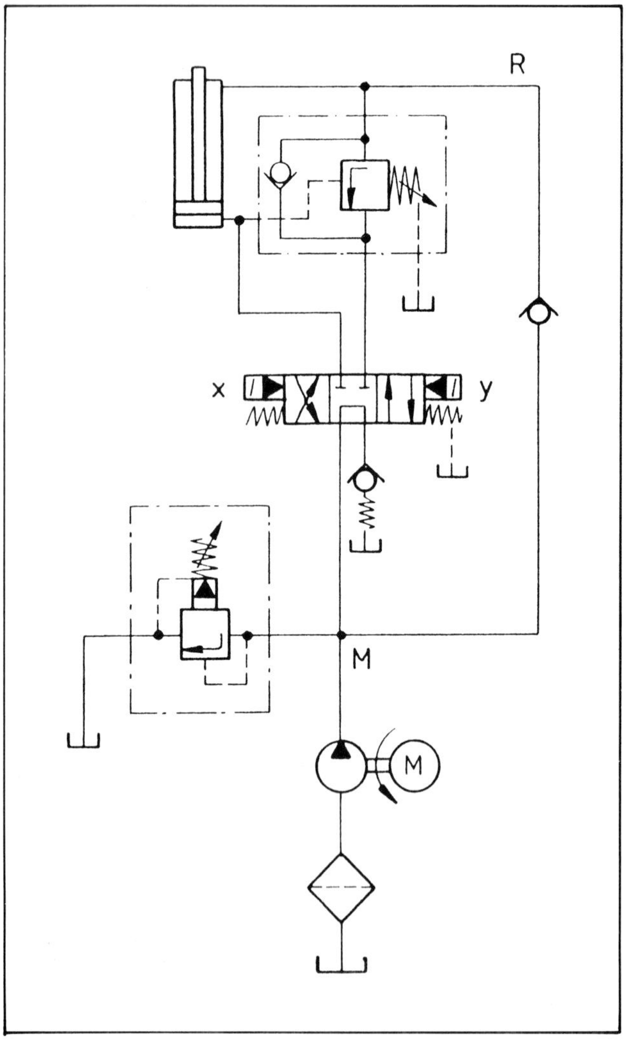

Fig. 262 Sequenced regenerative power saving system.

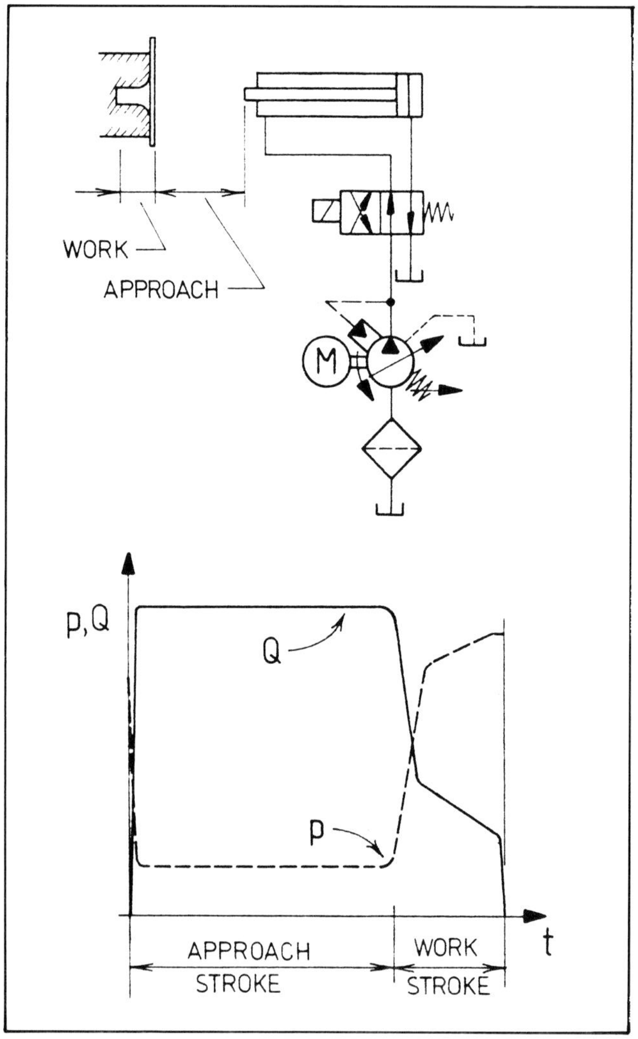

Fig. 263 Variable displacement pressure controlled pump provides power saving control.

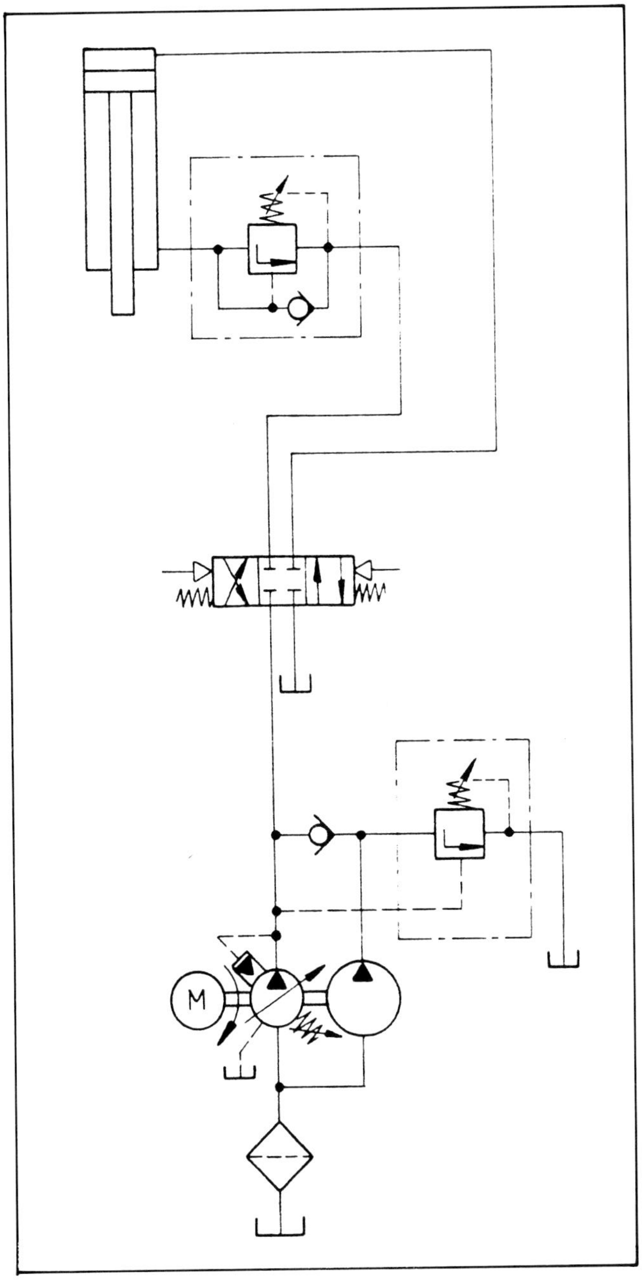

Fig. 264

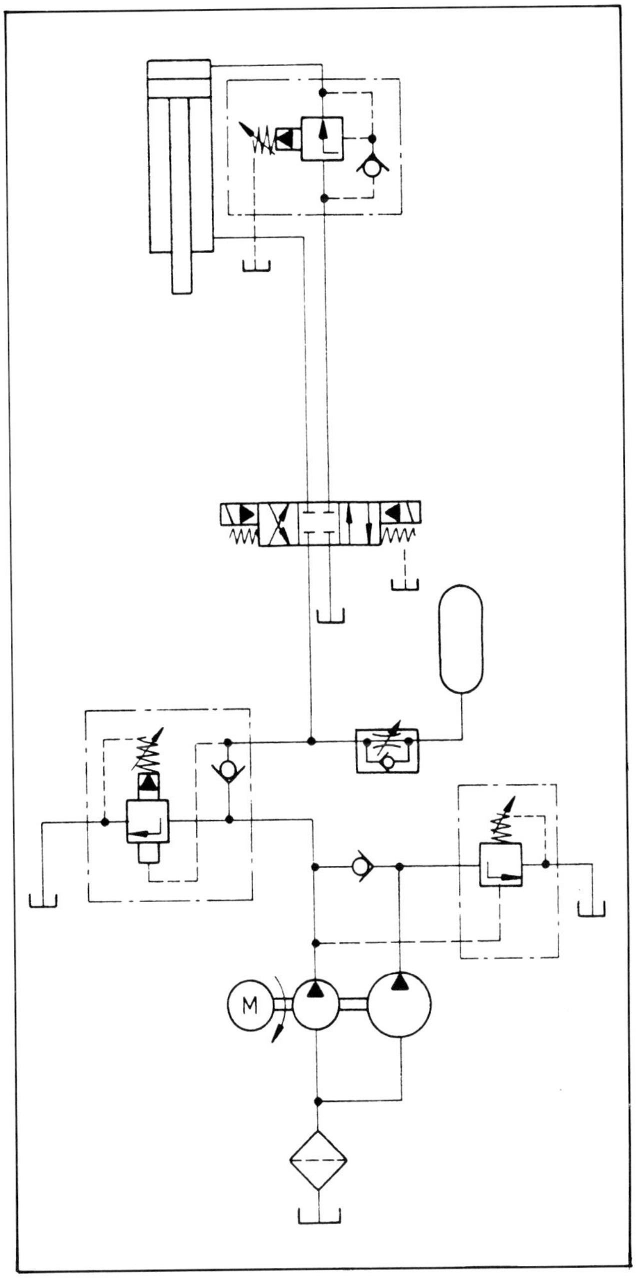

Fig. 265

With the circuit in fig. 268, the high volume pump is off-loaded with a vented compound relief valve. The compound relief valve also serves as a protection from over-pressurising the low pressure pump. The circuit in fig. 267 uses less components to achieve pump off-loading, but the remote-controlled pressure sequence valve does not protect the low pressure, high volume pump against excessive pressure build-up during the approach stroke.

Hydrostatic transmissions (motor circuits)

A hydrostatic transmission (hydrostatic drive) is a combination of two interconnected positive displacement units, a hydraulic pump, and a hydraulic motor. The pump may be located at some distance from the motor, or the two units (pump and motor) may be incorporated into a single housing.

Hydrostatic transmissions can give stepless changes of output speed, output torque, and output power. Other features are dynamic braking, smooth reversing, low inertia, fast response, and compact design. Figure 269 shows the four possible transmission output characteristics for combinations of fixed and variable motors and pumps (see also pumps, and rotary actuators, Chapters 3 and 7).

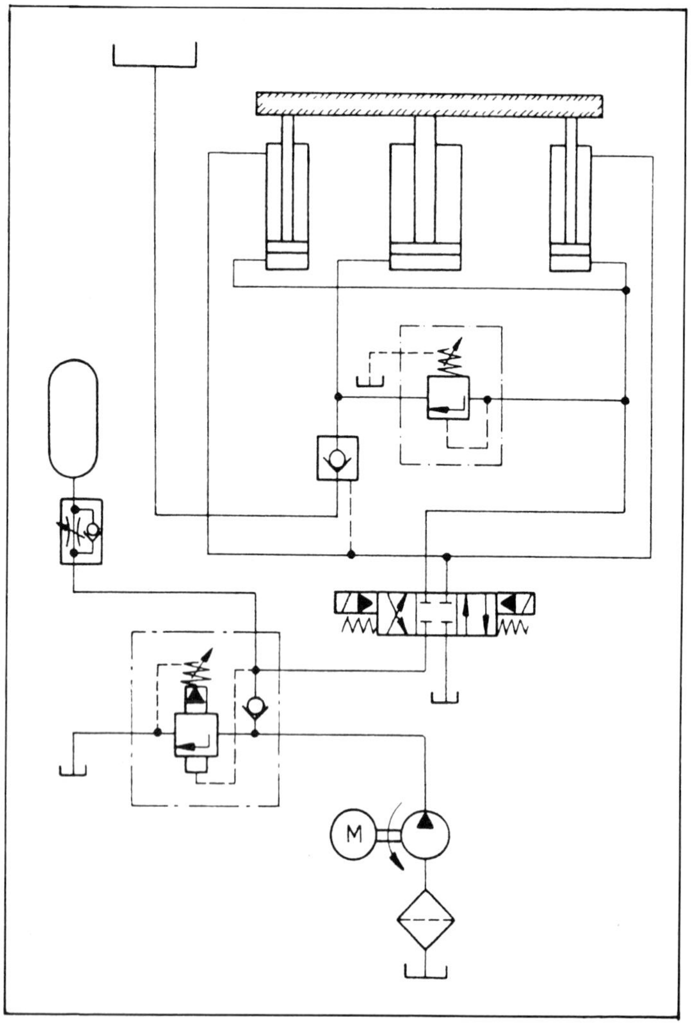

Fig. 266 Kicker actuator circuit with unloading valve and accumulator to maintain system pressure and unload the pump.

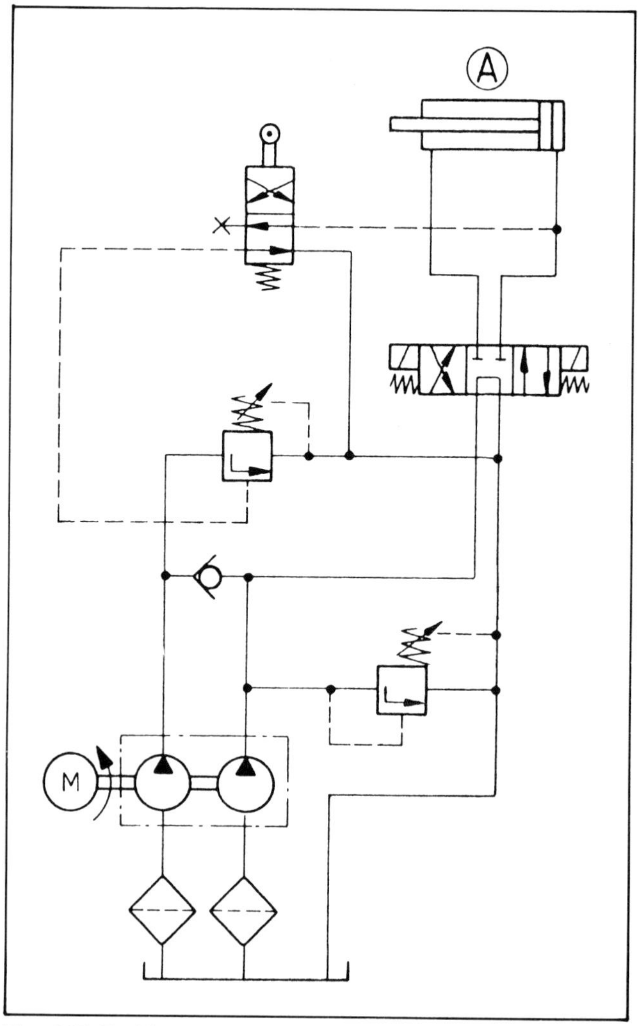

Fig. 267 Position sequenced pump offloading with remotely controlled pressure sequence valve.

Open circuit hydrostatic transmissions

In a hydrostatic transmission designed as an "open circuit" drive, the discharge fluid from the hydraulic motor returns to the reservoir before re-entering the pump inlet. Figure 270 illustrates an "open circuit" hydrostatic transmission for single direction rotation. Figure 271 illustrates an open circuit hydrostatic transmission for double direction rotation (reversible motor drive). The circuit in fig. 270 also has remote pressure control, pump unloading via the D.C.V., and free-wheeling provision via the check valve line. The motor is protected by a high pressure filter. The circuit in fig. 271 features pump unloading via open centre D.C.V.

An open circuit bi-directional drive with cross-line relief valves is depicted in fig. 272. These cross-line relief valves eliminate pressure shock when the motor is reversed or stopped. An open circuit uni-directional drive with selective torque control is depicted in fig. 273. A replenishing line provides flow to the inlet port during braking, and pump unloading is accomplished with a tandem centre directional control valve. An open circuit bi-directional drive with brake valve is shown in fig. 274. The brake valve provides controlled and adjustable braking when the D.C.V. is centred. Brake valves are used with high inertia loads to prevent the motor from overrunning during the braking phase. An open circuit bi-directional drive with three selectable speeds and cross-line relief valves is shown in fig. 275.

Motor braking

With a heavy load attached to the transmission motor shaft, an excessive pressure shock will be generated when the D.C.V. is centred or when the load is suddenly reversed. This results from the load inertia which pushes fluid from the motor outlet, and with the D.C.V. being fully reversed or having a closed centre, the discharge fluid has nowhere to go, and an

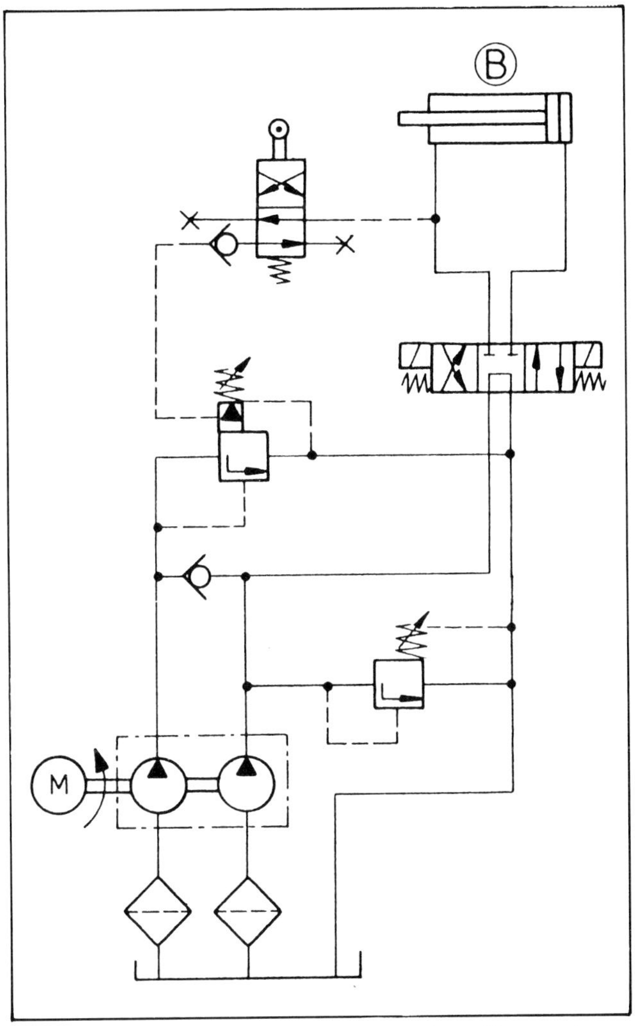

Fig. 268 Position sequenced pump offloading with vented compound relief valve.

excessive pressure shock will occur. To alleviate such shocks, brake valves and shock relief valves are used. Shock relief valves are simple or compound type relief valves which are teed-off from the motor discharge line. Shock relief valves should have a 25% higher pressure setting than the system relief valve (fig. 272).

Brake valves, as illustrated in fig. 274, automatically retard the motion of a spinning load as the pressure diminishes on the motor inlet side (see also Chapter 5, pressure controls). However, this is not the case with shock relief valves. Shock relief valves retard a motor load only during motion reversal and remove pressure shocks which could otherwise damage the system.

A reversible transmission circuit using shock relief valves for both directions of motor rotation, can be designed so that the discharge flow from the relief valve crosses over (or is connected) to the pressure line (fig. 272). At first glance it may appear that these "cross-line" relief valves would keep the motor inlet well supplied with hydraulic fluid. However, with externally drained motors replenishing fluid is still required, since leakage fluid depletes the closed circuit and motor cavitation can occur. Replenishing fluid is drawn into the system via low cracking pressure check valves (30 kPa or less). Such check valves are known as make-up checks or replenishing checks (figs. 272, 273 and 275).

Displacement		Transmission Output			
Pump	Motor	Speed	Torque	Power	Direction
Fixed	Fixed	Fixed	Fixed	Fixed	Single
Fixed	Variable	Variable	Variable	Variable	Single
Variable	Fixed	Variable	Fixed	Variable	Double
Variable	Variable	Variable	Variable	Variable	Double

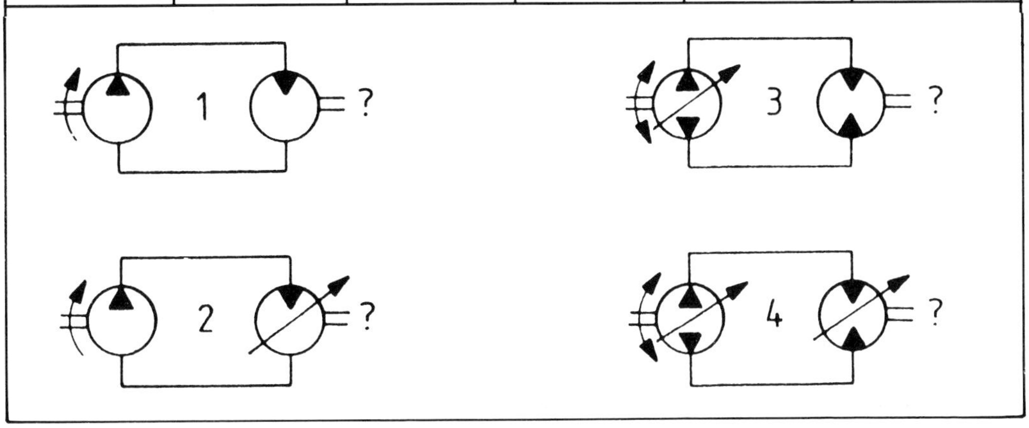

Fig. 269 Transmission output characteristics.

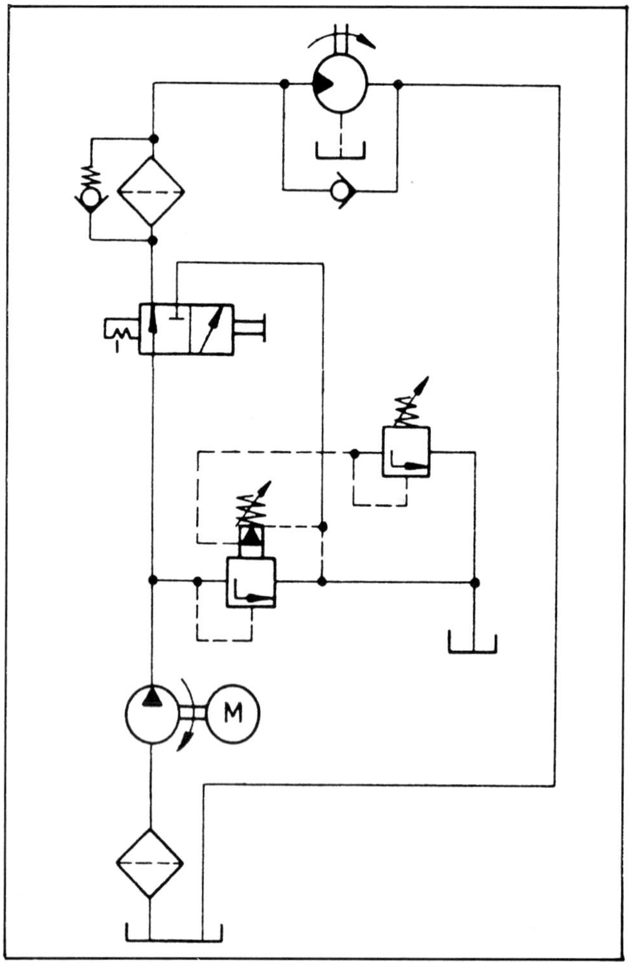

Fig. 270 Single direction, open circuit hydrostatic transmission.

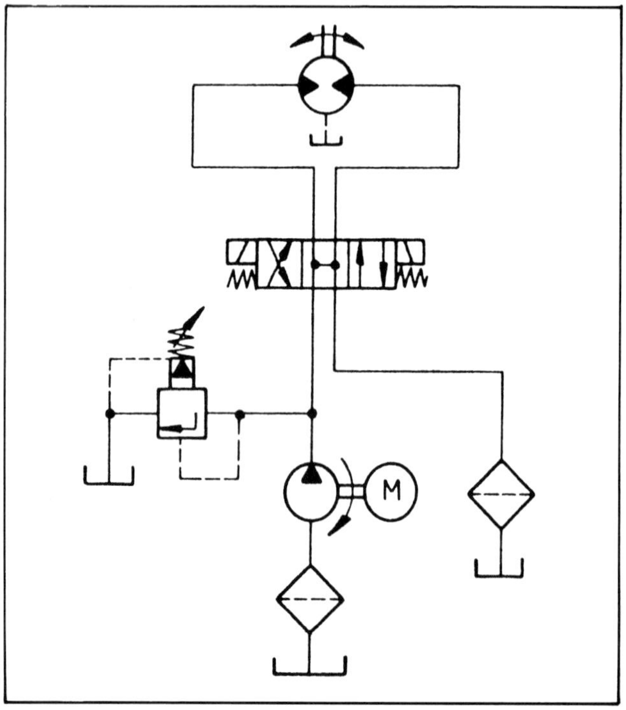

Fig. 271 Reversible, open circuit drive (hydrostatic transmission).

Closed circuit hydrostatic transmissions

Most hydrostatic drives built today are of the closed circuit type. Compared with the open circuit, this type of system has many advantages, like:

- positive boost to the transmission pump and motor;
- more positive control of motor speed and direction; and
- the use of a much smaller reservoir for a given size system.

In a hydrostatic transmission designed as a "closed circuit" drive, the discharge fluid from the hydraulic motor returns directly to the pump inlet, not to a reservoir. Such a closed circuit transmission is shown in fig. 276.

In commercially built transmissions, the make-up fluid is usually pumped into the low pressure side of the system by means of a replenishing pump which is driven in tandem with the transmission pump (fig. 72). In most closed circuit hydrostatic transmissions, the transmission pump has variable displacement and is designed for both flow directions (figs. 276–278).

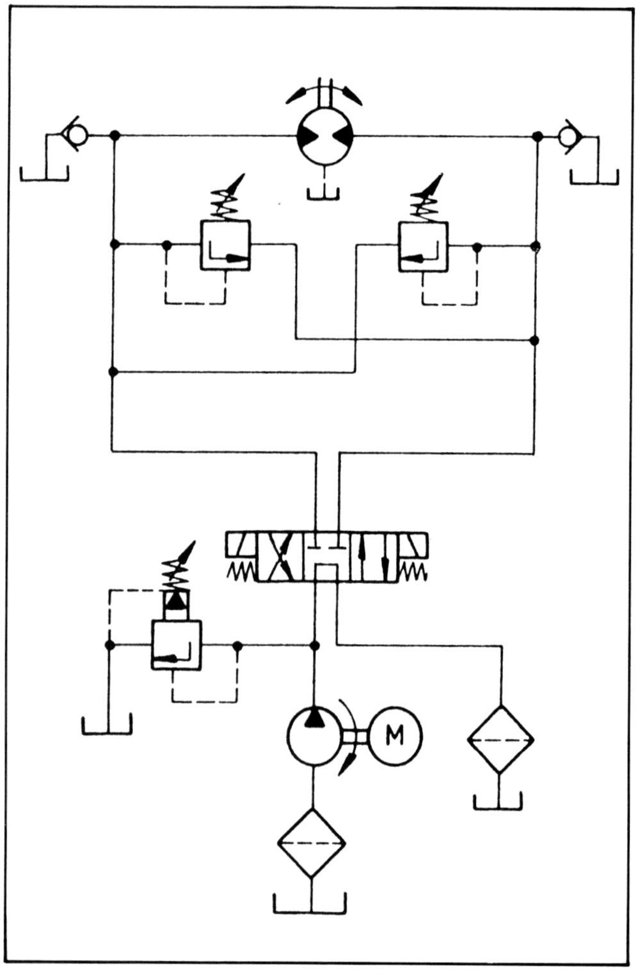

Fig. 272 Reversible, open circuit drive with cross-over relief valves.

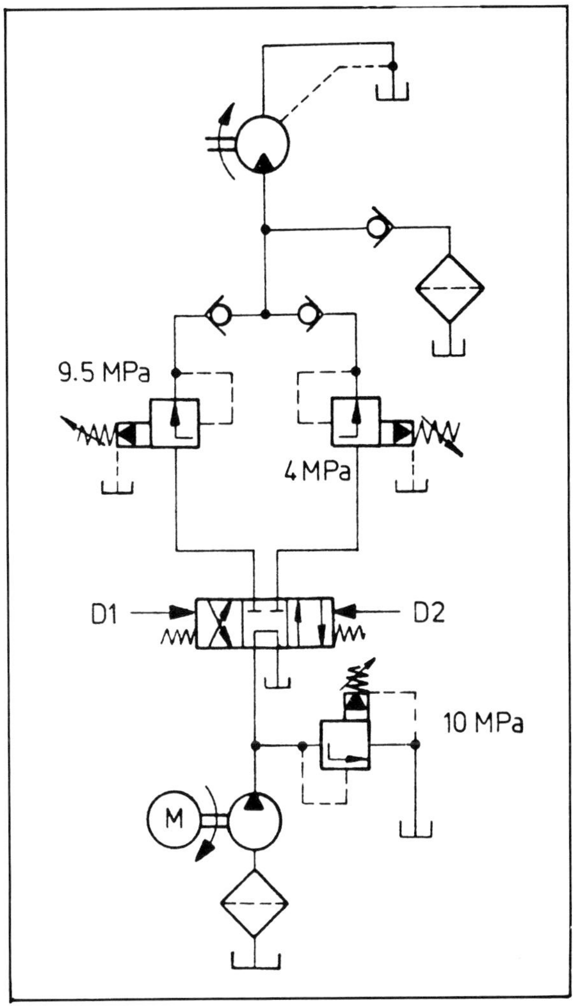

Fig. 273 Single direction, open circuit drive with selective torque control (D1/D2).

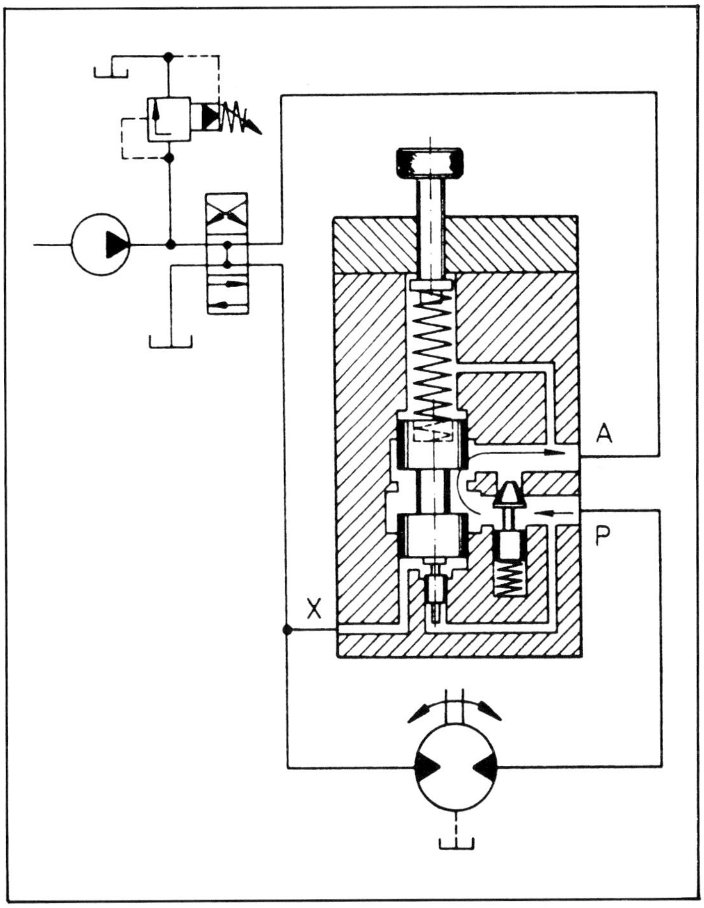

Fig. 274 Open circuit reversible drive with braking control for one direction only.

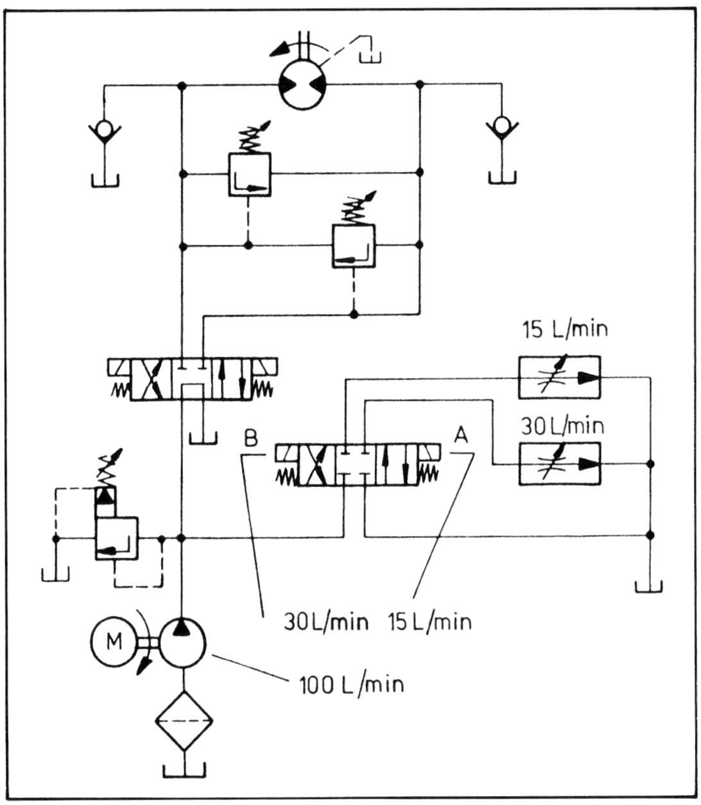

Fig. 275 Reversible, open circuit drive with three selectable speeds and cross-over relief valves to prevent shocks during motor reversal.

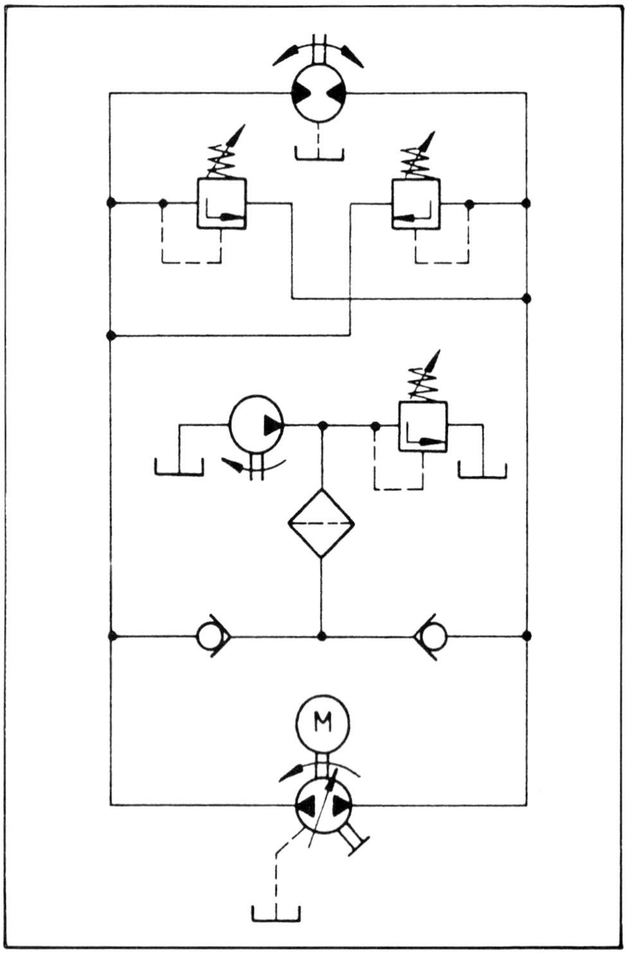

Fig. 276 Reversible, closed circuit hydrostatic transmission with cross-over relief valves.

If the hydraulic motor in a closed circuit hydrostatic transmission should stall either for a prolonged time, or for short but frequent intervals, then the pump flow will be pushed over the system relief valve. This causes the fluid to froth and subsequently overheat. Since the hot fluid does not return to a reservoir (closed circuit system) where it could dissipate heat and air, the fluid will soon overheat and deteriorate (oxidise), causing irreparable damage to the pump. Some methods currently used to prevent this are:

- Pressure switch with time delay to switch off the prime mover (electric motor), thus protecting the system if stalling should occur (fig. 277).
- Pumps with "pressure override" which automatically reduces the displacement at a pre-set maximum pressure.
- Pressure selected bleed-off from the low pressure side of the closed circuit. This diverts a predetermined flow to tank via an inbuilt cooler (figs. 277 and 278).
- Temperature switch in the reservoir to detect overheating.

To rely only on the leakage fluid from the pump and the motor to remove heat and contamination from the closed circuit would not suffice. In order to utilise all the fluid from the replenishing pump, which has a capacity in excess of the flow required to replace system leakage, a hydraulically pilot operated flushing valve is used. This diverts the excess of hot and contaminated fluid from the low pressure side of the circuit, via a cooler, and back to the reservoir. The replenishing pump relief valve must be set higher than the flushing circuit relief valve in order to ensure the transfer of fluid.

Example calculation for a high-low double pump system

A hydraulic system is required to control a vertically mounted single ended double acting actuator. The circuit must be designed as a "high-low" type control (double pump, fig. 260). The force output exerted onto the workpiece is to be held at a constant level for 30 minutes, with a maximum fluctuation of 15 per cent.

1. Draw the hydraulic circuit and label all pressure controlling valves for their pressure setting.
2. Size all required components and determine the flow output from both pumps, the pressure setting for all pressure controlling valves, and the pump input power for fast approach stroke and slow work stroke.

Specifications for fast approach stroke:
Force = 37.5 kN
Speed = 0.32 m/s

Specifications for slow work stroke:
Force = 100 kN
Speed = 0.13 m/s

General specifications:
Actuator dia. = 106 mm
Overall efficiency = 70%
Pressure settings 10% above calculated values
Cut-in to cut-out $\Delta p = 15\%$
Working volume of accumulator = 0.8 L

Rapid approach

Flow = Speed × Area

Therefore the flowrate required from both pumps $= 0.32 \times 0.106^2 \times 0.7854 \times 1000 = 2.824$ L/s

$$\text{Pressure} = \frac{\text{Force}}{\text{Area}}$$

$$= \frac{37.5 \times 10^3}{0.106^2 \times 0.7854 \times 10^6} = 4.249 \text{ MPa}$$

$$\text{Power} = \text{Pressure} \times \text{Flowrate} \times \text{Efficiency}$$

$$= \frac{4.249 \times 10^6 \times 2.824 \times 100}{10^3 \times 10^3 \times 70} = 17.14 \text{ kW}$$

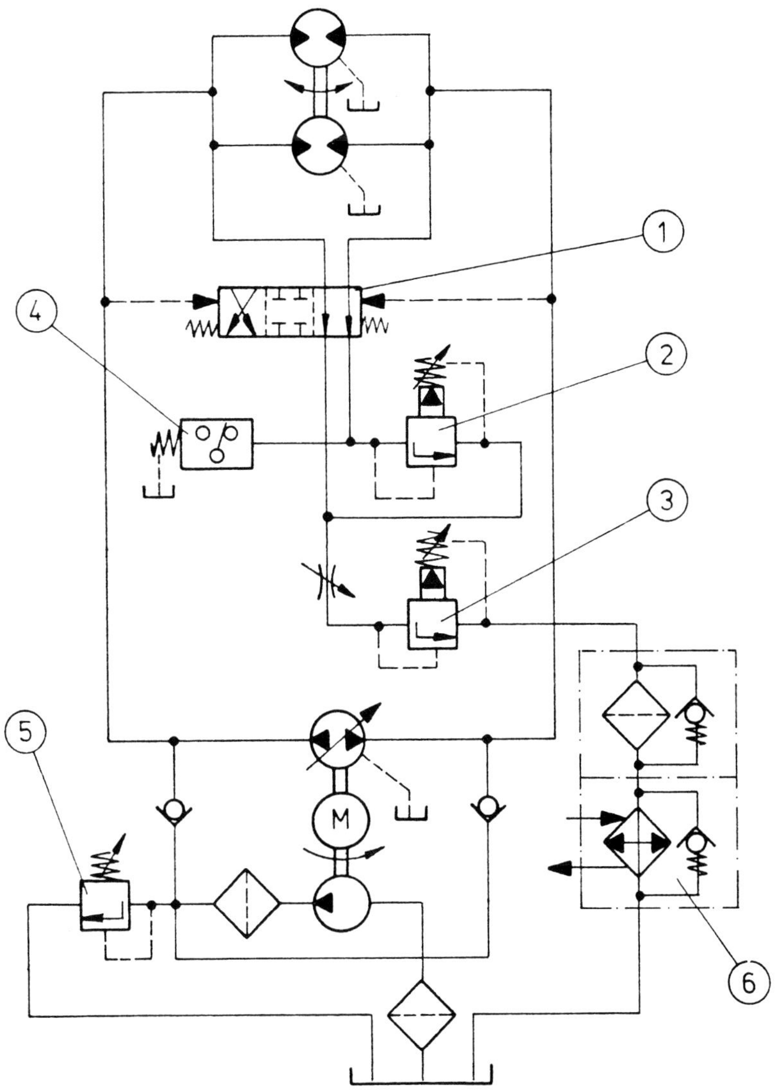

(1) Flushing valve to select the pressure switch and the system relief valve into the high pressure circuit side, and divert excessive hot fluid from the motor discharge via the filter and cooler back to tank.

(2) Relief valve to absorb shocks during motor reversal and limit maximum system pressure.

(3) Back pressure valve to maintain positive pump inlet and motor outlet pressure.

(4) Pressure switch to protect the system from overload or motor stalling. The pressure switch is set at a lower pressure than the shock relief valves.

(5) Replenishing pump relief valve to protect the pump from overload during motor reversal.

(6) Filter and cooler with bypass. The cooler ensures that only cool fluid is pumped into the replenishing circuit.

Fig. 277 Closed circuit hydrostatic transmission.

Work stroke

The flowrate required from the high pressure pump $= 0.13 \times 0.106^2 \times 0.7854 \times 10^3 = 1.147$ L/s

$$\text{Pressure} = \frac{100 \times 10^3}{0.106^2 \times 0.7854 \times 10^6} = 11.33 \text{ MPa}$$

$$\text{Power} = \frac{11.33 \times 10^6 \times 1.147 \times 100}{10^3 \times 10^3 \times 70} = 18.57 \text{ kW}$$

Low pressure pump size

Flowrate both pumps minus flowrate high pressure pump $= 2.824 - 1.147 = 1.677$ L/s

Valve setting pressures

Unloading valve $= 11.33$ MPa (cut-in pressure)
Offloading valve $= 4.25$ MPa $+ 10\% = 4.7$ MPa

Power input into electric motor

$\text{Power}_{(MAX)} + 15\%$ for unloading valve $= 21.85$ kW

Reservoir size

$3 \times \text{flow}_{(MAX)} = 3 \times 2.824 \times 60 = 500$ L

Accumulator size

Precharge pressure = 90% of minimum system pressure
p1 = 11.33 MPa × 0.9 = 10.2 MPa (102 bar)
p2 = 11.33 MPa = cut-in pressure (15% less than p3)

$$p3 = \frac{11.33 \text{ MPa} \times 100}{85} = 13.33 \text{ MPa (133.3 bar)}$$

Using the accumulator sizing chart (fig. 156) for the calculated parameters, the correct size of accumulator is 10 litres (nominal), with a working volume (V_w) of approximately 0.95 litres.

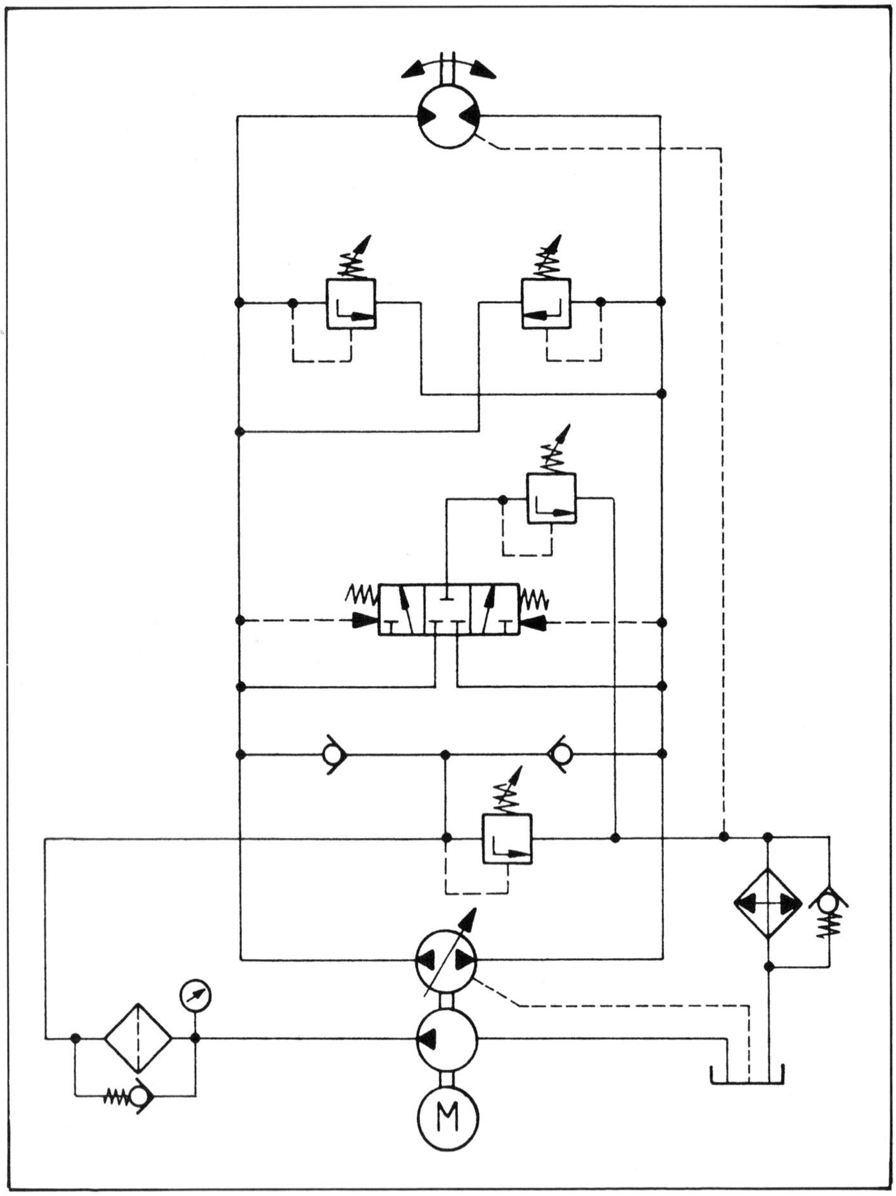

Fig. 278

16 Proportional and servo hydraulic control systems

Even long before the advent of servo controls and proportional controls, most industrial hydraulic systems were electrically sequence controlled. This meant that solenoid valves with distinct on/off characteristics caused the actuators to commence moving or stop moving. Load acceleration and deceleration was in most cases achieved with manually adjusted controls, or where the load acceleration and/or deceleration had to be progressive, circuit designers used cam-operated flow control valves. For many applications these systems have performed satisfactorily and they may, in a limited way, well continue to do so in the future.

Today, however, hydraulic controls for modern production machines and hydraulic controls used widely in mining, agriculture, transport, amusement, shipping and aerospace industries often demand an infinitely adaptable and quickly variable setting of actuator forces and their stroking speeds. For such applications, cam controlled or manually adjustable speed and force controls are not sufficient. The modern fast sequencing plastics injection moulding machine is a typical example of a machine which uses proportional controls. Its actuator motions require an infinitely variable stroke speed with progressively varying actuator forces within one stroke. Furthermore, for the next product to be run on the same machine, these parameters must often be changed quickly and with precise performance repetition to different settings. To make matters worse, on-the-job fine tuning may be required because of variations in environmental and machine temperature and inconsistencies between different batches of materials.

For such stringent conditions, machine designers have no choice but to use proportionally variable controls, which give the flexibility and infinitely variable performance of hydraulic actuators with regard to force, speed and stroke position (see fig. 279).

In the past, servo valves in a closed loop control arrangement were used to produce the sophisticated performance required by such machines. Servo valves, however, are high precision directional and flow control valves, which often create high installation and purchasing costs. Even the slightest fluid contamination causes the delicate feedback mechanisms in the servo valves to malfunction. To compound the problems, serviceability is often limited and complex to access and expensive.

For such reasons, major hydraulic component manufacturers developed a range of proportional valves, which filled the wide gap between simple on/off valves and sophisticated servo valves. The ability of proportional valves to be used for directional, pressure and flow control, and to adjust them electronically to an infinite number of positions within their range, makes them extremely useful and popular for a wide range of hydraulic applications.

With some manufacturers' products, proportional valves at the high performance end of their range approach the performance of servo valves. This chapter defines the two valve groups as follows:

- Proportional valves are valves producing a hydraulic output proportional to its solenoid input signal. This input signal can be adjusted remotely by electronic means. Its solenoid is a proportional solenoid causing proportional spool shift in proportional directional controls or proportional spring tension in proportional pressure controls and flow controls.
- Servo valves have a force or torque motor to achieve the variable output. The input is also electrical and is also adjustable by electronic means. Servo valves are predominantly used for proportional directional and flow control.

Basic definitions

The technology pertaining to servo and proportional valves involves a large range of more or less complicated concepts and unusual terminology. For this reason these concepts are briefly defined. Some of these concepts apply specifically to proportional or to servo valves only, while others apply to both.

Hydraulic control system concepts

Hydraulic power transmission systems in general are used to control the position of, the motion direction of, the speed of motion of, and the force against a resisting load. Conventional control systems, as explained in the chapter introduction, greatly limit the performance of such hydraulic power transmissions. Hence, servo and proportional valves and control concepts were developed. Proportional, as well as servo valve systems have one major concept in common: "They motion, position and force a resistive load proportionally to a provided input signal". Such an input signal may be manual, mechanical or electrical. The last is the most common in servo and proportional valve systems.

Open loop control systems

In an open loop control system, the actuator performance data with regard to its stroke position, operating pressure and required flowrate is provided as an analog electrical signal (voltage). This "input signal" describes the desired "set point value" or

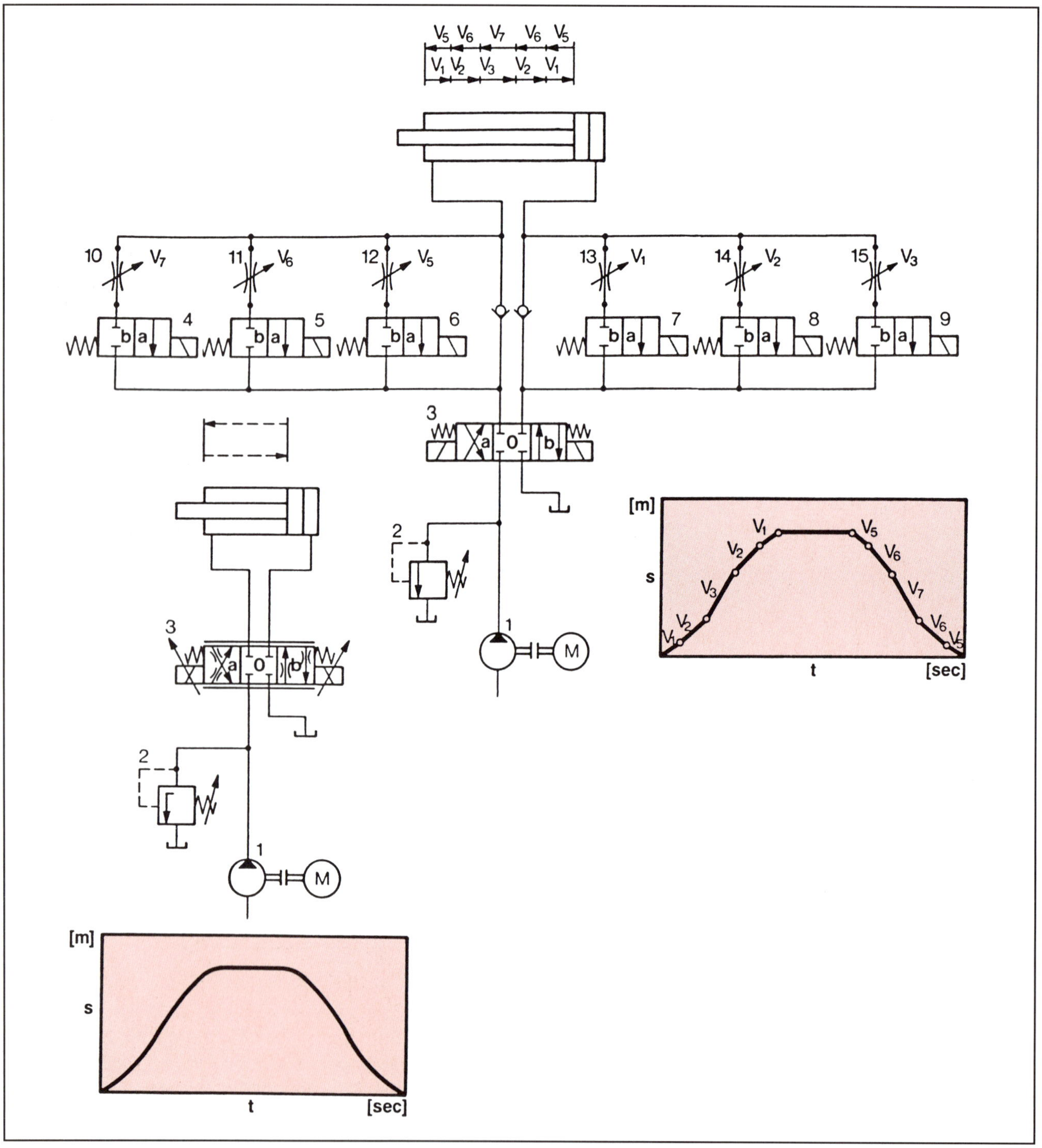

Fig. 279 Transport actuator with ramped acceleration and deceleration for both directions. With non-analog on/off technology the circuit uses 16 hydraulic components, whereas the solution with a proportional directional control valve reduces the control to a mere three component circuit and smoother ramping may be achieved.

performance data. From here on, for simplicity reasons, we use the term "input value signal". Transient functions, such as accelerating or decelerating the load opposing the hydraulic actuator, are achieved and controlled by means of electronic ramp generators.

Note: An open loop control system has no feedback provision from the load behaviour and no corrections are made, if the load behaviour falls short of the desired performance. There are, however, numerous control applications where an open loop control system is sufficient and extreme response accuracy does not warrant the extra costs of a closed loop system (see fig. 282).

Closed loop control system

In a closed loop control system, the "input value signal" also describes the desired system performance as explained for open loops. The system output in a closed loop system is, however, constantly monitored by electronic measuring devices and the "feedback signal" from these devices is then, at the summing point, compared with the input value signal. Any system deviation (error) resulting from this comparison is then fed back to the proportional magnet to cause corrective measures at the valve spool. Thus the control system is continually monitoring and correcting any minute deviation of load behaviour to maintain the desired input data (see fig. 282).

The relationship between the input value signal and the system response is called the "transfer response". This overall transfer response is dependent on the transfer response of all the individual control components in the control chain.

Follow-up versus regulated systems

In a "regulator type system", the hydraulic output must as precisely as possible maintain load position, load speed, or actuator force, or a combination of these parameters as dictated by the input signal. It maintains these output parameters independently of system disturbances and it regulates even more effectively if the control valve in use has an armature position sensor or a valve internal position feedback, and the control is designed as a closed loop system. In a regulator type system the input signal does not usually vary (see fig. 282).

In a "follow-up type system", the input signal frequently changes and therefore the output must automatically follow the input and maintain output parameters dictated by the input regardless of system disturbances (see fig. 281). A typical follow-up system is the power steering on a motor car.

Manual input hydraulic systems

A manually controlled valve with hand lever is shown in fig. 280. Movement of the lever in the clockwise direction, will move the spool to the right causing the pressure port to be connected to the rear end of the actuator, and the return port to the rod end. This will move the machine part in the same direction as the spool, taking the valve body with it, which closes the fluid ports again. As a result, the position of the machine part will always be proportional to the valve handle position. Thus, a partial movement on the hand lever will cause the actuator to respond with a proportional partial movement, and a full input movement on the hand lever will cause the actuator to move through its full stroke.

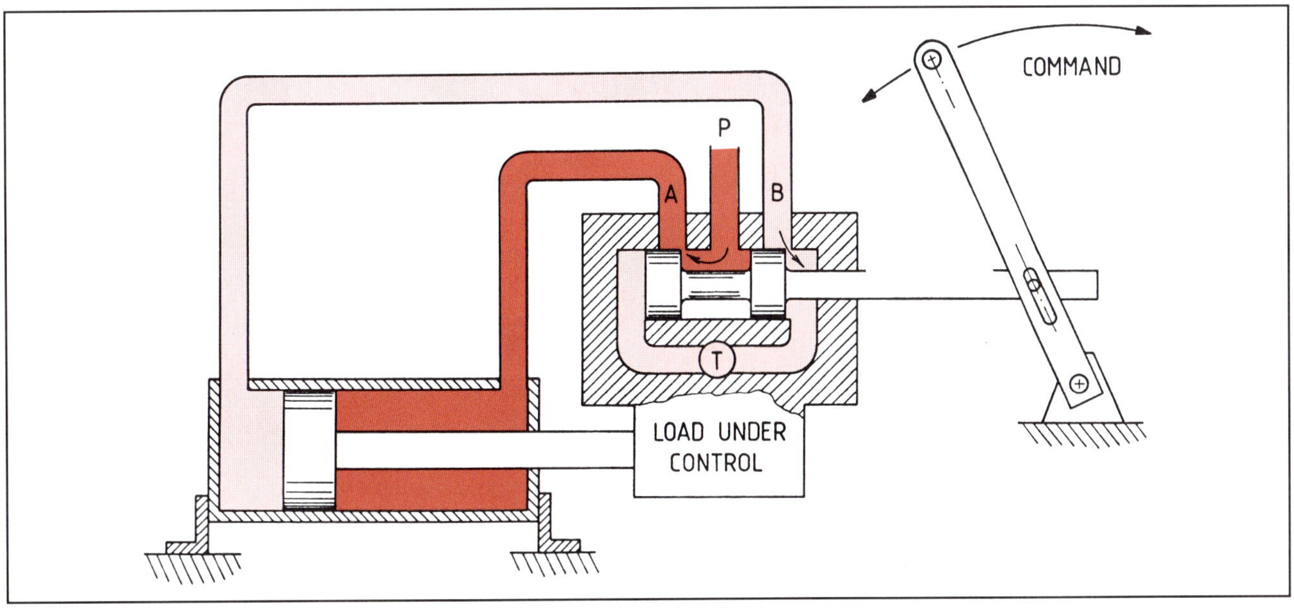

Fig. 280 Manual input hydraulic system of the follow-up type similar to automotive power steering.

Mechanical input systems

An example of a proportional or servo system with mechanical input is the hydraulically powered copying attachment for a lathe, shown in fig. 281.

A spring loaded stylus bears lightly against the outline of the master form, tracing it as the carriage is fed along the lathe bed. The profiles of the nose of the stylus and the tool must be the same, and cannot be larger than the smallest radius of the work if reproduction is to be accurate.

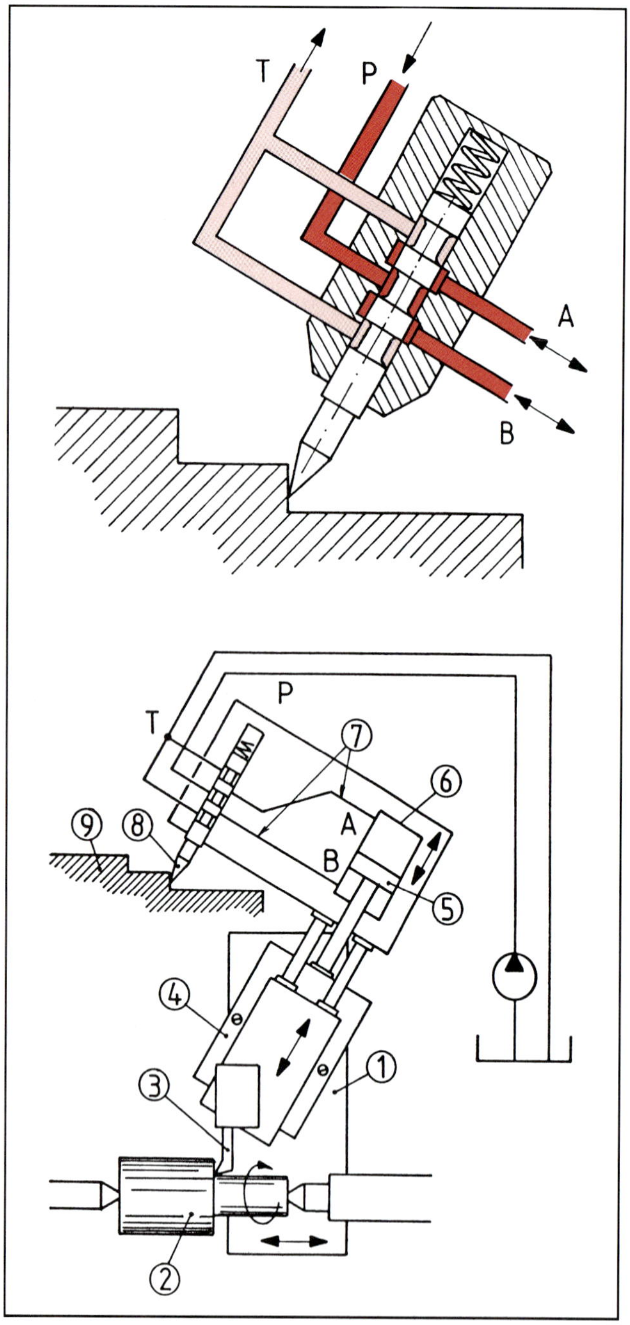

Fig. 281 Turning lathe with hydraulic control type servo copying attachment. The stylus provides the mechanical control input signal as it moves along the template (master form).

Any movement of the stylus operates the spool valve to which the stylus is directly connected. The spool valve controls the flow of oil from the pump to the cylinder. The oil flow causes the motion of the cylinder, which is directly coupled to the tool on the copy slide. The cylinder moves relative to the piston, which remains stationary. In some designs the motion is the reverse, with the piston moving and the cylinder remaining stationary. As soon as the stylus deviates from a path parallel to the lathe axis, the spool will uncover the oil ports.

Oil will enter the cylinder port A and leave at cylinder port B, or vice-versa, depending on the direction of the valve movement. This causes the cylinder, cutting tool and spool valve housing to move distances proportional to, and in the same direction as, the spool valve. Movement ceases when the cylinder has carried the spool valve housing to the point where the ports to the cylinder are covered, thus stopping the flow of oil to and from the cylinder.

Electrical input hydraulic systems

Figure 282 shows four typical systems with electrical input. In system Ⓐ the proportional pressure reducing valve maintains a precise, but variable, pressure to the actuator. In system Ⓑ a proportional pressure limiting valve is used to control pressure limits which must differ for extension or retraction stroke. In system Ⓒ the proportional directional valve controls motor rotation direction and motor rpm. Systems Ⓐ, Ⓑ and Ⓒ are so called open loop systems, which have no load behaviour feedback. System Ⓓ is a classical closed loop proportional control, which includes load feedback for precise load positioning. The electronic control includes gain, ramps and dead band adjustments with remote PLC analog output for setpoint and position controls.

Proportional valves

Proportional valves differ from conventional on/off valves through their ability to assume an infinitely varying switching position. This switching position is a step-less conversion of a analog electrical input signal into a proportional hydraulic actuator output. Proportional valves may be used to achieve:

- Proportional actuator position by using proportional directional control valves.
- Proportional actuator force (or torque) by using proportional pressure control valves.
- Proportional actuator motion velocity by using proportional flow control valves.

The fundamental principle of their operation is based on the use of a proportional solenoid, which moves the valve element into the desired position and maintains that position with great precision, if the input signal is not varied. For any variation of the input signal, the solenoid armature responds with a proportional linear movement or force (see fig. 284).

Proportional valves are available with or without a solenoid armature position sensor and may be used for closed or open loop hydraulic control systems. For closed loop hydraulic control systems, the proportional valve usually has a zero spool overlap (see fig. 31). To achieve greater flow capacity, manufacturers recommend the use of pilot operated proportional directional control valves to obtain higher positioning forces on the valve spool.

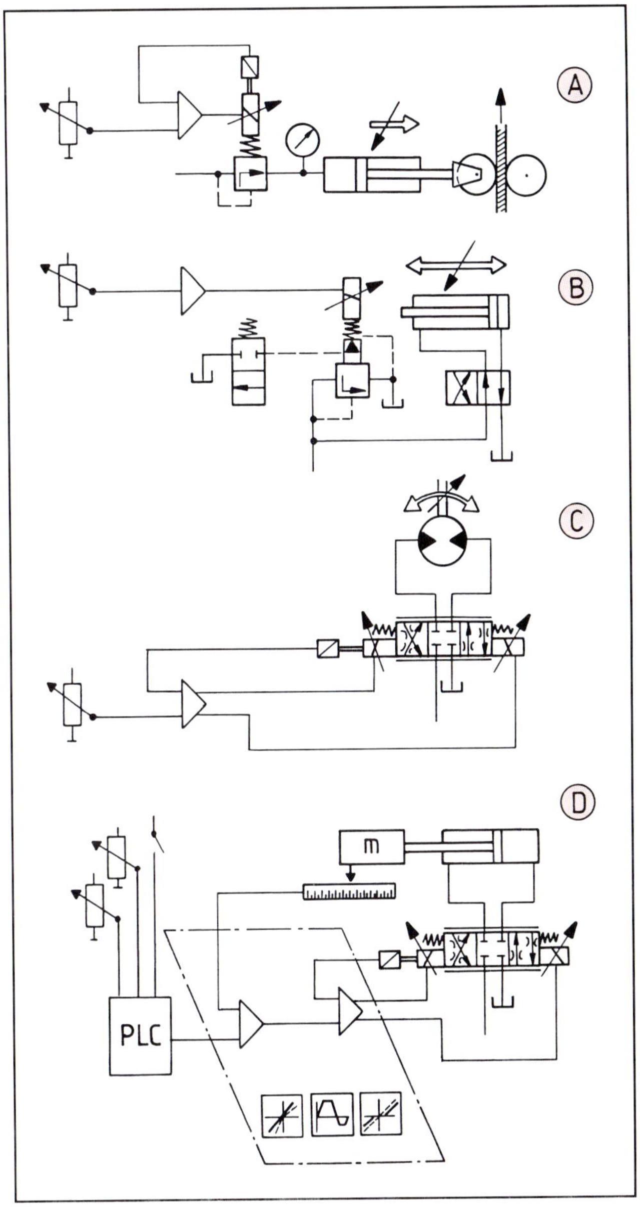

Fig. 282 Proportional hydraulic systems with electrical input.

The proportional solenoid

The proportional solenoid, unlike a conventional DC solenoid, generates an almost consistent force (F_M) over its working stroke (see comparison conventional solenoid/proportional solenoid fig. 283). Its solenoid armature chamber is filled with oil and is sealed for pressures up to 250 Bar. Such solenoids are usually referred to as wet armature or wet pin solenoids. The essential difference between a conventional on/off DC solenoid and the proportional solenoid is the unique design of the proportional armature, the adjoining pole piece and the gap between core tube and armature, which conducts the magnetic flux. When passing a current through the solenoid coil, a magnetic force is generated which pulls the armature and the attached push rod towards the pole piece. The magnitude of the created linear force on the push rod is proportional to the current. This means a 50% current creates 50% of maximum push force and a mere 25% current results in a 25% push rod force. Opposing this push rod force is a spring on the opposite end of the valve spool (see fig. 284). Hence, balancing of forces takes place. The spool moves partially to the right if the proportional magnetic force is greater, or it moves to the left if the magnetic force becomes reduced. The spool remains stationary if the two forces are in equilibrium. To show the spring/magnet relationship, the linear spring characteristics of the opposing spring is added in fig. 284.

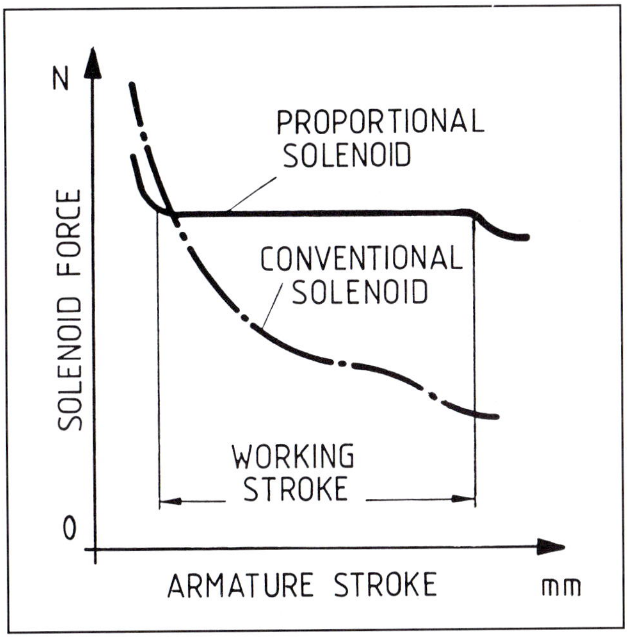

Fig. 283 Comparison of normal DC solenoid to proportional solenoid.

Proportional solenoid construction and design may differ depending on the particular application. For proportional directional controls, the solenoid has a linear distance control and the armature force is balanced against the valve spool spring (centring spring or return spring). The armature usually has a maximum working stroke distance of about 5 mm. To reduce its relatively high hysteresis (3 to 6%) a dither signal may be applied (see fig. 284).

For force controlled proportional solenoids, the magnetic force produced on the armature and attached push rod is directly proportional to the applied solenoid current and at an armature stroke of one to two millimetres (see fig. 285). Such proportional force solenoids are used on proportional pressure controls (proportional pressure limiting valves and proportional pressure reducing valves).

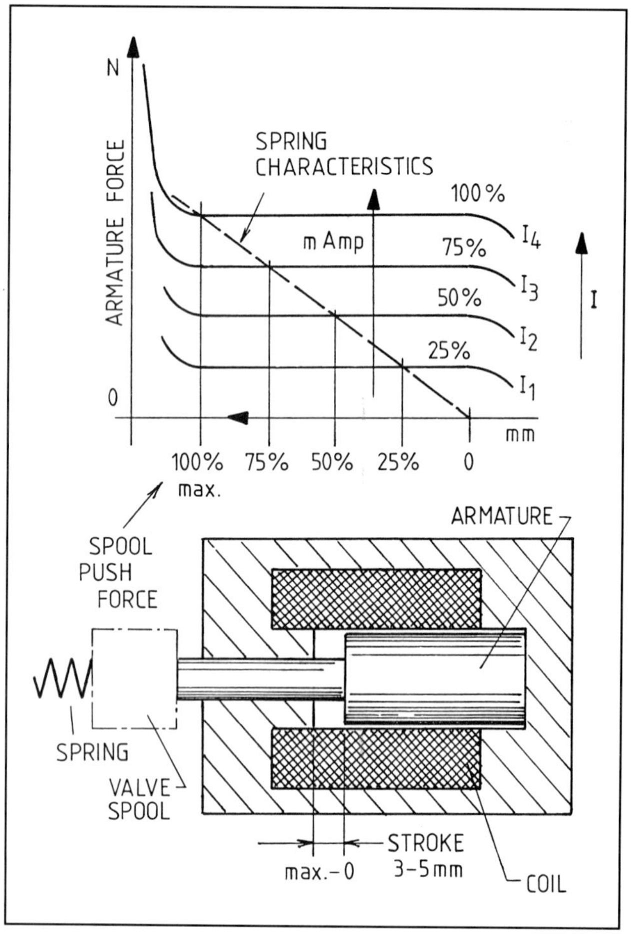

Fig. 284 Proportional linear distance control solenoid. This type of solenoid is predominantly used for directional control type proportional valves.

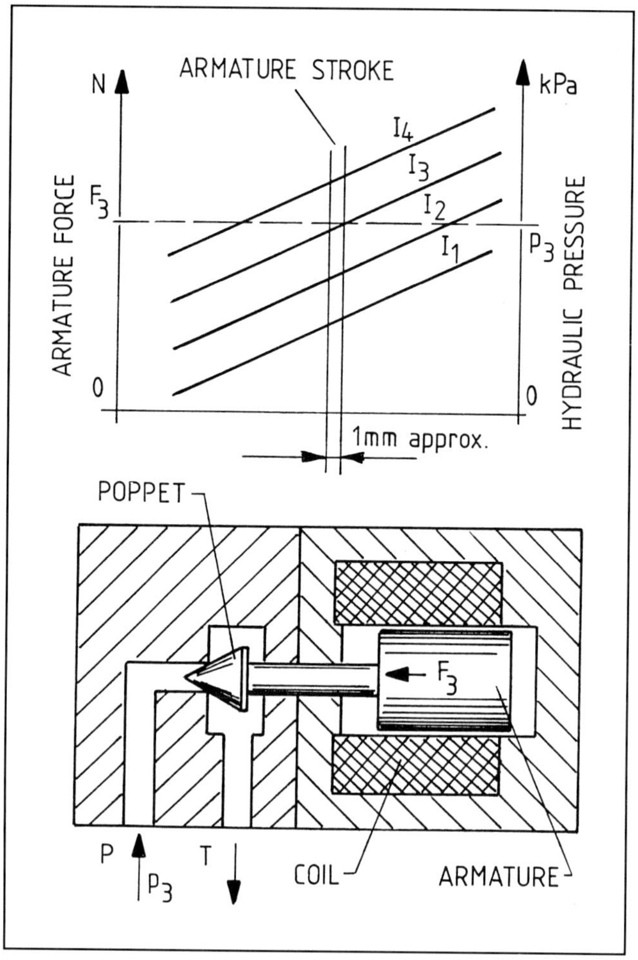

Fig. 285 Proportional force control solenoid. This type of solenoid is predominantly used for proportional pressure control valves.

Proportional valves with solenoid armature position sensor

Position sensors or so called LVDT (linear variable differential transducer) are used to ascertain precise spool position in proportional directional control valves. Since the spool is always in direct contact with the proportional solenoid armature and its connected push rod, one can say that the LVDT in reality senses the relative stroke position of the armature (see fig. 286).

Such position sensors (LVDT) are used on proportional directional valves, proportional pressure control valves, as well as proportional flow control valves (see fig. 289). Flow and system pressure fluctuations induced by varying load resistance conditions create a disturbance factor on the precise equilibrium between solenoid armature force and opposing spring force (see fig. 284). For any proportional valve having to regulate or maintain a desired hydraulic parameter, whether this be an actuator position, a precise system pressure or a desired flowrate to a motor or linear actuator, this equilibrium is of great importance. Should the armature, because of such disturbances, move from its stipulated equilibrium position, the LVDT will then immediately discover this movement (deviation) and cause the electronic proportional signal amplifier to change the current to the solenoid and thus cause the armature to correct its position. It may, therefore, be said that any minute deviation from the desired input value position for the armature is automatically corrected. Although this automatic correction circuit is a valve internal closed loop, it is not to be mistaken for a closed loop control system (see closed loop control systems, fig. 282). Therefore, position sensors, ascertaining armature position (LVDT) are used for open loop, as well as closed loop proportional control systems to enhance greater performance and response accuracy.

Figure 286 illustrates the function of an LVDT (Linear Variable Differential Transducer). The LVDT consists of two secondary coils and in between them a primary coil. A ferrit iron plunger, which is connected to the valve element (spool or poppet), moves inside these three coils. The primary coil is connected to a high frequency oscillator. The two secondary coils are connected in opposition. With the plunger in the precise centre of the three coils, the induced voltages by transformer action from the primary coil cancel each other out, thus causing zero output. If the plunger is moved out of the centre (by the proportional solenoid), the induced secondary voltage increases in one outer coil, and decreases accordingly in the other outer coil. The resulting net output voltage is proportional to the plunger movements and the phase indicates left-hand or right-hand motion.The output voltage is then demodulated into a DC signal. LVDTs are highly accurate for short movement recognition, and are therefore widely used for armature position sensing.

Proportional directional control valves

The proportional directional control valve with four ports and two or three switching positions is the most common type of proportional control. When a solenoid in a conventional on/off D.C.V. is energised, the armature moves over the full stroke distance and so does the attached spool. In a proportional directional control valve, the armature moves a percentage of its full stroke, according to the percentage of the maximum input signal. If the valve spool moves only a percentage of its maximum spool stroke, measured from the centre position, logically it will pass only a percentage of its maximum possible flowrate (Q) at a given pressure. Hence, proportional directional control valves do not only control actuator motion direction, but they also control the amount of flow passing to and from the actuator. Therefore, one may safely say that proportional directional control valves also control actuator speed (see fig. 287).

Proportional directional control valves, similar to conventional on/off D.C.V. can be direct operated valves or pilot operated valves. Both versions may operate with or without an armature position sensor. For pilot operated proportional D.C.V., an armature position sensor may be built into the pilot, as well as into the slave valve (see fig. 288).

Above NG10 most manufacturers recommend the use of pilot operation. For such applications, the pilot stage valve may be of size NG6 or smaller, whereas the slave valve may be as large as an NG32 passing 870 L/min or more!

A direct operated proportional valve with two solenoids and no armature position sensor is shown in fig. 287.

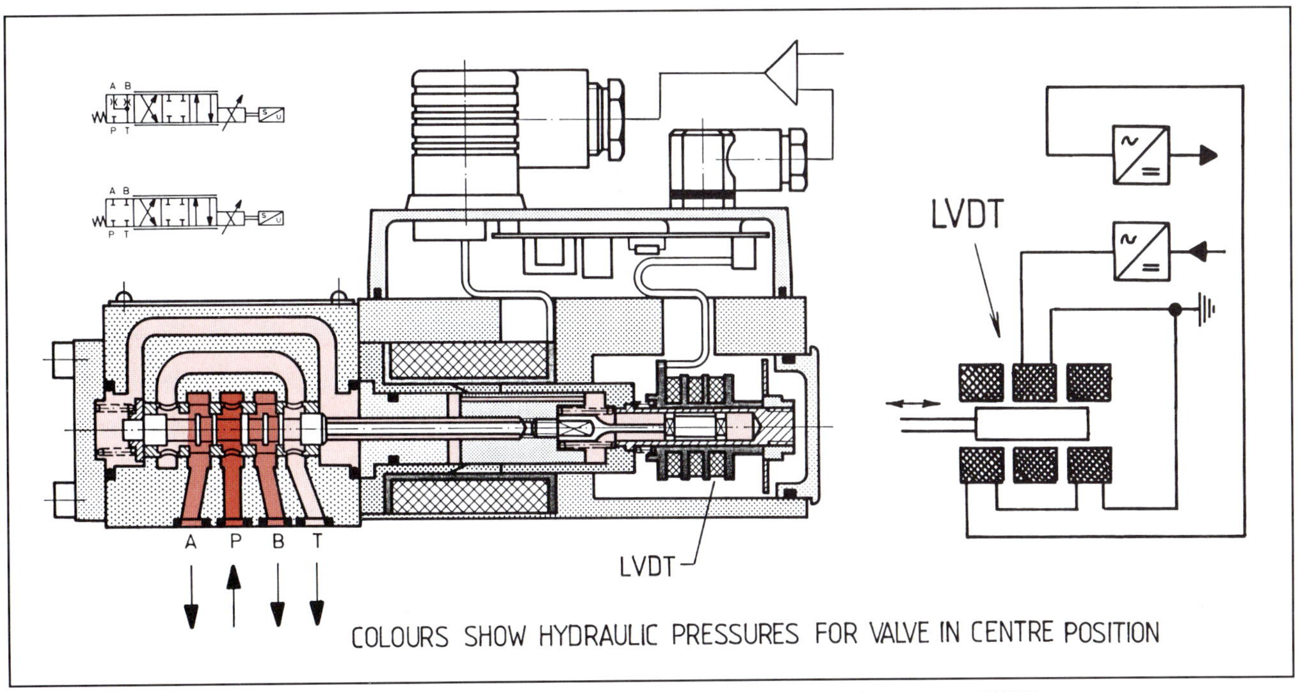

Fig. 286 Proportional closed loop directional control valve with solenoid armature position sensor (LVDT).

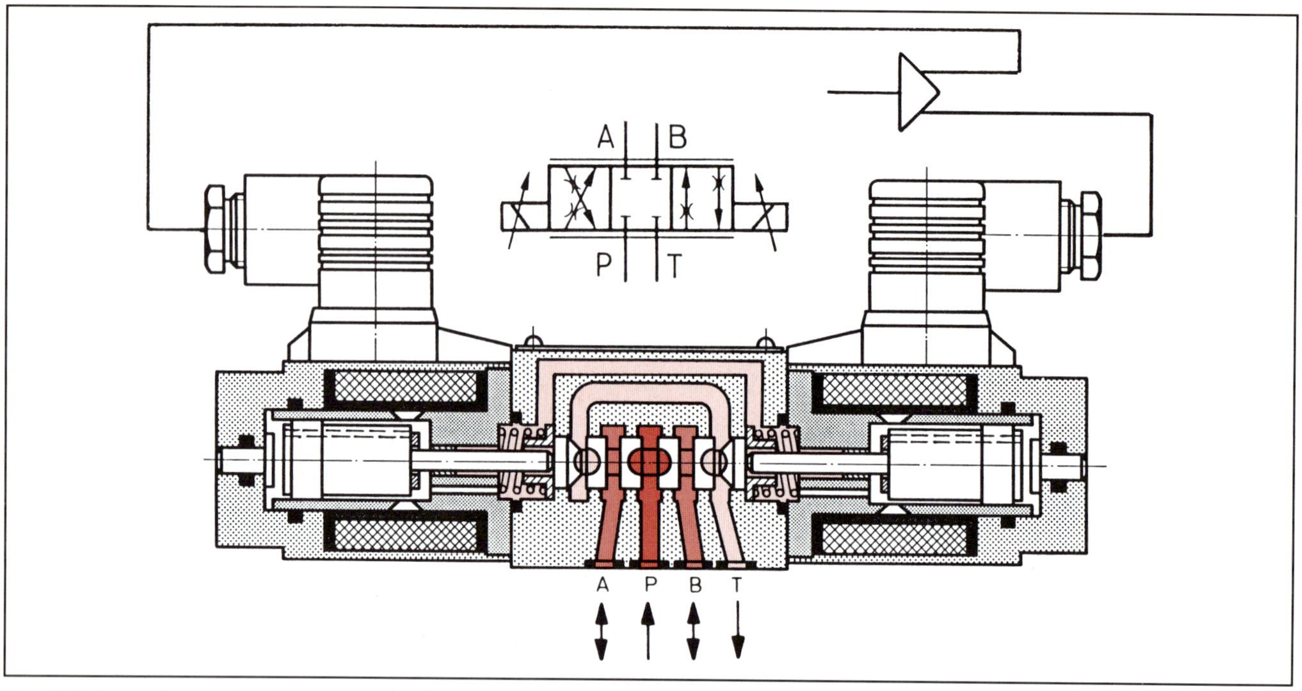

Fig. 287 Proportional directional control valve without armature position sensor.

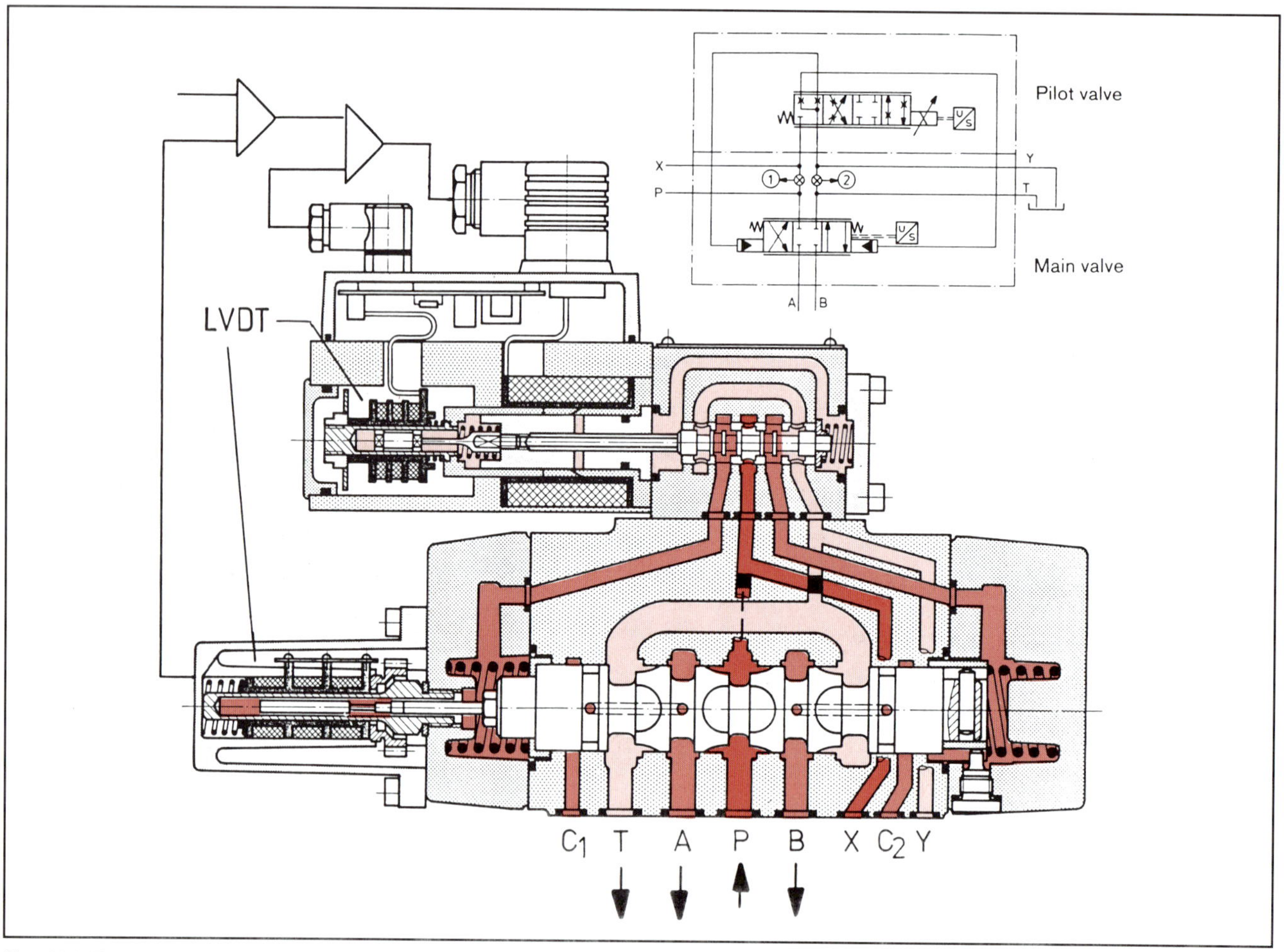

Fig. 288 Proportional directional control valve with pilot master valve and slave valve. Both pilot and slave valve have an armature position sensor.

Proportional pressure control valves

Proportional pressure relief valves (pressure limiting valves), as well as proportional pressure reducing valves, are explained in this category. For both types, the manual adjusting mechanism (usually a threaded plunger or a hand wheel spindle) has been replaced with a proportional force control solenoid. Both valve types are available as pilot operated compound type valves or as cartridge type valves for manifold block assembly (see fig. 282 A and 282 B.

Proportional pressure relief valves limit system pressure by remote control, whereas proportional pressure reducing valves may be applied to remotely obtain a reduced pressure for a defined section of the control circuit.

Proportional pressure relief valves (pressure limiting valves)

Proportional pressure relief valves are available with a variety of design principles. **Nozzle type valves** have a push rod (spindle) acting directly onto a nozzle opening without a spring. Applying a defined, but proportional, current to the solenoid coil pushes the spindle against the nozzle opening with a force precisely proportional to the solenoid current. To open the valve and obtain an oil flow to tank, sufficient pressure must build up at the valves P port to force the spindle off the nozzle and towards the solenoid. With **poppet type** relief valves, the solenoid armature acts onto a spring. The spring forces a cone shaped poppet onto the nozzle opening. Opening the poppet is based on the same principles as described in nozzle type valves without a spring.

For pilot operated (compound type) relief valves, the proportionally adjusted pilot stage valve creates a minute oil flow to the tank. This in turn induces a pressure drop (Δp) across the main stage valve spool or cartridge, which causes the main stage valve to open and pass all pump flow to the tank. Due to the proportionality of the pilot stage closure force, the main stage valve can with ease, remotely, electrically and infinitely be adjusted or varied (see fig. 289).

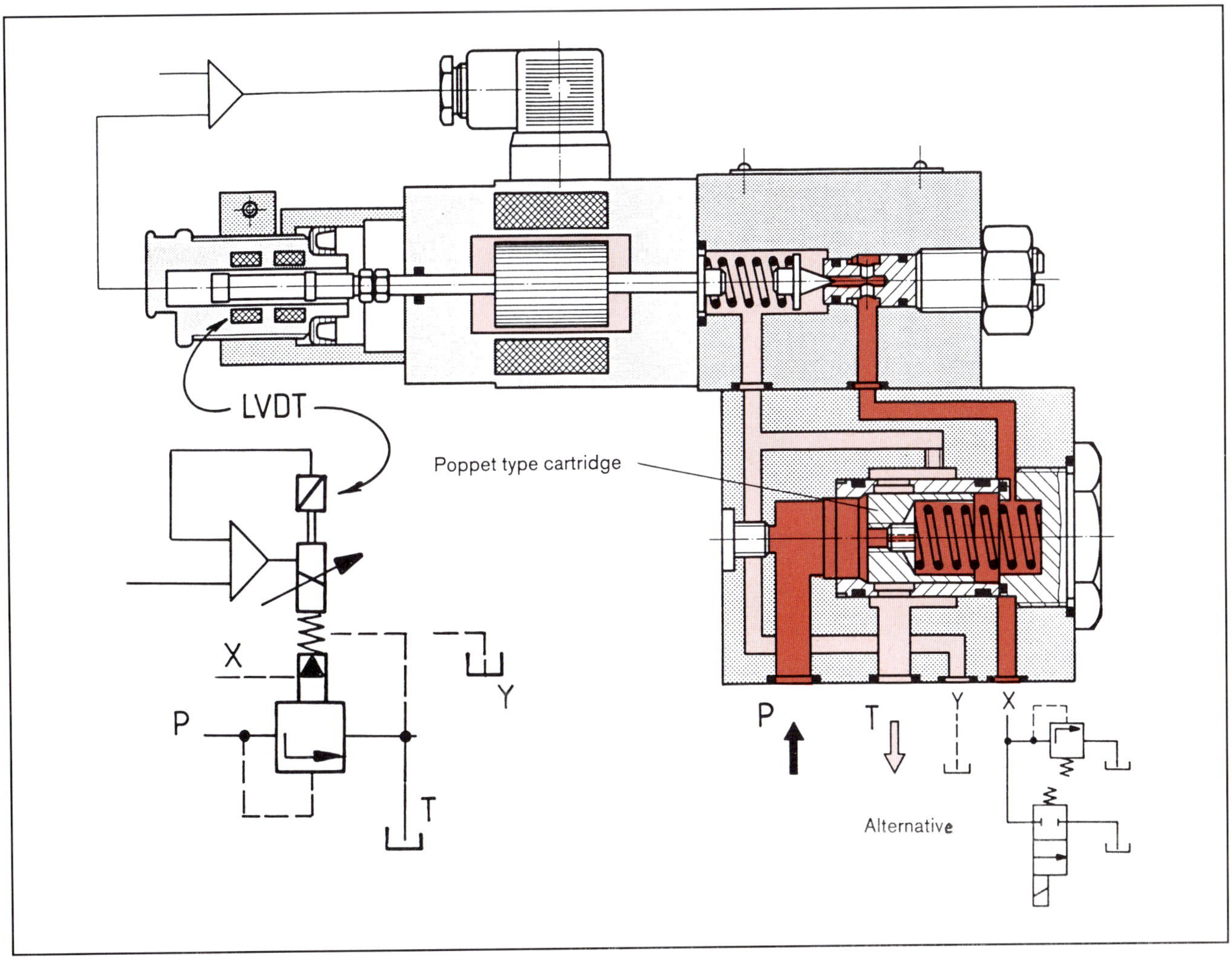

Fig. 289 Pilot operated cartridge type proportional pressure relief valve with armature position sensor.

Proportional pressure reducing valves

Proportional pressure reducing valves are used to reduce pressure from the valves input port to the valves output port and maintain a predetermined pressure at the exit port (downstream), regardless of pressure fluctuations at the inlet port. The valve shown in fig. 290 has a poppet valve for the pilot stage, a hydrostat type flow controller for the pilot oil flow and a spool type main stage valve. Port B remains blocked. The valve is a Bosch NG6 proportional pressure reducer with a Q_{max} of 40 L/min, and is available in three pressure ranges (0–75 bar, 0–175 bar, 0–300 bar). The valve also has an inbuilt secondary system relief provision (see also figs. 114 and 115).

Proportional flow control valves (throttle valve)

With proportional flow control valves the armature force acts directly onto the valve's flow metering spool. The spool is biased to the left by the spring opposing the armature pull-in force. Depending on the electrical input signal and the resulting armature stroke, different flow throttling results may be achieved (see fig. 291).

Flow regulating valves are essentially a variable orifice. Thus, for all flow regulating valves, the flow rate (Q) through the metering orifice(s) depend(s) essentially on the:

- size and shape of the orifice (mm^2)
- pressure drop across the orifice (Δp).

As the flowrate is approximately proportional to the square root of the pressure differential (Δp), nominal flow is usually given in manufacturer's catalogue information, in reference to a Δp of 8 bar (see fig. 297). Where other pressure differentials apply, flow is calculated according to the formula:

$$Q_X = Q_{nom} \bullet \sqrt{\frac{\Delta p}{8}}$$

In general, efforts are made not to exceed the stipulated pressure drop of 8 bar across the flow control valve, as ensuing **Bernoulli forces** on the valve spool could then affect the throttling accuracy. To avoid such unavoidable spool forces, the use of pressure compensators is recommended (see fig. 292).

Pressure compensation for proportional valves

Both proportional directional control valves and proportional flow control valves are used to control actuator speed. Where the actuator speed must be held constant with precision or regulated proportionally with accuracy, it is essential to add pressure compensation to the proportional valve.

If pressure fluctuations caused by hydraulic load change occur, these may affect the pressure differential (Δp) across the flow metering orifices in the proportional valve and flowrate variations are inevitable. Such flowrate variations are the direct result of varying Bernoulli forces tending to upset the adjusted equilibrium between spring force plus Bernoulli force to armature force (see fig. 292).

$$F_{SPRING} + F_{BERNOULLI} = F_{ARMATURE}$$

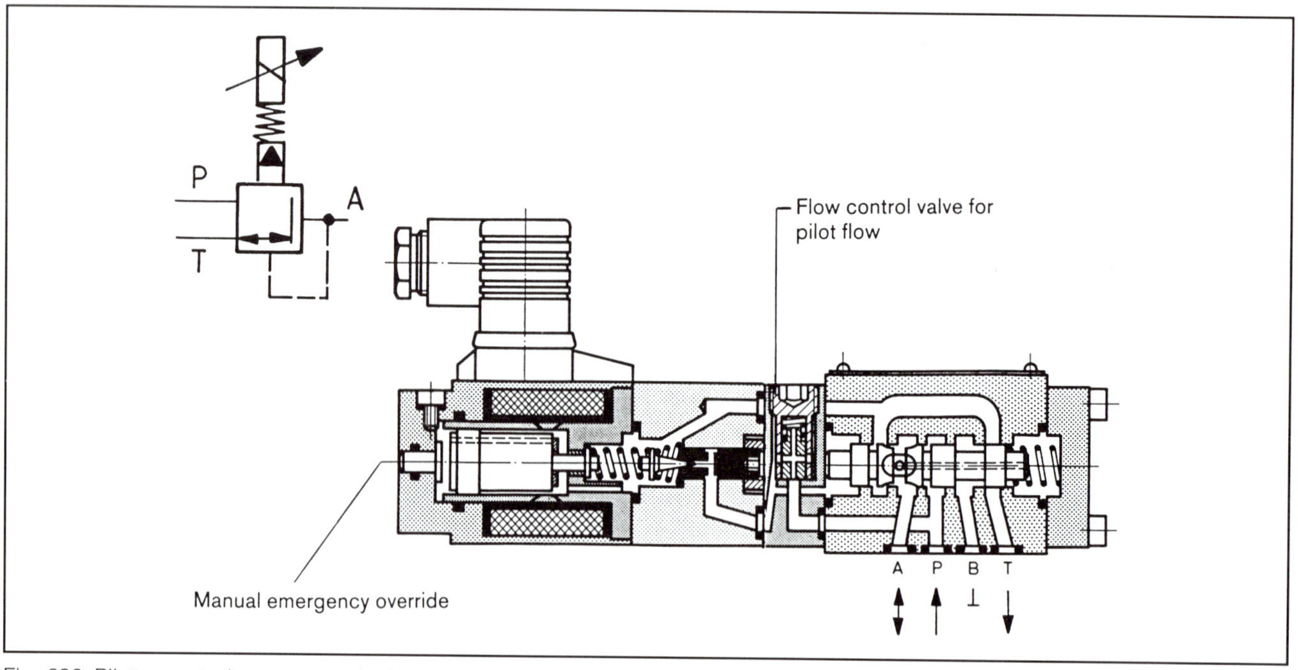

Fig. 290 Pilot operated pressure reducing valve with armature position sensor. The main stage is a spool valve.

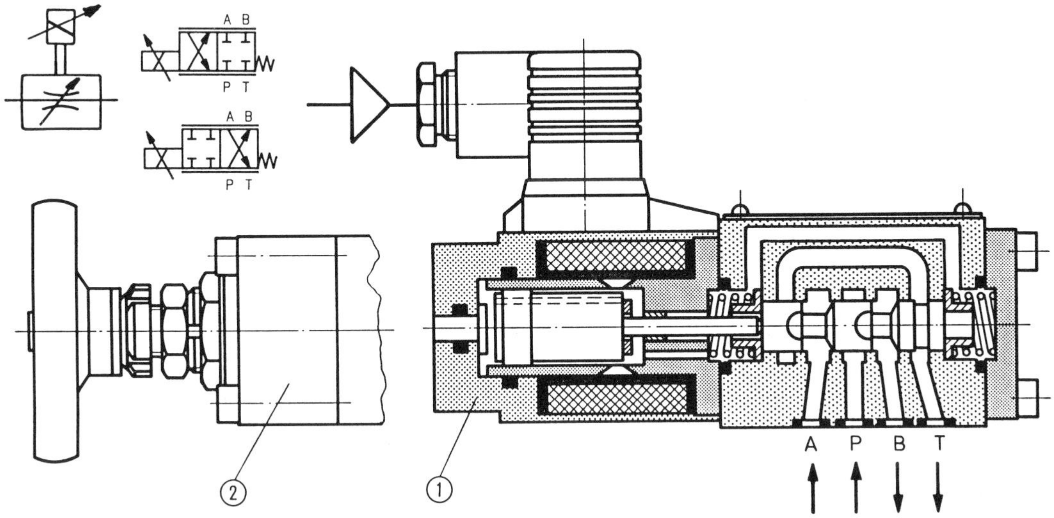

① Proportional throttle valve without position control
② optional with manual emergency override

Fig. 291 Proportional flow control valve with optional emergency manual override.

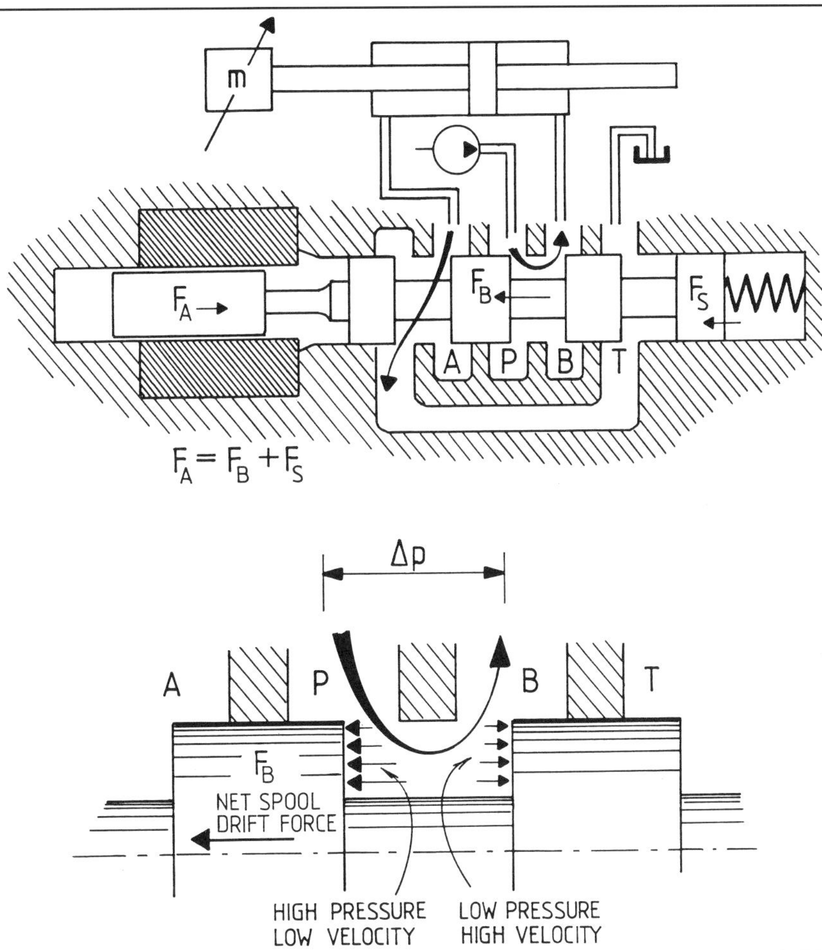

Fig. 292 Bernoulli's principle (theorem) rendering net spool drift force against armature.

As long as the pressure differential (Δp) is kept stable, the net spool drift force ($F_{BERNOULLI}$) also remains stable and, added to the spring force (F_{SPRING}), the two forces will be equal to the armature force ($F_{ARMATURE}$). Changing the Δp results in different net spool drift forces. The spool/armature equilibrium is then disturbed and the spool moves to a different position (error drift), thus the metering orifice is no longer the same and as a result the flowrate is changed. To avoid this chain reaction, pressure compensated valves must be used, where load pressures are expected to vary (see fig. 293).

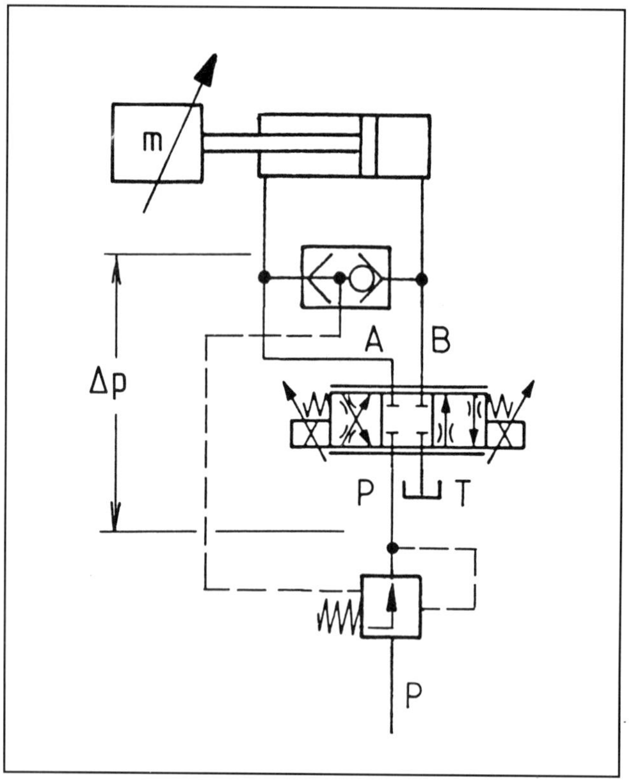

Fig. 293 Proportional directional/flow control valve with load pressure fluctuation compensation to maintain Δp constant.

Depending on the actuator motion direction in linear actuators, or clockwise or anticlockwise motion for hydraulic motors, load pressure must be obtained (tapped off) at all relevant actuator flow ports. The sensed load pressure is then directed to the pressure compensator, which essentially is a normally open hydrostat. The hydrostat is a spool valve, which is spring biased to be normally open. When the Δp across the varying throttling orifice of the proportional valve reaches the spring bias value of the hydrostat, the hydrostat spool starts to close the pressure compensator valve orifice. The resulting pressure drop across the compensator valve reduces the inlet pressure on the proportional valve and thus maintains the essential Δp across the proportional valve at a precise and predetermined level.

Measuring devices

Those hydraulic control systems which require a high degree of accuracy in system response when compared with the input value signal at the summing point, need precise measuring devices to measure partial spool stroke position, partial actuator stroke position, varying load velocity, varying system pressure, varying revolutions, and so on.

Furthermore, the accuracy of the measured parameters is also of importance, as the controlled parameter can never be more accurate than the measuring device's performance! This also applies to the frequency at which the measurements taken are transferred to the electronic controller, which in turn directly influences the system's response and dynamic behaviour.

Devices used to measure position or displacement

- Potentiometers
- Inductive voltage dividers
- Photoelectric digital sensors
- Linear variable differential transducers (LVDT)

Of these, the LVDT is probably the most widely used for measuring distance (see fig. 286). The LVDT is also used to measure spool position within the proportional valve (see fig. 291).

Devices used to measure speed or velocity

- Rotational speed tachometers with permanently exited DC generators are used to measure rotational speed or revolutions.
- Linear speed is measured with a rod shaped permanent magnet attached to the moving load. The permanent magnet slides to and fro within adjoining tube-like coils and generates velocity dependent voltages in these coils.

Devices used to measure pressure

- Strain gauge principal sensors are ideally suited to measure static system pressure. By application of the obtained pressure value, multiplied by the actuator piston area, one can compute the force exerted by the investigated linear actuator. Thus, measuring its operating pressure relates directly to obtaining its operating force.
- Piezo resistive pressure sensors produce an electrical signal through mechanical deformation of quartz crystals or special plastic foils. To make the piezo electric effect useful to measure pressures,

the charge obtained by the deformation must be amplified and transformed. Piezo resistive pressure sensors are used to measure fast dynamic pressures, such as pressure spikes and pulsating pressures. They are not suitable for extended static pressure measurements.

Electronic amplifier cards

Different proportional valves, even from the same manufacturer, have different electronic amplifier cards (see manufacturer's catalogue). The sources of input signals to these electronic control amplifiers may be a variety of input components, such as:

- potentiometers
- PLC controllers
- temperature sensors
- pressure sensors
- tacho generators
- microprocessors.

These input devices produce a low power signal in terms of voltage or current. In order to produce a high voltage or current output signal, necessary to power the proportional solenoid, these input signals must be amplified. This amplification is accomplished by the electronic amplifier card (see fig. 294). An amplifier card may include adjustment for dead band correction, gain variation, ramp profile generation, and so on.

Amplifier gain

Amplifier gain is an expression describing the relationship of an input signal voltage to its corresponding output signal voltage. The output voltage is always directly proportional to the input voltage.

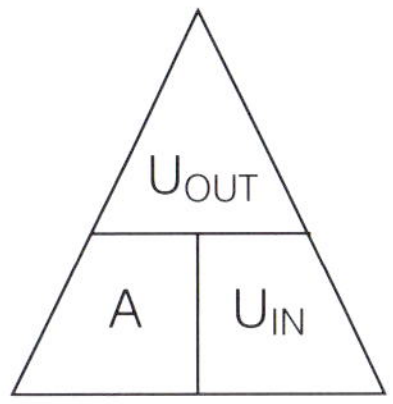

$$\text{Gain (A)} = \frac{\text{Output voltage}}{\text{Input voltage}}$$

Input signals, usually derived from low voltage potentiometers, various hydraulic actuator performance sensors or PLC controllers, are amplified into a current which is powerful enough to drive the proportional valves solenoid and convert this signal into a spool position or push rod force.

Speed ramp generation

If the electrical input signal to a proportional solenoid valve is gradually increased or decreased, the valve's

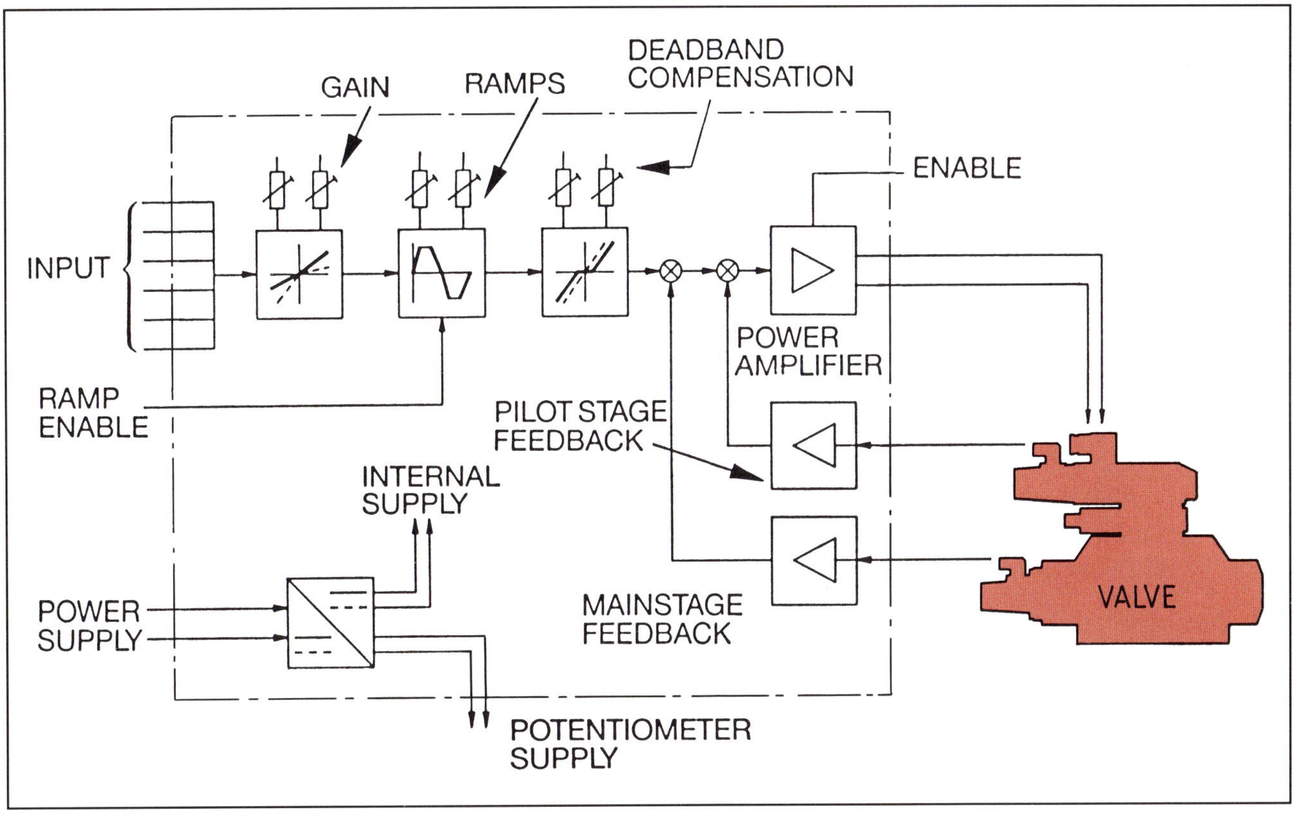

Fig. 294 Electronic amplifier card for a proportional directional control valve. Adjustments on the card may be made for gain, ramps and dead band.

spool starts to move gradually and the valve's hydraulic output is also varied gradually. In hydraulic terms this means actuator speed increase or decrease. Such gradual electrical signal increase or decrease is called ramping.

Ramping is significant for accelerating or breaking of cylinder loads or motor rotations, or for increasing or decreasing of pressures and their resulting forces on actuators or torque on hydraulic motors.

The transit time of the ramp depends on the measured difference between the initial and final signal value (U) and the inclination or the declination of the ramp. In practice one adjusts the ramp angle (α) and not the ramp time (see fig. 295). The steepness (angle α) of the ramp is controlled by a gradual charging of a capacitator in the electronic ramp generator circuit. This in turn creates a corresponding gradual increase of the solenoid signal (see fig. 295 A). The final speed of the actuator with its attached load, when ramping stops, is determined by the signal level U (see fig. 295 B). Ramp inclination angle or declination angle may be identical or may be individually adjusted (see fig. 295 C).

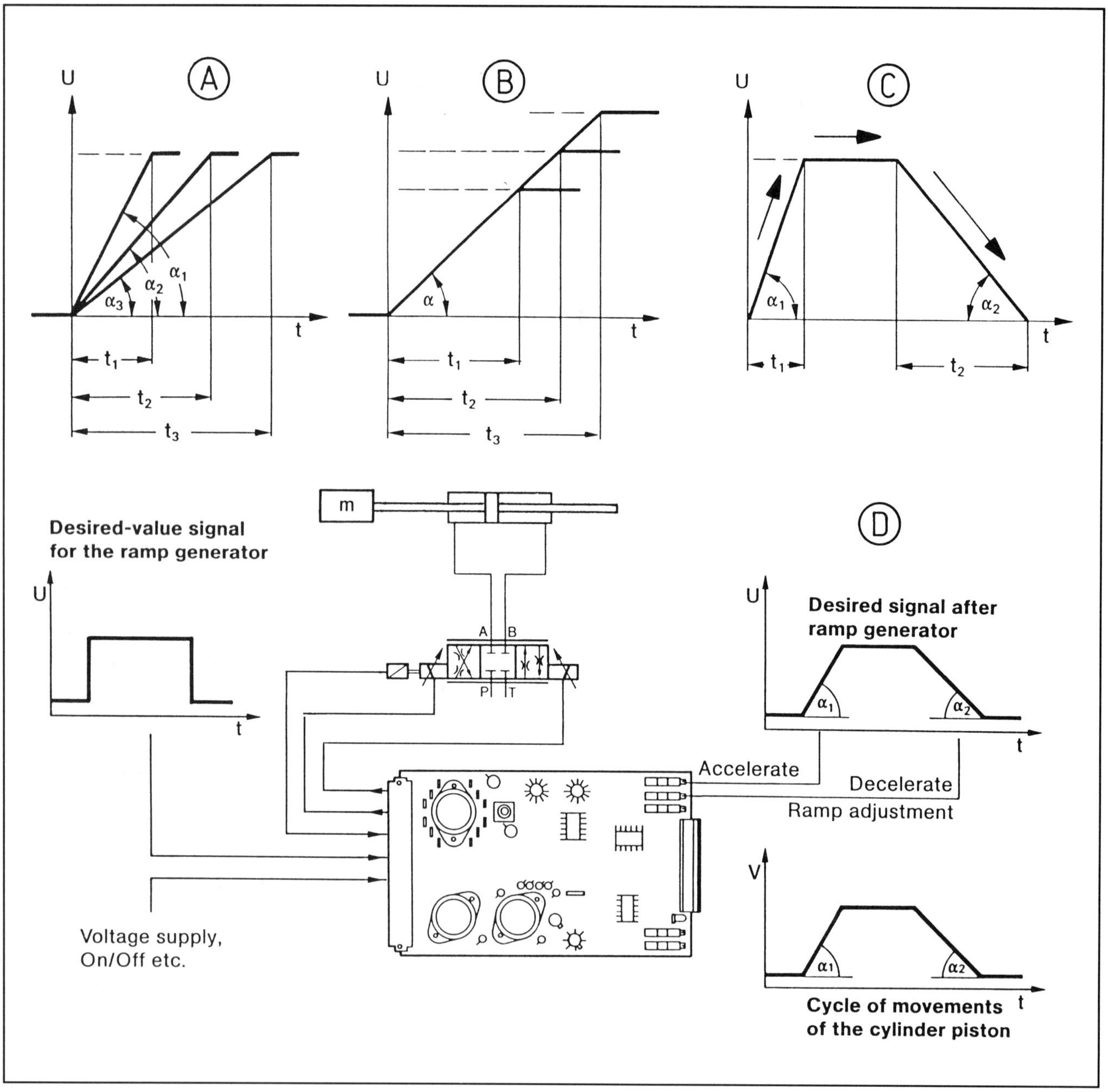

Fig. 295 Various speed ramps to achieve actuator acceleration or deceleration, or both.

Quadrant identification

For ramping on proportional directional control valves, when the valve spool moves through its centre position (zero position), the spool motion direction is maintained (left to right or right to left), but the actuator, being a cylinder or a motor, changes motion direction (for a motor clockwise to anticlockwise rotation). Thus, breaking from the first quadrant when moving to the adjoining quadrant becomes acceleration. To make sure that the acceleration remains the same in both directions, the ramp needs to be switched over when the spool moves through the perfect centre (quadrant one to quadrant two border). This takes place automatically, when ramp generators with quadrant identification are used (see fig. 296).

Valve characteristics data

For hydraulic system designers, to be able to make an informed choice for selecting the correct components for a hydraulic control system, manufacturers of proportional valves and servo valves provide precise catalogue data for their products (see fig. 297). Such data may be broadly grouped into

- static performance data
- dynamic performance data
- electrical requirement data
- hydraulic data
- general engineering data.

This chapter describes only the first two data categories: **static data** and **dynamic data**. However, these data categories apply to servo, as well as proportional, valves.

Static data

Static data describes the connection between the input signal and the response output of the system to that input signal. Valve behaviour characteristics listed under static data are:

- valve gain
- linearity
- dead band
- valve hysteresis
- response sensitivity
- reversal error
- repeatability.

Valve gain

Valve gain is adjusted at the valve's electronic control card potentiometer "labelled valve gain". Valve gain adjustment is carried out after completion of dead band adjustment. Valve gain adjustment alters the gradient of the valve flow versus input signal curve, see figs 297 and 298.

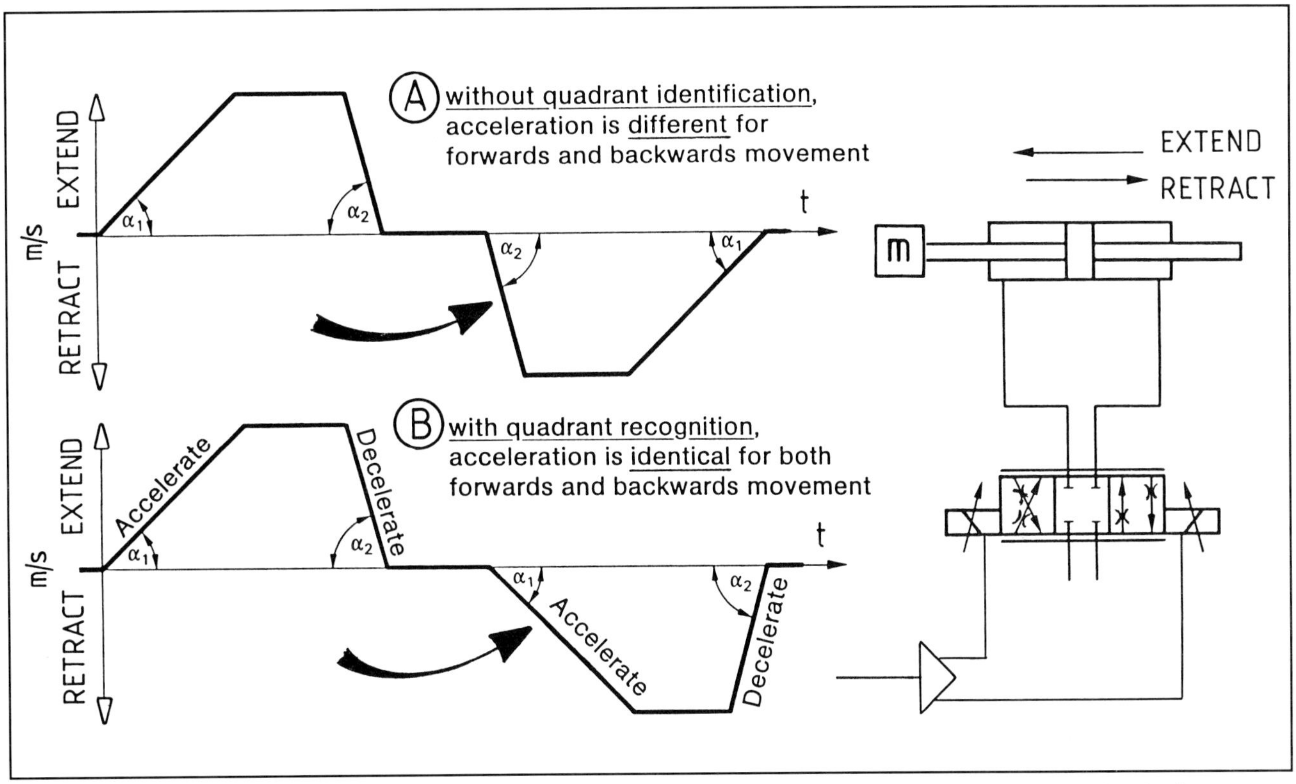

Fig. 296 Quadrant identification automatically takes place for most amplifier cards as used on modern proportional valves.

General Characteristics		
Construction	Spool type valve	
Actuation	Proportional solenoid without position control	
Connection type	Sub-plate, mounting hole configuration NG 6 (ISO 4401)	
Mounting position	optional	
Ambient temperature range	−20...+50°C	
Hydraulic		
Pressure medium	Hydraulic oil as per DIN 51 524...535 Other fluids after prior consultation	
Viscosity, recommended max. permitted	20...100 mm²/s 10...800 mm²/s	
Pressure medium temperature	−20...+80°C	
Filtration	Permissible contamination class of pressure medium as per NAS 1638	Achieved using filter βx = 75
In line with operational reliability and service life	8	X = 10
	9	20
	10	25
Flow direction	cf. symbol	
Nominal flow *	18 or 35 l/min (Δp = 8 bar)	
Max. working pressure Port P, A, B Port T	315 bar 250 bar	
Electrical		
Cyclic duration factor	100 % (9 V DC)	
Degree of protection	IP 65	
Solenoid connector	Connector DIN 43 650 / ISO 4400	
Solenoid current	max. 2.7 A	
Coil resistance R_{20}	2.5 Ω	
Power consumption	max. 25 W	
Static / dynamic		
Hysteresis	≦ 4 %	
Range of inversion	≦ 3 %	
Manufacturing tolerance	≈ 10 %	
Response time for 100 % signal change	70 ms	

Operating limits

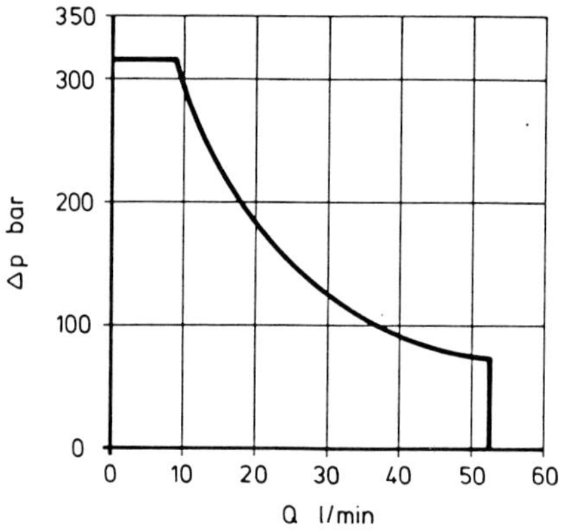

* **Nominal flow**
This always refers to a pressure differential of Δp = 8 bar at the throttle point.
Where other pressure differentials are involved, flow is calculated according to the following formula:

$$Q_x = Q_{nom} \cdot \sqrt{\frac{\Delta p_x}{8}}$$

However, the **operating limits** must be borne in mind here.
When the operating limits are exceeded, the ensuing flow forces lead to uncontrollable spool movements.
To achieve effective limitation of Δp, use is made of **pressure compensators**.

[1]) Zero adjustment
[2]) Gain adjustment
[3]) U_E = 0...8 V when using plug amplifier B 830 303 114

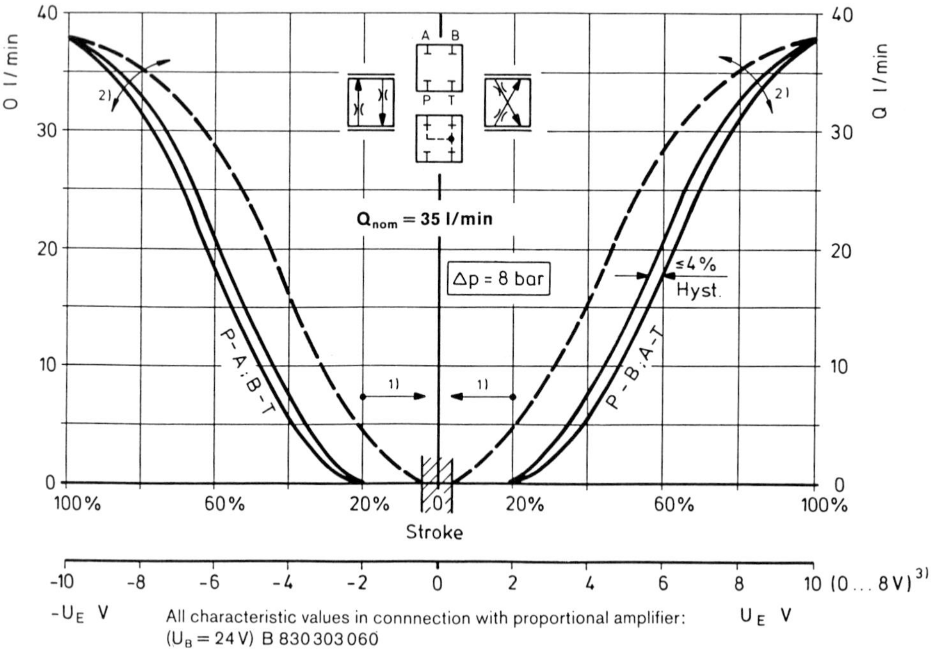

Fig. 297 BOSCH catalogue data for the proportional directional control valve shown in fig. 287.

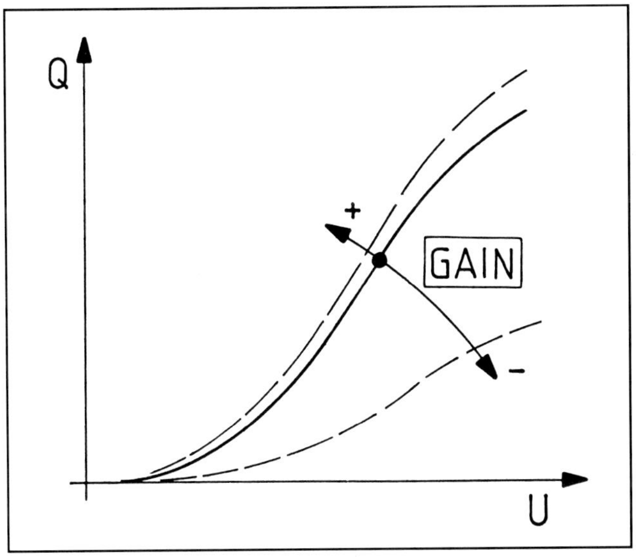

Fig. 298 Valve gain adjustment alters the gradient of the valve flow versus the electrical input signal curve.

Linearity

The linearity graph shows the interrelationship between the amplifier input signal U_E (voltage) and the proportional valves output flowrate Q in L/min. The "S" shaped curve indicates that proportional valves do not produce a strict linear output. Servo valves, however, do produce a near perfect linear output, proportional to the input. Fig. 299 shows a characteristics graph for the proportional directional control valve shown in fig. 287. A 40% input voltage produces a flowrate of 4 L/min, whereas an 80% input signal renders a flowrate of 16 L/min. Hence, doubling the signal voltage produces four times the flowrate.

Dead band

Proportional valves for open loop control functions, in the valve's centre position, have a distinct positive overlap of the spool land, compared with the flow passage port inside the valve. This overlap minimises internal valve spool leakage in the null position (for example, when the valve spool is centred or passes through the centre). To obtain flow from the P port to an output port (A, B), and from an output port (A, B) to the tank port (T), the valve spool must move a certain distance depending on the amount of overlap (usually + 20% of spool travel). Hence, an electrical input signal is necessary to cause this unproductive initial spool shift (see fig. 297). This initial unproductive spool shift region is called "dead band".

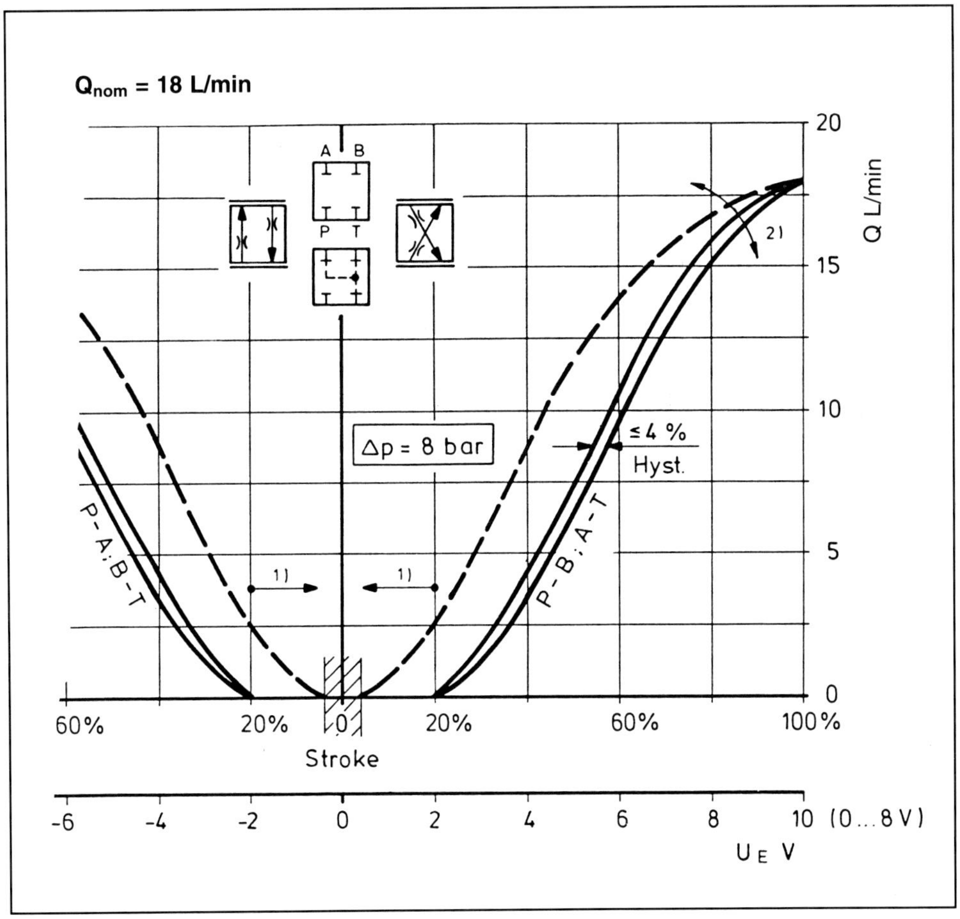

Fig. 299 Valve linearity.

Dead band applies to the left-hand as well as right-hand spool movement out of the perfect centre position (left-hand/right-hand quadrant). Most proportional valve amplifiers provide an inbuilt electronic dead band compensation adjustment (zero adjustment). Such electronic dead band compensation often produces excellent dead band region input signal versus output flowrate linearity characteristics.

Valve hysteresis

Valve hysteresis describes the valve's lagging effect by producing a different amount of spool movement, and hence flowrate for a specific electrical input signal target level when the input signal is increased or decreased to that target level. Valve hysteresis is the result of frictional forces between spool and valve body not being evenly distributed over the entire spool stroke. A dither signal applied to the proportional solenoid reduces hysteresis remarkably, and so does the armature position sensor used in conjunction with some proportional solenoids. Figure 300 shows the hysteresis of a servo valve, compared with a proportional valve. The servo valve performance line is almost straight, whereas the proportional valve's performance is depicted by the curve. Hysteresis affects the performance of all types of proportional and servo valves. Manufacturing catalogue information usually shows hysteresis for various valve sizes and types (see fig. 297) or hysteresis may be expressed in a percentage. For the average proportional valve, this percentage may be less than 1%.

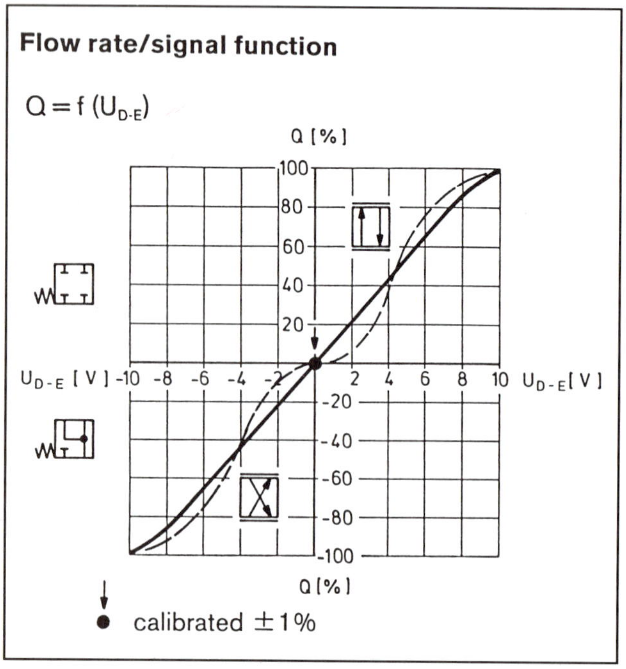

Fig. 300 Valve hysteresis.

Response sensitivity

Response sensitivity describes the amount of electrical input signal change needed to obtain a measurable valve output change (Q) in the same direction from which stopping occurred. Such a signal change to the negative or positive is measured and expressed in milliamps (see fig. 301).

$$E_{(mA)} = I_2 - I_1$$

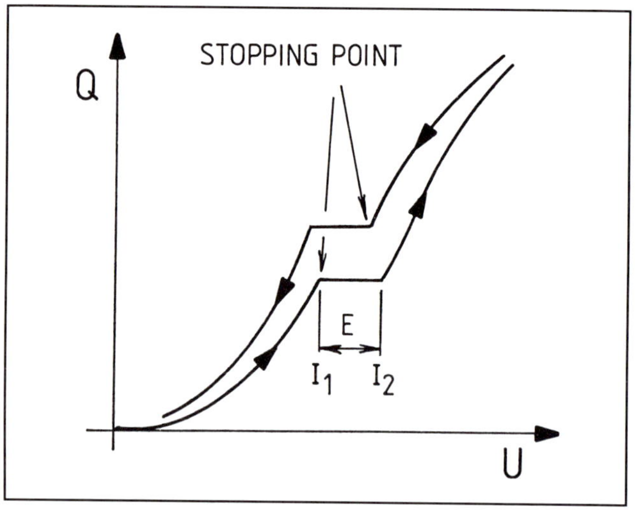

Fig. 301 Response sensitivity.

Reversal error

Reversal error describes the amount of electrical input signal change needed to obtain a measurable valve output change (Q) in the opposite direction from which stopping occurred. This necessary opposite signal change is also measured and expressed in milliamps (see fig. 302).

$$R_{(mA)} = I_2 - I_1$$

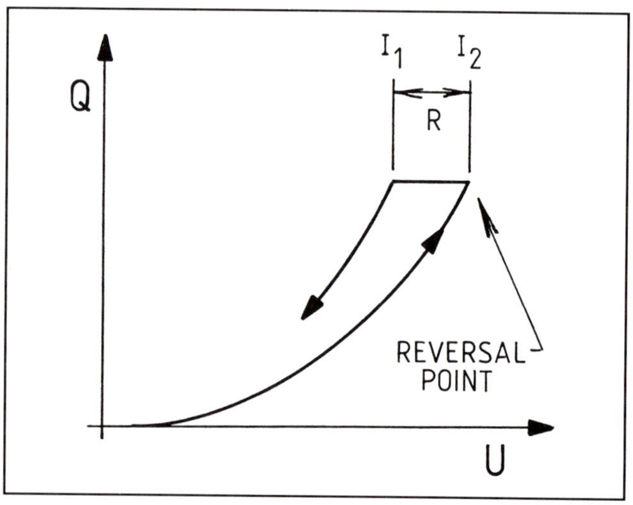

Fig. 302 Reversal error.

Repeatability

Repeatability describes the valve's ability to produce an identical flowrate when the same input signal is repeated (see fig. 303). Since this perfection is not possible, manufacturers of proportional valves provide data of the range of output (Q). This range depends also on the use of an armature position sensor. Without an armature position sensor, the repeatability range is approximately 2 to 3%; with an armature position sensor it is often better than 1%.

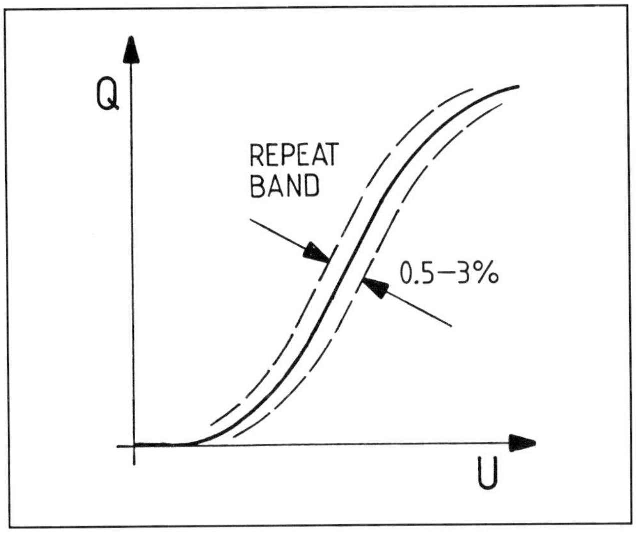

Fig. 303 Valve output repeatability.

Dynamic data

Dynamic data or dynamic response describes the system's or valve's behaviour over a period of time in response to the changes of the input signal. Dynamic response information from the manufacturer of components is important for the precise analysis of a complete control system.

There may be any kind of input signal change in the real technical system (slow or fast signal change, sinusoidal signal change, increasing or decreasing signal change), but it is standard procedure to group these changes into:

- step input change
- oscillating input change with different frequencies.

The system's response to the step input change is called the **step response**. The system's response to an oscillating input change is called the **frequency response**.

Step response

Step response describes the proportional valve's or system's ability to respond to a sudden input signal change. The valve's behaviour to such a sudden input signal change for a 100% signal step on a NG6 proportional directional valve may be in the vicinity of 30 ms, whereas a mere 10% signal step has a 25 ms response delay. Ideally, the valve's or system's step response should be one with a zero delay and it should be non-oscillatory and stable (see fig. 304).

As mentioned before, step response applies to individual valves as well as to entire proportionally controlled systems. Curve 1 in fig. 304 shows the ideal, but physically unattainable, step response. Curve 3 depicts an oscillatory, but ultimately stable and steady system. Such a system is said to be "under damped". Curve 6 shows a step response which is non-oscillatory and too slow. Such a system is said to be "over damped". Curve 2 depicts a system with a step response which is far too fast. The system is heavily oscillatory and unstable. Curves 4 and 5 show systems with the best damping response. Such step responses are highly desirable, but often difficult to achieve.

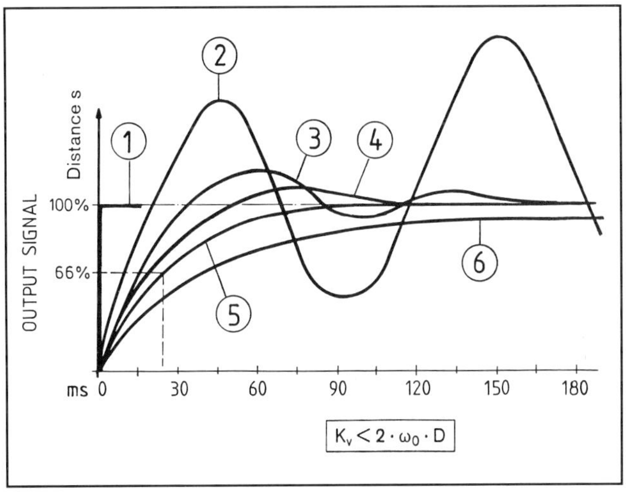

Fig. 304 Various step response curves.

Frequency response

A valve's step response is often insufficient to define all of its dynamic characteristics. For this reason, the frequency response characteristics of a system, or a singular proportional or servo valve, are important and desirable data. Frequency response is usually used in connection with closed loop proportional or servo systems. It refers to a system's or valve's dynamic behaviour (response) to a sinusoidal input signal. To obtain the desired frequency response data, the valve's sinusoidal input signal is maintained at a constant amplitude, but initially at low frequency. The frequency is then gradually increased and the valve's output response measured. The valve's hydraulic output response (Q) is also sinusoidal but starts to lag behind the input sinusoidal signal. This is called phase lag (φ). Whilst the frequency is gradually increased, the output amplitude responds with a gradual decrease to a parameter called: "modified amplitude value" ($\bar{A}$). Manufacturers of servo and closed loop proportional valves provide such dynamic data for their products. The "Bode diagram" is one of these data provided (see fig. 305) and it refers to a real valve at a stipulated and tested operating pressure. In the Bode diagram, phase shift or phase lag (φ) is shown in degrees for a complete period of 360° and amplitude damping (dB), is depicted by a logarithmic scale to the equation:

$$dB = 20 \bullet \log \frac{H_{output}}{U_{input}}$$

In the above formula the expression dB stands for decibel, H_{output} refers to the hydraulic output signal and U_{input} refers to the electrical amplifier input signal.

The frequency limit (a valve's working frequency limit) is defined as a point in the increasing frequency, where a minus 90° phase lag (φ) appears (see point A in fig. 305). A further definition points to an amplitude damping figure of (minus) –3 dB (see point B in fig. 305). Both definitions render an approximately equal frequency limit, which means, the –90° phase lag and the –3 dB amplitude dampening occur at nearly the same frequency.

Because of the non-linearity of the valve's dynamic response behaviour, the frequency response is a function of the input signal amplitude, which in turn is defined as a parameter. A valve's dynamic behaviour is particularly interesting in the small input signal range, and manufacturers therefore provide a Bode diagram with an amplitude ratio curve and phase characteristics curve for the 5% of U_{max} signal, the 25% of U_{max} signal and the 100% U signal.

Servo valves

A servo valve is a device, which converts a low power electrical input signal into a proportionally much higher hydraulic output with high accuracy. Servo valves in general consist of:

- a torque motor (on some servo valves)
- a flapper jet valve (first hydraulic stage on other servo valves)
- a spool valve for the second hydraulic stage
- an internal feedback mechanism (link between the spool displacement and the torque motor)

Torque motor

This is a small-deflection rotary-motion electric motor. The angular deflection of the rotor is proportional to the level of electrical current supplied to it. The normal maximum deflection for these motors is about seven degrees.

The rotational movement of the motor is transmitted to the valve spool via a specially designed crank (commonly called a scotch yoke) or by means of flats ground onto the rotor, with the valve spool off-centre from the shaft. In small flow size valves, a direct motor-to-valve drive combination is sufficient. But in the larger flow ranges the motor drives a primary stage (master) valve which provides the hydraulic force to drive a secondary (slave) spool.

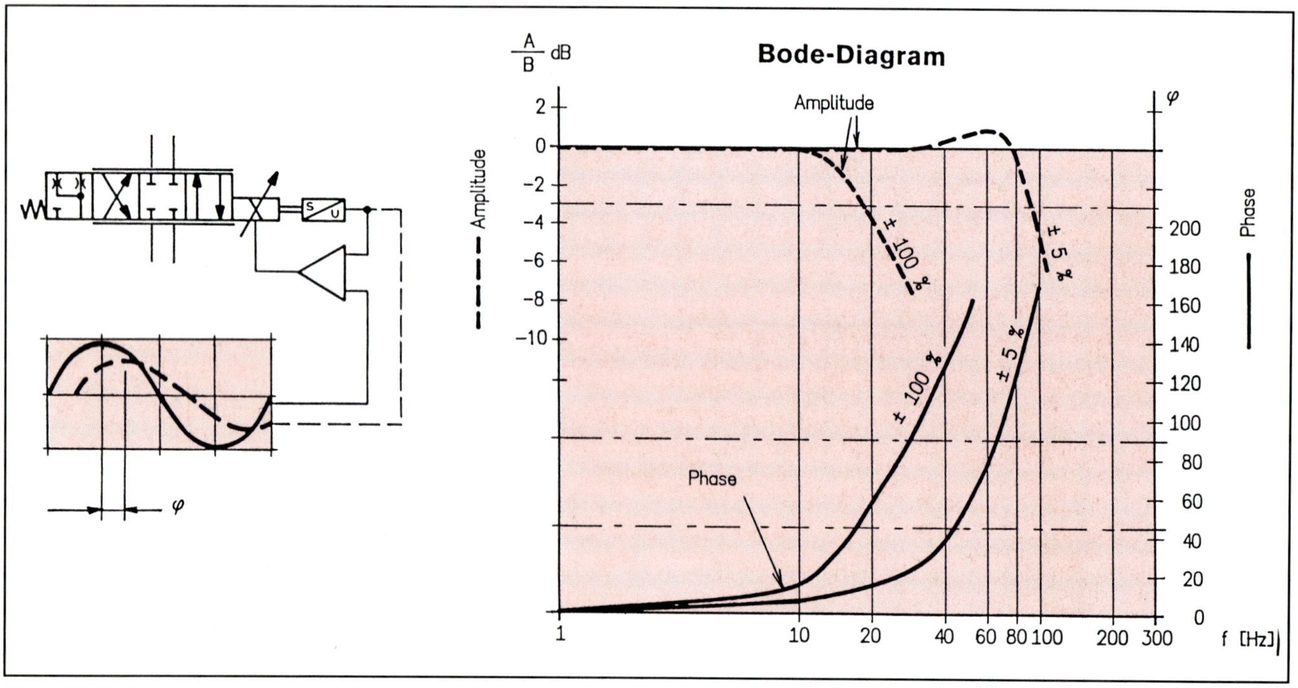

Fig. 305 Bode diagram.

A direct spool drive is shown in fig. 306, which also shows how an electrohydraulic servo valve is linked into a closed loop with its signal amplifier and feedback signal from the load positioning potentiometer.

Flapper valves

So called "flapper" valves (figs 307 and 308) are currently the most commonly used type of servo valves in industry, with the most popular version being manufactured by MOOG.

Operation

The armature is held in a balanced position by a permanent magnetic field generated by the upper and lower pole plates surrounding it. This field can be unbalanced by an opposing field generated by the armature coils. This unbalanced field causes the armature to rotate about the base of the beryllium copper flexure tube which is anchored to the valve body. As shown in fig. 307, a tube — which acts as the flapper and also contains the cantilever spring — is fitted inside the flexure tube. Assuming that the electrical signal to the force motor will cause a clockwise torque onto the armature, this torque will rotate the flapper tube in a clockwise direction. It should be pointed out here, that these movements are very small. This rotation of the flapper changes the clearance between the flapper and the nozzle faces, increasing the leakage from the right-hand nozzle and decreasing it from the left-hand one. These nozzles are supplied with fluid via fixed orifices in a filter located at the valve's pressure connection, and the fluid channel to the nozzles is also connected to the ends of the main valve spool (fig. 307). Change to the nozzle leakage changes the gallery pressure behind the nozzles. So, as the leakage from the right-hand gallery is reduced, similarly the pressure in the left-hand gallery is increased. The resulting force imbalance is towards the right, causing the spool to move in that direction. This spool movement connects the right-hand piston proportionally to the pressure port, and hydraulically connects the left-hand piston to return (tank). The spool movement also takes with it the feedback cantilever spring, which reacts back onto the armature in opposition to the generated magnetic field. This reduces the resultant clockwise torque on the flapper, causing it to move to the right and re-adjust the nozzle leakages, which in turn change the gallery pressures. When the net forces of

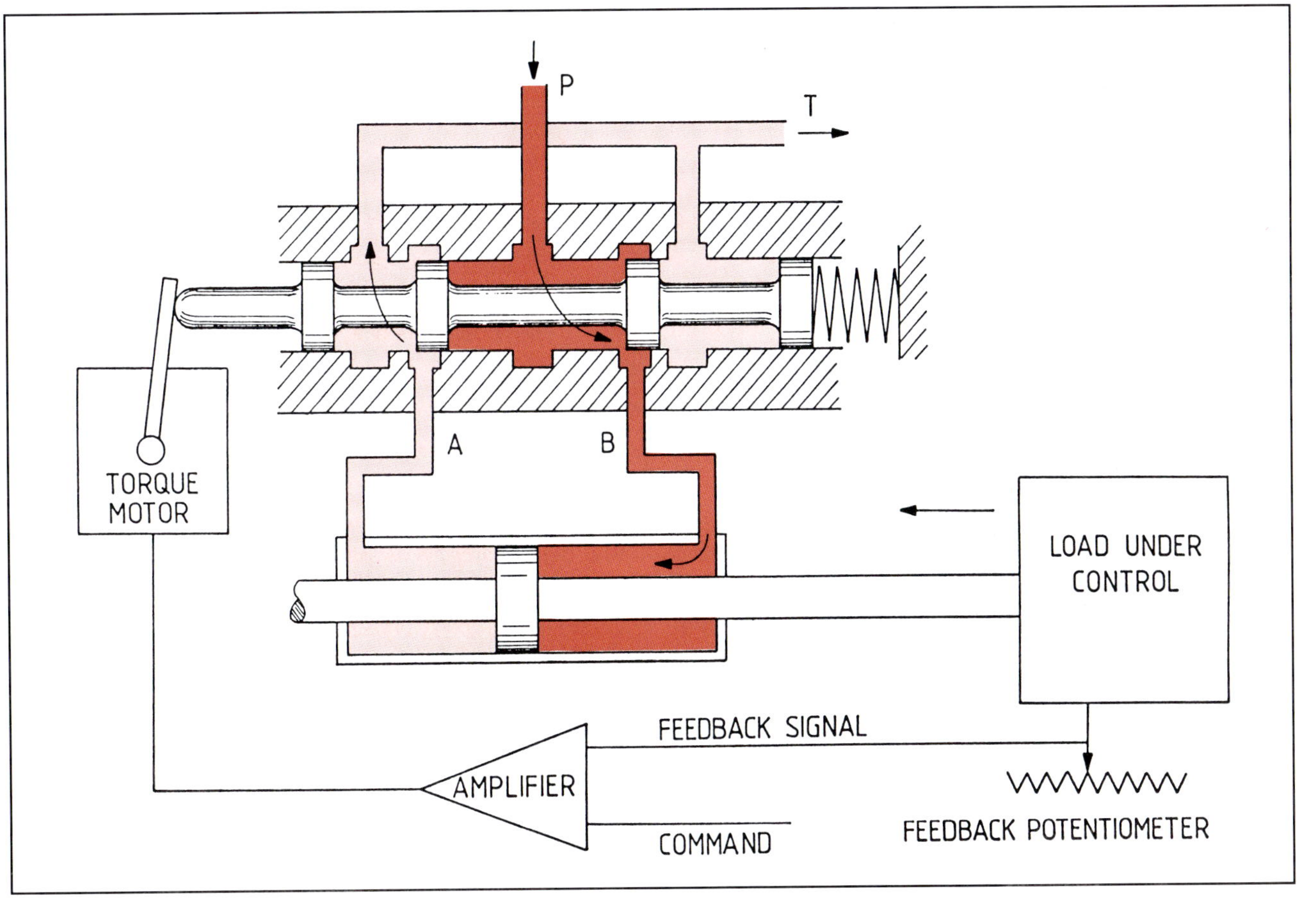

Fig. 306 Servo valve with torque motor.

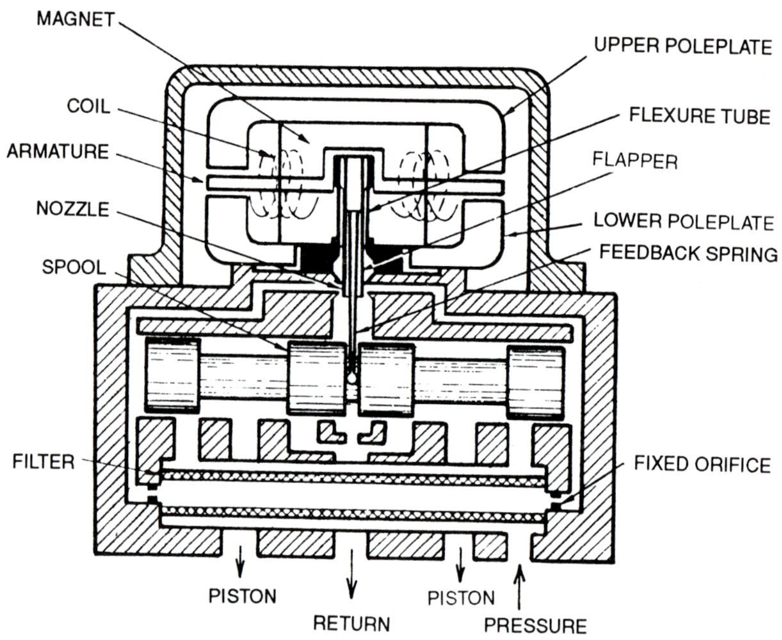

Fig. 307 Flapper type servo valve.

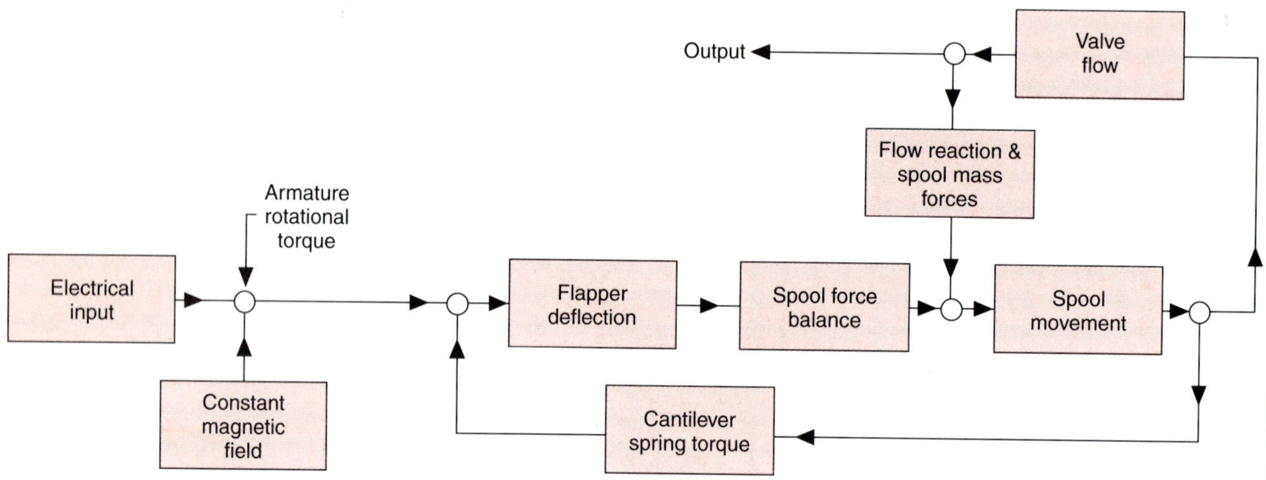

Fig. 308 Block diagram for the action of a flapper type servo valve.

flow reaction, armature torque, cantilever spring and gallery pressures balance, the spool will stop moving in the spool position required by the input electrical signal. The block diagram (fig. 307) illustrates the closed loop of the valve, which causes any extraneous disturbance to be corrected.

Circuit example 1 with proportional valves

Figure 309 shows the conventional circuit design approach and the proportional valve design approach for a machine, where actuator A requires six different retraction speeds and one extension speed. If an electrical signal closes valve 4, one may select any or all six of the previously adjusted speed control valves into the circuit. If valve 4 is opened, the actuator is switched into fast retraction mode by eliminating (bypassing) the meter-out speed control valves. The clamping actuator B requires three separate pressure settings for the extension stroke. Valves 17 to 20 provide these pressure settings and they are selectable by switching D.C.V. 20 in either a, b or centre position. Maximum system pressure for actuator B is set on valve 16. Using proportional valve technology reduces the hardware requirements considerably and makes the speed and pressure adjustments remotely and infinitely variable (compare the two circuits).

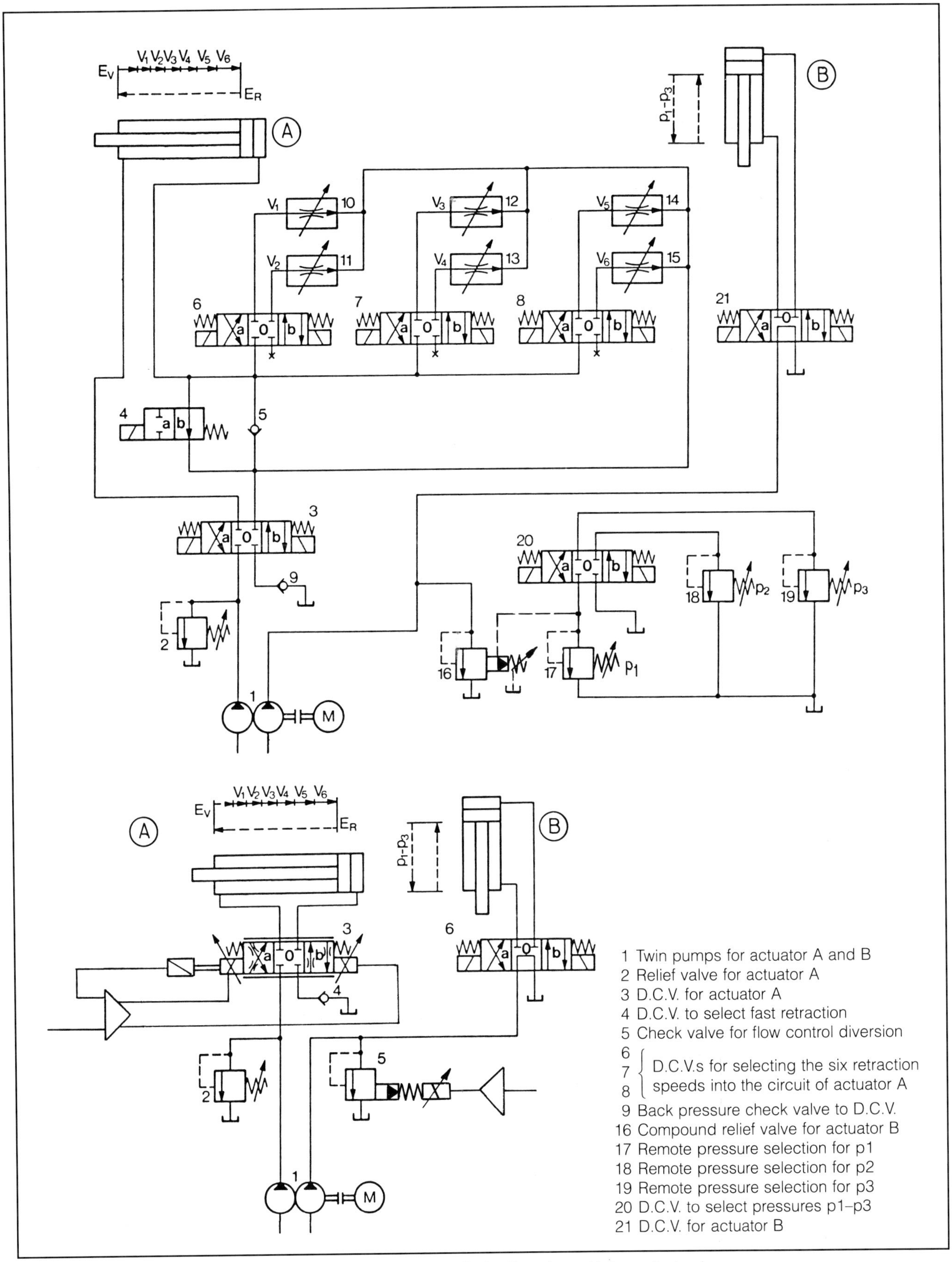

Fig. 309 Automated manufacturing machine with clamping cylinder B and machining cylinder A.

Circuit example 2 with proportional valves

A typical application where proportional valves are used to control injection pressure and injection speed for a plastics injection moulding machine. The flow from the variable displacement pump (1) is controlled by proportional valve (4). The maximum pump pressure is controlled by proportional pressure limiting valve (2). Valve (3) would only operate in an emergency, if the electronic control on valve (2) should fail. Hydraulic motor operation is achieved by a directional control valve (8). Actuator extension stroke and velocity are controlled by a proportional directional control valve (7) and its retraction stroke is controlled by a conventional directional control valve (6). Since the extension stroke velocity and force of the linear actuator must vary from product to product produced on the injection moulding machine, valve (7) is required to throttle the flow for actuator extension and maintain this flow regardless of system fluctuations. For this reason, a back pressure valve (5) is used as a counter-balance valve to create a variable back pressure. Thus, the net thrust force on the actuator extension stroke may be kept constant, regardless of load fluctuations and therefore the throttled flow across valve (7) is also kept constant. As a result, the governing factor, the extension speed of the linear actuator is kept constant.

Summary proportional valves

The circuits shown in figs. 309 and 310 clearly demonstrate the advantage gained by using proportional control valves. Not only are speeds and pressure settings for the controls in the given two circuit examples infinitely variable, any need to increase speed or pressure level settings may be simply achieved electrically by adding further input set-points on the PLC controller.

It is beyond the scope of this chapter to present calculation examples, since such calculations for the design of proportional control circuits requires extensive design knowledge, experience and command of calculus. However, various manufacturers of proportional and servo valves produce excellent information regarding design and use of their products.

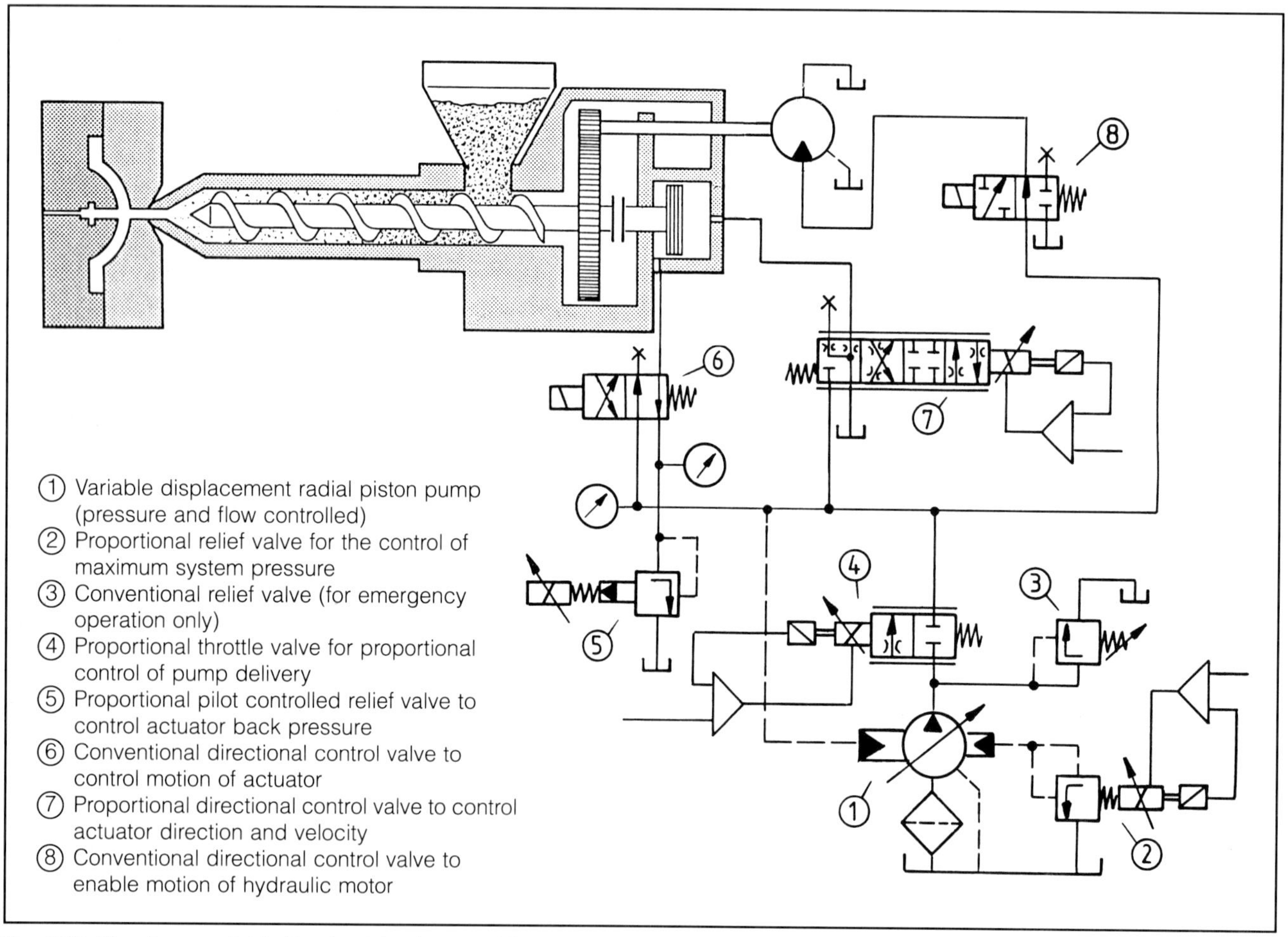

Fig. 310 Circuit example 2. Plastics injection moulding machine controlled by proportional valves to achieve infinitely adjustable pressure settings and injection speed.

17 Faultfinding of hydraulic circuits

Note
When a fluid other than mineral oil is used in a hydraulic system, identical fluid should always be used for faultfinding.

Noisy pump

Cause	*What to do*
Air leaking into system.	Be sure that the oil reservoir is filled to the normal level and that the oil intake is below the oil surface. Check the pump shaft seal or packing, pipe and tubing connections, and all other points where air might leak into the system. One good way to check a suspected leak on the intake side is to pour oil over it. If the pump noise stops, you have found the leak.
Air bubbles in intake oil.	If the oil level is low, or the return line to the reservoir is installed above the oil level, air bubbles will appear in the reservoir. Check the oil level and return line position.
Cavitation. (The formation of vacuum in a pump when it does not get enough oil.)	Check for a clogged or restricted intake line, or a plugged air vent in the reservoir. Check that the lining of the suction hose has not become detached from the wire reinforcement. Check the strainers in the intake line. The oil viscosity may be too high: check the manufacturer's recommendations.
Loose or worn pump parts.	Check the manufacturer's maintenance instructions first. Tightening every nut in sight may not be the way to stop leakage. Look for worn gaskets and packings, and replace if necessary. There is usually no way to compensate for wear in a part: it is always better to replace it. The oil may be of improper grade or quality: check the manufacturer's recommendations.
Stuck pump vanes, valves, pistons, etc.	Parts may be stuck by metallic chips, bits of lint, etc. If so, dismantle and clean thoroughly. Avoid the use of files, emery cloth, steel hammers, etc. on machined surfaces. Products of oil deterioration such as gums, sludges, varnishes, and lacquers may be the cause of sticking. Use a solvent to clean parts, and dry thoroughly before reassembling. If parts are stuck through corrosion or rust, they will probably have to be replaced. Be sure the oil has sufficient resistance to deterioration and provides adequate protection against rust and corrosion.
Filter or strainer too dirty; filter too small.	Filter and strainers must be kept clean enough to permit adequate flow. Check the filter capacity. Be sure that the original filter has not been replaced by one of a smaller capacity. Use oil of a quality high enough to prevent rapid sludge formation.
Pump running too fast.	Determine the recommended speed. Check pulley and gear size. Make sure that no one has installed a replacement motor with another than the recommended speed.
Pump out of line with driving motor.	Check alignment. Misalignment may be caused by temperature distortion.

Pump not pumping

Cause	*What to do*
Pump shaft turning in the wrong direction.	Shut down immediately. Some types of pump can turn in either direction without causing damage; others are designed to turn in one direction only. Check belts, pulleys, gears, and motor connections. Reversed leads on 3-phase motors are the commonest cause of wrong direction.
Intake clogged.	Check line from reservoir to pump. Make sure that filters and strainers are not clogged.
Low oil level.	Make sure that the oil is up to the recommended level in the reservoir. Intake line must be below the oil level.
Air leak in intake.	If any air at all is going through the pump, it will probably be quite noisy. Pour oil over suspected leaks; if the noise stops, you have found the leak.
Pump shaft speed too low.	Some pumps will deliver oil over a wide range of speeds; others must turn at the recommended speed to give appreciable flow. Find out first what the manufacturer's recommended speed is, then, check the speed of the pump, with a speed counter if possible. If speed is too low, look for trouble in the driving motor.
Oil too heavy.	If the oil is too heavy, some types of pump cannot pick up prime. You can make a very rough check of viscosity by using some oil, that is known to have the right viscosity, for comparison. Then, with both oils at the same temperature, pour a quart of each oil through a small funnel. The heavier oil will take a noticeably longer time to run through. Oil that is too heavy can do great harm to hydraulic systems. Drain and re-fill with oil of the right viscosity.
Mechanical trouble (broken shaft, loose coupling, etc.).	Mechanical trouble is often accompanied by a noise that can be very easily located. If you find it necessary to dismantle, follow the manufacturer's recommendations to the letter.

Leakage around pump

Cause	*What to do*
Worn packing.	Tighten the packing gland or replace the packing. The trouble may be caused by abrasives in the oil. If you suspect this sort of trouble, make a thorough check of points where abrasives may enter the system.
Head of oil on suction line.	Usually it is better to have slight pressure on the suction side of the pump, although it may not be necessary. With more than a slight head, leakage may result. If a head is not required and components can be arranged, do so. Otherwise, don't worry about the leakage. Just wipe off the pump periodically, and, if feasible, install a drip pan. Don't let oil leak onto the floor!

Overheating

Cause	*What to do*
Oil viscosity too high.	Check oil recommendations. If you are not sure of the viscosity of the oil in the system, it may be worth your while to drain the system and re-fill with oil of proper viscosity. Unusual tempeature conditions may cause oil of proper viscosity for "working temperature" to thicken too much on the way to the pump. In this case, using oil with a higher viscosity index may cure the trouble.

Internal leakage too high.	Check for wear and loose packings. Oil viscosity may be too low. Check the recommendations. Under unusual working conditions, the temperature may go high enough to reduce the viscosity of the recommended oil too much. Proceed with caution if you are tempted to try a higher viscosity oil.
Excessive discharge pressure.	If oil viscosity is found to be okay, the trouble may be caused by a high setting of the relief valve. If so, re-set.
Poorly fitted pump parts.	Poorly fitted parts may cause undue friction. Look for signs of excessive friction and make sure that all parts are in alignment.
Oil cooler clogged.	On any machine equipped with an oil cooler, high temperatures are to be expected. If temperatures normally run high, they'll go even higher if oil cooler passages are clogged. If you find a clogged cooler, try blowing it out with compressed air. If this won't work, try solvents.
Low oil.	If the oil supply is low, less oil will be available to carry away the same amount of heat. This will cause a rise in oil temperature, especially in machines without oil coolers. Be sure the oil is up to the right level.

Erratic action

Cause	*What to do*
Valves, pistons, etc., sticking or binding.	First, check the suspected part for mechanical deficiencies such as misalignment of a shaft, worn bearings, etc. Then, look for signs of dirt, oil sludge, varnishes, and lacquers caused by oil deterioration. You can make up for mechanical deficiencies by replacing worn parts, but don't forget that these deficiencies are often caused by using the wrong oil.
Sluggishness when a machine is first started.	Sluggishness is often caused by oil that is too thick at starting temperatures. If you can put up with this for a few minutes, the oil may thin out enough to give satisfactory operation. But if the oil does not thin out or if the surrounding temperature remains relatively low, you may have to switch to oil with a lower pour point, lighter viscosity, or, perhaps, higher V.I. Under severe conditions, immersion heaters are sometimes used to pre-heat the oil.

Low pressure in system

If an "in-line" hydraulic tester is available it should be fitted between the relief valve and the system. This will enable the measurement of pump flow rate to be made at various pressures up to the setting of the system relief valve. The results should then be compared with the manufacturer's specifications.

Cause	*What to do*
Relief valve setting too low.	If the relief valve setting is too low, oil may flow from the pump through the relief valve and back to the oil reservoir without reaching the point of use. To check the relief setting, block the discharge line beyond the relief valve and check the line pressure with a pressure gauge.

Relief valve stuck open.	Look for dirt or sludge in the valve. If the valve is dirty, dismantle and clean. A stuck valve may be an indication that the system contains dirty or deteriorated oil. Be sure, therefore, that the oil has a high enough resistance to deterioration.
Leak in system.	Check the whole system for leaks. Serious leaks in the open are easy to detect, but leaks often occur in concealed piping. One routine in leak testing is to install a pressure gauge in the discharge line near the pump and then block off the circuit progressively. When the gauge pressure drops, the leak is between this point and the check point just before it.
Broken, worn, or stuck pump parts.	Install a pressure gauge and block the system just beyond the relief valve. If no appreciable pressure develops, and the relief valve is okay, look for mechanical trouble in the pump. Replace worn and broken parts.
Incorrect control valve setting; oil "short circuited" to reservoir.	If open-centre directional control valves are unintentionally set at the neutral position, oil will return to reservoir without meeting any appreciable resistance, and very little pressure will develop. Scored control valve pistons and cylinders can cause this trouble. Replace worn parts.

18 Pneumatic step-counter circuit design for hydraulic power systems

It is puzzling, and to a certain extent regrettable, that modern industrial hydraulic systems with their relatively high sequence content have been up till now almost entirely neglected by the pneumatic application engineer. This is in spite of the fact that for many years hydraulic directional control valves with pneumatic pilot actuation have been readily available on the market. This phenomenon can be mainly blamed on the fluid power teaching programs. The pneumatic circuit designer has a more or less good command of his own area of circuit control, but has up till now hardly bothered to gain the knowledge to integrate pneumatics with hydraulic controls.

On the other hand, the hydraulics application engineer or circuit designer deserves no compliment either. As a rule, he designs his systems in accordance with the force, speed, and power conditions defined in the technical specifications, and lets the electrician bother with the remaining sequence control. This automatism should be interrupted, and pneumatic control logic promoted as an ideal alternative of controlling sequential and combinational hydraulic power systems. The reasons are:

- Pneumatically pilot operated directional control valves have a much longer service life than solenoid valves. "Burning-out" is a common complaint with solenoids and their replacement can be expensive because of lost production and solenoid replacement costs.
- Pneumatically pilot operated valves have a response time of 5–10 milliseconds, whereas the response time of comparable solenoid valves only reaches the range of 30–50 milliseconds.
- Pneumatic control circuits are explosion, and fire-hazard free and for most applications pneumatic machine control as a total package is 20–30% less expensive than electrical circuits with solenoids and contactor relays.
- Pneumatic installations require no wire ducts or conduits, and push-in line connections for the rigging-up of pneumatic valves have minimised labour costs.
- Most pneumatic component manufacturers offer a complete control package specifically designed and built to comply with job specifications. All valves are neatly arranged in a control cabinet. Component lists, circuit drawings, and commissioning manuals are generally included in such packages.
- Today's available variety of robust switching components including non-contact type sensors, interface elements, high speed pneumatic counting equipment, and integrated logic modules make pneumatics the ideal control choice for many industrial hydraulic systems.
- Design, installation, and maintenance of pneumatic control circuits require no specially licensed tradesmen. Maintenance personnel servicing the hydraulic circuit can also be used to maintain and service the pneumatic control. Thus, demarcation problems are eliminated.

What pneumatic circuit designers should know about hydraulic systems

Industrial hydraulic power systems, unlike pneumatic systems, are constantly power absorbing. Hydraulic energy is produced as long as the prime mover drives the pump, and pressure within the hydraulic system is developed by resistance to pump flow. Hence, the hydraulic system could suffer damage if such pump flow would not be stopped or recirculated to tank during non-action periods of the hydraulically powered machine. Such non-action periods arise from stalling an actuator, or by reaching the end of the control sequence, or during time-delay periods of the control sequence. To eliminate power wastage, excessive fluid heat and the resulting system damage, hydraulic control designers use one of the four pump flow control methods depicted in fig. 230. In such power saving hydraulic systems one differentiates between systems where the maximum pressure during non-action periods:

- must be maintained (That means the pressure in the actuators is maintained, see fig. 311a and d.);
- must fall off. (That means pressure in the system drops back to zero, see fig. 311b and c.)

In both cases the pump flow is either stopped or recirculated back to tank, and pump input power is almost zero. In the control systems of fig. 311a and d, pump flow and system pressure are automatically controlled, and therefore require no attention by the pneumatic sequence control. The control systems fig. 311b and c operate with directional control valves and must therefore receive pneumatic signals.

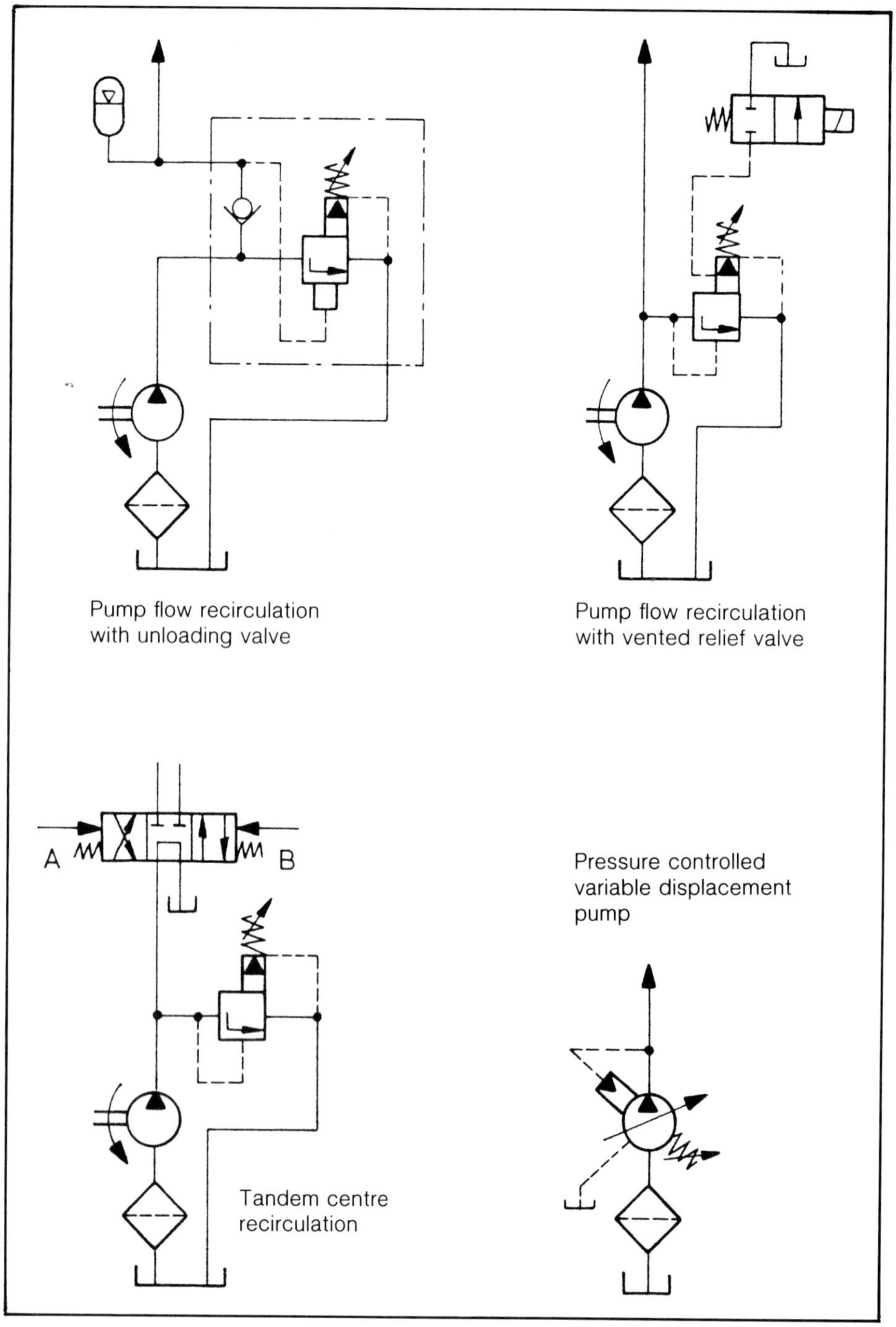

Fig. 311 Methods of system pressure control and pump offload control.

Three position valves

Three position pilot operated control valves can only select the centre position when neither of the two pilot signals are present. These valves will therefore automatically select the centre position whenever the pilot signal disappears. Pneumatic pilot signals leading to such hydraulic directional control valves (with three positions) must therefore be carefully timed and maintained for a precise duration to match the stringent requirements of the hydraulic circuit (see figs. 324–326).

Sequence control systems designed with the step-counter method cancel the pilot signals as soon as the next sequence step is switched. Where step-counter systems are used to control pilot operated three position valves, additional memory valves (to retain the pneumatic pilot) are required, and cancellation of such memorised pilot signals must not occur until an opposing pilot signal is imminent or centre position selection of the D.C.V. is required (fig. 324).

Industrial hydraulic systems (especially controls for machine tools) require almost exclusively three position directional control valves. To merge hydraulics and pneumatics successfully, pneumatic system designers must therefore acquire at least some knowledge of the function and operation of the more commonly used pressure and flow control valves.

Traverse-time diagram (step-motion diagram)

The traverse-time diagram is a simple, easily understood graphic representation of the motion sequence (step sequence) at which linear fluid power actuators operate. The machine designer defines the motion sequence and draws the traverse-time diagram, thus establishing a communication and design basis between all parties involved in the design and implementation of the total control system.

The following points are to be observed when setting up a traverse-time diagram:

- Step commands are represented by vertical lines (timelines), the direction and speed rate of the actuators is shown by inclined lines (fig. 312).
- Sequence steps are numbered above the step command lines. Time delay periods are labelled T (fig. 312), and stroke length is regarded as equal since it has no bearing on the sequencing of the machine.
- Horizontal lines represent stroke limits and are labelled according to the name given to the limit valves located at the stroke end (position sequencing). ing).
- Circuit and signal labelling is based on the binary system, whereby actuator extension is denoted the logic 1, and retraction the logic 0. For example, signal command A1 causes actuator A to extend. Limit valve a_1 confirms this extension. Signal command A0 causes actuator A to retract, and limit valve feedback a_0 confirms the completion of this motion.

Step-counter circuit design

The step-counter circuit is a modular circuit design concept consisting of a chain of integrated circuit blocks or modules, whereby each sequence step (depicted in the traverse-time diagram) is allocated a

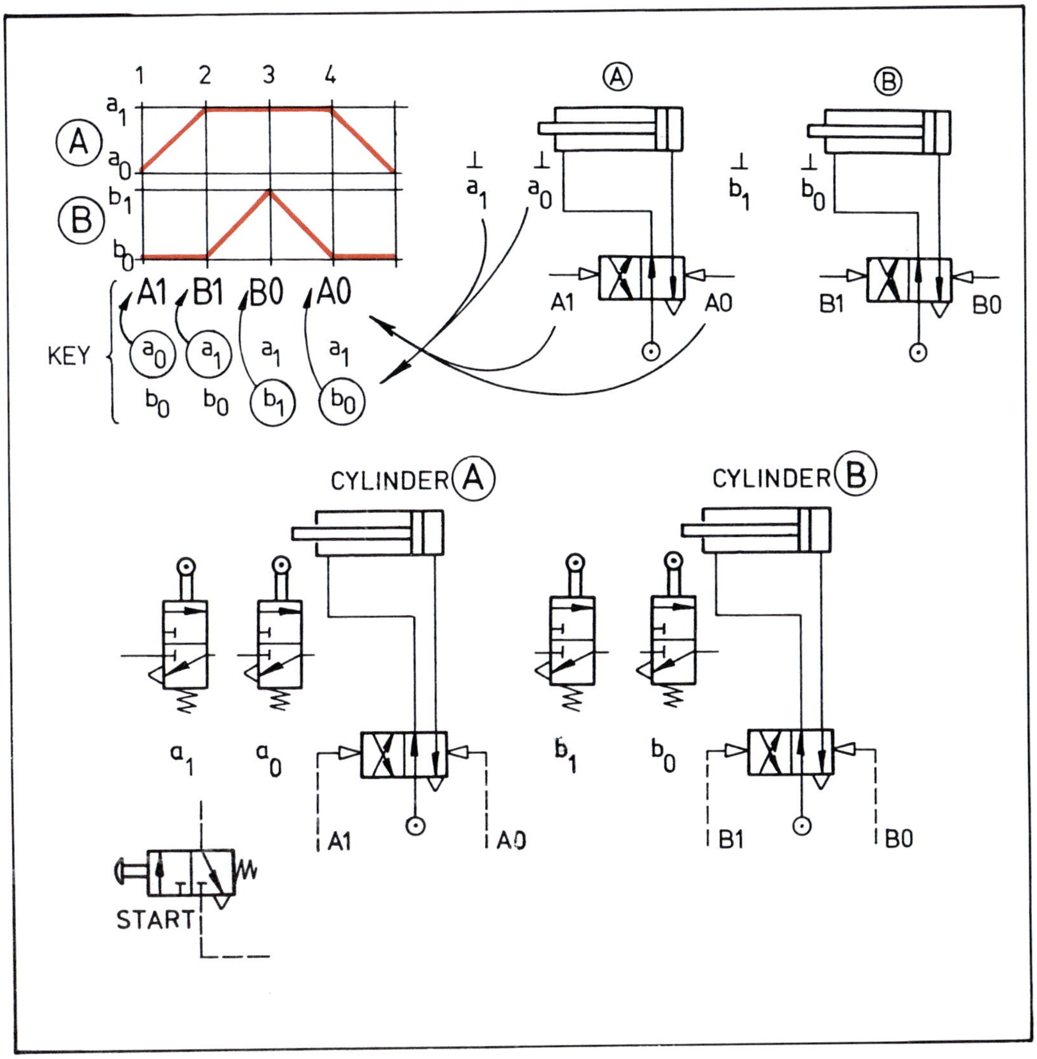

Fig. 312 Traverse-time diagram and circuit labelling.

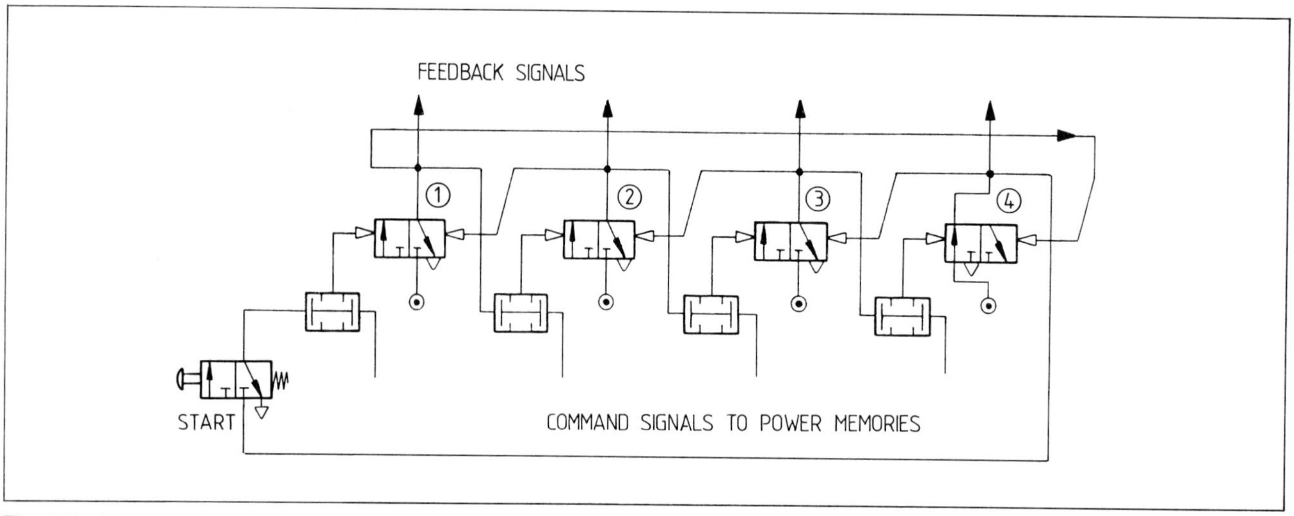

Fig. 313 Chain of integrated step-counter modules.

module. The simplicity of the step-counter (sometimes called shift-register or step-sequencer) stems from the construction of its modules and the integration of these modules into a sequence chain (fig. 313). Each module is built from a "memory" valve and a pre-switched "and" function valve. The two signals required to make the "and" function, are the feedback signal, confirming the completion of the previous sequence step, and the preparation signal from the memory valve of the previous step-counter module. The "and" function sets the memory into the "on" position. The memory must accomplish three functions. It

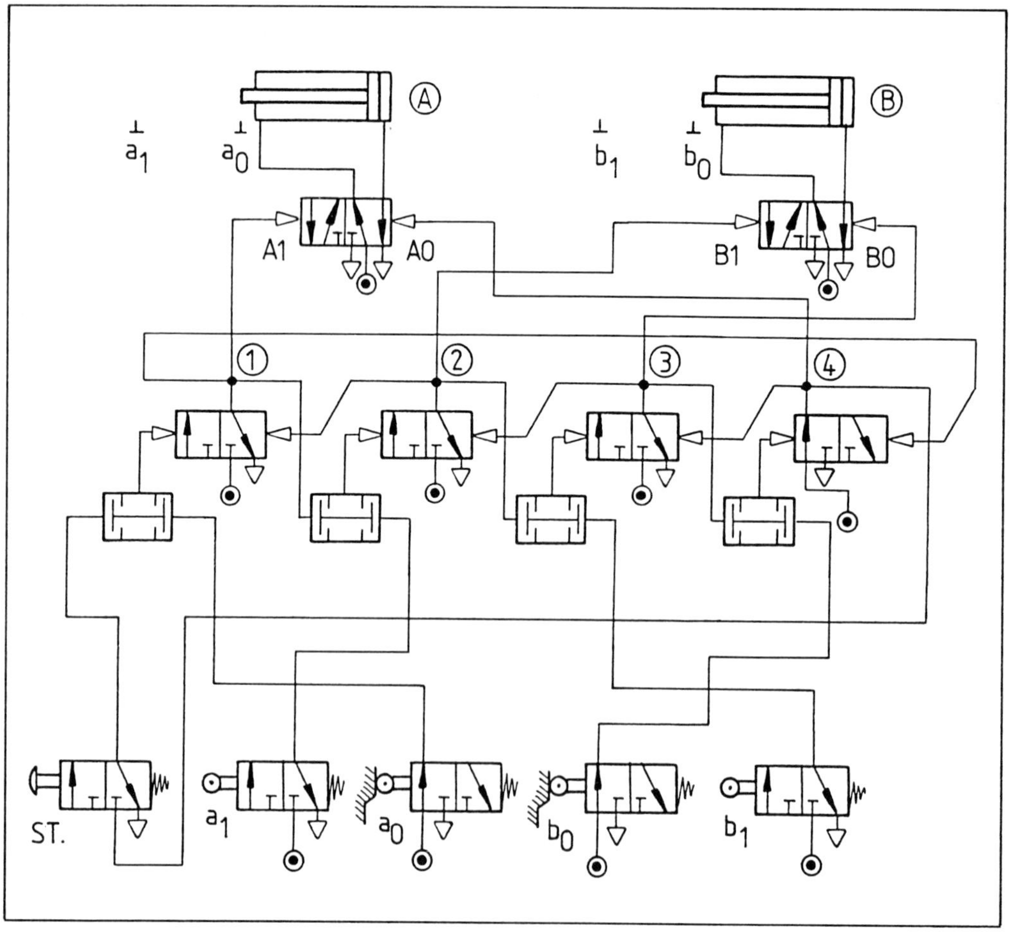

Fig. 314 Complete step-counter control circuit.

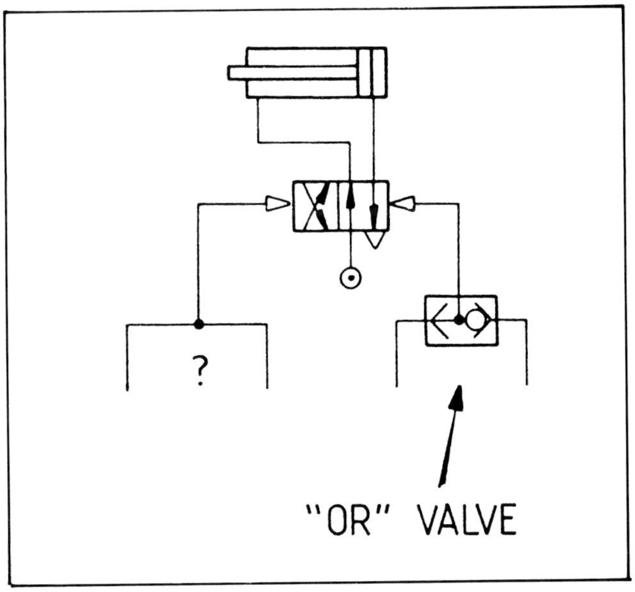

Fig. 315 A T-connector is no substitute.

provides the power memory valves (D.C.V.) with a signal command to extend or retract an actuator; it re-sets the memory valve in the pre-vious step-counter module; and it provides the next step-counter module in the chain with a preparation signal.

After the sequence program is completed, the last module in the chain remains set until the first module (after the start signal for a new cycle has been given) re-sets the last module. The preparation signal for the "and" function in module 1 is fed through the start valve. By this, continuous cycling is interrupted until the start command is given. A circuit for a clamp and drill application is presented in fig. 314.

Where an actuator is to extend and retract more than once during a control cycle, the appropriate output signals from the step-counter chain are connected by "or" function before being connected to the signal port on the D.C.V. It should be noted that T-connectors do not substitute for "or" function valves, since the pilot signals from the step-counter chain would exhaust through the valve on the other end of the T-connection. This is illustrated in figs. 315 and 316, and a circuit application is presented in fig. 317.

Simultaneous actuator movements

Where a sequence step module actuates more than one actuator (that is, two or more actuators traverse simultaneously), the output signal from the appropriate step-counter module is simply branched out (see arrows in fig. 317), and connected to the signal ports of the power memories involved in the simultaneous movement. The next sequence step, which precedes the step with simultaneous actuator movements, must not switch until all these movements are confirmed. This is accomplished with "and" function valves or, where possible, by connecting the appropriate limit valves in series order. (See the arrows in fig. 317, the additional "and" valve in fig. 318, and the traverse-time diagram in fig. 319.)

Time-delayed sequencing for three position valves

Time-delayed sequencing is often encountered in hydraulic machine tool circuits. The pneumatic timer may either be built into the confirmation signal-line leading to the step-counter module (see arrows in fig. 320) or into the switching line leading to the directional control valve. Timer placement in the switching line delays only the switching of the D.C.V., but resetting of the previous step-counter module is not delayed. With the timer in the confirmation signal line, both the D.C.V. switching function and the step-counter module switching function are delayed.

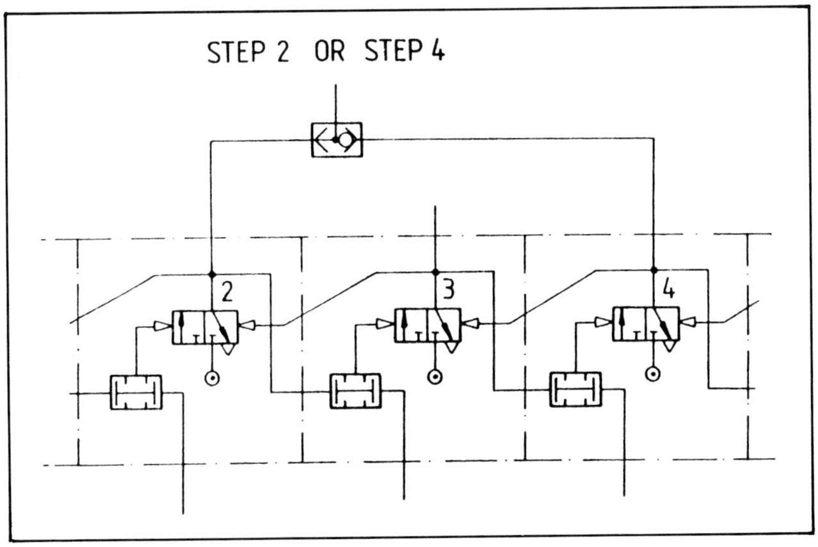

Fig. 316 "Or" valve application.

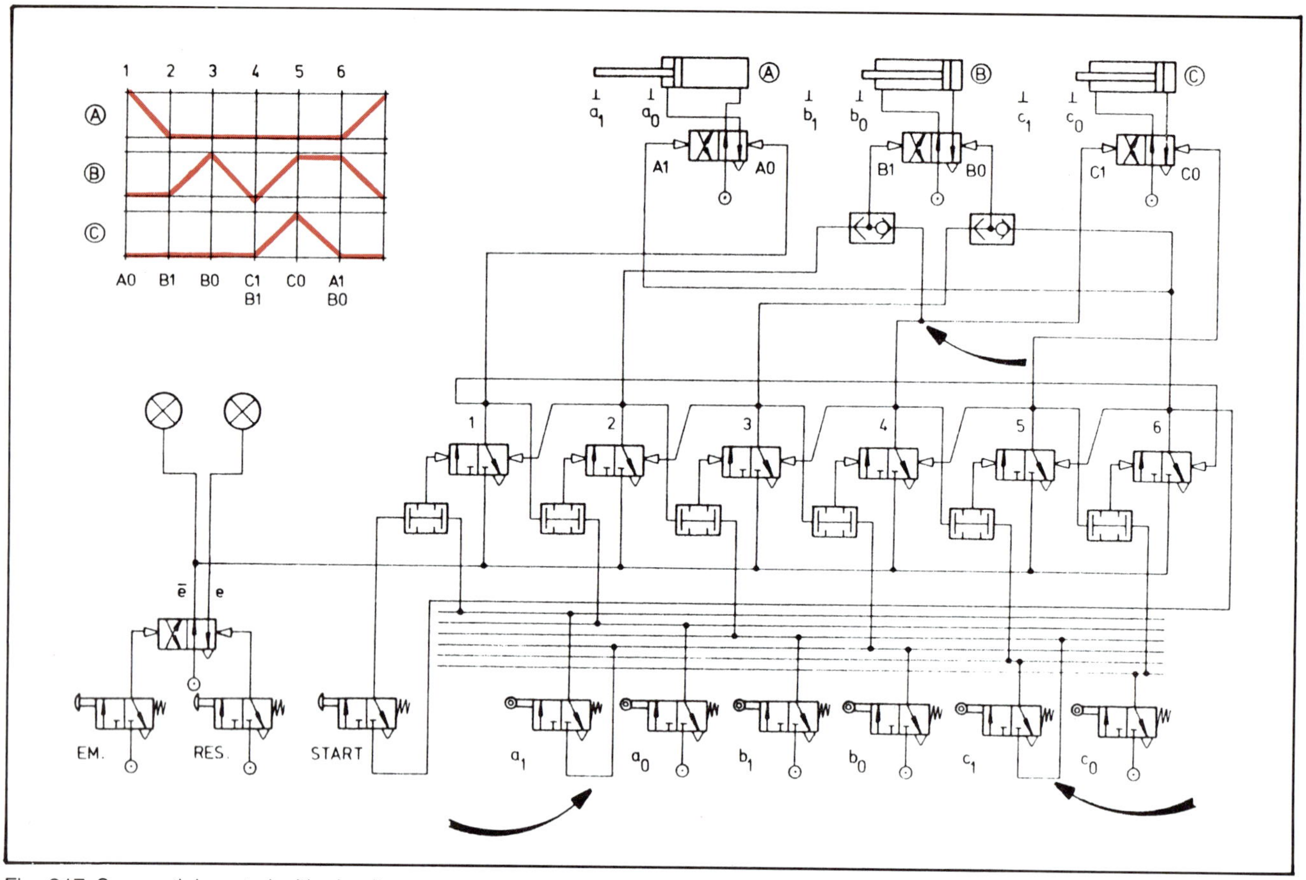

Fig. 317 Sequential control with simultaneous actuator movements at steps 4 and 5.

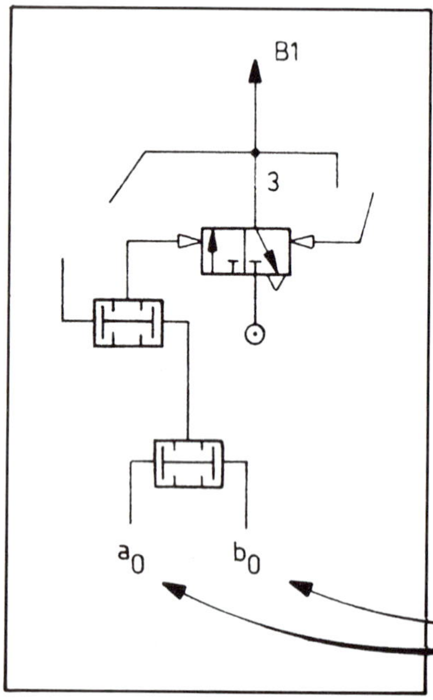

Fig. 318 Multiple confirmation of limit valve signals after simultaneous actuator movement.

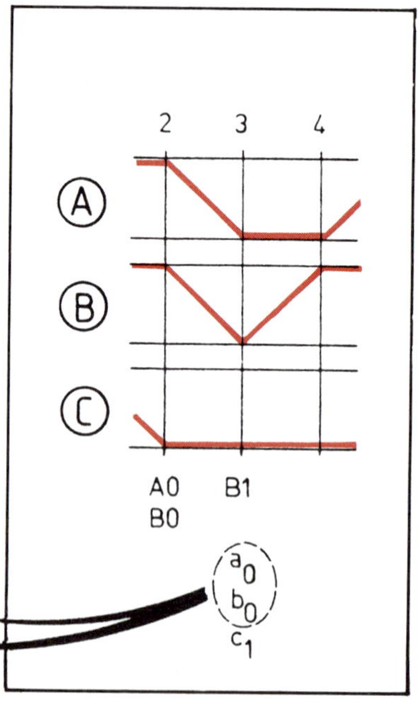

Fig. 319 Key shows multiple confirmation.

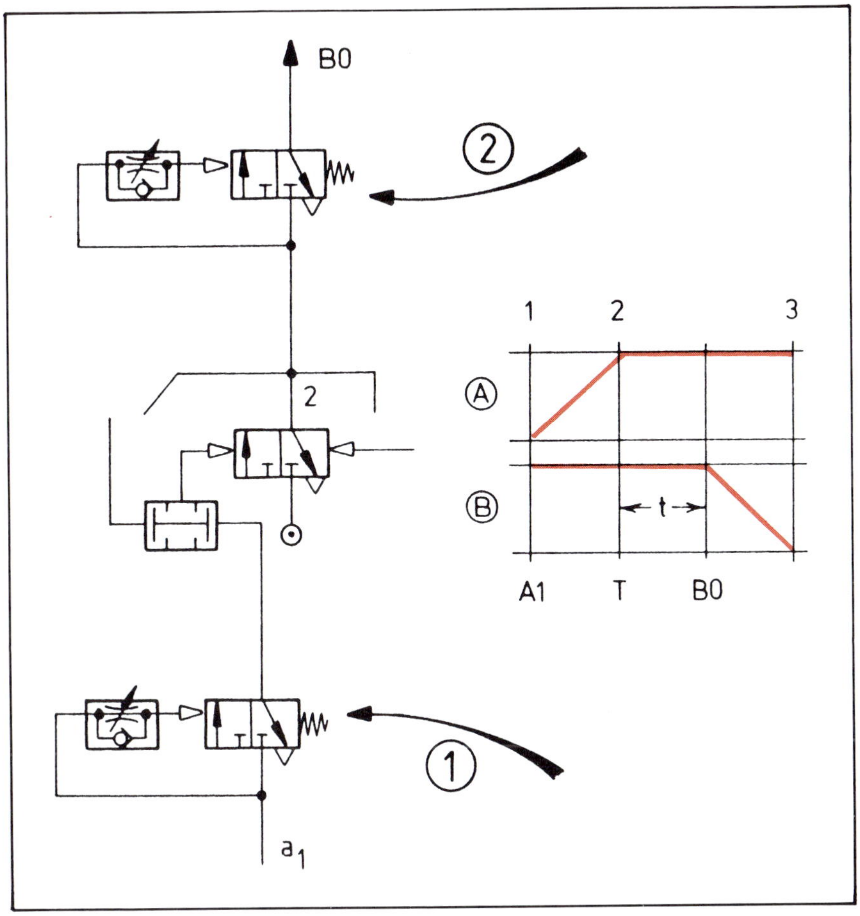

Fig. 320 Timer placement options.

Pneumatic circuit designers therefore have to evaluate both timer positions for their merits and select the correct position according to circuit application.

Circuit example 1

Pneumatic step-counter control, integrated with two hydraulic actuators. If the pneumatic timer were wrongly placed (in the position indicated by the arrow) then actuator B would overshoot the limit valve b_1, at the completion of sequence step 3 (fig. 321).

Circuit example 2

A machine tool comprises actuator A for clamping, actuator B for drilling, and actuator C to off-set the table. Its sequence A1-B1-B0-C1-B1-B0-C0/A0 is controlled with a pneumatic step-counter circuit (fig. 322). A vented compound relief valve controls the offloading function at the end of the cycle. An additional step-counter module is used to control pump offloading (see arrow in fig. 322), which must occur as soon as sequence step 7 is confirmed by signals $a_0 \bullet c_0$. By switching step 1, step-counter module 8 re-sets and pump flow is directed into the system.

Circuit example 3

A time delay phase is added to the control circuit shown in fig. 322. The machine sequence now reads A1-B1-B0-T-B1-B0-C0/A0. As previously explained, fixed displacement hydraulic pumps require pump offloading whenever non-action periods occur (end-of-cycle or time delay periods). Hence, pump offloading is required in this example at the end of sequence step 3 (time delay period), and at the end of sequence step 7 (end of cycle).

A simple and economic switching solution may be achieved by inserting an additional step-counter module between modules 3 and 4 (see arrow in fig. 323). This module switches the pump into offloading as soon as sequence step 3 is completed, and it also operates the pneumatic timer. The timer (after switching) provides the confirmation signal for step-counter module 4 which, when switched, re-sets step-counter module U, and pump flow is again redirected into the system.

Circuit example 4

This control problem is identical to circuit example 2 except for the offloading method. Here, pump offloading is achieved by switching the tandem centre valve for actuator A into its centre position. To prevent premature cancellation of pilot signal A1, a memory valve is added. Sequence step 7 cancels the memory,

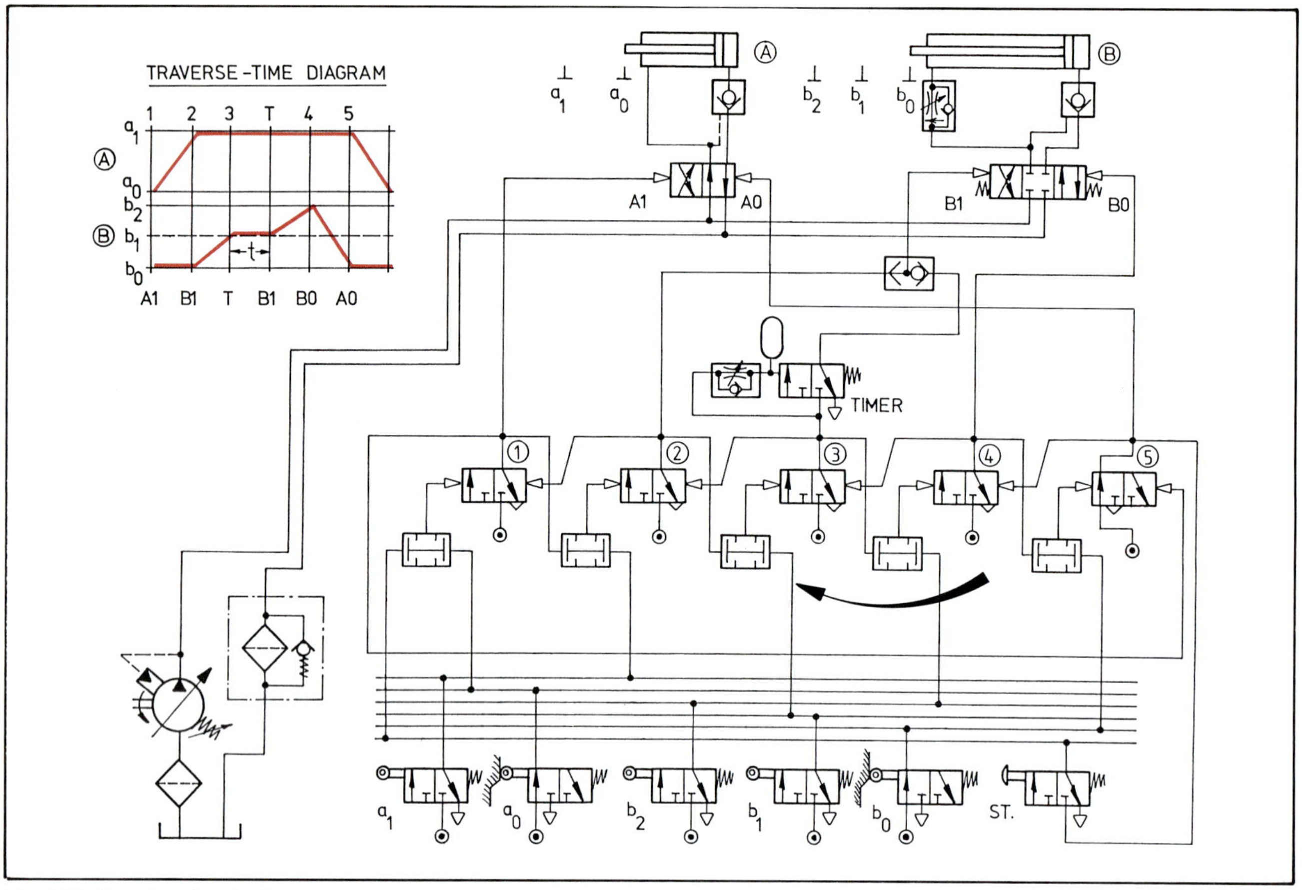

Fig. 321 Circuit example 1.

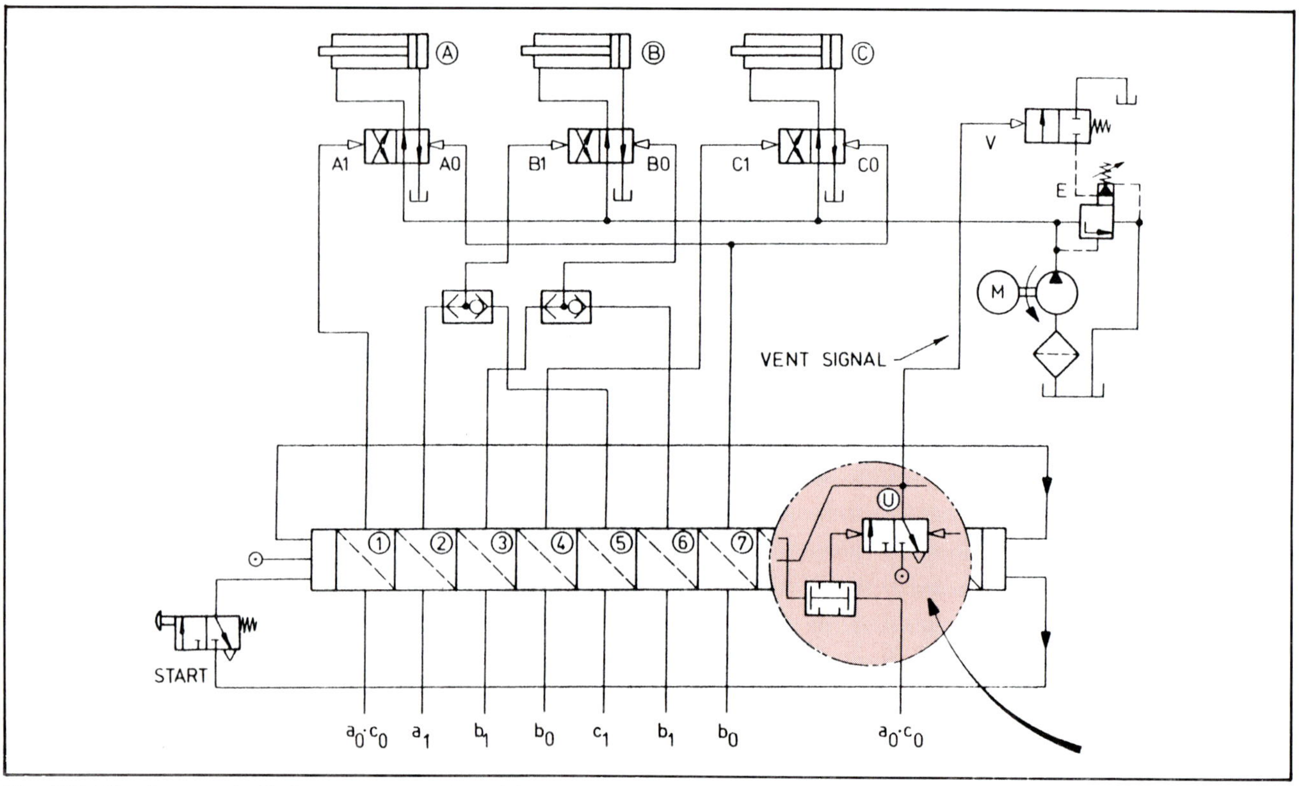

Fig. 322 Circuit example 2: the step-counter modules are presented by logic modules but the last module shows the symbolic valve arrangement inside the circle (enlarged).

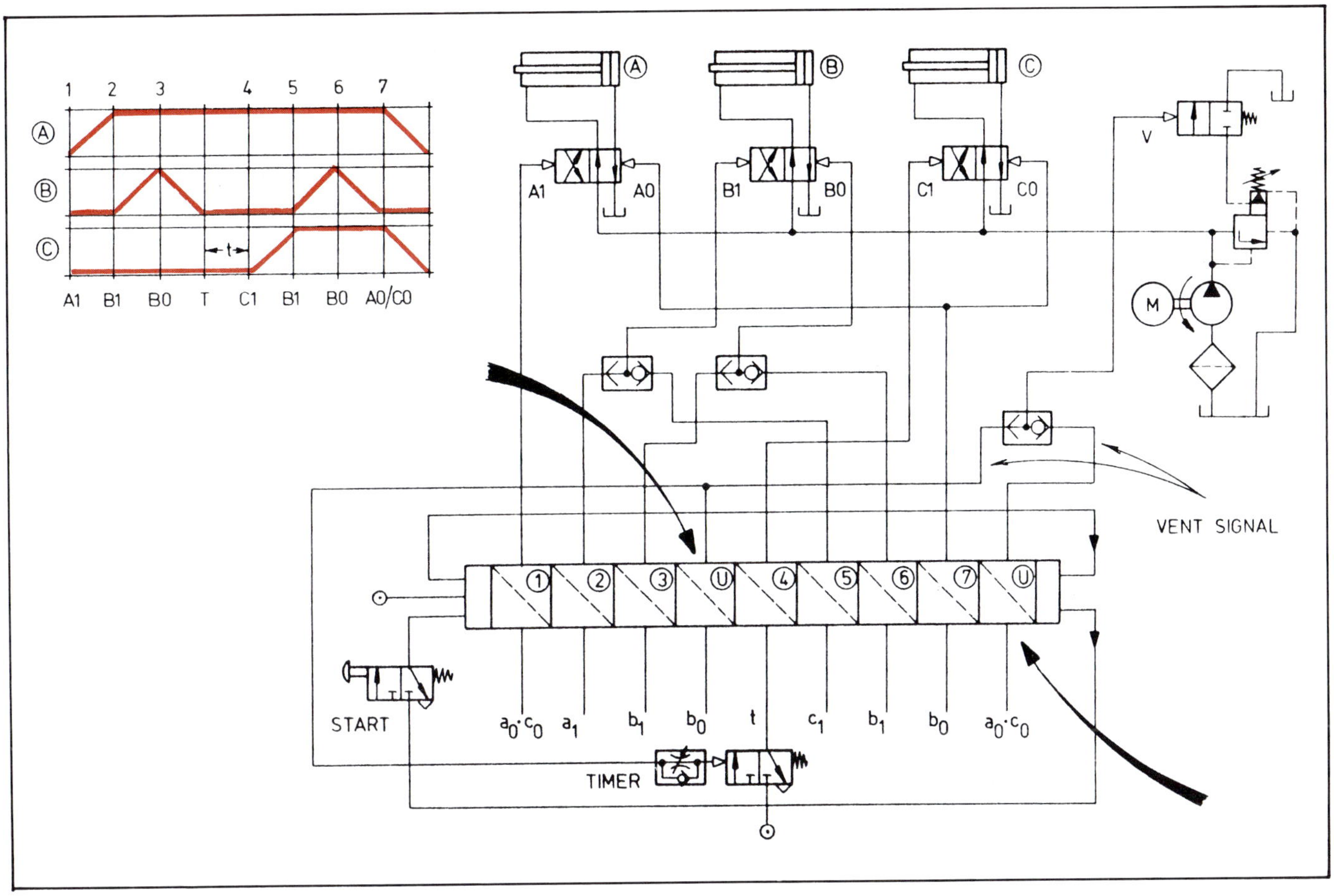

Fig. 323 Circuit example 3.

and thus signal A1 switches the counteracting signal A0 with its partner signal C0. Without this memory valve, signal A1 would disappear on the D.C.V. as soon as step-counter module 2 re-sets step-counter module 1. This would let the three position valve make centre, and pump flow to the actuators would cease.

With some step-counter types, the last step-counter module remains set until step-module 1 re-sets it. Therefore, an inhibition valve is required which cuts signal A0 short as soon as the last sequence function is confirmed. This lets the three position valve make centre and offloads the pump to tank. With step-counter types where the last step-module re-sets itself, this inhibition valve is not required (see fig. 324).

Circuit example 5

This control problem is identical to circuit example 3. However, pump offloading is achieved by means of a tandem centre valve, and has to occur at the end of sequence steps 3 and 7. This requires signal A1 to be maintained from the beginning of step 1 to the end of step 7, but not for the duration of the time delay at the end of step 3 where the pump must be offloaded (see fig. 325).

The switching equations for this circuit are:

Step module 1 $= a_0 \cdot c_0 \cdot$ start
Step module 2 $= a_1$
Step module 3 $= b_1$
Step module 4 $= b_0$
Step module 5 $= c_1$
Step module 6 $= b_1$
Step module 7 $= b_0$

A1 = step 1 + step 4
A0 = step 7 $\cdot \overline{a_0 \cdot c_0}$ (inhibition)
B1 = step 2 + step 5
B0 = step 3 + step 6

C1 = step 4 · timer
C0 = step 7

Memory 1 (set) = step 1
Memory 1 (re-sets) = step 4 (offload)

Memory 2 (set) = step 4 · timer
Memory 2 (re-set) = step 7 (offload)

Circuit example 6

In one single pass, a milling machine machines three faces of an aluminium casting. All machining functions are hydraulically powered and all sequence steps are pneumatically controlled.

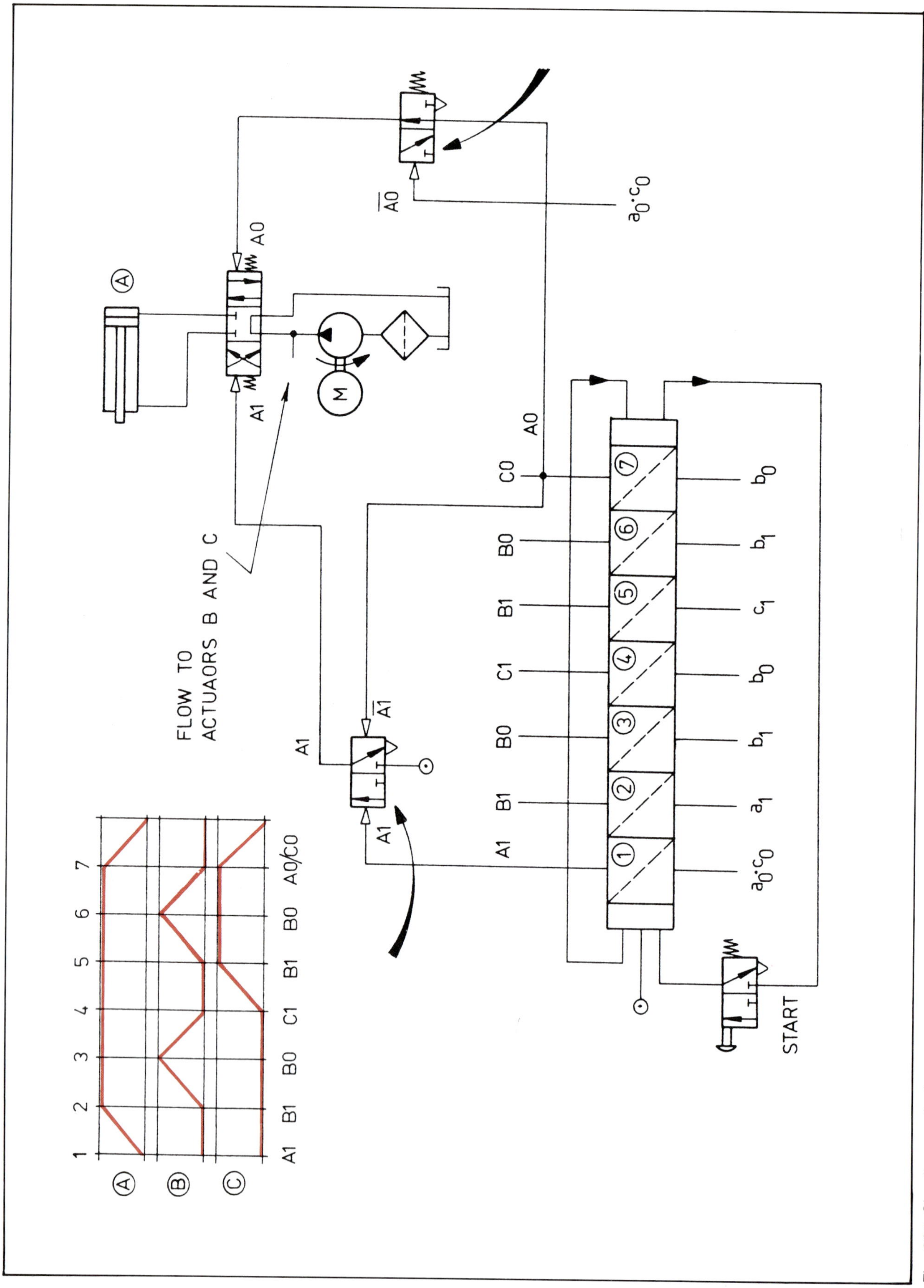

Fig. 324 Circuit example 4.

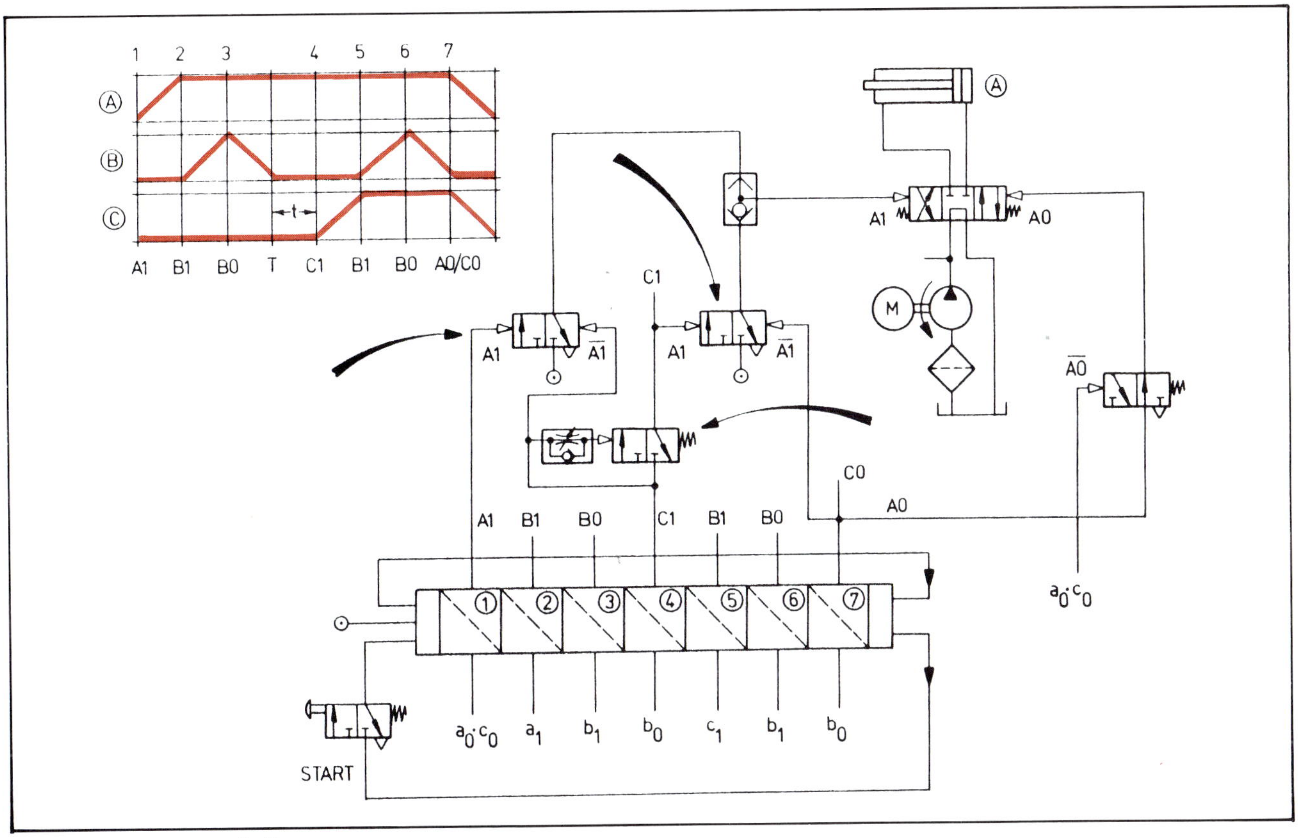

Fig. 325 Circuit example 5.

Workcycles

The sequence program is shown in the traverse-time diagram of fig. 326. The clutch/brakes for the three milling heads are pneumatically operated, but only one unit is shown in the circuit. The clutch engages and disengages the milling cutters, and all three electric motors, which run continuously, are switched on and off by means of pneumatically operated electrical switches (see fig. 326).

Actuator A clamps the workpiece onto the table and locking actuator C prevents the hydraulically driven table from creeping. The three milling heads are moved against their fixed limit stops by actuators B, and the hydraulic motor drives the table. The table approaches the milling position with fast advance speed (G3) until it reaches limit valve d_2. It then advances with slow milling speed (G1) to position d_3. After milling is completed, the table returns with fast speed (G3) to position d_1, and changes to creeping speed (G2) to prevent system shocks.

Safety precautions

If the hydraulic or the pneumatic system should fail, clamping pressure can be maintained from the accumulator. The three milling heads will automatically retract to position b_0 and the clutch/brake actuators disengage and brake the milling cutters respectively.

Pneumatic circuit

Memory E performs a dual function. It serves as a D.C.V. for the clutch/brake (signal E1), and provides the step-counter circuit with confirmation signals (e_1 and e_0). Signal B1 must be memorised because the D.C.V. which controls actuators B is not a memory valve, and without the additional memory the milling heads would retract when step-module 3 re-sets step module 2.

To prevent premature cancellation of signal D1, a further memory is required. This memory sustains signal D1 until signal D0 for table reversing (step 5) must appear. Signal D0 is sustained by step-modules 5 and 6 via an "or" function valve. The two extremity signals, G3 and G1, are alternating between steps 3 and 6, and therefore require no additional memory valve.

Hydraulic circuit

The dual pump delivers the system with two different flowrates. System pressure for the milling head movements is reduced. The unloading valve controls maximum system pressure of the right-hand pump, and unloads the pump during non-action periods. The vented compound relief valve limits the pressure from the left-hand pump and offloads the pump when no flow is required.

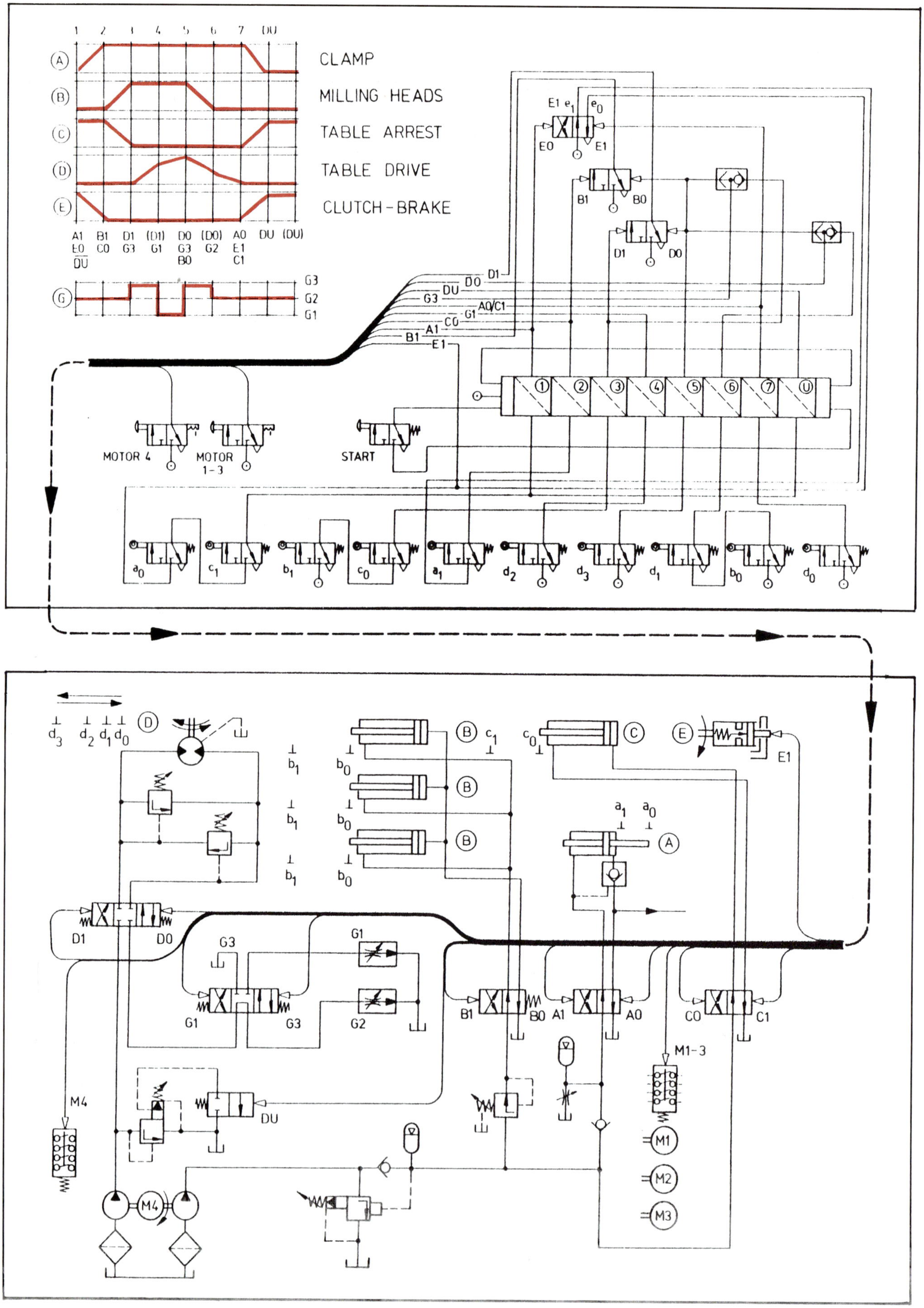

Fig. 326 Circuit example 6 (milling machine).

19 PLC electronic programmable controllers for hydraulic systems

Over recent years electronic programmable logic controllers have sprung up like mushrooms, and over 300 different types and brands are now available. Electronic programmable logic controllers have become an indispensable feature of the advances in industrial automation.

These controllers evolved as industry sought more economical ways of automating production lines, particularly those involved in the manufacturing of equipment, consumer goods and heavy industry products. Thus the electronic programmable controller has replaced relay based, hard wired electrical systems, and more recently it has also made significant inroads into the traditional domain of pneumatic logic circuitry. This encroachment has brought about some problems:

- programming difficulties caused through non-standardised and varying programming techniques
- programming difficulties caused through the use of varied logic element names and element description for hardware and software components within the controller
- demarcation problems in factories between electrical and metal-worker unions when maintenance on machines with fluid power control and electronic control is required
- extreme shortage of skilled personnel who can program and install such electronic controllers and interface them with fluid power circuitry.

What is an electronic programmable controller?

Electronic programmable logic controllers are often abbreviated with the letters PC. This abbreviation is misleading, since the letters PC are also used for "personal computers". It is therefore suggested that the abbreviation PLC (programmable logic controller) should be used instead.

Programmable logic controllers (PLC) operate by monitoring input signals from such sources as push-button switches, proximity sensors, heat sensors, liquid level sensors, limit switches, pressure and flow-rate sensors. When binary logic changes are detected from these input signals, the PLC reacts through a user-programmed internal logic switching network and produces appropriate output signals. These output signals may then be used to operate external loads and switching functions of the attached fluid power control system (see fig. 327).

Programmable logic controllers eliminate much of the wiring and rewiring that was necessary with conventional relay-based systems. Instead, the programmed "logic network" replaces the previously "hard wired" network. This logic network may be altered as required by simply programming or reprogramming the PLC. Thus, the automated processes of a production line or complex manufacturing machine can be controlled and modified at will for highly economical adaptability to a rapidly changing manufacturing environment.

How does an electronic programmable controller work?

A typical programmable electronic controller (PLC) has four separate yet interlinked components. These are:

- an input/output section which connects the PLC to the outside world (the machine with its sensors, solenoid valves and electric motors
- a central processing unit which is microprocessor based
- a programming device which may be a hand-held programming console or a computer
- a power supply to power input sensors and output signals leading to solenoids on the fluid power valves (usually 24 volt DC).

The CPU is microprocessor based and may be regarded as the "brain" of the controller. It reads all the on/off conditions of all input terminals and memorises them in its input image memory before executing the program. The CPU then processes that information according to the control plan programmed into it. Such an internal control plan may include numerous "memory" functions, logic "and", "or", as well as "inhibition" functions, arithmetic computations, timers and counter functions. Furthermore, the CPU continuously scans (monitors) the status of all output signals and thus constantly updates the content of the image memory according to changes made to the input image memory. Apart from the tasks already mentioned, the CPU also organises its internal operation (watchdog timer, initialising program, etc.).

The program is entered with ladder diagram or graphic logic symbols via a computer and monitor. It may also be entered by statement list or so-called mnemonics via a hand-held or on-board programming console. It then remains in the RAM memory of the CPU. Execution results based upon the logic combinations producing internal or external output signals are then written internally (electronically) into the element image memory. The element image memory then drives the output relays of the PLC.

When a programmable logic controller (PLC) operates, that is, when it executes its program to control an external system with, for example, fluid power valves and their actuators, a series of operations is automatically performed within the PLC. These internal operations can broadly be grouped into four categories:

1. Common internal processes such as resetting the watchdog scan-cycle timer and checking the user program memory
2. Data input/output refreshing
3. Instruction execution
4. Peripheral device command servicing.

From these four internal operations, the only one easily visible is number 3 — instruction execution (driving valves and their actuators and starting or stopping motors).

Summary

A programmable controller is an aggregate control mechanism made up of multiple electronic relays, timers and counters used to execute the internal logical wiring by the programming panel. A conventional hard wired relay panel basically differs from a PLC in the sequence execution method. Its sequences are executed in parallel while the PLC executes its sequences in the order of the program and cyclically as the scanning reveals any input changes.

This chapter is not intended to promote any particular brand or type of programmable controller and, therefore, great care has been taken to present the necessary information in as unbiased a way as possible. Nevertheless, to make the applications meaningful, some practical programming procedures are given, and have been based on the "Omron C20" programmable controller which is basically identical to the "Festo FPC 201" and closely related to the

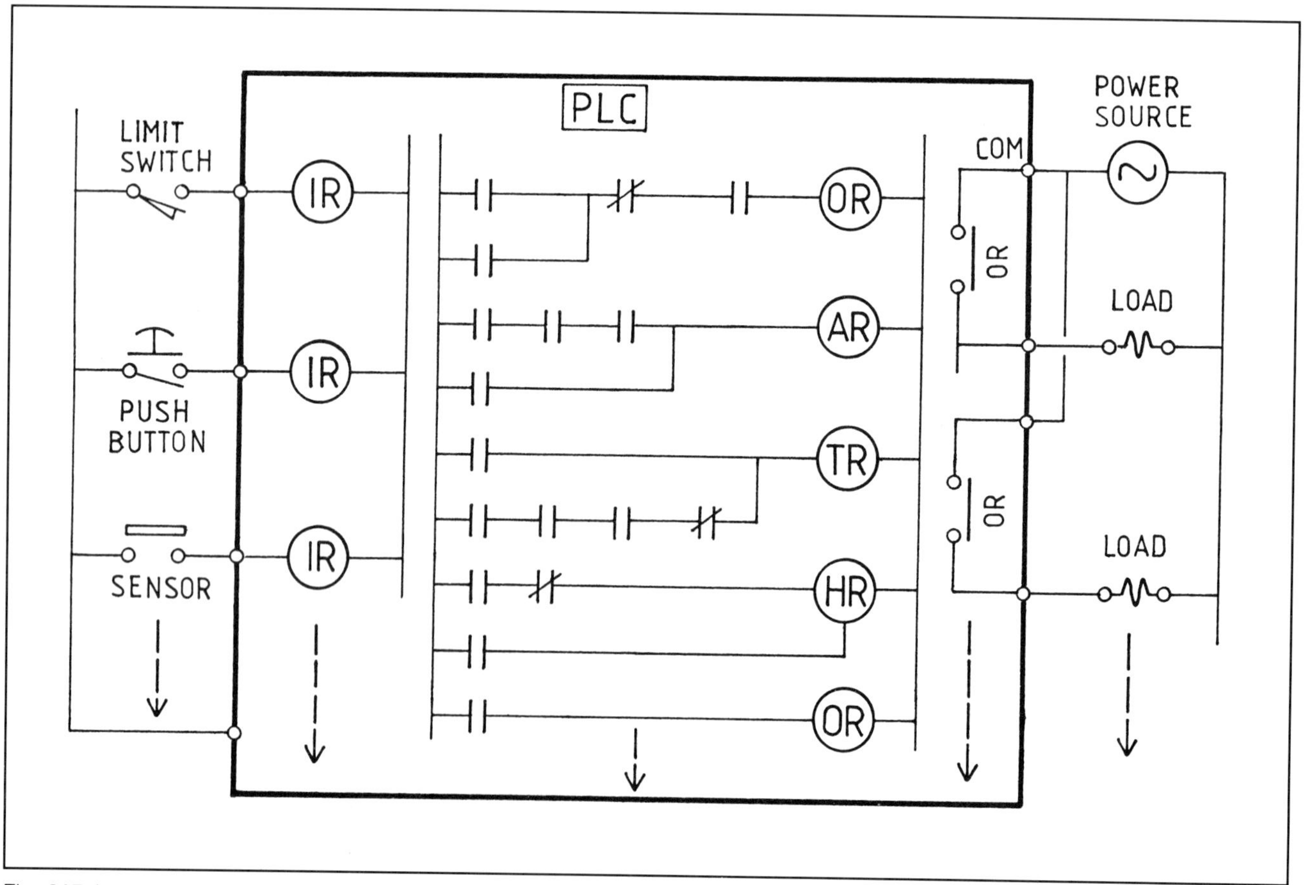

Fig. 327 Internal configuration of a PLC.

"Mitsubishi Melsec F series". These are all small types of controllers, ideally suited for fluid power sequential and combinational control.

Programming the electronic controller (PLC)

A most misleading impression for many control engineers and electricians when confronted with a programmable controller is that these machines (PLC) can easily be programmed with the "ladder logic" method. Ladder logic is *not a method of circuit design*; it is a method of circuit presentation, like a pneumatic or hydraulic circuit. For this reason numerous amateur programmers have had insurmountable difficulties and frustration whilst programming their little electronic "wizard"; some to such an extent that they gave up before they ever saw their controller work! In the same way as one has to learn to design a properly functioning hydraulic or pneumatic control circuit one needs to know how to design a PLC ladder logic circuit. It is therefore mandatory that the designer of PLC circuits possesses knowledge of general switching logic, knows and understands the function of the five basic logic concepts: "not", "yes", "and", "or" and "inhibition". Other equally essential switching concepts include: timer function, counter function, shift-register function, auxiliary relay function, memory function as applied to flip-flops, BCD number systems and data conversion.

Today, most machinery used in manufacturing, farming and mining is controlled by fluid power actuators and valves and is often of a highly sequential nature. This chapter therefore presents a programming and design method which works for most types and brands of PLC. This method is of course not new to the experienced circuit designer of pneumatic sequential controls. It is called the "step-counter design method". It is extensively explained and applied in chapter 8 of the book *Pneumatic Control for Industrial Automation* by Peter Rohner and Gordon Smith. In fact most principles outlined in that book can be directly applied to PLC programming.

Electronic switching peculiarities

Compared with hydraulic and pneumatic valves, the only electronic counterpart to a memory valve is the "keep relay". The keep relay is sometimes also called "holding relay" or "retentive relay", and maintains its logic status (set or reset) even during power failure. For hydraulic and pneumatic memory valves, this memory or retentive behaviour is achieved with the mechanical friction between the spool and the valve body. For electronic memories (relays) the "latch-in" must be maintained with a battery back-up when power fails. This demands that every logic switching function which in a pneumatic or hydraulic circuit would normally end up as a pilot signal to a memory valve, when electronically achieved, also has to end up driving an electronic memory which is in fact a keep relay or holding relay (see figs. 329 and 339).

Another peculiarity is the terminology used for normally open and normally closed hydraulic valves compared to electronic contacts and switches. A normally closed valve does not pass any flow in its non-actuated position. A normally closed contact, however, or a normally closed switch, does pass a signal in its normally closed state (chapter 2, fig. 20; also fig. 328).

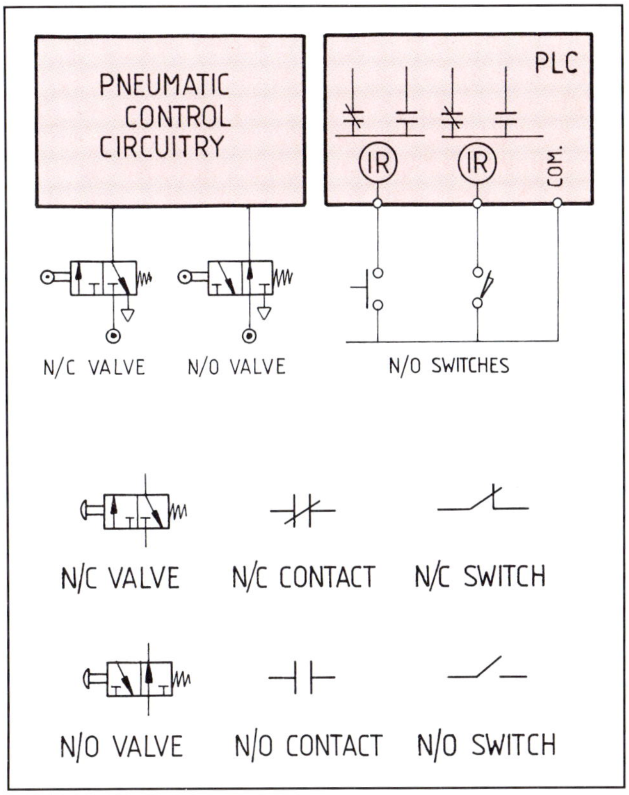

Fig. 328 A normally closed valve does not pass a signal but a normally closed contact does. Electronic controllers make all relay signals available in their logic 1 and also logic 0 form (normally closed and normally open contacts; see also fig. 22).

The third peculiarity is related to series connection of hydraulic valves, as explained in chapter 2, figs. 48 and 49. The electronic and electric ladder diagram presentation shows only series connection and no "and" connection by means of "and" gates. The PLC always connects these inputs by means of "and" gates and not by series connection. Electronic input relays as well as internal relays drive a vast number of normally closed and normally open contacts. Therefore, one never has to worry about reducing a complex logic input equation to its fully minimised form, as would be necessary for a pneumatic or hydraulic control. Thus, when programming an electronic controller, one may safely assume that each relay has an unlimited number of contacts which can always be "and" connected or "or" connected. These contacts may be called upon at any time. External inversion of signal (within the limit switch) is no longer required as input signals may be examined as normally open or normally closed (see fig. 328).

Relay types

A Programmable Logic Controller (PLC) has basically four different types of relays. Figures 329 and 360 list these relays in conjunction with their word (channel) allocation number. PLC relays with their respective acronym can be grouped into the following four categories:

1. Input relays [IR]
2. Output relays [OR]
3. Internal auxiliary relays [AR]
4. Holding relays [HR]

- Input relays (IR) are used to receive incoming signals and distribute them wherever these input signals are required. Input relays may be compared to a limit valve signal or push-button start signal or stop valve signal in a pneumatic circuit. The signal emerging from such relays may be processed in normally open (N/O) or normally closed (N/C) form (see figs. 328 and 20). The "Omron C20" has 80 input relay if expanded to its full input-output relay capacity.
- Output relays (OR) are the only relays which can be used to drive a load outside the PLC (for example a solenoid, a siren, start a motor or turn on an oven). The "Omron C20" has 60 output relays, if expanded to its maximum capacity (fig. 360). Output relays are not memory retentive during power failure (fig. 329).
- Internal auxiliary relays (AR) are only used for internal logic signal processing and cannot be used to drive an outside load. The "Omron C20" has 136 internal auxiliary relays. Internal auxiliary relays are not memory retentive during power failure (fig. 329) and may be compared to an air pilot operated, spring reset pneumatic valve. Internal auxiliary relays are an integral part of the CPU and are not affected by the input/output relay number of the PLC (expanded or not expanded).
- Holding relays (HR) are also used for internal logic signal processing, in much the same way as internal auxiliary relays. Holding relays are memory retentive during power failure if they are combined with the Fun. 11 instruction (see fig. 329). Holding relays are basically set–reset type relays or flip-flops and may be compared to a pilot operated hydraulic or pneumatic memory valves (see fig. 329 and chapter 2, fig. 19). The "Omron C20" has 160 holding relays (fig. 360), which are convertible to "Keep relays" (see keep function). Holding relays are an integral part of the CPU and are not affected by the input/output relay number of the PLC (expanded or not expanded).

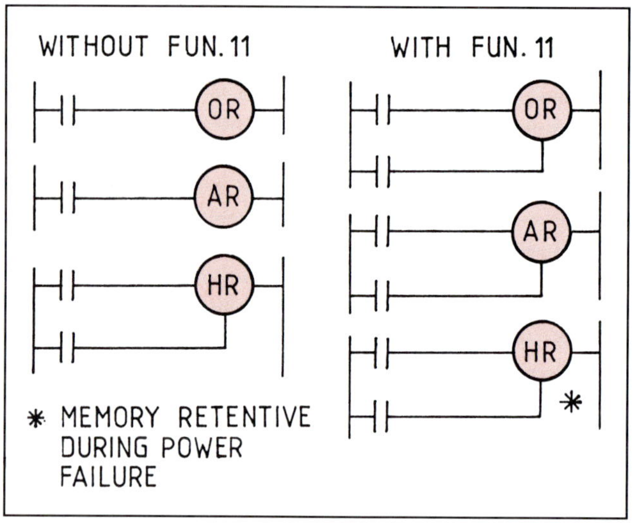

Fig. 329 Relay types and integration of "Fun. 11" for conversion of relays to set/reset type flip-flops.

Latching relay

The latching function is often also called self-holding function. It may be used to create an R-S flip-flop, but one must be aware that such a relay is not memory retentive during power failure. Hence, latching relays (on some PLCs) no longer return to their logic state when the power to the PLC is restored after a power interruption. If the flip-flop (latching relay) had the "set" status prior to power failure it will then be in the "reset" state when power is restored. For this reason most PLCs also possess a number of keep relays which maintain switching state with the aid of a back-up battery (see figs. 329 and 360).

Keep function (Fun. 11)

The "Omron C20" has 190 holding relays (fig. 360). Only "keep relays" have true memory capacity which makes them memory retentive during power failure (see fig. 328). Keep relays must not be compared to latching relays (see latching relays). With "Omron" PLCs the keep function (Fun. 11) may also be assigned to output relays (OR), as well as internal auxiliary relays (AR) to turn them into set–reset relays with a separate set and reset input signal (see fig. 328). However, this does not make the output relay or the internal auxiliary relay memory retentive during power failure.

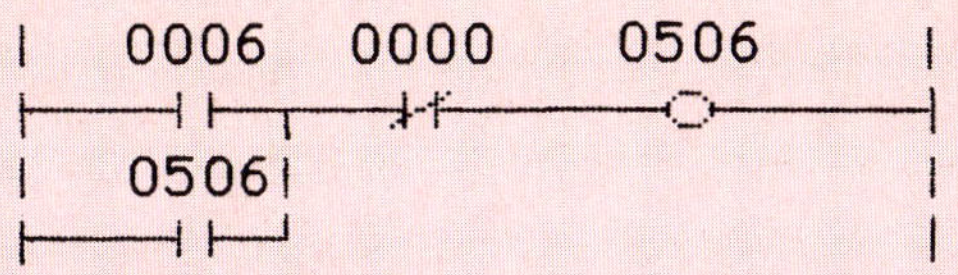

MNEMONIC		OPERAND
LD		0006
OR		0506
AND	NOT	0000
OUT		0506

Fig. 330 Latching relay. Input 0006 is used to "set" the relay 506: Input 0000 is the "reset" signal and input 506 acts as the latch (self-holding).

PLC programming and integration concepts

There are several basic steps involved in designing and writing a satisfactory and economically working PLC program. It is therefore recommended that, prior to programming, the person endeavouring to write such a program should study:

- the concepts of Boolean logic functions with their ladder diagram interpretation (see fig. 338).
- the concept of sequential machine control notation by means of traverse-time diagrams (step-motion diagrams figs. 312 and 317), and last but not least
- the concept of step-counter sequencing as a means of efficient, economical and effective programming of asynchronously operating sequential machine controls (see figs. 313 and 314, as well as fig. 341).

All these concepts are explained in detail in chapter 18 of this book and more extensively in the book *Pneumatic Control for Industrial Automation* by Peter Rohner and Gordon Smith (chapters 7, 8 and 10). Designing a PLC circuit and programming the PLC should ideally be attempted in the following steps:

Step 1

If the circuit to be controlled by the PLC is of a predominantly sequential nature (and most industrial circuits are), then draw a traverse–time diagram for all actuators involved (hydraulic linear actuator, hydraulic motors, electric motors, pneumatic vacuum generators etc.). For a typical traverse–time diagram, including a timer, see figs. 353 and 356.

Step 2

Draw the electro-hydraulic circuit or the electro-pneumatic circuit with all its directional control valves, linear actuators, motors etc. Assign PLC output point numbers to all these hydraulic, pneumatic and electric elements which require a PLC signal for their operation (valve solenoids, prime movers, buzzers, control panel indication lamps, heating elements etc. — see fig. 354 and control problem 2).

Step 3

Now transfer the assigned output point numbers to the sequence step timelines of the traverse–time diagram developed before (fig. 353).

Step 4

Establish and assign input points and their corresponding PLC input point numbers for all input elements (limit switches, push button switches, heat sensors, pressure switches, flowrate sensors and proximity sensors etc.) which contribute to the essential sequencing and functioning of the hydraulically or pneumatically powered machine (see fig. 354).

Step 5

Transfer all these input numbers, which directly affect the sequencing of the machine, into the traverse–time diagram established in step 1. Some input signals such as program selection, machine interrupt, automatic or manual cycling do not directly affect the normal sequencing. These signals are required in step 8 (see fig. 353).

Step 6

Design and draw a ladder diagram (or system flow chart, if this is preferred), first for the sequential circuit only, and then also for all output commands, leading to the fluid power and electrical devices on the machine (see fig. 355).

Step 7

Design and draw a ladder diagram for all fringe condition modules, if such modules are required (machine interrupt module, cycle selection module, stepping module, program selector module etc.). Place the ladder diagram developed in step 7 in front of the sequential circuit diagram of step 6. This makes the circuit orderly and grouped into segments for ease of troubleshooting (see figs. 355 and 357).

Step 8

Design and draw a ladder diagram for all other machine control functions which are not typically of a sequential or modular nature. Such machine control functions are usually of pure or semi-combinational nature and may require a knowledge of Boolean algebra, truth tables and Karnaugh Veitch maps, and they often require extensive circuit design experience. This is particularly true if special interrupt sequences, sequence jumps and alternative sequence programs are required to control the automated machine (see chapters 8–10 in *Pneumatic Control for Industrial Automation* by Peter Rohner and Gordon Smith).

Step 9

Input the program established in steps 6–8 into the CPU. When using a programming console, this will involve converting the ladder diagram to mnemonic form. Alternatively, if a PC with the appropriate software is used, the ladder diagram is established via the keyboard and the monitor screen, and then transferred by the host link into the PLC.

Step 10

Check the program for syntax errors and if the PLC gives an error message, correct such errors. To check for syntax errors on an "Omron C20", one may press the keys: CLR, CLR, FUN, MONTR in any of the three selectable modes (RUN, MONITOR, PROGRAM). For further debugging checks and error messages, it is best to consult the operation manual of the PLC in use.

Step 11

Execute the program with simulation input switches to check for basic programming and execution errors and correct these immediately. These errors do not fall into the syntax error category. Such basic programming errors might creep in relatively easily. For example, an actuator is meant to extend and reach a particular limit switch, but instead of extending, the actuator retracts and hence does not reach the limit switch the sequential logic is looking for. The step-counter therefore stops at this point until the programming error is corrected.

Step 12

Once the PLC is programmed, the syntax checked and execution errors eliminated, one may run the PLC in conjunction with the fluid power controlled machine. Any time-dependent functions can now be fine tuned, if such fine tuning is required (see changing set value of timer or counter in operation manual of PLC).

Step 13

Once the machine runs with satisfactory performance, it is wise to make a final back-up copy of the program. This copy contains all logic, timing and counting data for future reference. The program may now be stored on disk, tape or paper, or one may wish to transfer it onto an EEPROM for safe-keeping.

Step 14

Finally, it is necessary to document all input and output numbers in an input/output assignment list. It is also advisable to document these keep relays, which need to be "set" for program start-up.

Logic functions for PLC

The five most important basic logic functions are shown in fig. 331. From these five basic functions one can construct further logic compound functions and even a latching type R–S flip-flop. The left-hand column of fig. 331 shows the name of the logic function, the truth table is shown in column 2, the internationally standardised logic symbol in column 3, and the equivalent fluid power symbol in column 4. Column 5 shows the equivalent ladder diagram symbol circuit, as used in PLC ladder diagram programming; and in column 6 is the Boolean switching equation for that logic function. The letter A stands for signal input 1, the letter B for signal input 2 and the letter S is the output signal of the logic function. Since various PLC brands use differing input and output signal numbering concepts, names for input and output signals have been generalised in fig. 331.

A dash above the signal label means the inverse of the signal. The inverse of a signal means the signal is "not on", or the signal is "not present". The logic Boolean expression for connecting inputs by the "and" function is denoted by a dot sign (•), the logic Boolean expression for connecting inputs by "or" function is denoted by a plus sign (+). These two unique signs may also be used to logically connect switching expressions enclosed by brackets. For example:

(0006 • 0003) + (1000 • 502)
for "and" connecting within brackets or

(0011 + 1003) • (507 + T 03)
for "or" connecting within brackets.

Switching equations may be used directly to program a PLC controller by translating them into mnemonic codes. Alternatively, one may start with a ladder diagram and then enter instructions into the PLC by translating the ladder diagram graphics into mnemonic codes. Experienced programmers however, program the PLC directly from ladder, by making the translation a mental process. Most PLCs offer the facility to translate the program with a so-called printer interface module and transfer the PLC program directly into the buffer of an attached printer. The printer then prints the ladder diagram, the mnemonic list and, if required, also a cross-reference list.

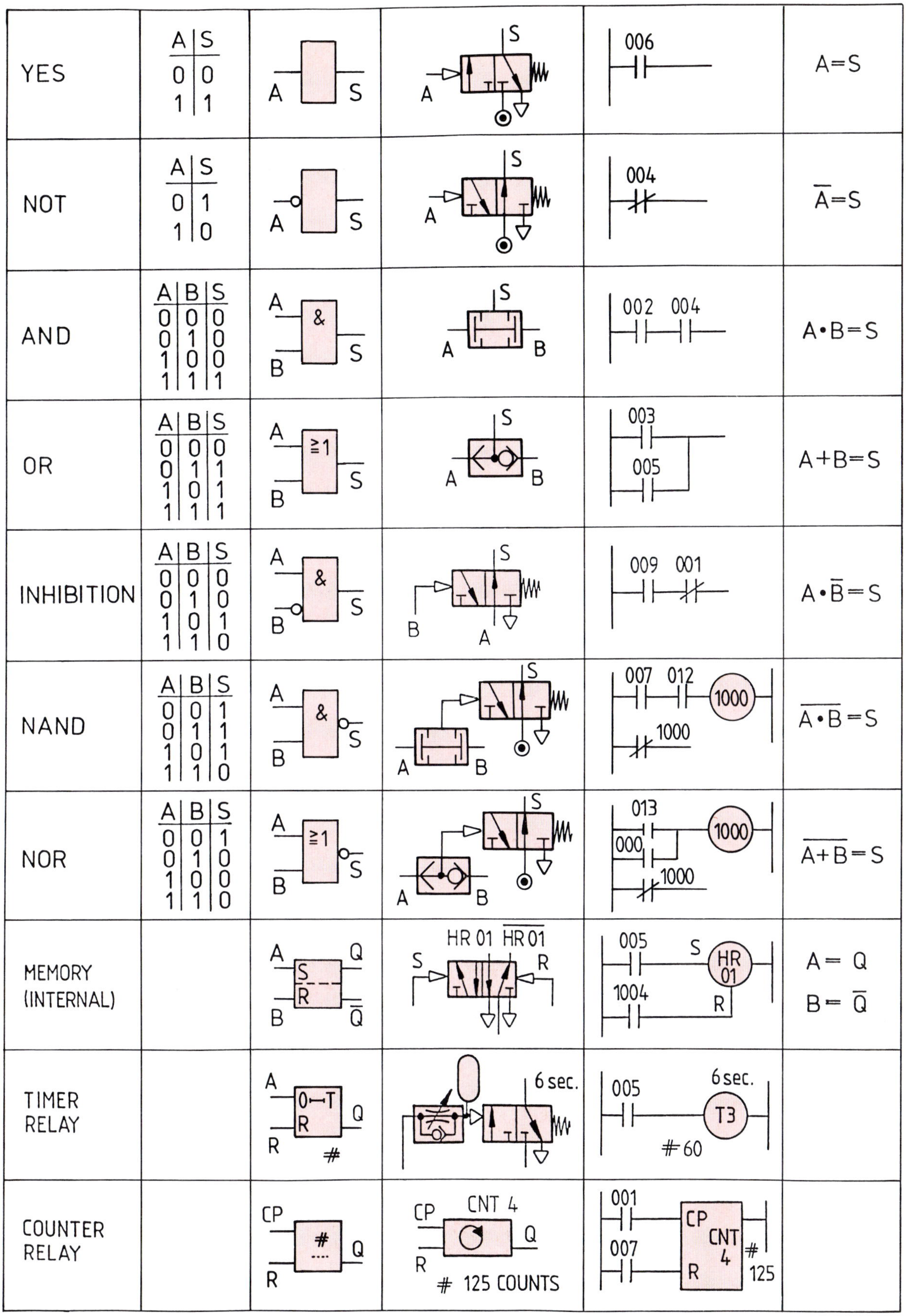

Fig. 331 Basic logic functions for PLC.

Logic function description

"Yes" function

With the "yes" function (one of the five basic logic functions, see fig. 332), the central processing unit (CPU) of the PLC interrogates a designated signal input point for the presence of its signal. For example, if input 0006 is "on", then output 504 is turned "on".

Here, for the purpose of illustration, output relay 504 is used to show the presence of input 0006. The "yes" function itself often requires no immediate output, as it may also be embedded within an "or" logic connection or an "and" logic connection or may contribute to an "inhibition" function (fig. 338). Most PLCs have LED input and output monitoring lights, which may be used for output monitoring of logic functions.

Conclusion

Output relay 504 is "on" when input point 0006 receives a signal from its connected sensor or switch.

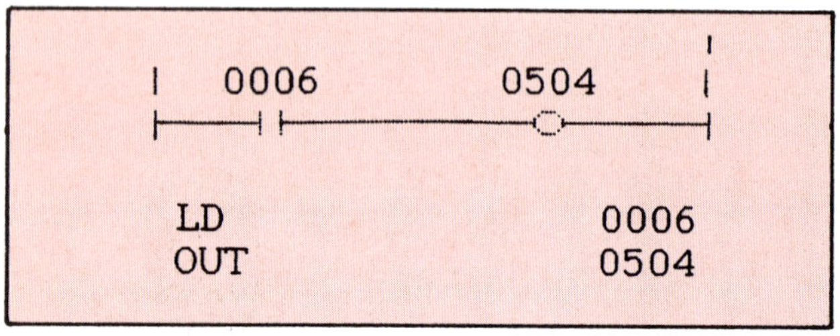

Fig. 332 Ladder diagram and mnemonic list for the "yes" function.

"Not" function

With the "not" function, the central processing unit of the PLC interrogates a designated signal input point for the "absence" of its signal. For example, if input 0013 is not "on", then output 511 is turned "on".

Here, for the purpose of illustration, output relay 511 is used to show the absence of input 0013. The "not" function itself often needs no immediate output, as it may also be embedded within an "or" logic connection or an "and" logic connection or may contribute to an "inhibition" function.

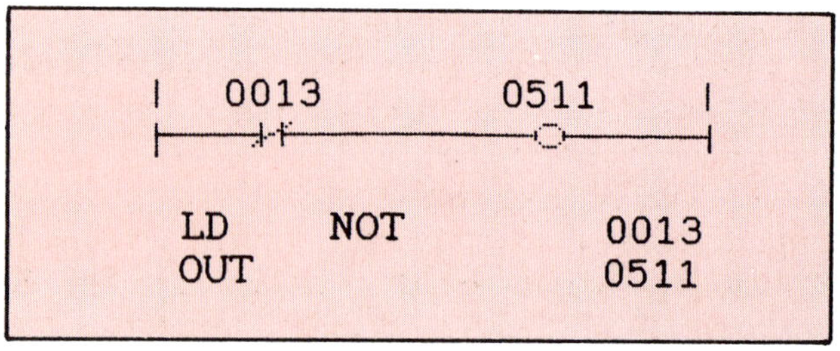

Fig. 333 Ladder diagram and mnemonic list for a "not" function.

Conclusion

Output relay 511 is "on" when input point 0013 receives no signal from its connected sensor or switch.

"And" function

With the "and" function, the central processing unit interrogates two or more designated input points for the simultaneous presence of their signals. For example, if input 0007 is "on" and also input 0015 is "on" at the same time, then output 503 is turned "on". The order in which the two inputs are presented in the ladder diagram and the mnemonic list is of no importance, as the PLC in reality does not connect them in series, but only by means of "and" gates.

Here again, for the purpose of logic function illustration only, output relay 503 is used to display the achievement of the "and" function (or simultaneous presence of signals 0007 and 0015). The "and" function often requires no direct or immediate output as it may also be connected to and followed by other logic functions, thus terminating in an output further along the logic line (see logic line).

Conclusion

Output relay 503 is "on" when both inputs are "on" within one scan (0007 and 1115) and therefore receive a signal from their connected switches or sensors.

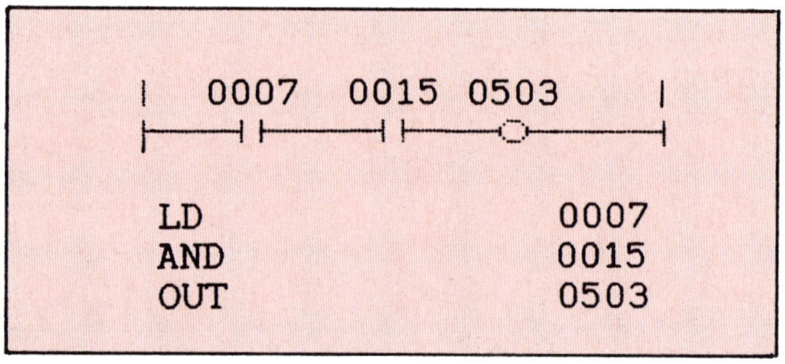

Fig. 334 Ladder diagram and mnemonic list for the "and" function. Although the "and" function in the electronic ladder diagram is shown as a series function, inside the PLC these input signals are "and" connected by electronic "and" gates and not in series order.

"Or" function

With the "or" function (inclusive "or" function), the CPU interrogates two or more designated input points logically connected in parallel for the presence of their signals. For example, if input 0000 is "on" or input 0008 is "on", then output 501 is turned "on".

The order in which the two inputs are listed in the ladder diagram and the mnemonic list is of no importance. Output 501 is "on" regardless of whether only one or both inputs are "on" within the same scan. This "or" function is therefore often also called an inclusive "or" function. The "or" function itself often requires no direct or immediate output as it may also be connected to and followed by another logic function, thus terminating in an output further along the logic line (see logic line).

Conclusion

Output relay 501 is "on" when input point 0000 or alternatively input 0008 or alternatively both inputs receive a signal from their connected limit switches or sensors.

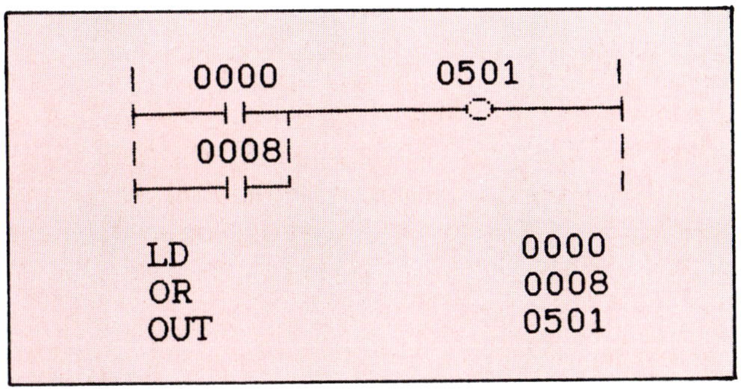

Fig. 335 Ladder diagram and mnemonic list for an inclusive "or" function.

"Inhibition" function

With the "inhibition" function the CPU interrogates two designated input points for their simultaneous logic state. One of them for its "presence", the other for its "absence". For example, if input 0004 is "on" and input 0005 is not "on", then output 507 is turned "on".

Conclusion

Output relay 507 is "on" when a signal is present on input 0004 and no signal is present on input 0005. One may also say output relay 507 is "on" when input 0004 is present and input 0005 is absent.

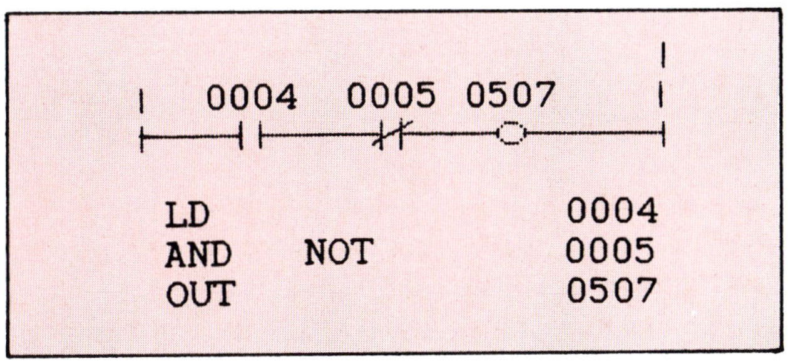

Fig. 336 Ladder diagram and mnemonic list for the "inhibition" function.

The order in which the two signals are listed in the ladder diagram and in the mnemonic list is of no importance. The "inhibition" function often requires no direct or immediate output as it may also be connected to and followed by other logic functions further along the logic line (see logic line). The "not" function, as an integral part of the "inhibition" function, may also be applied to inhibit preceding logic functions, such as "or" or "and" (see examples, fig. 337).

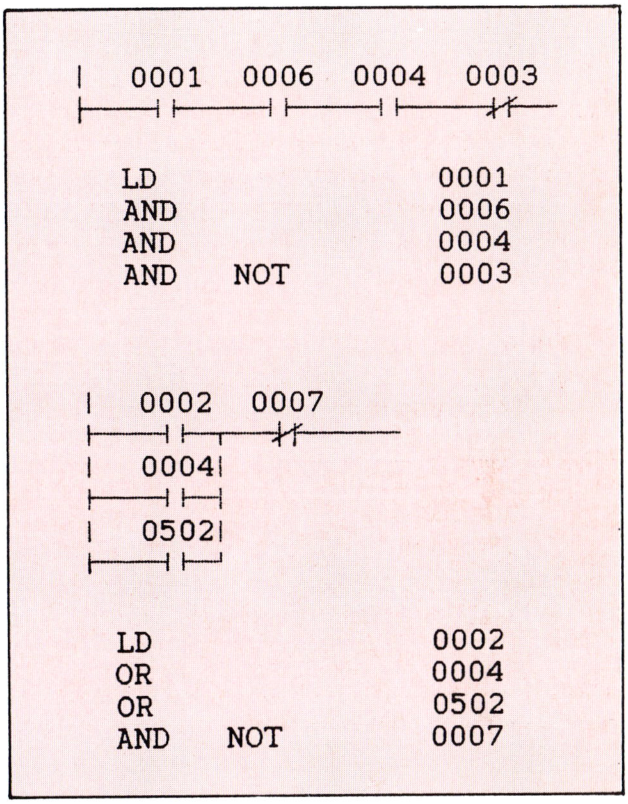

Fig. 337 Ladder diagram and mnemonic list for logic functions preceding the "not" function as part of a compound "inhibition" function.

Logic line (network)

Ladder diagrams consist of one or a number of individual logic lines (sometimes also called networks or function blocks). Each logic line starts from the left-hand bus bar of the ladder diagram and ultimately terminates with a relay coil, a timer coil, a counter coil or a special instruction on the right-hand bus bar. It is not absolutely essential to draw the right-hand bus bar in a ladder diagram; it is, however, customary to do so.

The number of contacts in "and" function arrangements or in "or" function arrangements is not limited for use within a logic line.

All contacts used within a logic line may be normally open or normally closed, or a combination of both (e.g. with an inhibition function; see fig. 338 and fig. 336). For Omron PLCs the beginning of a logic line is always denoted by the load instruction (LD). This may differ for other PLC brands.

Examples

When a logic line starts with, say, input 0015 as a normally open contact, the programming instruction in the mnemonic list will be LD 0015. If the logic line starts with a normally closed contact, say 0007, the instruction in the mnemonic list would then be LD NOT 0007. The load instruction (LD) is also used to start a new block within a logic line (see fig. 338). Illustration 338 shows six typical logic lines as they may frequently be found in ladder diagrams for PLC.

- Logic line A in fig. 338 drives output relay coil 500 if its logic is one (on).

$$(\overline{0007} \bullet 0001) + (0004) \bullet \overline{0008} = 500$$

- Logic line B is a keep relay (memory retentive relay).

$$0007 \bullet 0005 \bullet \overline{0009} = \text{SET HR001}$$
$$0003 = \text{RESET HR001}$$

- Logic line C drives output relay coil 509, if its logic, consisting of a complex "or" as well as an "and" function, including an inhibition function, is one (on).

$$(0001 \bullet 0003) \bullet [(\overline{1000} \bullet 0004) + (502 \bullet 0006)] = 509$$

- Logic line D illustrates how a complex logic including a timer contact drives several output coils.
- Logic line E illustrates how an extremely complex logic, consisting of an "or" block (B1) inhibited by the presence of HR006 and then "and" connected to two more "or" blocks (B2 and B3), ultimately

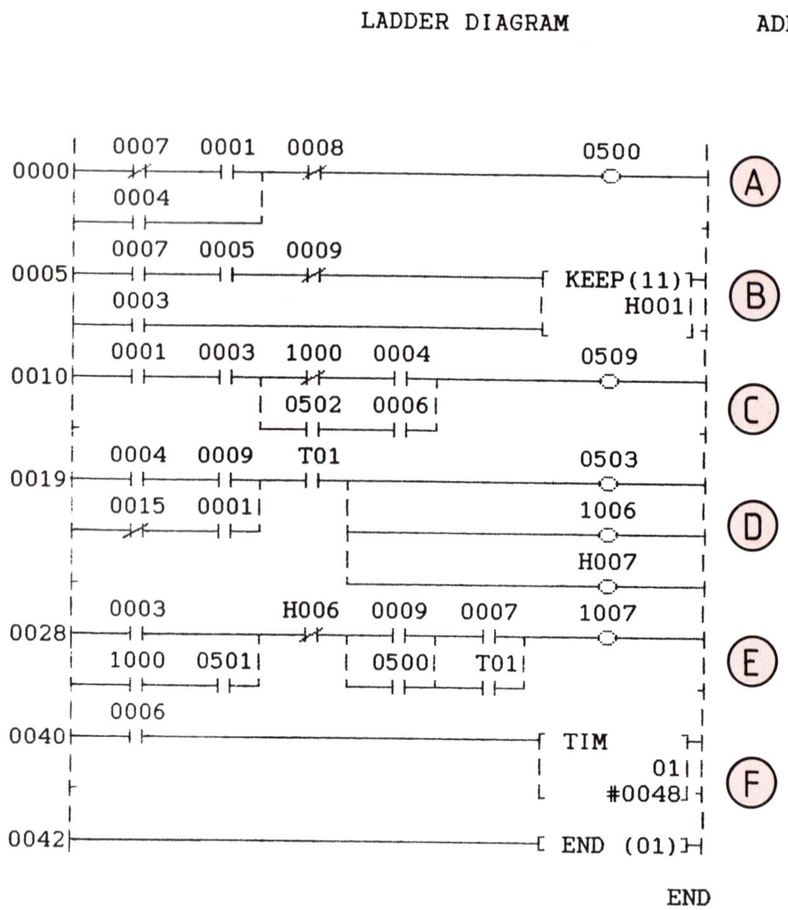

ADDRESS	MNEMONIC			OPERAND
0000	LD	NOT		0007
0001	AND			0001
0002	OR			0004
0003	AND	NOT		0008
0004	OUT			0500
0005	LD			0007
0006	AND			0005
0007	AND	NOT		0009
0008	LD			0003
0009	KEEP	(11)	HR	001
0010	LD			0001
0011	AND			0003
0012	LD	NOT		1000
0013	AND			0004
0014	LD			0502
0015	AND			0006
0016	OR	LD		
0017	AND	LD		
0018	OUT			0509
0019	LD			0004
0020	AND			0009
0021	LD	NOT		0015
0022	AND			0001
0023	OR	LD		
0024	AND		TIM	01
0025	OUT			0503
0026	OUT			1006
0027	OUT		HR	007
0028	LD			0003
0029	LD			1000
0030	AND			0501
0031	OR	LD		
0032	AND	NOT	HR	006
0033	LD			0009
0034	OR			0500
0035	AND	LD		
0036	LD			0007
0037	OR		TIM	01
0038	AND	LD		
0039	OUT			1007
0040	LD			0006
0041	TIM			01
			#	0048

Fig. 338 Six typical logic lines, as used in ladder diagrams for PLC programming.

drives auxiliary relay coil 1007, if its complex logic is one (on). It must be noted that inhibiting contact HR006 could also be placed in front of block 1 or behind block 3, or between blocks 2 and 3, giving the same logic effect.

- Logic line F shows a timer coil (T01) with its driving contact 0006. The timer starts to time out when contact 0006 is logic one (on). See also fig. 341.

Allocation of relay numbers for sequential controls

For programming convenience, labelling of output relay points to hydraulic solenoids and input points for signals from fluid power actuator limit switches is based on the same logic principle as explained in chapter 18, under the heading "Traverse-time diagram".

Actuator extension commands are labelled with "even" relay numbers: 502, 504, 506, 508 etc. Retraction commands are labelled with "odd" numbers: 501, 503, 505, 507 etc. (see figs. 339 and 354).

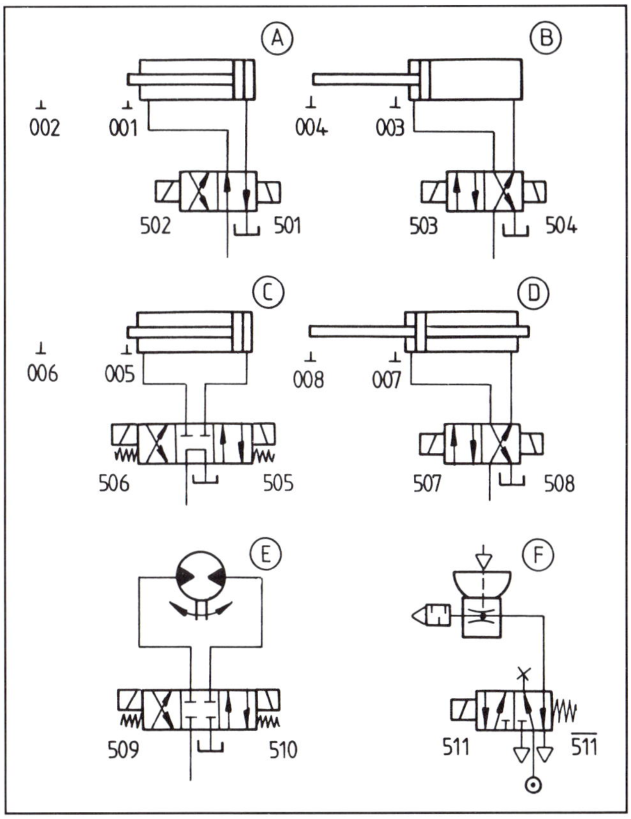

Fig. 339 Allocation of relay numbers for solenoid output signals and limit switch input signals when integrating the hydraulic control system with a PLC.

Limit switches placed in the extended position on the hydraulic actuators are given an "even" number: 002, 004, 006, 008 etc. and limit switches placed in the retracted position of the hydraulic actuator are given an "odd" number: 001, 003, 005, 007 etc. (see figs. 339 and 354). Hence, if actuator B for example must extend, the PLC actuates output relay 504. Solenoid 504, which is connected to PLC output point 504 causes the actuator to extend, and when actuator B is fully extended, limit switch 004 signals to the PLC that the initiated action is completed. Thus, a 504 solenoid command causes limit switch 004 to be actuated as a feedback signal (see application of this principle in figs. 341, 353, 354, 356 and 357).

Step-counter programming for a PLC

The step-counter circuit design method, as explained and illustrated in chapter 18, may also be used to program a PLC (see figs. 313 and 321). Each step-counter module is "built" from a keep relay (HR) and a pre-switched "and" function. The two signals required to make the "and" function are the essential confirmation signal, confirming the completion of the previous sequence motion, and the preparation signal given by the keep relay of the previous step-counter module (or sequence step).

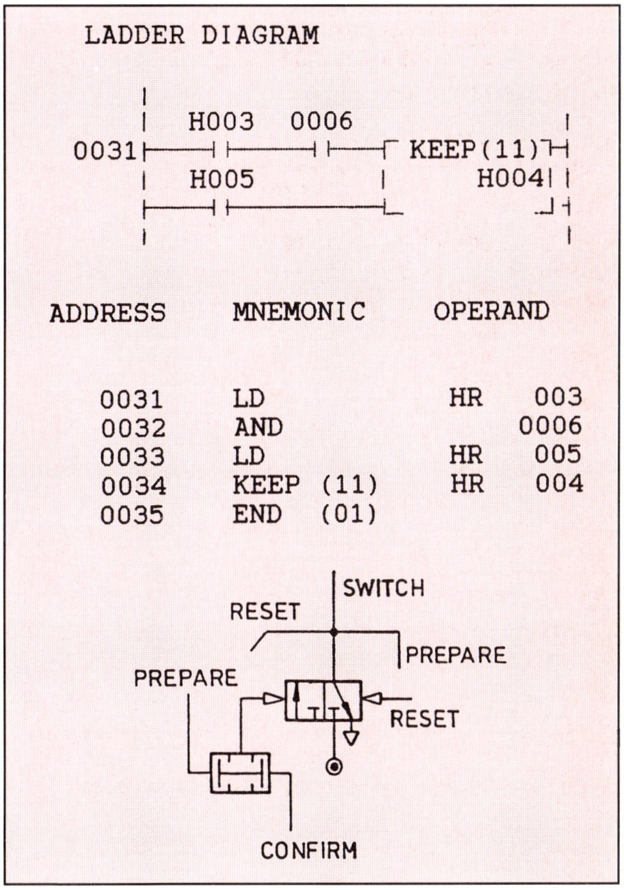

Fig. 340 Step-counter module comparison for pneumatic circuit diagram, PLC ladder logic diagram and PLC mnemonic list. The designation HR stands for holding relay (memory retentive keep relay). For the purpose of illustration, programming starts at random on address 31.

The essential confirmation signal is, of course, only a contact which is driven by the input relay (IR). The input relay is actuated by the limit switch on the machine (see figs. 340–342). The preparation signal is also only a contact which is driven by the keep relay (HR) of the previous step-counter module (compare fig. 314 with fig. 341).

The PLC integration of the individual step-counter modules into the step-counter chain follows the same pattern as established in pneumatic circuit design. The first rung in the ladder diagram includes the start switch or a contact from a cycle selection module (figs. 313, 314 and 341).

Control problem 1

To demonstrate the correlation and integration of the individual step-counter modules, a ladder diagram is now established for a clamp and drill machine with the sequence A1–B1–B0–A0. The pneumatic or hydraulic power circuit is given and all necessary limit switches and solenoid signal commands from the output relays leading to the solenoids are allocated their appropriate relay numbers (see fig. 341).

It must be noted that the keep relay (HR) in a PLC step-counter program performs the same functions as the memory valve in a pneumatic step-counter. These functions are:

- to switch the output relay. The output relay then drives the solenoids on the power valves (fig. 341).
- to reset the keep relay (HR) of the previous electronic step-counter module (figs. 313 and 341).
- to prepare the "and" function which switches the keep relay of the next electronic step-counter module (compare fig. 314 with fig. 341).

At this stage it may be appropriate to mention that immediately after the last sequential address (address 024 in the given exercise) an "end" instruction must also be programmed into the PLC program. For the "Omron C20" this instruction is function 01 (see fig. 341). An error message will usually be displayed on the PLC if this end instruction has been omitted. The CPU of the programmer scans the program data from address 0000 to the address with the end instruction in the sequence, but does not scan unused memory capacity.

Testing the PLC program with the "dry-run" monitoring method

Once the program is entered into the central processing unit of the PLC, it is highly recommended to check the entered program not only for syntax errors, but also for program design errors and program entering errors. Such a test may save later machine problems or, in the worst case, damage to the machine and harm to operators. A "dry-run" test is given below for control problem 1 in a step-by-step explanation.

Step 1

Set the last holding relay in the step-counter chain in the "on" state. For this, use the "force set" facility of the PLC controller in use. For an "Omron C20" PLC, forcing of outputs (output contacts) is accomplished in the "monitor" mode. To force the lasting holding relay of the PLC step-counter (HR004), the following keys on the programming console need to be pressed in the following order: CLR, CLR, SHIFT, CONT, HR, 4, MONTR, SET.

For the given example of control problem 1, HR004 needs to be set in the same way as one would set the last memory valve in a pneumatic step-counter. For control problem 1 see fig. 341. *Note:* HR004 is the last step-counter holding relay in the ladder diagram step-counter chain (compare with fig. 314).

Step 2

Actuating (forcing) HR004 causes output 501 to turn "on". This is displayed with the output 501 LED (light emitting diode) on the diagnostics panel of the PLC. Now briefly actuate in "and" connection the two switches circled in the traverse-time diagram of control problem 1 (inputs 000 and 001, under timeline 1). This causes output 501 to disappear and output 502 will light up. Output 502 will cause clamping action on the machine (actuator A extension).

Step 3

Now briefly actuate input 002 (timeline 2, step 2 in traverse-time diagram). This will cause output 504 to light up. Output 504 is the solenoid signal for sequence step 2 causing drill extension (actuator B extension).

Step 4

Now briefly actuate input 004, as circled in step 3 of the traverse-time diagram. This causes output 503 to light up (drill retraction, actuator B retraction).

Step 5

Now briefly actuate input 003, as circled in step 4 of the traverse-time diagram. This causes output 501 to light up (unclamping action on the machine, actuator A retraction).

The "dry-run" test is now complete and HR004 should remain "on", whereas all other holding relays from HR001 to HR003 will be "off". Should the entered control program in the PLC have any hidden programming errors, the "dry-run" test would either not perform as shown above, or the step-counter could not proceed beyond certain steps. In that case, the control circuit would need corrections.

TRAVERSE-TIME DIAGRAM

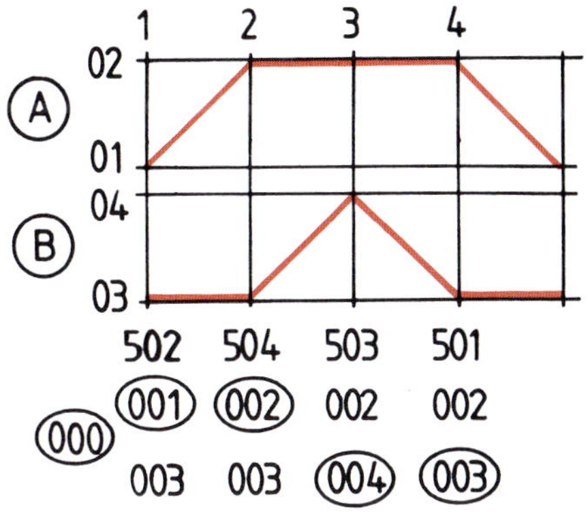

ELECTRO-HYDRAULIC CIRCUIT

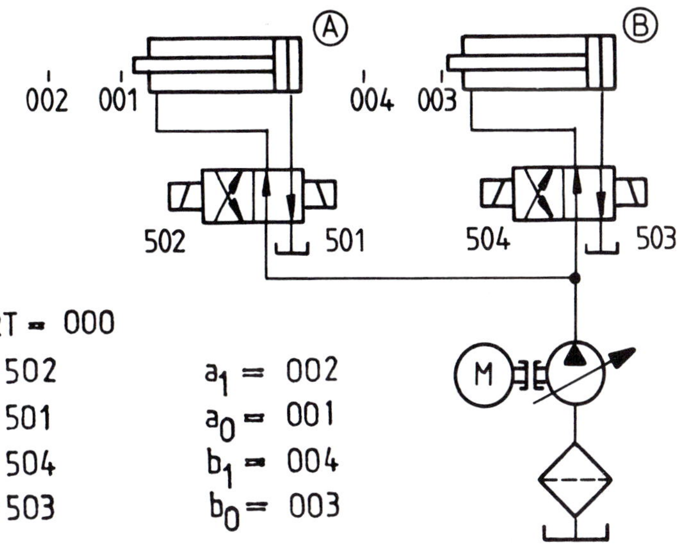

START = 000
A1 = 502
A0 = 501
B1 = 504
B0 = 503

a_1 = 002
a_0 = 001
b_1 = 004
b_0 = 003

LADDER DIAGRAM

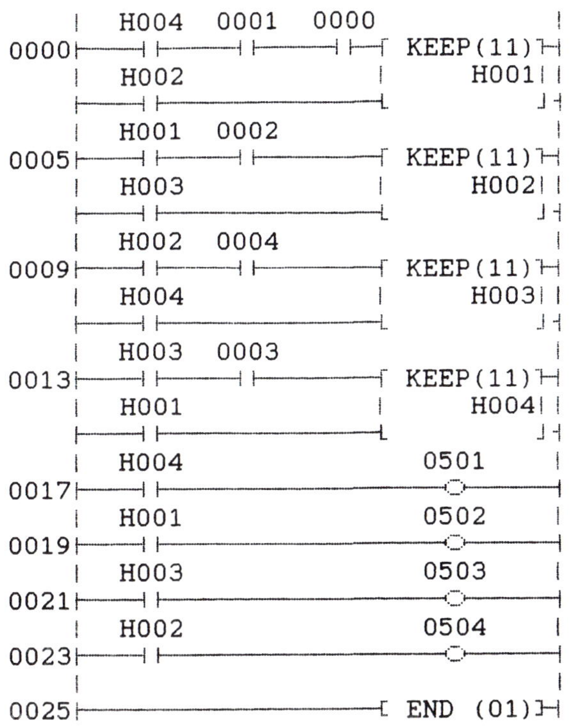

MNEMONIC LIST

ADDRESS	MNEMONIC	OPERAND
0000	LD	HR 004
0001	AND	0001
0002	AND	0000
0003	LD	HR 002
0004	KEEP (11)	HR 001
0005	LD	HR 001
0006	AND	0002
0007	LD	HR 003
0008	KEEP (11)	HR 002
0009	LD	HR 002
0010	AND	0004
0011	LD	HR 004
0012	KEEP (11)	HR 003
0013	LD	HR 003
0014	AND	0003
0015	LD	HR 001
0016	KEEP (11)	HR 004
0017	LD	HR 004
0018	OUT	0501
0019	LD	HR 001
0020	OUT	0502
0021	LD	HR 003
0022	OUT	0503
0023	LD	HR 002
0024	OUT	0504
0025	END (01)	

Fig. 341 Traverse-time diagram, electro-hydraulic circuit, ladder diagram and mnemonic list for clamp and drill control sequence (control problem 1).

Timer functions

With timer functions one may again follow the structure given for internal auxiliary relays. A timer may be programmed for a value ranging between 0 and 999.9 seconds. Timer number allocations for the "Omron C20" are from 00 to 47. Timers and counters must not be allocated the same number (fig. 360).

The timer works by clock pulses in 0.1 second decrements and produces an output signal when the set value time has elapsed to 0000 s. The timer starts when its timer "on" signal is logic 1 and resets when its "on" signal is cancelled. The timer output is transmitted externally through an output relay as shown in fig. 342 or it may be used as a contact within a logic line (fig. 338 E). Such a timer contact may be normally open or normally closed and its contact may be used as often as required.

The timer and its activating signal ("on" signal) must be separately programmed in the ladder diagram and its contact (or contacts if used several times) appear in the ladder diagram wherever required (see figs. 338 and 351).

The timer is in reality a counter in which an internal clock pulses the timer from its "set value" to count down to zero. This internal clock works on a pulse rate of 0.1 per second. Thus a 10.6 second delay is programmed as # 106.

Fringe condition modules

Machine routines (better known as fringe condition modules), are extensively explained and illustrated in figs. 343 to 350. Since such modules are usually attached on the fringe of a sequential control circuit, they are often also called fringe conditions or fringe condition modules. Fig. 358 shows a pneumatic circuit, and fig. 359 the equivalent ladder diagram plus a corresponding mnemonic list, where all these commonly used fringe condition modules are integrated with each other and merged to a pneumatic step-counter (for PLC step-counter integration of such modules see figs. 355 and 357).

Simple cycle selection module

This module is used to furnish a sequentially controlled machine with a permanent start signal when "auto" cycling is selected. When a "manual" cycling start is selected, the operator is required to actuate the start valve each time a cycle is to begin (see

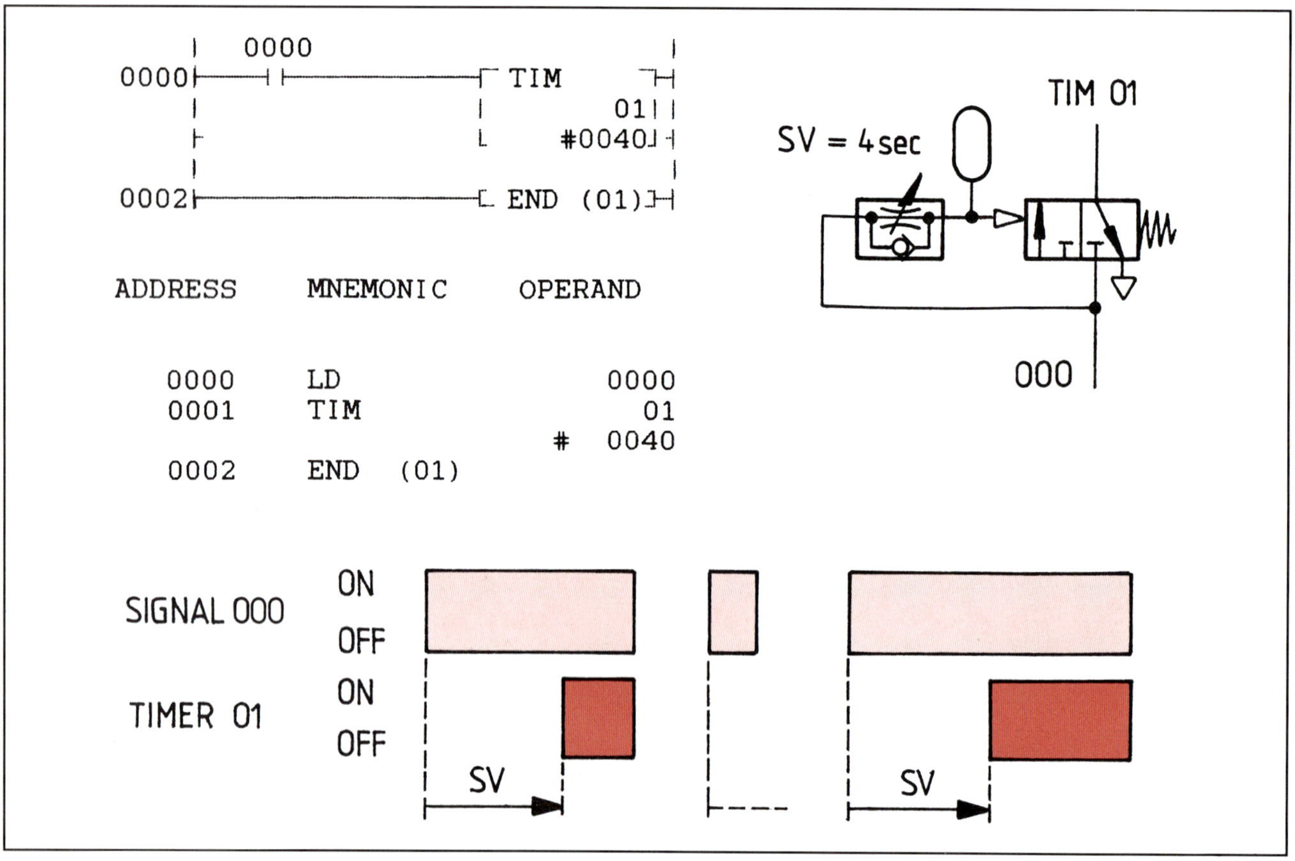

Fig. 342 On-delay timer. The timer requires an output relay when used externally to drive a load. Timer set value does not require an address. Timer values are set in increments of 0.1 seconds (pulse rate of the internal clock). Thus a 2 second time delay is programmed as value 20 (# 20).

figs. 343 and 344). An application circuit which includes a time-delay function and a complex cycle selection module is given in fig. 351.

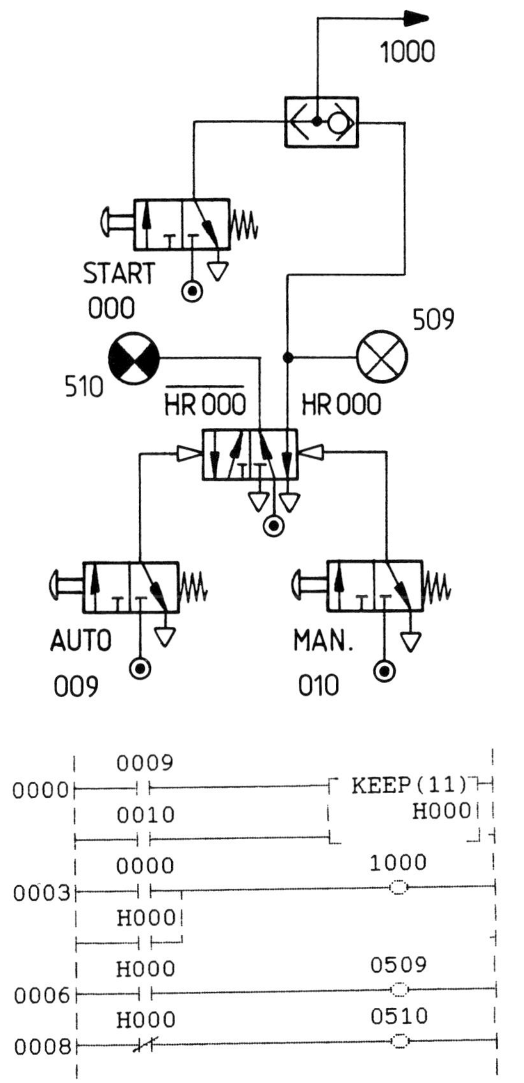

ADDRESS	MNEMONIC		OPERAND	
0000	LD			0009
0001	LD			0010
0002	KEEP	(11)	HR	000
0003	LD			0000
0004	OR		HR	000
0005	OUT			1000
0006	LD		HR	000
0007	OUT			0509
0008	LD	NOT	HR	000
0009	OUT			0510

Fig. 343 Simple cycle selection module to provide either automatic or manual cycling. To achieve a true memory function, a holding relay must be used. The output relay (OR) is driven either by the contact of the holding relay (HR) or by the start switch.

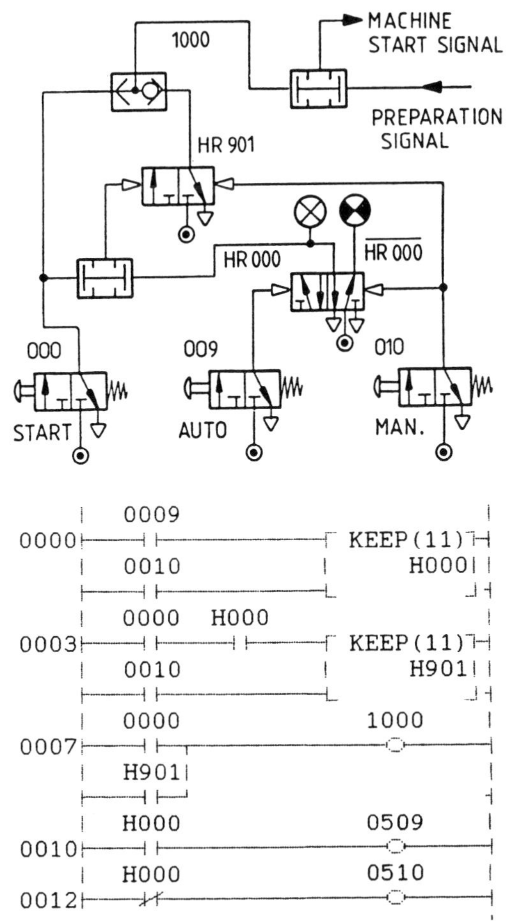

ADDRESS	MNEMONIC		OPERAND	
0000	LD			0009
0001	LD			0010
0002	KEEP	(11)	HR	000
0003	LD			0000
0004	AND		HR	000
0005	LD			0010
0006	KEEP	(11)	HR	901
0007	LD			0000
0008	OR		HR	901
0009	OUT			1000
0010	LD		HR	000
0011	OUT			0509
0012	LD	NOT	HR	000
0013	OUT			0510

Fig. 344 Extended cycle selection module. This module provides separate selection and start for both cycling modes. For appli-cation of this module see fig. 351.

Extended cycle selection module

The extended cycle selection module is an extension to the simple cycle selection module (fig. 344). It eliminates the disadvantage of the simple cycle selection module by providing separate selection as well as separate start for both cycling modes. The additional costs, when pneumatically constructed, are: one "memory" valve and one "and" valve. When programmed on a PLC controller, no extra costs arise, since the extra components are within the CPU of the PLC memory.

Emergency stop for PLC step-counter

The majority of automated and PLC controlled machines operate according to a programmed sequence. Due to mishaps such as a damaged tooling, a misaligned workpiece in an assembly line, an

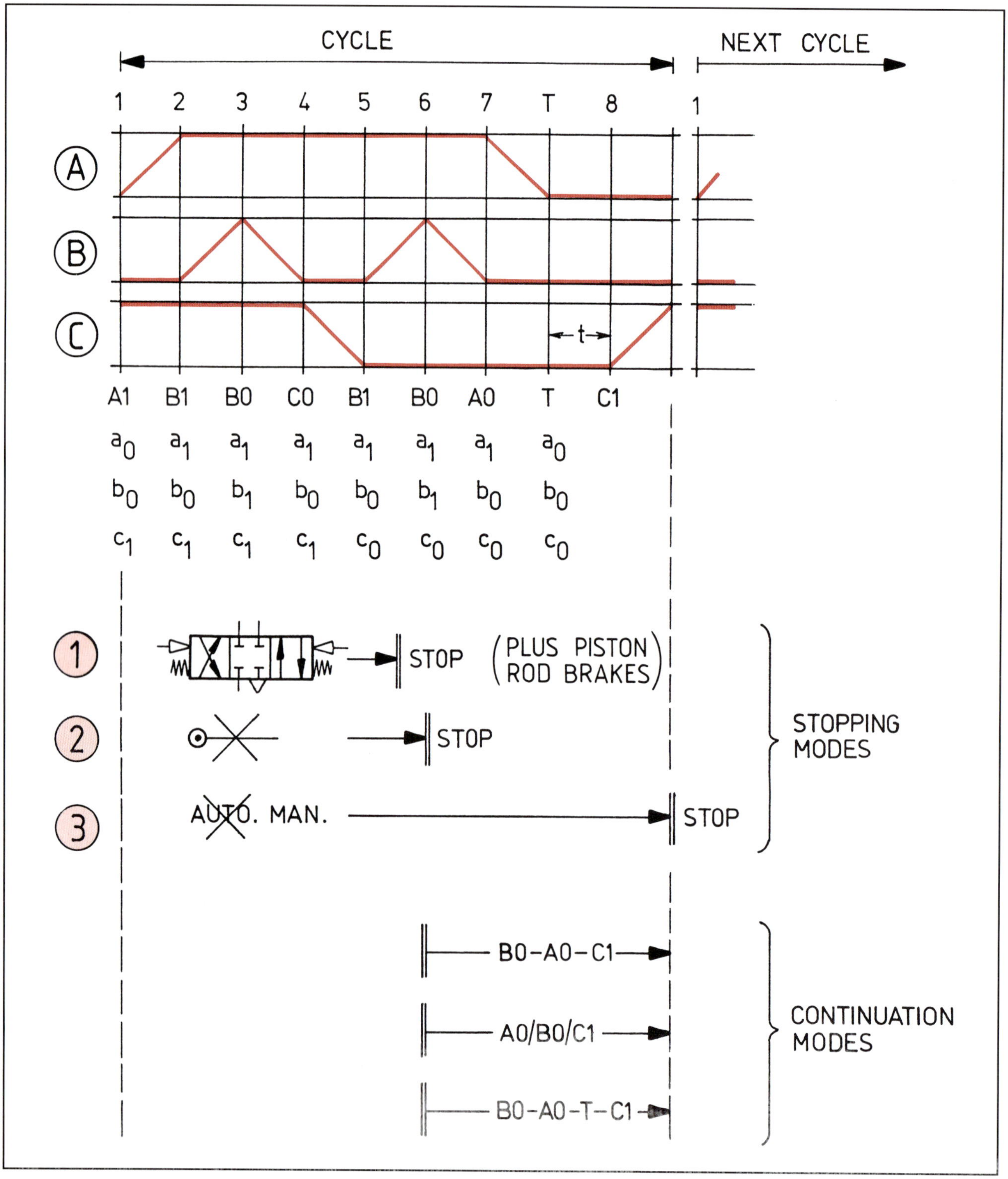

Fig. 345 Emergency stop modes (machine sequence interrupt concepts).

empty workpiece magazine, a sudden drop in supply pressure or an endangered operator, the sequence must be interrupted or the start of a new cycle prevented if automatic cycling has been selected.

The nature of such an interruption varies with the anticipated degree of danger and the inherent safety equipment already built into the machine (safety guard, remote control intermittent operation, various unskilled operators, etc.). The choice is basically from the following three emergency stopping modes:

1. Stop instantaneously (use piston rod brakes if necessary). Piston rod brakes are available for pneumatic as well as hydraulic linear actuators (see fig. 345).
2. Stop at the end of the sequence step (that step which is in action at the time when the emergency is signalled; fig. 345).
3. Stop at the end of the sequence cycle (if the machine cycles automatically and does not need a start command for each new cycle).

Once the machine cycle is interrupted and actuators have stopped, one has several options to choose from for further sequence or machine action. It is, therefore, the circuit designer's responsibility to decide what this action is to be. The choice is crucial, since it can prevent serious harm to the machine operator or avoid costly damage to workpieces, tooling and machinery.

To illustrate: in a machine tool, one has to stop the drilling head and retract it completely before the workpiece may be unclamped and the automatic workpiece ejectors finally permitted to clear the machine. It would be disastrous if all these movements were to happen simultaneously. However, in another machine, all actuators may quite safely act in unison and move into the position required to start a new cycle once the emergency situation is cleared. In another machine application the machine cycle may be brought to a halt with either stopping mode 1 or 2, as depicted in fig. 345, and when the emergency situation is cleared the sequence could be permitted to continue to the end of the cycle. Such machine applications are quite common. They may be regarded as classical "continuation modes" for machine sequence behaviour after emergency stop has been signalled. For more extensive machine interrupt and emergency stopping information see chapter 8 of *Pneumatic Control for Industrial Automation* by Peter Rohner and Gordon Smith.

Basic sequencing interrupt module

For maximum operator and machine safety, all emergency or machine interrupt signals should be stored in a memory. In PLC control the memory function is performed by a holding relay (fig. 346). It is recommended that the reset switch is placed inside the control cabinet, or if possible that a key operated reset switch be used, which is only accessible to authorised personnel.

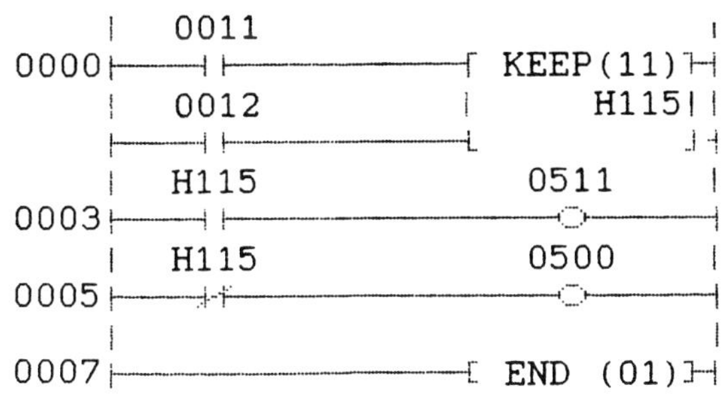

ADDRESS	MNEMONIC		OPERAND	
0000	LD			0011
0001	LD			0012
0002	KEEP (11)		HR	115
0003	LD		HR	115
0004	OUT			0511
0005	LD	NOT	HR	115
0006	OUT			0500
0007	END (01)			

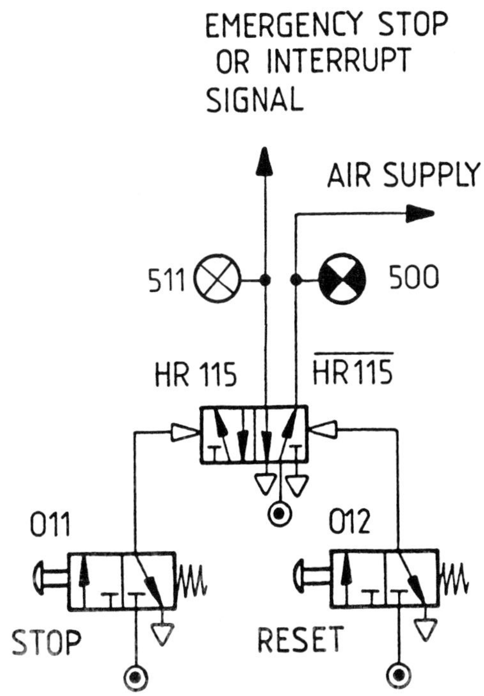

Fig. 346 Basic sequencing interrupt module (emergency stop module).

To make the machine sequence interrupt functional at all times, its memory (holding relay) "set" signal must never be "and" connected to any other signals.

Complex sequencing interrupt module

With the basic sequencing interrupt module, the machine sequence is permitted to continue as soon as the reset switch is actuated. For some machines, this feature may prove to be confusing and even dangerous. To eliminate this, the complex sequencing interrupt module has an additional memory (Keep flip-flop), which inhibits machine restart, unless "reset" has been actuated and (in "and" connection) "start" is actuated. The "start" signal may come from a single start valve (fig. 347) or from a cycle selection module (fig. 348).

Cycle selection module and sequencing interrupt module combination

Where a sequential machine control needs a cycle selection module as well as a sequencing interrupt module, it is advantageous to link the two modules with an auto-selection reset link (see fig. 348). This link causes the cycle selection module to be shifted into the manual mode, hence a fresh auto-selection signal needs to be given if automatic machine operation is to resume. For integration of these combined modules see ladder diagram in fig. 351.

Control problem 2

To demonstrate the integration of a cycle selection module and a machine interrupt module, the following circuit is presented.

A hydraulically powered machine with seven sequential steps is to operate with the following machine sequence:

A1-B1-B0-C0-B1-B0-T-A0/C1.

The confirmation from sequence step 7 must be an "and" function consisting of the signals 001 • 005 (a_0 and c_0). The timer delaying sequence step 7 is to be placed into the step-counter input (the confirmation line leading to the step-counter module;

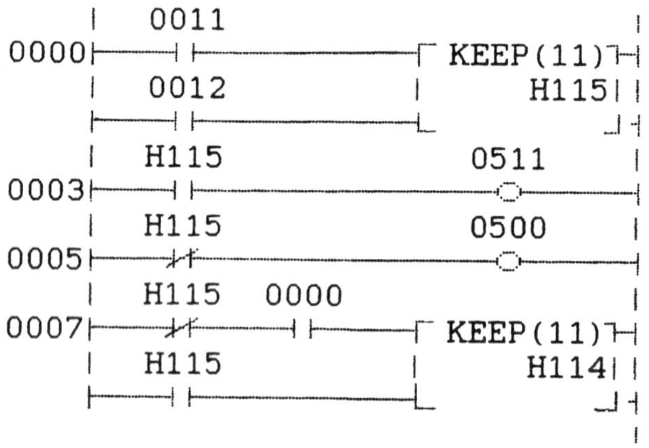

ADDRESS	MNEMONIC		OPERAND	
0000	LD			0011
0001	LD			0012
0002	KEEP (11)		HR	115
0003	LD		HR	115
0004	OUT			0511
0005	LD	NOT	HR	115
0006	OUT			0500
0007	LD	NOT	HR	115
0008	AND			0000
0009	LD		HR	115
0010	KEEP (11)		HR	114

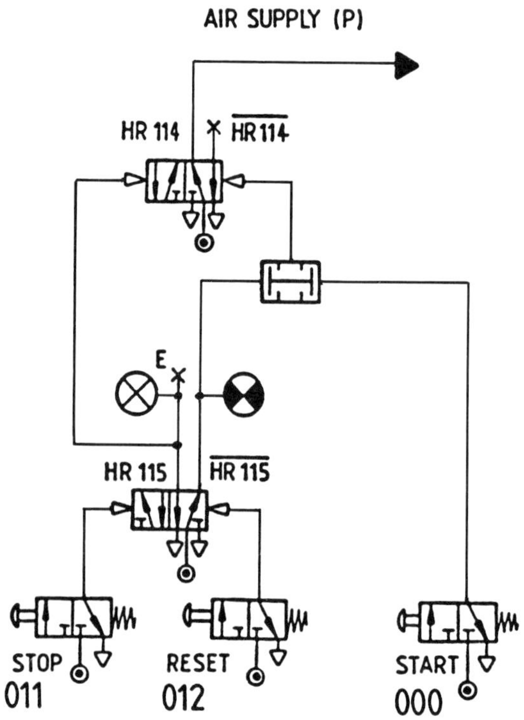

Fig. 347 Complex sequencing interrupt module with extra safety features for machine restart.

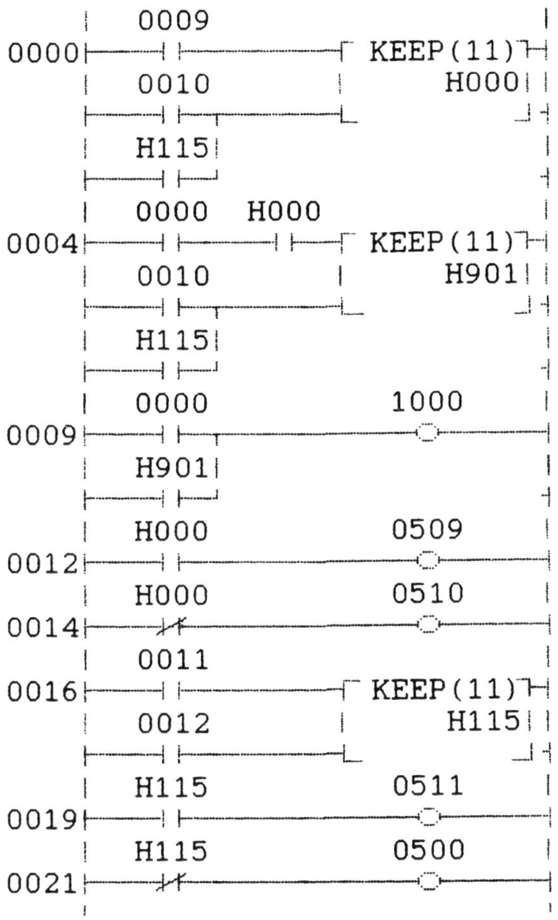

ADDRESS	MNEMONIC		OPERAND	
0000	LD			0009
0001	LD			0010
0002	OR		HR	115
0003	KEEP	(11)	HR	000
0004	LD			0000
0005	AND		HR	000
0006	LD			0010
0007	OR		HR	115
0008	KEEP	(11)	HR	901
0009	LD			0000
0010	OR		HR	901
0011	OUT			1000
0012	LD		HR	000
0013	OUT			0509
0014	LD	NOT	HR	000
0015	OUT			0510
0016	LD			0011
0017	LD			0012
0018	KEEP	(11)	HR	115
0019	LD		HR	115
0020	OUT			0511
0021	LD	NOT	HR	115
0022	OUT			0500
0023	END	(01)		

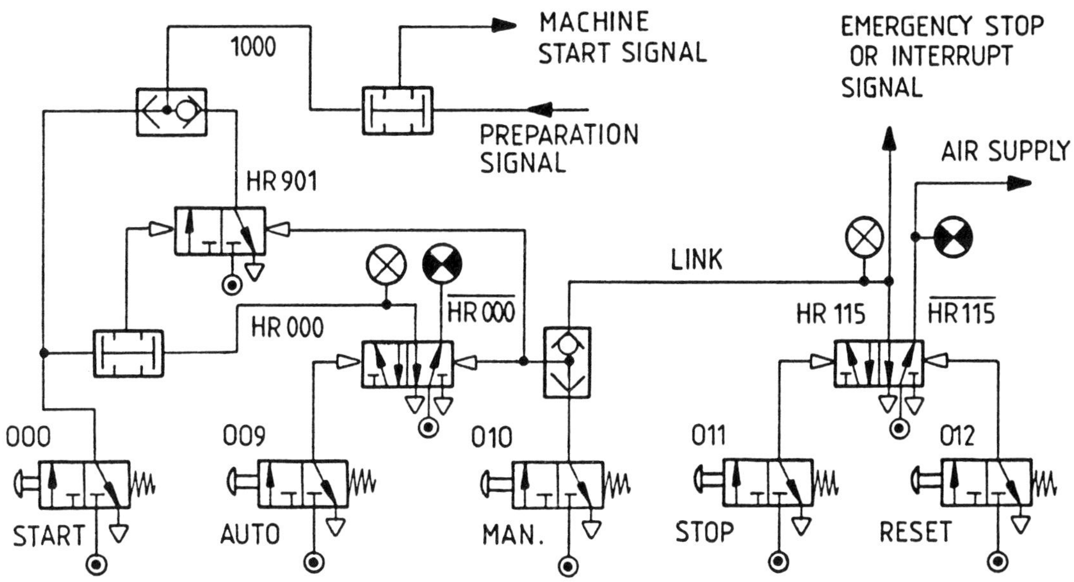

Fig. 348 Cycle selection module and sequencing interrupt module combination. The link causes the cycle selection module to be automatically shifted into manual mode when machine interrupt is signalled.

see figs. 340 and 351). Allocation of relay numbers is as shown in the following input/output assignment list:

Input/output assignment list

A	Ret. Sol.	= 501	Auto Switch	=	009
A	Ext. Sol.	= 502	Man. Switch	=	010
B	Ret. Sol.	= 503	Stop Switch	=	011
B	Ext. Sol.	= 504	Res. Switch	=	012
C	Ret. Sol.	= 505	Start Switch	=	000
C	Ext. Sol.	= 505	Auto Lamp	=	509
A	Ret. Ls.	= 001	Man. Lamp	=	510
A	Ext. Ls.	= 002	Stop Lamp	=	511
B	Ret. Ls.	= 003	Res. Lamp	=	500
B	Ext. Ls.	= 004	Start Relay	=	1000
C	Ret. Ls.	= 005	Auto Relay	=	HR000
C	Ext. Ls.	= 006	Stop Relay	=	HR115
TIM 01 # 40			Aux. Relay	=	HR901
(4 seconds)			Aux. Relay	=	HR114

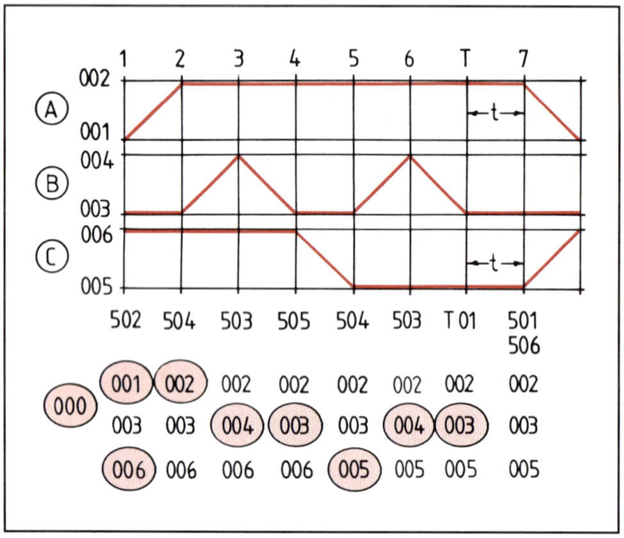

Fig. 349 Traverse-time diagram for control problem 2. See also circuit in fig. 351. Note time delay period at end of sequence step 6.

Stepping module

For initial machine or tool set-up, it is often advantageous to run a machine sequence for trial purposes on a step-by-step basis. Once the machine is proved to function correctly, one can then select automatic step advancing. For the stepping operation, the initial select "step" push button switch actuation causes holding relay 900 (HR900) to reset, thus giving signal HR900. Further short actuations of the "step" switch cause the step-counter to advance in a step-by-step manner.

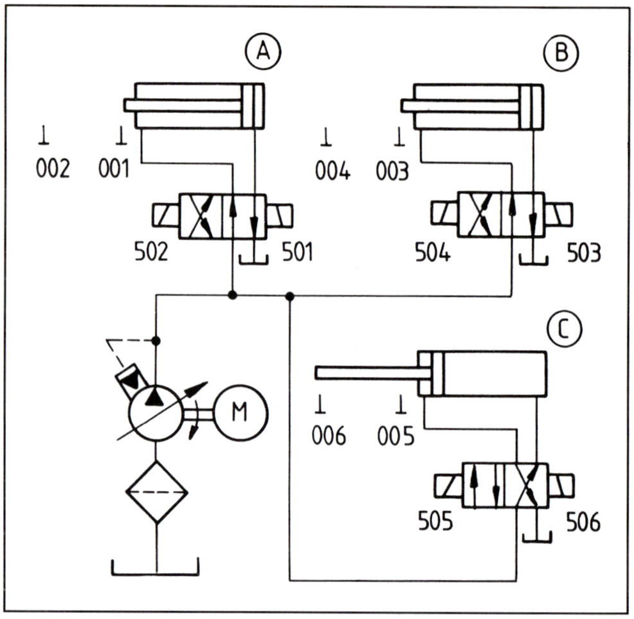

Fig. 350 Electro-hydraulic circuit with PLC input/output assignment labels for control problem 2.

To commence stepping at sequence step 1, the initial stepping selection does not cause the step-counter to start if the step selection push button is actuated for a period of less than one second. Hence the initial operation of this button merely causes stepping mode selection. Further actuations of the "step" button cause the step-counter to advance on a step-by-step basis. This arrangement saves an additional stepping push button switch and is in line with the policy of simplifying the control panel on machines. For integration of a stepping module see fig. 358.

Three position hydraulic valves

Electrically piloted three position hydraulic valves (usually closed centre, open centre or tandem centre) with spring selected centre position, can only select into the centre when neither of the two electrical pilot signals is present. Conversely, in the absence of these pilot signals, a three position valve automatically selects the centre position. Electrical pilot signals from PLC controllers leading to such hydraulic directional control valves must therefore be carefully timed and maintained for a precise duration to match the stringent requirements of the hydraulic machine sequence (see figs. 353–357).

PLC sequence control systems, designed by the step-counter method, automatically cancel the pilot signals of the previous sequence step whenever the present step is actuated (see figs. 340, 341 and 357). Therefore, where step-counter PLC control systems are used to control electrically pilot operated three position valves or spring returned two position valves,

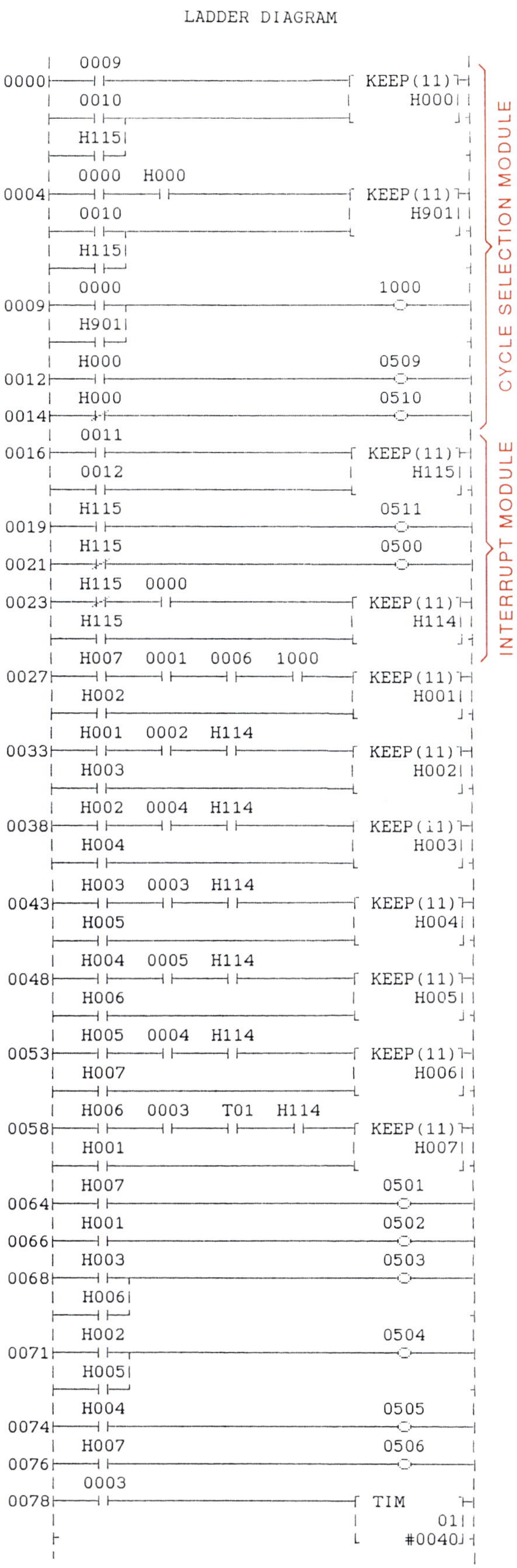

ADDRESS	MNEMONIC			OPERAND
0000	LD			0009
0001	LD			0010
0002	OR		HR	115
0003	KEEP (11)		HR	000
0004	LD			0000
0005	AND		HR	000
0006	LD			0010
0007	OR		HR	115
0008	KEEP (11)		HR	901
0009	LD			0000
0010	OR		HR	901
0011	OUT			1000
0012	LD		HR	000
0013	OUT			0509
0014	LD	NOT	HR	000
0015	OUT			0510
0016	LD			0011
0017	LD			0012
0018	KEEP (11)		HR	115
0019	LD		HR	115
0020	OUT			0511
0021	LD	NOT	HR	115
0022	OUT			0500
0023	LD	NOT	HR	115
0024	AND			0000
0025	LD		HR	115
0026	KEEP (11)		HR	114
0027	LD		HR	007
0028	AND			0001
0029	AND			0006
0030	AND			1000
0031	LD		HR	002
0032	KEEP (11)		HR	001
0033	LD		HR	001
0034	AND			0002
0035	AND		HR	114
0036	LD		HR	003
0037	KEEP (11)		HR	002
0038	LD		HR	002
0039	AND			0004
0040	AND		HR	114
0041	LD		HR	004
0042	KEEP (11)		HR	003
0043	LD		HR	003
0044	AND			0003
0045	AND		HR	114
0046	LD		HR	005
0047	KEEP (11)		HR	004
0048	LD		HR	004
0049	AND			0005
0050	AND		HR	114
0051	LD		HR	006
0052	KEEP (11)		HR	005
0053	LD		HR	005
0054	AND			0004
0055	AND		HR	114
0056	LD		HR	007
0057	KEEP (11)		HR	006
0058	LD		HR	006
0059	AND			0003
0060	AND		TIM	01
0061	AND		HR	114
0062	LD		HR	001
0063	KEEP (11)		HR	007
0064	LD		HR	007
0065	OUT			0501
0066	LD		HR	001
0067	OUT			0502
0068	LD		HR	003
0069	OR		HR	006
0070	OUT			0503
0071	LD		HR	002
0072	OR		HR	005
0073	OUT			0504
0074	LD		HR	004
0075	OUT			0505
0076	LD		HR	007
0077	OUT			0506
0078	LD			0003
0079	TIM			01
			#	0040

Fig. 351 Ladder diagram for control problem 2 with time delay function, extended cycle selection module and complex sequencing interrupt module.

ADDRESS	MNEMONIC	OPERAND
0000	LD	0007
0001	LD	0008
0002	AND NOT	TIM 10
0003	KEEP (11)	HR 900
0004	LD	0008
0005	AND NOT	TIM 10
0006	AND	TIM 09
0007	OR	HR 900
0008	OUT	1003
0009	LD	0008
0010	AND NOT	TIM 10
0011	AND	TIM 09
0012	LD	HR 900
0013	AND	0000
0014	OR LD	
0015	OUT	1004
0016	LD	HR 900
0017	OUT	0507
0018	LD NOT	HR 900
0019	OUT	0508
0020	LD NOT	HR 900
0021	TIM	09
		# 0010
0022	LD	0008
0023	TIM	10
		# 0005

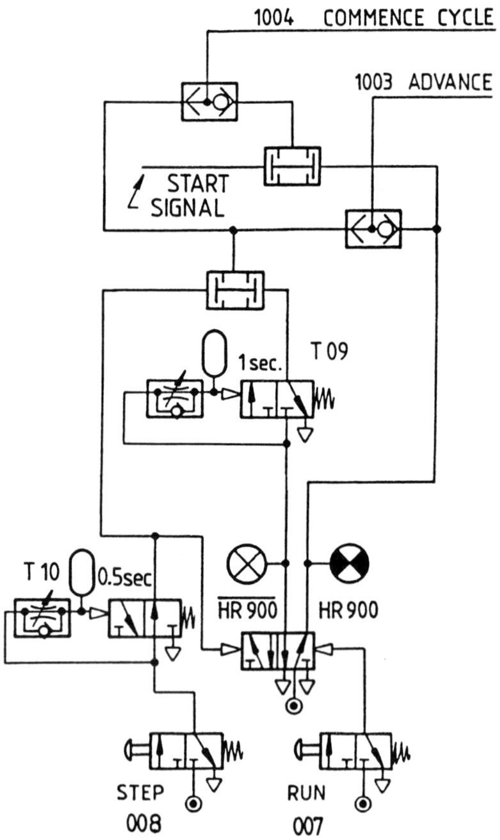

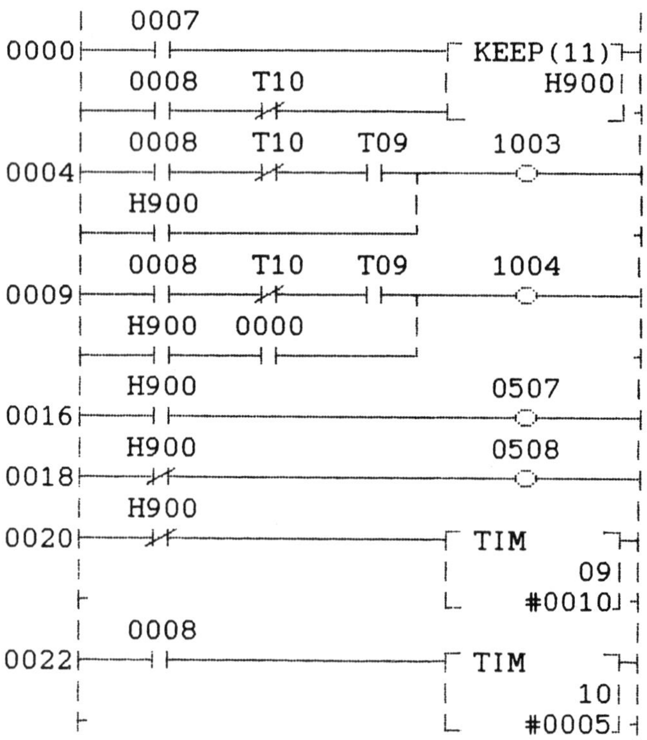

Fig. 352 Stepping module to provide sequential machines with the option of manual stepping or automatic step advancing.

the circuit designer has three options for maintaining (sustaining) step-counter output signals. These options are as follows:

1. The "mono-flop" type output relay, leading to the non-memory type hydraulic valve, may be converted into an R/S flip-flop type set/reset relay which requires a specific set and a specific reset instruction (see "Keep function" and fig. 329).
2. Additional holding relays may be used to maintain the PLC output signal and thus the position of such non-memory type hydraulic valves (see chapter 18, circuit example 5).
3. The step-counter output signals may be "or" connected in the ladder diagram, so that each consecutive step-counter step sustains the output relay until cancellation is required (see fig. 355, output relay 502 and output relay 507).

Of these three methods, the last one is the best and the simplest for designing circuits and maintaining step-counter output signals to non-memory type hydraulic valves.

Time-delayed sequencing for three position hydraulic valves with PLC

Time delay sequencing is often encountered in hydraulic sequential control circuits. The PLC timer may be built into the confirmation signal line leading to the step-counter holding relay, or into the switching line leading to the PLC output relay and eventually to the directional control valve solenoid (see arrows in figs. 320 and 357). Timer contact placement in the switching line of the output relay delays the switching of the DCV solenoid, but resetting of the previous PLC step-counter module is not delayed. With the timer in the confirmation signal line, leading to the step-counter holding relay, both the DCV solenoid switching function and the step-counter module switching function are delayed. PLC control circuit designers, therefore, have to evaluate and contemplate both timer positions for their merits and select the correct position according to circuit application (see also description of "Circuit example 1" in chapter 18 and fig. 321).

Control problem 3

This is a control problem for a hydraulically powered machine with two linear actuators (actuators A and B) and a hydraulic motor (actuator C). The sequence of this machine is illustrated in the traverse-time diagram in fig. 353.

Actuator A is controlled by a spring return type DCV (see "Electro-hydraulic circuit", fig. 354). Actuators B and C require a closed centre, three position directional control valve and the fixed displacement pump recirculates its flow back to tank via the vented compound relief valve during non-action stages of the machine sequence. The venting valve is also a spring return valve. Solenoid actuation of this venting valve causes the pump to come on stream and supply flow into the system, whereas the absence of the PLC output signal 507 causes venting and thus flow recirculation back to tank (see "Electro-hydraulic circuit", fig. 354).

For actuators A and B the sequencing action is obvious from the traverse-time diagram. The motions of hydraulic motor C are not so obvious. The motor starts to rotate anti-clockwise at sequence step 5 for 9 seconds. Anti-clockwise rotations are terminated and confirmed by timer 2. At sequence step 8, linear actuator A and motor C operate simultaneously. The motor, however, now rotates clockwise; its clockwise rotations are to be counted by counter 3 and must amount to 7 full rotations. Motor rotations at step 8 must stop as soon as 7 rotations are completed, regardless of the position of actuator A. For this reason, the PLC output signal 505 leading to solenoid 505 of DCV C must be terminated as soon as the counter has fallen to zero (7 rotations completed) and its termination signal must not be tied ("and" connected), to the completion signal of the retraction stroke of cylinder A (see fig. 355).

In order to achieve minimal power consumption during non-action machine stages, venting of the compound relief valve must occur during the time-delay period, starting with sequence step 4, and at the completion of sequence step 8. Step-counter steps 4 and 9 (HR004 and HR009) signal the venting action. Venting must also occur when sequencing interruption by the interrupt module is signalled (HR115 "on" and HR114 "off").

The bar chart in the key of the traverse-time diagram for this machine shows when the various solenoid signals from the PLC are switched "on", and when they are to be switched "off". Solenoid signal 502 from the PLC controller must deliberately be held "on" (sustained) during sequence steps 1 to 7, because actuator A must remain extended (spring return valve). This is achieved with a multiple "or" function, where holding relay 1 initially switches output signal 502 "on" and holding relays 2 to 7 are sustaining the output signal 502 (see ladder diagram, fig. 355, addresses 56–63). The same principle is applied to the venting valve, which also requires sustaining of its signal (see addresses 74 to 82 of the ladder diagram). As an alternative, one could use an auxiliary relay (HR) as is shown in fig. 324 of chapter 18 (circuit example 4).

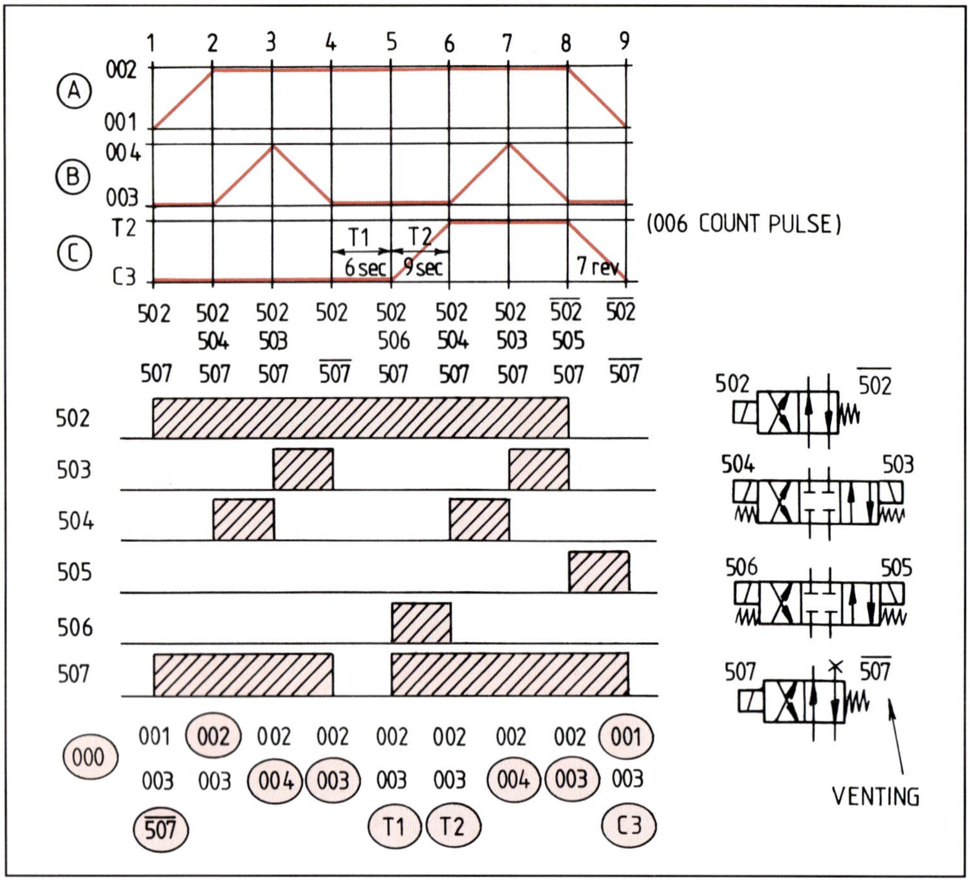

Fig. 353 Traverse-time diagram for control problem 3 with inclusion of signal bar chart for PLC output signal display.

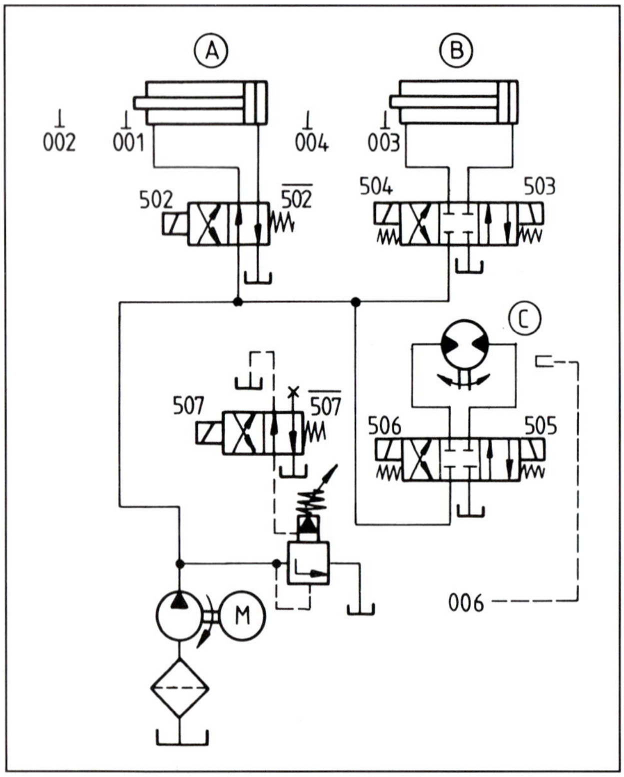

Fig. 354 Electro-hydraulic control circuit with PLC input-output assignment number labels for control problem 3.

Control problem 4

This control problem has the same machine sequence as control problem 3. However, to achieve minimal power consumption during non-action periods, pump offloading is achieved by the pump offloading method with a tandem centre D.C.V. for actuator B. For this reason, the PLC must not send any signals to D.C.V. B during the time-delay period, starting at the end of sequence step 3 and the machine "off" period, starting at the end of sequence step 7.

Whenever pump flow is required, D.C.V. B must receive either a signal 504 or a signal 503 to switch out of the centre position. For this reason signal 503 is given at sequence step 1, 4 and 7, even if actuator B is already in the retracted position.

Step-counter module 4 does not reset step-counter module 3 until timer 1 has elapsed (timed out). Pump offloading, however, must occur as soon as actuator B has fully retracted. Limit switch 0003 is therefore used to inhibit signal HR003 and thus cancels PLC output signal 503 (see addresses 51 and 52 in ladder diagram of fig. 357). The same principle is applied at the completion of sequence step 7, where signal 503 is not cancelled until counter contact C03 signals motor rotations completed and limit switch 0001 signals actuator A fully retracted. If one of the two

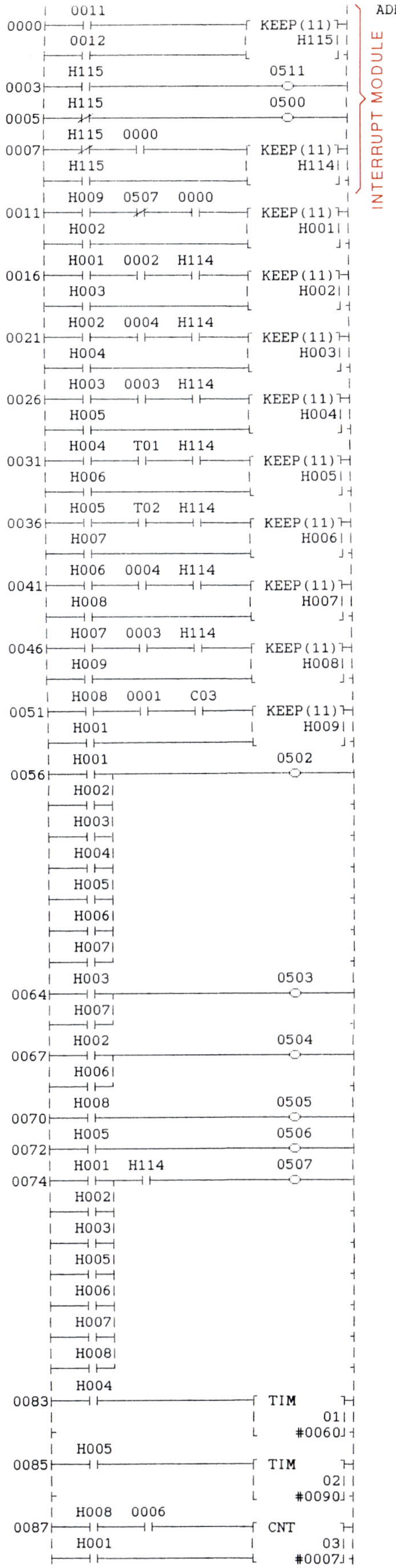

ADDRESS	MNEMONIC	OPERAND
0000	LD	0011
0001	LD	0012
0002	KEEP (11)	HR 115
0003	LD	HR 115
0004	OUT	0511
0005	LD NOT	HR 115
0006	OUT	0500
0007	LD NOT	HR 115
0008	AND	0000
0009	LD	HR 115
0010	KEEP (11)	HR 114
0011	LD	HR 009
0012	AND NOT	0507
0013	AND	0000
0014	LD	HR 002
0015	KEEP (11)	HR 001
0016	LD	HR 001
0017	AND	0002
0018	AND	HR 114
0019	LD	HR 003
0020	KEEP (11)	HR 002
0021	LD	HR 002
0022	AND	0004
0023	AND	HR 114
0024	LD	HR 004
0025	KEEP (11)	HR 003
0026	LD	HR 003
0027	AND	0003
0028	AND	HR 114
0029	LD	HR 005
0030	KEEP (11)	HR 004
0031	LD	HR 004
0032	AND	TIM 01
0033	AND	HR 114
0034	LD	HR 006
0035	KEEP (11)	HR 005
0036	LD	HR 005
0037	AND	TIM 02
0038	AND	HR 114
0039	LD	HR 007
0040	KEEP (11)	HR 006
0041	LD	HR 006
0042	AND	0004
0043	AND	HR 114
0044	LD	HR 008
0045	KEEP (11)	HR 007
0046	LD	HR 007
0047	AND	0003
0048	AND	HR 114
0049	LD	HR 009
0050	KEEP (11)	HR 008
0051	LD	HR 008
0052	AND	0001
0053	AND	CNT 03
0054	LD	HR 001
0055	KEEP (11)	HR 009
0056	LD	HR 001
0057	OR	HR 002
0058	OR	HR 003
0059	OR	HR 004
0060	OR	HR 005
0061	OR	HR 006
0062	OR	HR 007
0063	OUT	0502
0064	LD	HR 003
0065	OR	HR 007
0066	OUT	0503
0067	LD	HR 002
0068	OR	HR 006
0069	OUT	0504
0070	LD	HR 008
0071	OUT	0505
0072	LD	HR 005
0073	OUT	0506
0074	LD	HR 001
0075	OR	HR 002
0076	OR	HR 003
0077	OR	HR 005
0078	OR	HR 006
0079	OR	HR 007
0080	OR	HR 008
0081	AND	HR 114
0082	OUT	0507
0083	LD	HR 004
0084	TIM	01
		# 0060
0085	LD	HR 005
0086	TIM	02
		# 0090
0087	LD	HR 008
0088	AND	0006
0089	LD	HR 001
0090	CNT	03
		# 0007
0091	END (01)	

Fig. 355 Ladder diagram and mnemonic list for PLC programming of control problem 3.

actuators finishes its task earlier than the other, the PLC signal 503 must still be maintained to cause pump flow to go to the system. But as soon as both tasks are completed (step 7 is completed), pump flow must stop. Hence, PLC signal 503 must disappear and DCV B must centre. Here, De Morgan's theorem of switching theory is applied, which yields:

$$\overline{0001 \bullet C03} = \overline{0001} + \overline{C03}$$

As an incomplete Boolean equation, one could write:

$$\ldots + (HR007 \bullet \overline{0001}) + (HR007 \bullet \overline{C03}) = 503$$

For an explanation of De Morgan's theorem, see *Pneumatic Control for Industrial Automation* by Peter Rohner and Gordon Smith, page 139. An application of De Morgan's theorem is given in the previous chapter in figs. 324 and 325, where pneumatic signal A0 from step-counter module 7 is inhibited when the confirmation signals a_0 and c_0 in "and" connection are present. This inhibition function of signal A0 causes the DCV to switch into tandem centre and thus offloads the pump to tank. The equation for that application reads:

$$A0 = \text{Step } 7 \bullet (\overline{a_0 \bullet c_0})$$

Here, De Morgan's equation would read:

$$A0 = \text{Step } 7 \bullet (\overline{a_0 \bullet c_0}) = \text{step } 7 \bullet (\overline{a_0} + \overline{c_0})$$

In the ladder diagram for the PLC controller we use the last section of the De Morgan's theorem formula:

$A0 = \text{step } 7 \bullet (\overline{a_0} + \overline{c_0})$ but for simplicity reasons in the PLC ladder diagram, we use the non-minimised form, reading:

$$A0 = (\text{step } 7 \bullet \overline{a_0}) + (\text{step } 7 \bullet \overline{c_0})$$

Conclusion

The complexity of hydraulic control systems with their three-position, non-memory type valves and their sometimes elaborate pump control circuits, necessitates that designers of PLC control circuits have extensive knowledge of the peculiarities of industrial, mining and agricultural hydraulics. Only then will they be able to create functional and economical PLC circuitry to do justice to both engineering faculties, that of PLC electronic control and that of hydraulic control. It is therefore desirable that university engineering schools include PLC programming and fluid power control subjects in their engineering curriculum.

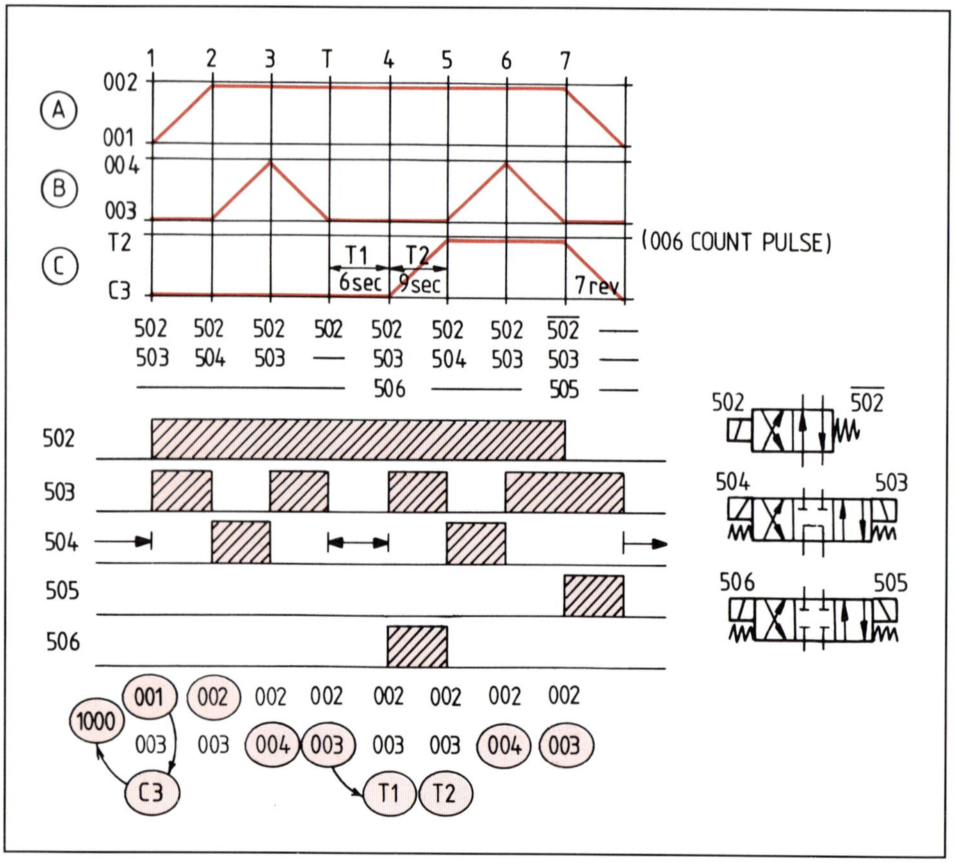

Fig. 356 Traverse-time diagram with inclusion of signal bar chart for PLC output signal display for control problem 4.

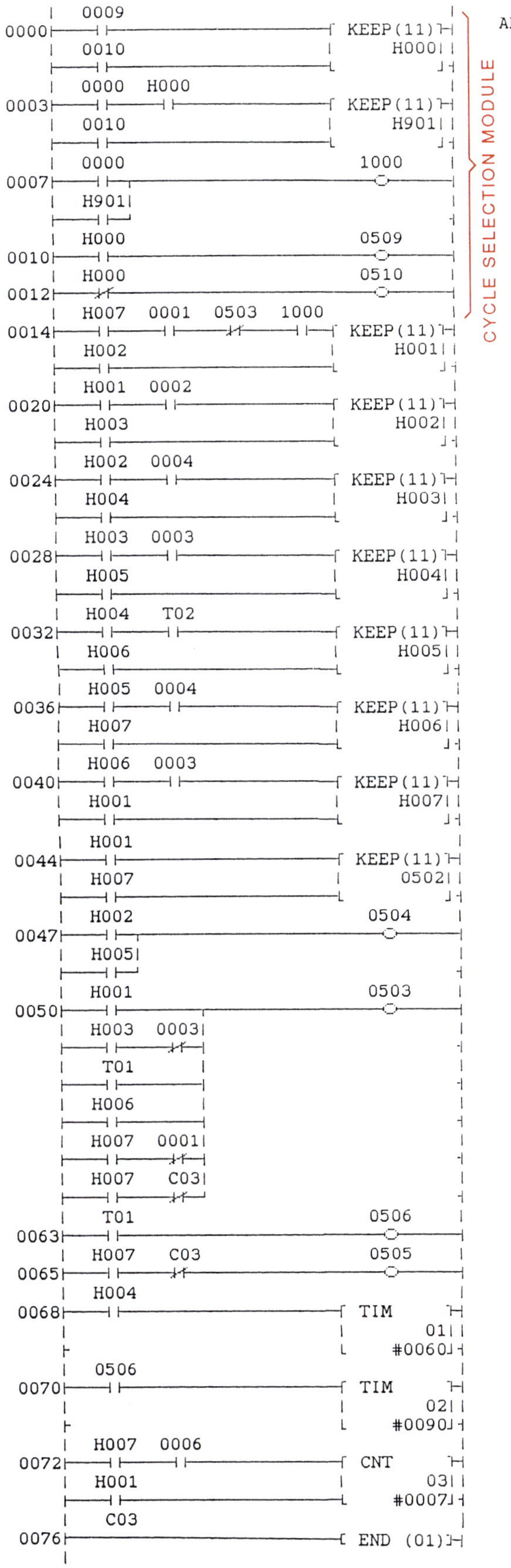

ADDRESS	MNEMONIC	OPERAND
0000	LD	0009
0001	LD	0010
0002	KEEP (11)	HR 000
0003	LD	0000
0004	AND	HR 000
0005	LD	0010
0006	KEEP (11)	HR 901
0007	LD	0000
0008	OR	HR 901
0009	OUT	1000
0010	LD	HR 000
0011	OUT	0509
0012	LD NOT	HR 000
0013	OUT	0510
0014	LD	HR 007
0015	AND	0001
0016	AND NOT	0503
0017	AND	1000
0018	LD	HR 002
0019	KEEP (11)	HR 001
0020	LD	HR 001
0021	AND	0002
0022	LD	HR 003
0023	KEEP (11)	HR 002
0024	LD	HR 002
0025	AND	0004
0026	LD	HR 004
0027	KEEP (11)	HR 003
0028	LD	HR 003
0029	AND	0003
0030	LD	HR 005
0031	KEEP (11)	HR 004
0032	LD	HR 004
0033	AND	TIM 02
0034	LD	HR 006
0035	KEEP (11)	HR 005
0036	LD	HR 005
0037	AND	0004
0038	LD	HR 007
0039	KEEP (11)	HR 006
0040	LD	HR 006
0041	AND	0003
0042	LD	HR 001
0043	KEEP (11)	HR 007
0044	LD	HR 001
0045	LD	HR 007
0046	KEEP (11)	0502
0047	LD	HR 002
0048	OR	HR 005
0049	OUT	0504
0050	LD	HR 001
0051	LD	HR 003
0052	AND NOT	0003
0053	OR LD	
0054	OR	TIM 01
0055	OR	HR 006
0056	LD	HR 007
0057	AND NOT	0001
0058	OR LD	
0059	LD	HR 007
0060	AND NOT	CNT 03
0061	OR LD	
0062	OUT	0503
0063	LD	TIM 01
0064	OUT	0506
0065	LD	HR 007
0066	AND NOT	CNT 03
0067	OUT	0505
0068	LD	HR 004
0069	TIM	01
		# 0060
0070	LD	0506
0071	TIM	02
		# 0090
0072	LD	HR 007
0073	AND	0006
0074	LD	HR 001
0075	CNT	03
		# 0007

Fig. 357 Ladder diagram and mnemonic list for PLC programming of control problem 4.

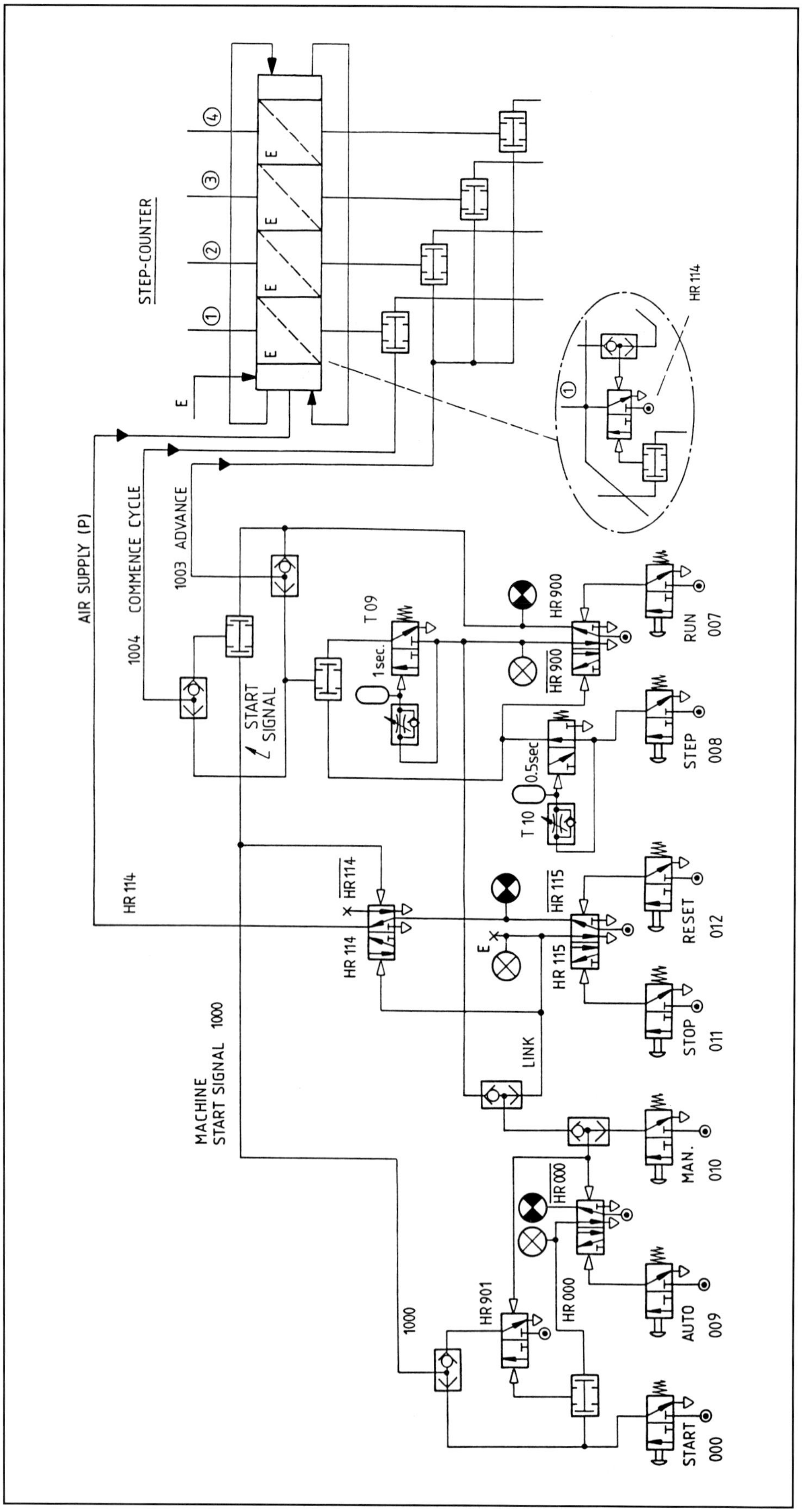

Fig. 358 Fluid power circuit with four step-counter modules (four sequence steps) and integration of three fringe condition modules (complex cycle selection module, complex sequencing interrupt module and a stepping module).

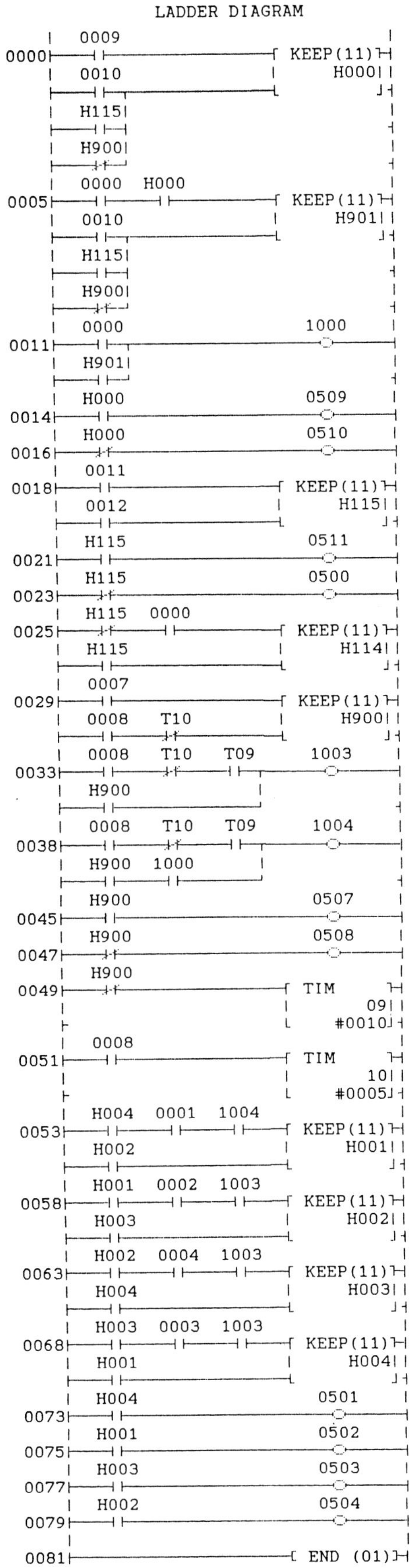

ADDRESS	MNEMONIC		OPERAND	
0000	LD			0009
0001	LD			0010
0002	OR		HR	115
0003	OR	NOT	HR	900
0004	KEEP	(11)	HR	000
0005	LD			0000
0006	AND		HR	000
0007	LD			0010
0008	OR		HR	115
0009	OR	NOT	HR	900
0010	KEEP	(11)	HR	901
0011	LD			0000
0012	OR		HR	901
0013	OUT			1000
0014	LD		HR	000
0015	OUT			0509
0016	LD	NOT	HR	000
0017	OUT			0510
0018	LD			0011
0019	LD			0012
0020	KEEP	(11)	HR	115
0021	LD		HR	115
0022	OUT			0511
0023	LD	NOT	HR	115
0024	OUT			0500
0025	LD	NOT	HR	115
0026	AND			0000
0027	LD		HR	115
0028	KEEP	(11)	HR	114
0029	LD			0007
0030	LD			0008
0031	AND	NOT	TIM	10
0032	KEEP	(11)	HR	900
0033	LD			0008
0034	AND	NOT	TIM	10
0035	AND		TIM	09
0036	OR		HR	900
0037	OUT			1003
0038	LD			0008
0039	AND	NOT	TIM	10
0040	AND		TIM	09
0041	LD		HR	900
0042	AND			1000
0043	OR	LD		
0044	OUT			1004
0045	LD		HR	900
0046	OUT			0507
0047	LD	NOT	HR	900
0048	OUT			0508
0049	LD	NOT	HR	900
0050	TIM			09
			#	0010
0051	LD			0008
0052	TIM			10
			#	0005
0053	LD		HR	004
0054	AND			0001
0055	AND			1004
0056	LD		HR	002
0057	KEEP	(11)	HR	001
0058	LD		HR	001
0059	AND			0002
0060	AND			1003
0061	LD		HR	003
0062	KEEP	(11)	HR	002
0063	LD		HR	002
0064	AND			0004
0065	AND			1003
0066	LD		HR	004
0067	KEEP	(11)	HR	003
0068	LD		HR	003
0069	AND			0003
0070	AND			1003
0071	LD		HR	001
0072	KEEP	(11)	HR	004
0073	LD		HR	004
0074	OUT			0501
0075	LD		HR	001
0076	OUT			0502
0077	LD		HR	003
0078	OUT			0503
0079	LD		HR	002
0080	OUT			0504
0081	END	(01)		

Fig. 359 Ladder diagram and mnemonic list for circuit given in fig. 358.

Name	No. of points	Relay number									
Input relay	80	0000 to 0415									
		00CH		01CH		02CH		03CH		04CH	
		00	08	00	08	00	08	00	08	00	08
		01	09	01	09	01	09	01	09	01	09
		02	10	02	10	02	10	02	10	02	10
		03	11	03	11	03	11	03	11	03	11
		04	12	04	12	04	12	04	12	04	12
		05	13	05	13	05	13	05	13	05	13
		06	14	06	14	06	14	06	14	06	14
		07	15	07	15	07	15	07	15	07	15
Output relay	60	0500 to 0915									
		05CH		06CH		07CH		08CH		09CH	
		00	08	00	08	00	08	00	08	00	08
		01	09	01	09	01	09	01	09	01	09
		02	10	02	10	02	10	02	10	02	10
		03	11	03	11	03	11	03	11	03	11
		04	12	04	12	04	12	04	12	04	12
		05	13	05	13	05	13	05	13	05	13
		06	14	06	14	06	14	06	14	06	14
		07	15	07	15	07	15	07	15	07	15
Internal auxiliary relay	136	1000 to 1807									
		10CH		11CH		12CH		13CH		14CH	
		00	08	00	08	00	08	00	08	00	08
		01	09	01	09	01	09	01	09	01	09
		02	10	02	10	02	10	02	10	02	10
		03	11	03	11	03	11	03	11	03	11
		04	12	04	12	04	12	04	12	04	12
		05	13	05	13	05	13	05	13	05	13
		06	14	06	14	06	14	06	14	06	14
		07	15	07	15	07	15	07	15	07	15
		15CH		16CH		17CH		18CH			
		00	08	00	08	00	08	00			
		01	09	01	09	01	09	01			
		02	10	02	10	02	10	02			
		03	11	03	11	03	11	03			
		04	12	04	12	04	12	04			
		05	13	05	13	05	13	05			
		06	14	06	14	06	14	06			
		07	15	07	15	07	15	07			
Holding relay (retentive relay)	160	HR000 to 915									
		00CH		01CH		02CH		03CH		04CH	
		00	08	00	08	00	08	00	08	00	08
		01	09	01	09	01	09	01	09	01	09
		02	10	02	10	02	10	02	10	02	10
		03	11	03	11	03	11	03	11	03	11
		04	12	04	12	04	12	04	12	04	12
		05	13	05	13	05	13	05	13	05	13
		06	14	06	14	06	14	06	14	06	14
		07	15	07	15	07	15	07	15	07	15
		05CH		06CH		07CH		08CH		09CH	
		00	08	00	08	00	08	00	08	00	08
		01	09	01	09	01	09	01	09	01	09
		02	10	02	10	02	10	02	10	02	10
		03	11	03	11	03	11	03	11	03	11
		04	12	04	12	04	12	04	12	04	12
		05	13	05	13	05	13	05	13	05	13
		06	14	06	14	06	14	06	14	06	14
		07	15	07	15	07	15	07	15	07	15

Name	No. of points	Timer/counter number						
Timer/counter	48	TIM/CNT00 to 47						
		00	08	16	24	32	40	
		01	09	17	25	33	41	
		02	10	18	26	34	42	
		03	11	19	27	35	43	
		04	12	20	28	36	44	
		05	13	21	29	37	45	
		06	14	22	30	38	46	
		07	15	23	31	39	47	

Fig. 360 Assignment of input-output channels (16 bit words) and relay numbers for an "Omron C20" programmable controller.

Appendices

1 Units of measurement and their symbols

Mechanical oscillations are commonly expressed in cycles per unit of time, and rotational frequency in revolutions per unit of time. Since "cycle" and "revolution" are not units, they do not have internationally recognised symbols. However, they are often expressed as abbreviations, and in English the common expressions for them are r.p.m. (revolutions per minute), and c/min (cycles per minute).

The bar and Pascal are given equal status as pressure units. The American and Australian fluid power industries still do not agree on one preferred unit. However, the Pascal is the SI unit for pressure. Thus, common current usage in industry opts for the multiples kiloPascal (kPa) and megaPascal (MPa) and for the bar, which is equivalent to 100 kPa. The European fluid power industry predominantly uses the bar as the preferred unit, and both equipment and product information from European countries has its pressure ratings specified in bars.

Symbols used in this book (base units)

Quantity	Symbol	SI unit	Other recognised units
Area	A	m^2	mm^2, km^2, cm^2
Acceleration	a	m/s^2	
Displacement	V	m^3	mL, cm^3
Flow rate	Q	m^3/s	L/min
Force	F	N	kN, MN
Frequency	f	Hz	1/s
Circle constant	π	3.1416	
Length	l	m	mm, cm, km
Mass	m	kg	
Moment	M	Nm	kNm, MNm
Power	P	W	kW, MW
Pressure	p	Pa	kPa, MPa, BAR
Radius	r	m	mm, cm
Revolutions	n	1/s	1/min
Temperature	T	K	°C
Torque	M	Nm	kNm, MNm
Velocity	v	m/s	m/min, km/h
Time	t	s	ms, min, h, d
Viscosity (DYN)	η	Pa.s	$\frac{N.s}{m^2}$
Viscosity (KIN)	ν	m^2/s	mm^2/s = 1cSt
Volume	V	m^3	mL, cm^3
Work	W	J	kJ, mJ

Commonly used area and volume conversions

$1\ m^2 = 10\,000\ cm^2 = 1\,000\,000\ mm^2$

$1\ m^3 = 1000\ dm^3 = 1000\ L = 1\,000\,000\ cm^3 = 1\,000\,000\,000\ mm^3$

Commonly used pressure conversions

1 bar = 100 000 Pa 1 bar = 100 kPa 1 bar = 0.1 MPa	1 MPa = 10 bar 1 MPa = 1 000 kPa 1 MPa = 1 000 000 Pa
1 kPa = 1000 Pa 1 kPa = 0.01 bar 1 kPa = 0.001 MPa	1 bar = 14.5 psi 100 kPa = 14.5 psi 1 MPa = 145 psi

Prefixes for fractions and multiples of base units

Fraction	Prefix	Symbol	
10^{-18}	atto	a	
10^{-15}	femento	f	
10^{-12}	pico	p	
10^{-9}	nano	n	
10^{-6}	micro	µ	0.000 001
10^{-3}	milli	m	0.001
10^{-2}	centi	c	0.01
10^{-1}	deci	d	0.1
Multiple	**Prefix**	**Symbol**	
10^{1}	deca	da	10
10^{2}	hecto	h	100
10^{3}	kilo	k	1 000
10^{6}	mega	M	1 000 000
10^{9}	giga	G	1 000 000 000
10^{12}	tera	T	

Examples for using fractions and multiples of base units
***units used in Europe**

Fract./Mult.	Symbol	Pa	m	L	N	W
10^{-3}	m		mm	mL		mW
10^{-2}	c		cm			
10^{-1}	d		dm*	dL*		
10^{1}	da				daN*	
10^{2}	h			hL		
10^{3}	k	kPa	km			kW
10^{6}	M	MPa			MN	MW

2 Fluid power formulae

	Symbols	Coherent SI units	Commonly used units
Pump	$P=\frac{p\times Q\times 100}{\eta_0}$	$W=\frac{Pa\times m^3/s\times 100}{\eta_0}$	$kW=\frac{MPa\times L/min\times 100}{60\times\eta_0}$
	$V=\frac{Q\times 100}{n\times\eta_v}$	$m^3=\frac{m^3/s\times 100}{n/s\times\eta_v}$	$mL=\frac{L/min\times 10^5}{rpm\times\eta_v}$
	$Q=\frac{n\times V\times\eta_v}{100}$	$m^3/s=\frac{n/s\times m^3\times\eta_v}{100}$	$L/min=\frac{rpm\times mL\times\eta_v}{10^5}$
	$n=\frac{Q\times 100}{V\times\eta_v}$	$n/s=\frac{m^3/s\times 100}{m^3\times\eta_v}$	$rpm=\frac{L/min\times 10^5}{mL\times\eta_v}$
Motor	$P=\frac{2\times\pi\times n\times M\times\eta_0}{100}$	$W=\frac{6.28\times n/s\times Nm\times\eta_0}{100}$	$kW=\frac{6.28\times rpm\times Nm\times\eta_0}{10^3\times 60}$
	$Q=\frac{V\times n\times 100}{\eta_v}$	$m^3/s=\frac{m^3\times n/s\times 100}{\eta_v}$	$L/min=\frac{mL\times rpm}{10\times\eta_v}$
	$n=\frac{Q\times\eta_v}{V\times 100}$	$n/s=\frac{m^3/s\times\eta_v}{m^3\times 100}$	$rpm=\frac{L/min\times\eta_v\times 10}{mL}$
	$V=\frac{Q\times\eta_v}{n\times 100}$	$m^3=\frac{m^3/s\times\eta_v}{n/s\times 100}$	$mL=\frac{L/min\times\eta_v\times 10}{rpm}$
Linear Actuator	$F_{Ext}=\frac{p\times A\times\eta_{hm}}{100}$	$N=\frac{Pa\times d^2\times\pi\times\eta_{hm}}{4\times 100}$	$kN=\frac{MPa\times mm^2\times 0.7854\times\eta_{hm}}{10^5}$
	$F_{Ret}=\frac{p\times A\times\eta_{hm}}{100}$	$N=\frac{Pa\times(d_p^2-d_r^2)\times\pi\times\eta_{hm}}{4\times 100}$	$kN=\frac{MPa\times(mm^2-mm^2)\times 0.7854\times\eta_{hm}}{10^5}$
	$p=\frac{F\times 100}{A\times\eta_{hm}}$		$kN=\frac{bar\times mm^2\times 0.7854\times\eta_{hm}}{10^6}$
	$Q=v\times A \quad A=\frac{Q}{v}$	$m^3/s=m/s\times m^2$	$L/min=\frac{mm/s\times mm^2\times 0.7854\times 60}{10^6}$
	$v=\frac{Q}{A}$	$m/s=\frac{m^3/s}{m^2}$	$mm/s=\frac{L/min\times 10^6}{mm^2\times 0.7854\times 60}$
	$P=\frac{p\times Q\times\eta_{hm}}{100}$	$W=\frac{Pa\times m^3/s\times\eta_{hm}}{100}$	$kW=\frac{MPa\times L/min\times\eta_{hm}}{60\times 100}$
	$V=A\times S$	$m^3=d^2\times 0.7854\times m$	$L=\frac{mm^2\times 0.7854\times mm}{10^6}$
Pipe diameter	$d=\sqrt{\frac{Q}{v\times 0.7854}}$	$m=\sqrt{\frac{m^3/s}{m/s\times 0.7854}}$	$mm=\sqrt{\frac{L/min}{m/s\times 0.7854\times 10^6\times 60}}$
Reservoir volume	$V=3\times Q$	$m^3=m^3/s\times 60\times 3$	$L=3\times L/min$

3 Industrial hydraulic symbols

These symbols are based upon the international I.S.O. 1219 fluid power symbols. Only the most common symbols have been included.

Composite symbols can be devised for any fluid power components by combining the relevant basic symbols.

Pumps, motors, and drives

	Fixed	Variable
Single direction pump	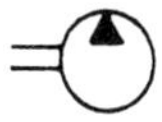	
Double direction pump	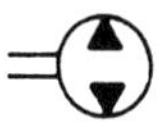	
Single direction motor	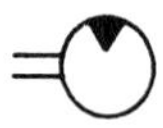	
Double direction motor	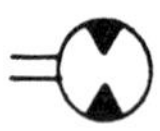	
Single direction pump/motor with reversal of flow direction	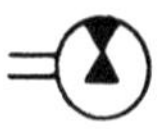	
Single direction pump/motor with single flow direction		
Double direction pump/motor with two directions of flow	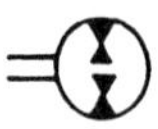	

Hydrostatic drive, split system type	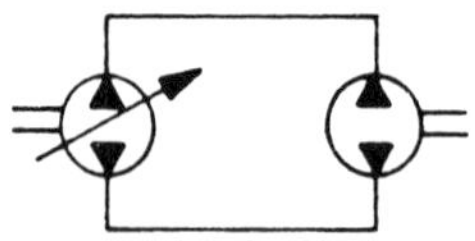
Hydrostatic drive, compact, reversible output	
Semi rotary actuator	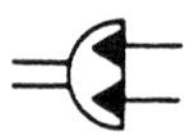

Linear actuators

Single acting ram (load returns the ram)	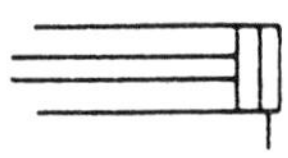
Single acting actuator (load returns the piston)	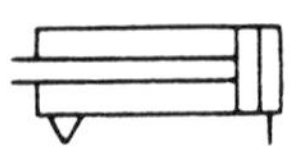
Single acting actuator (spring returns the piston)	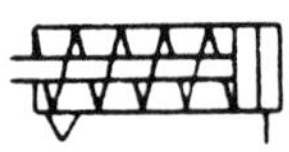
Double acting actuator	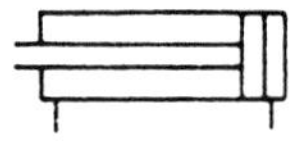
Differential actuator with oversize rod	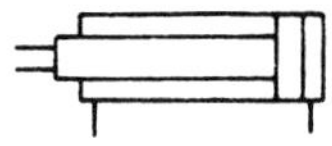
Double acting actuator with double ended rod	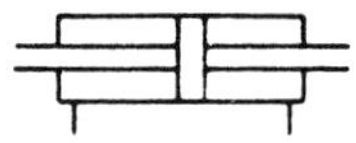
Piston with adjustable end cushioning	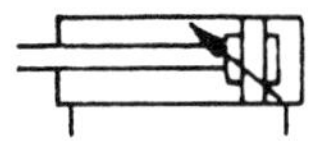
Piston with fixed end cushioning	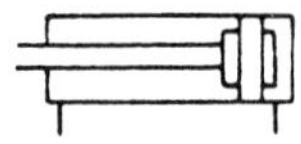
Telescopic, single acting actuator	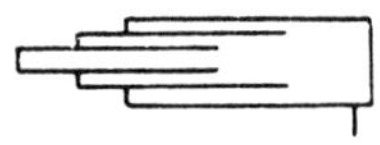
Telescopic, double acting actuator	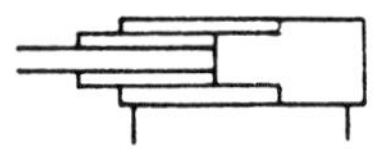
Pressure intensifier	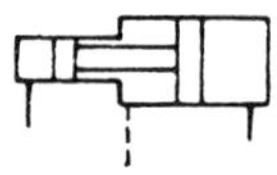

Valve control mechanisms

Undefined control	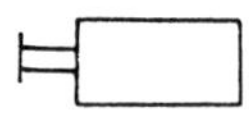
Hand lever (rotary or linear)	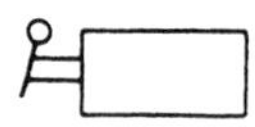
Push button	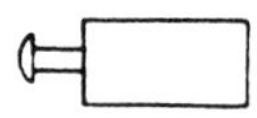
Foot lever	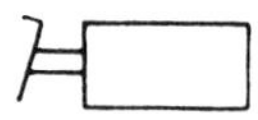
Cam roller	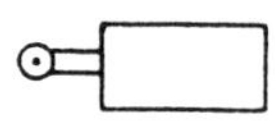
Plunger (piston or ball)	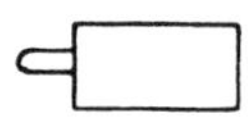
Spring	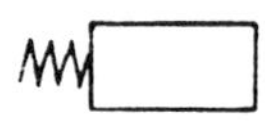

Detent mechanism 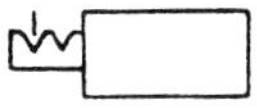

Pressure relief 

Pressure applied 

Pneumatic pilot 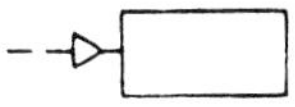

Hydraulic pilot 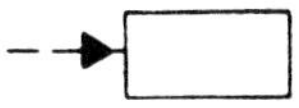

Solenoid 

Solenoid/hydraulic pilot 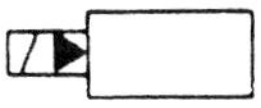

Pneumatic/hydraulic pilot 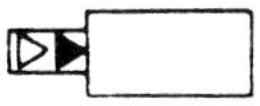

Spring centred 

Directional control valves

Directional control valve with two discrete positions 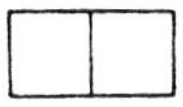

Directional control valve with three discrete positions 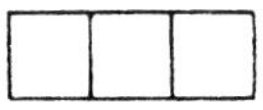

Directional control valve with significant cross-over positions 

Valve with two discrete and an infinite number of intermediate throttling positions 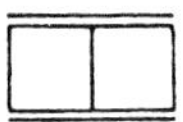

Valve with three discrete and an infinite number of intermediate throttling positions 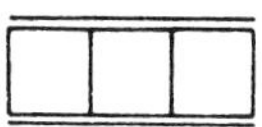

Two position, two port valve 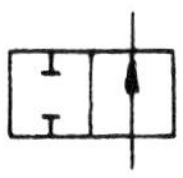

Two position, three port valve 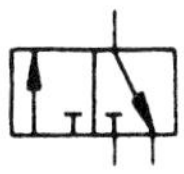

Two position, four port valve 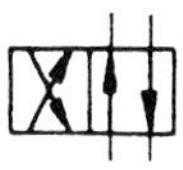

Two position, five port valve 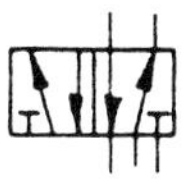

Three position, four port valve with fully closed centre configuration 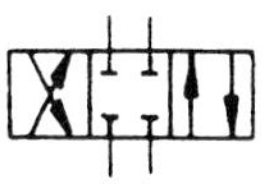

Port labelling:
- Working lines A, B
- Pilot lines X, Y
- Pressure line P
- Tank line T

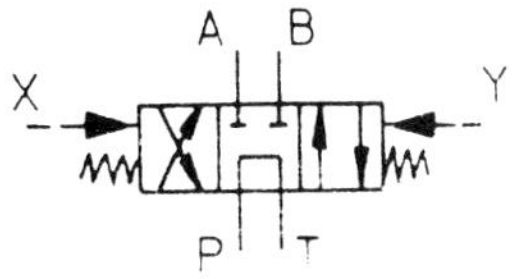

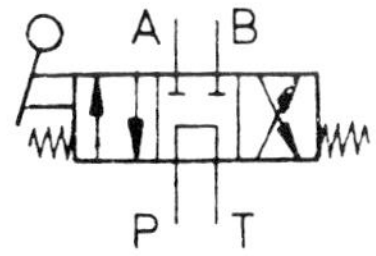

Check valve 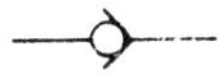

Spring loaded check valve 

Pilot operated check valve 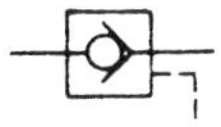

OR function valve 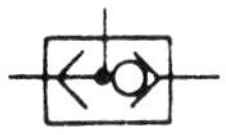

AND function valve 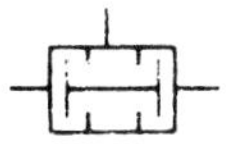

Deceleration valve 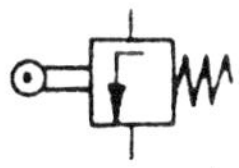

Deceleration valve 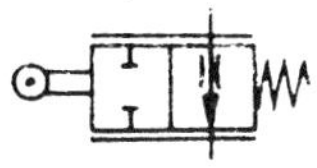

Servo and proportional valve

Proportional control pressure relief valve (with integral max. pressure limitation) 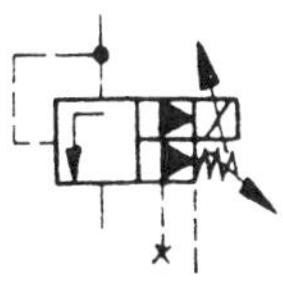

Pilot operated directional proportional valve 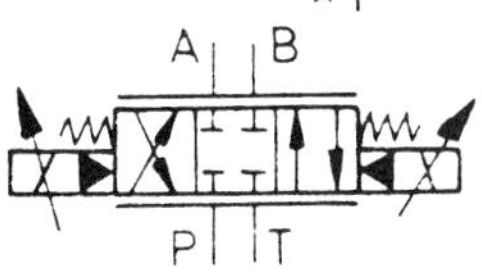

4-way servo valve with mechanical feedback, standard overlapping and hydraulic zero 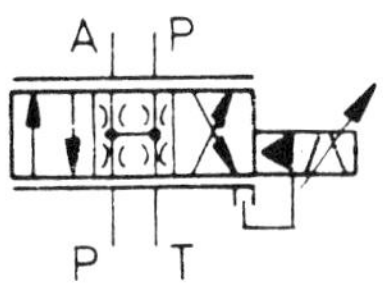

Pressure controls

Throttling orifice normally closed or normally open (*optional)

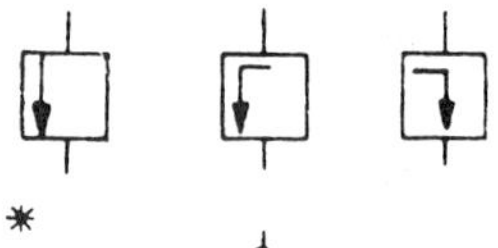

Pressure relief valve (fixed)

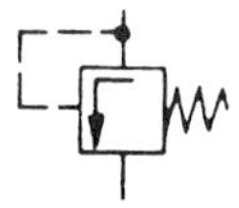

Pressure relief valve (adjustable)

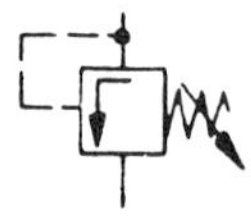

Detailed symbol of pilot operated pressure relief valve (compound relief valve)

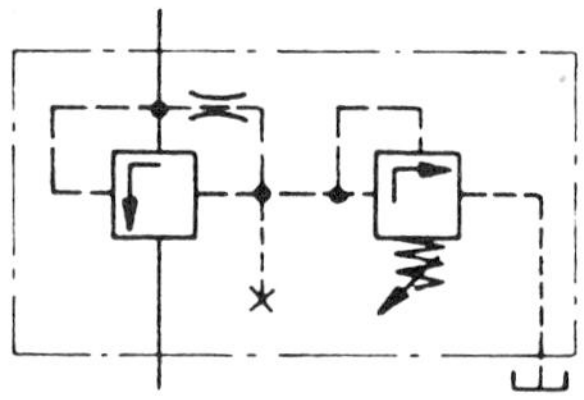

Simplified symbol of compound relief valve
(Pilot flow externally drained)

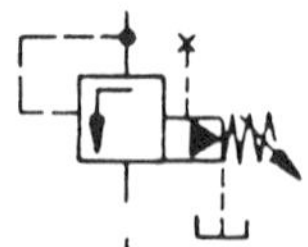

(Pilot flow internally drained)

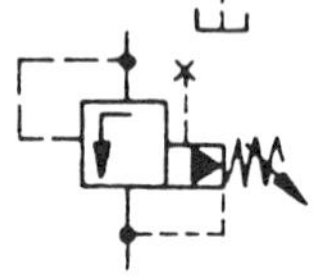

Brake valve

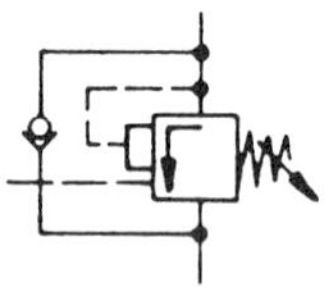

Unloading valve (accumulator charging valve)

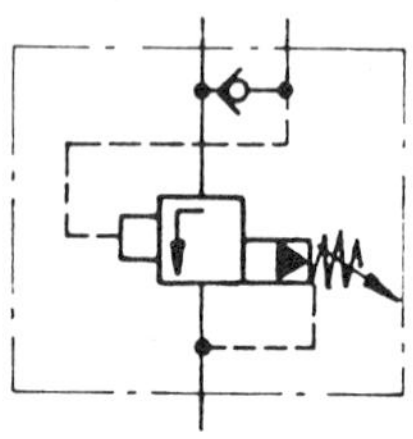

Counter balance valve (back pressure valve)

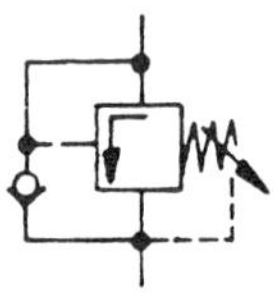

Sequence valve with remote control (external pilot)

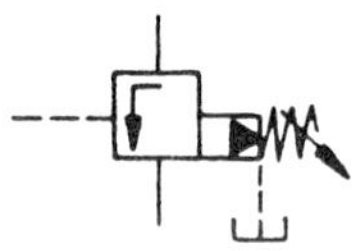

Sequence valve with direct control (internal pilot)

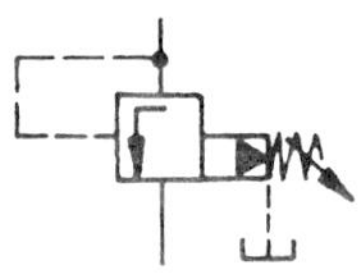

Offloading valve

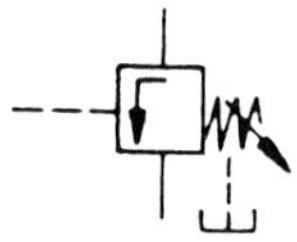

Pressure reducing valve (fixed)

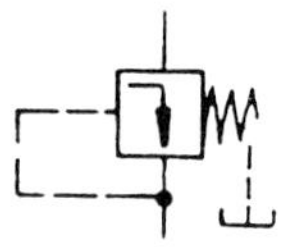

Pressure reducing valve (adjustable)

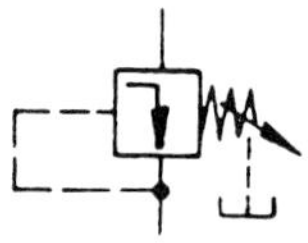

Pilot operated pressure reducing valve

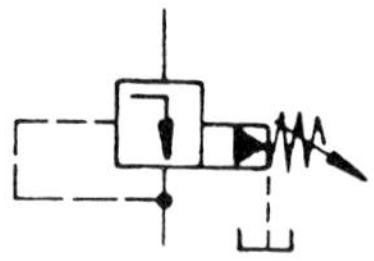

Pressure reducing valve with secondary system relief

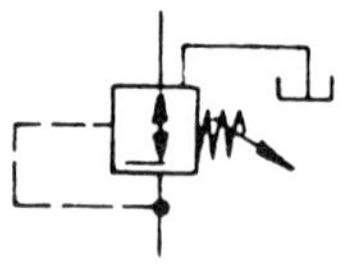

Flow controls

Throttle valve not affected by viscosity

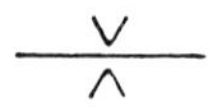

Throttle valve (fixed)

Throttle valve (adjustable)

Flow control valve, pressure and temperature compensated

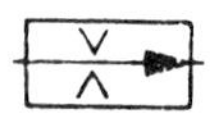

Flow control valve with reverse free flow check

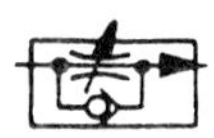

By-pass flow control valve

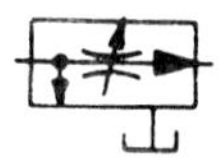

Flow divider

Fluid plumbing and storage

Pressure source	
Working line, return line, feed line	
Pilot control line	
Drain line	
Enclosure line	
Flexible line	
Electric line	
Pipeline connections	
Cross pipeline (not connected)	
Air vent	
Reservoir with inlet below fluid level	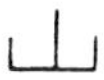
Reservoir with inlet above fluid level	

Miscellaneous symbols

Electric motor	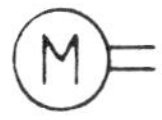
Heat engine	
Electric motor with pump and drive coupling	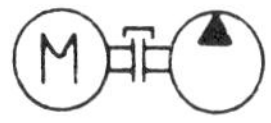
Plugged line	
Plugged line with take-off line	
Quick connect coupling	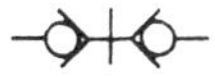
Rotary connection	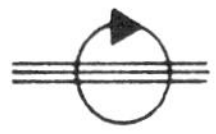
Accumulator	
Filter, strainer	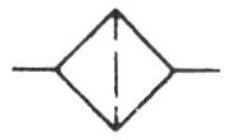
Cooler with coolant lines	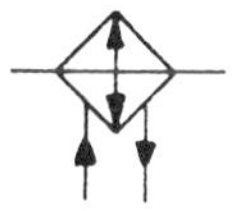
Heater	
Pressure gauge, pressure indicator	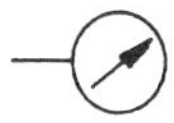
Flow meter	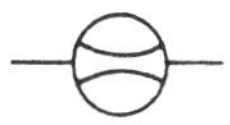
Thermometer	
Pressure switch (electrical)	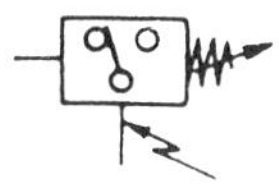
Shut-off valve	

4 Conversion table

Dots indicate SI units.

	TO CONVERT →		TO →	MULTIPLY BY
FORCE				
10^5	dyne		newton (N)	10^{-5}
0.101 97	kilogram-force (kgf)	•	newton (N)	9.806650
7.233	poundal (pdl)		newton (N)	0.1383
0.2248	pound-force (lbf)		newton (N)	4.448
2.2046	pound-force (lbf)		kilogram-force (kgf)	0.4536
0.1004	ton-force (UK)		kilonewton (kN)	9.964
32.174	poundal (pdl)		pound-force (lbf)	0.0311
1.000	kilopond (kp)		kilogram-force (kgf)	1.000
TORQUE				
0.7376	pound-force foot (lbf ft)	•	newton metre (Nm)	1.356
0.1020	kilogram-force metre (kgf m)		newton metre (Nm)	9.807
8.851	pound inch (lb in)		newton metre (Nm)	0.1130
1.39×10^{-3}	ounce inch (oz in)		gram-force centimetre (gf cm)	72.01
PRESSURE and STRESS				
9.869×10^{-3}	atmosphere (atm)		kilopascal (kPa)	101.30
0.1450	pound-force/in^2 (psi)		kilopascal (kPa)	6.894 76
0.010 20	kilogram force/cm^2 (kgf/cm^2)		kilopascal (kPa)	98.0665
0.068 04	atmosphere (atm)		pound-force/in^2 (psi)	14.70
20.89	pound-force/ft^2 (lbf/ft^2)		kilopascal (kPa)	0.047 88
0.01	bar	•	kilopascal (kPa)	100
10.000	millibar (mbar)		kilopascal (kPa)	0.1000
33.86	millibar (mbar)		inches mercury (inHg)	0.029 53
68.95	millibar (mbar)		pound-force/in^2 (psi)	0.014 50
7.501	torr or mm mercury (mmHg)		kilopascal (kPa)	0.1333
0.2953	inches mercury (inHg)		kilopascal (kPa)	3.386
760.0	torr or mm mercury (mmHg)		atmosphere (atm)	1.316×10^{-3}
4.015	inches water (inH_2O)		kilopascal (kPa)	0.2491
13.60	inches water (inH_2O)		inches mercury (inHg)	0.073 55
1.000	newton/metre^2 (N/m^2)	•	pascal (Pa)	1.000
0.0648	ton-force (UK)/inch^2		megapascal (mPa)	15.44
6.895×10^{-3}	megapascal (mPa)		pound-force/inch^2 (psi)	145.0
DENSITY				
0.062 428	pound/foot (lb/ft^3)		kilogram/metre^3 (kg/m^3)	16.0185
0.010 022	pound/gal (UK)		kilogram/metre^3 (kg/m^3)	99.776
10^{-3}	gram/centimetre^3 (g/cm^3)	•	kilogram/metre^3 (kg/m^3)	1000
7.5248×10^{-4}	ton/yard^3		kilogram/metre^3 (kg/m^3)	1328.94
0.160 544	pound/gal (UK)		pound/foot^3 (lb/ft^3)	6.228 84
ENERGY, WORK, and HEAT				
10^7	erg		joule (J)	10^{-7}
0.7376	foot pound-force (ft lbf)	•	joule (J)	1.3558
0.2388	calorie (cal)		joule (J)	4.1868
0.1020	kilogram-force metre (kgf m)		joule (J)	9.8066
MULTIPLY BY ←	TO ←		TO CONVERT	

	TO CONVERT →	TO →	MULTIPLY BY
9.478×10^{-4}	British thermal unit (Btu)	joule (J)	1055.1
0.3725	horsepower hour (hph)	megajoule (MJ)	2.6845
1.3410	horsepower hour (hph)	kilowatt hour (kWh)	0.7457
0.2778	kilowatt hour (kWh)	● megajoule (MJ)	3.600
3412.1	British thermal unit (Btu)	kilowatt hour (kWh)	2.931×10^{-4}
9.478×10^{-3}	therm	megajoule (MJ)	105.51
MASS			
0.035 274	ounce (oz)	gram (g)	28.3495
2.204 62	pound (lb)	● kilogram (kg)	0.453 592
1.102 31	ton US (short ton)	tonne (t)	0.907 185
0.984 207	ton UK (long ton)	tonne (t)	1.016 05
19.6841	hundred weight (cwt)	tonne (t)	0.050 80
1.429×10^{-4}	pound (lb)	grain	7000
0.031 08	slug	pound (lb)	32.174
0.01	quintal	kilogram (kg)	100
0.001	kip	pound (lb)	1000
5.0	carat	gram (g)	0.200
LENGTH			
0.039 370 1	inch (in)	millimetre (mm)	25.40
3.280 84	feet (ft)	metre (m)	0.3048
1.093 61	yard (yd)	● metre (m)	0.914 400
0.621 371	mile	kilometre (km)	1.609 34
0.0497	chain	metre (m)	20.1168
4.97	link	metre (m)	0.21168
0.5468	fathom	metre (m)	1.8288
39.370	thou or mil	millimetre (mm)	0.025 40
0.001	millimetre (mm)	micron (µm)	1000
10^{10}	angstrom (A)	metre (m)	10^{-10}
1.644×10^{-4}	mile UK nautical	feet (ft)	6080
5.396×10^{-4}	mile UK nautical	metre (m)	1853.2
5.399×10^{-4}	mile International nautical (n mile)	metre (m)	1852.00
1.894×10^{-4}	mile	feet (ft)	5280
AREA			
2.471	acre	hectare (ha)	0.4047
2.066×10^{-4}	acre	sq. yard (yd^2)	4840
2.471×10^{-4}	acre	sq. metre (m^2)	4047
1.973×10^{3}	circular mil	sq. millimetre (mm^2)	5.067×10^{-4}
1.2732	circular mil	square mil	0.7854
0.1550	sq. inch (in^2)	sq. centimetre (cm^2)	6.4516
10.764	sq. feet (ft^2)	● sq. metre (m^2)	0.092 90
1.562×10^{-3}	sq. mile	acre	640
3.861×10^{-3}	sq. mile	hectare (ha)	259.0
VOLUME			
0.2200	gal (UK)	● litre (l)	4.546
0.2642	gal (US)	litre (l)	3.785
0.0238	barrel (US)	gal (US)	42
0.035 32	cu. feet (ft^3)	litre (l)	28.32
0.1605	cu. feet (ft^3)	gal (UK)	6.229
MULTIPLY BY ←	TO ←	TO CONVERT	

	TO CONVERT →	TO →	MULTIPLY BY
35.31	cu. feet (ft^3)	● cu. metre (m^3)	0.028 32
1.308	cu. yard (yd^3)	cu. metre (m^3)	0.764 56
0.061 026	cu. inch (in^3)	cu. centimetre (cm^3) or millilitre (ml)	16.39
0.0352	fluid ounce (fl oz)	millilitres (ml)	28.41
1.76×10^{-3}	pint (UK) (pt)	millilitre (ml)	568.2
1×10^{-3}	litre (l)	millilitre (ml)	1000
1×10^{-3}	cu. metre (m^3)	litre (l)	1000

VELOCITY

0.039 370 1	inch/sec (in/s)	centimetre/sec (cm/s)	2.540
3.280 84	feet/sec (ft/s)	● metre/sec (m/s)	0.304 800
1.9685	feet/min (ft/min)	centimetre/sec (cm/s)	0.5080
0.621 371	miles/hr (mph)	kilometre/hr (km/h)	1.609 344
2.236 94	miles/hr (mph)	metre/sec (m/s)	0.447 04
0.5397	knot (UK)	kilometre/hr (km/h)	1.853
0.5400	knot International (kn)	kilometre/hr (km/h)	1.852

ACCELERATION

3.280 84	feet/sec² (ft/s^2)	metre/sec² (m/s^2)	0.304 800
0.101 97	gravitational acceleration (g)	● metre/sec² (m/s^2)	9.806 650

POWER and HEAT FLOW RATE

3.4121	British thermal unit/hr (Btu/h)	watt (W)	0.2931
0.8598	kilocalorie/hour (kcal/h)	● watt (W)	1.163
0.7376	foot pound-force/sec (ft lbf/s)	watt (W)	1.3558
0.1020	kilogram-force metre/sec (kgf m/s)	watt (W)	9.807
1.360	metric HP	kilowatt (kW)	0.7355
1.3410	horsepower (hp)	kilowatt (kW)	0.7457
7.457×10^{-4}	megawatts (MW)	horsepower (hp)	1341
0.2843	ton of refrigeration	kilowatt (kW)	3.517
8.333×10^{-5}	ton of refrigeration	British thermal unit/hr (Btu/h)	12000
1.000	joule/sec (J/s)	watt (W)	1.000
1.818×10^{-3}	horsepower (hp)	foot pound-force/sec (ft lbf/s)	550

ILLUMINATION

0.0929	foot candles	lux (lx)	10.764
0.0929	lumen/foot² (lm/ft^2)	lux (lx)	10.764
0.0929	candela/sq. foot (cd/ft^2)	● candela/sq. metre (cd/m^2)	10.764

ANGULAR MEASURE

57.296	degree (...°)	radian (rad)	0.017 453
9.5493	revs per min (rpm)	radian/sec (rad/s)	0.104 72

ELECTROMAGNETIC

2.778×10^{-4}	ampere hour (Ah)	coulomb (C)	3600
10^4	gauss	weber/sq metre (Wb/m^2)	10^{-4}
2.5407	microhms/cm³	microhms/in³	0.3937
10^5	gamma	gauss	10^{-5}

MULTIPLY BY ← TO ← TO CONVERT

TO CONVERT	→	TO	→	MULTIPLY BY

KINEMATIC VISCOSITY

10^6	centiStoke	square metre/sec (m^2/s)	10^{-6}
10.764	square feet/sec	square metre/sec	0.09290
10^6	square millimetre/sec	square metre/sec (m^2/s)	10^{-6}
10^4	Stoke (St) (cm^2/s)	square metre/sec	10^{-4}

DYNAMIC VISCOSITY

10^3	centipoise (cP)	pascal second (Pa s)	10^{-3}
2.419	pound/ft hr (lb/ft h)	centipoise (cP)	0.4134
0.02089	pound-force sec/foot^2 (lbf s/ft^2)	pascal second (Pa s)	47.88
1.000	gram/metre sec (g/m s)	centipoise (cP)	1.000

HEAT TRANSFER

0.1761	Btu/ft^2 h°F	watts/metre^2 K (W/m^2 K)	5.678
6.933	Btu in/ft^2 h°F	watts/metre K (W/m K)	0.1442
2.388×10^{-4}	Btu/lb°F	joule/kilogram K (J/kg K)	4186.8
4.299×10^{-4}	Btu/lb	joule/kilogram (J/kg)	2326
2.388×10^{-4}	kilocalories/kg	joule/kilogram (J/kg)	4187

SOUND LEVEL

0.1151	neper	decibel (Db)	8.686

MULTIPLY BY	← TO	←	TO CONVERT

PHYSICAL CONSTANTS

3.141593	π
2.718282	e
2.302585	$\log_e 10$

Index